Wide Bandgap Semiconductors

As Moore's law approaches its physical limit, research results beyond Moore's law based on novel materials, structures and devices is providing a new development direction for the semiconductor industry. GaN, SiC and other wide bandgap semiconductor materials are the third generation of semiconductor materials, following silicon, germanium, GaAs and InP. They are the preferred materials for manufacturing solid-state light sources, power electronic devices, and microwave RF devices due to their superior frequency, efficiency, voltage resistance, temperature resistance, and radiation resistance. The study of new materials can meet national strategic requirements for energy conservation and emission reduction, intelligent manufacturing and information security among others. It has become the forefront of global semiconductor technology research and a new industrial focus, and has a significant impact on industrial growth.

The book series "Wide Bandgap Semiconductors" is dedicated to clarifying the fundamental frontiers, key scientific and technological issues of wide bandgap semiconductors. It provides a comprehensive and prospective discussion of the characteristics of new generation semiconductor materials, fundamental physical properties, device fabrication, international R&D trends, key scientific and technological problems encountered, major R&D results achieved, device applications, and development prospects. This series represents the latest findings in fundamental semiconductor theory research and has significant application value.

The book series "Wide Bandgap Semiconductors" brings together the research experience and achievements of more than twenty of China's leading experts in this field. It is a creative exploration of the "synergy between fundamental and applied disciplines". This series fills a gap in the publication of monographs in this field both domestically and internationally, and can serve as a resource for scientific and technical workers engaged in the research and development of wide bandgap semiconductor electronic materials and devices, as well as for enterprise engineers and technicians seeking to strengthen their fundamental knowledge and investigate the problems encountered in practical scientific research.

Shijie Xu
Editor

Optical Characterization of Microstructures and Optoelectronic Devices Based on Wide Band Gap Semiconductors

Springer

Editor
Shijie Xu
Department of Optical Science
and Engineering, College of Future
Information Technology
Fudan University
Shanghai, China

ISSN 2948-2607 ISSN 2948-2615 (electronic)
Wide Bandgap Semiconductors
ISBN 978-981-95-1927-9 ISBN 978-981-95-1928-6 (eBook)
https://doi.org/10.1007/978-981-95-1928-6

Jointly published with Xidian University Press
The print edition is not for sale in China (Mainland). Customers from China (Mainland) please order the print book from: Xidian University Press.
ISBN of the Co-Publisher's edition: 978-7-5606-6283-1

Translation from the Chinese Simplified language edition: "宽禁带半导体微结构与光电器件光学表征" by Shijie Xu, © Xidian University Press 2022. Published by Xidian University Press. All Rights Reserved.

This Springer imprint is published by the registered company Springer Nature Singapore Pte Ltd.
The registered company address is: 152 Beach Road, #21-01/04 Gateway East, Singapore 189721, Singapore

Preface

Compared with traditional semiconductors such as silicon and gallium arsenide, wide bandgap semiconductors of gallium nitride (GaN), zinc oxide (ZnO), and silicon carbide (SiC) possess quite a few material advantages, such as large bandgap, high breakdown voltage, large saturated electron mobility, and high thermal conductivity. Among them, GaN, ZnO, etc., are direct bandgap semiconductors, which can be used to manufacture short-wavelength, high-efficiency light-emitting diodes (LEDs) and laser diodes (LDs) and other high-performance optoelectronic devices. At present, GaN-based LEDs are rapidly expanding toward full spectrum, deep ultraviolet, and micro-LED arrays. At the same time, other wide bandgap semiconductors such as gallium-doped ZnO and gallium oxide (Ga_2O_3) are also attracting extensive research interest across the world. Therefore, it is time to invite leading researchers to summarize and review their latest research results in the field of wide bandgap semiconductors as chapters of this book. On the one hand, this book is supposed to provide a frontier course for senior undergraduates and graduate students in semiconductor optoelectronics and related majors in our country's colleges and universities; on the other hand, it is intended to provide a reference and basis for our country's researchers in the cutting-edge strategic frontier field of wide bandgap semiconductors and their functional devices.

Since the breakthrough of GaN-based blue LEDs, which was awarded the Physics Nobel Prize in 2014, the science and technology of wide bandgap semiconductors has developed rapidly for over two decades. Numerous electronic and optoelectronic devices of wide bandgap semiconductors with high power, high speed, high-temperature resistance, and short wavelength have been rapidly industrialized. However, unlike the development path of traditional semiconductor technology, the technological development of wide bandgap semiconductor is much faster than the progress of basic research on wide bandgap semiconductors. This mismatch between the science and technology progresses of wide bandgap semiconductors has seriously hindered the further improvement of the performance of wide bandgap semiconductor electronic and optoelectronic devices. In addition to nitride wide bandgap semiconductors, research on oxide wide bandgap semiconductor materials and devices has also shown a booming trend. Therefore, publishing a series of books reflecting the current frontier of wide bandgap semiconductor technology is a good thing timely and naturally. Under the leadership and hosting of Academician Yue Hao from Xidian University, managers and colleagues of Xidian University Press decided to publish a series of books on wide bandgap semiconductors. Publishing this series of books could be the first in China and even in the world, which is indeed an outstanding event.

In advanced optoelectronic devices of wide bandgap semiconductors, laser diodes (LDs) could be at the top due to their extremely strict requirements for material quality, device design, and manufacturing process. In this book, the first two chapters (Chaps. 1 and 2) summarize and discuss the device structure design, growth, manufacturing process, and performance characterization of GaN-based LDs, letting readers to develop an overall and deep understanding of the current status of research, working mechanism, recent progress, and existing problems and difficulties of GaN-based ultraviolet and silicon-based nitride blue-green LD devices, respectively. The two chapters may also lay a good foundation for future research in this area by graduate students and scientific and technological workers and for further improving the performance of GaN-based LDs. It is worth mentioning that in the indium gallium nitride (InGaN) ternary alloy, which serves as the active layer of GaN-based LDs, there often exist local In-rich nanostructures formed by In component segregation. These In-rich nanostructures can lead to formation of localized electronic states. For the localized-state ensemble (LSE) luminescence, Professor Shijie Xu and his team members have developed a quantitative model—the LSE model—for 20 years. This model not only gives a quantitative explaination to the relevant experimental results, but also gains unprecedented deep understanding of the physical mechanisms in LSE luminescence, such as enabling us to determine several key parameters governing the luminescence processes. Since its first publication in 2001, the LSE model has been widely employed by different research groups worldwide to interpret their experimental results. In the three chapters (Chaps. 1, 10, and 11) of this book, application examples of the LSE model are presented. Another kind of advanced optoelectronic device based on GaN is the micro-LED array. Its process preparation, optoelectronic characteristics, key factors restricting its external quantum efficiency, and possible solutions are discussed in detail in another two chapters (Chaps. 4 and 5) in this book. In addition, the efficiency characterization of GaN-based LEDs is an important topic. It has been experimentally found that the luminous efficiency of GaN-based LEDs decreases after the injection current increases to a certain extent, which is the so-called "efficiency droop" challenging problem. Accurate characterization and in-depth study of the quantum efficiency of GaN-based LEDs will help solve this problem. Chapter 3 provides a beneficial research and exploration on this important issue. Due to the spontaneous polarization and piezoelectric polarization characteristics of the GaN with hexagonal wurtzite lattice, there often exists a significant built-in electric field in the multiquantum well structure of GaN-based LEDs. Such a large built-in electric field can lead to a noticeable quantum-confined Stark effect, causing spatial separation of electron and hole wave functions, and hence resulting in low luminous efficiency. A current method to suppress or solve this problem is to grow the LED multiquantum well structure on nonpolar or semipolar crystal faces. There is a dedicated chapter (Chap. 6) in this book to discuss the growth and optical characterization of nonpolar and semipolar GaN-based LEDs. In addition to the two-dimensional quantum well structures, the optical properties of lower-dimensional GaN nanowires and quantum dots are also a current eye-catching topic. Chapter 11 is arranged to describe the optical characterization work of quantum dots in GaN-based nanowires, especially focusing on the collective luminescence mechanisms of localized carriers and the issues of surface state passivation and temperature effects. The authors of this chapter use the LSE model to quantitatively explain their experimental results, obtaining a clear and profound understanding of the physical mechanism. For oxide wide bandgap semiconductors and their micro/nanostructures, the construction and in-depth investigation of ZnO micro/nano lasers are discussed in Chap. 7, while Chap. 8 is devoted to the preparation and characterization of Ga_2O_3 solar-blind deep-UV photodetectors. Optoelectronic properties and characterization of Ga-doped ZnO thin films are summarized and reviewed in detail in Chap. 9, while nonlinear optical properties of ZnO single crystals and micro/nanostructures are discussed in the last chapter of the book.

This book focusing on wide bandgap semiconductor lasers, including LDs, micro-LEDs, solar-blind deep-UV photodetectors, etc., and optical characterization of wide bandgap semiconductor micro/nanostructures is written mainly for senior undergraduates and graduate students in semiconductor and optoelectronics, as well as researchers in related fields. The book contents are arranged according to the principle of "devices, materials, and mechanisms," striving to describe the optoelectronic characteristics and characterization of short-wavelength LDs, LEDs, photodetectors, photovoltaic cells, etc., based on wide bandgap semiconductors of GaN, ZnO, and Ga_2O_3, etc. Although the book is prepared in a simple and understandable manner, its plentiful content highlights frontier and bottleneck issues in the field. This book can be used as a teaching reference book for courses such as "Wide Bandgap Semiconductor Optoelectronic Devices" in colleges and universities and as a reference material for researchers. The authors of this book come from the Institute of Semiconductors of the Chinese Academy of Sciences, Suzhou Institute of Nano-tech and Nano-bionics of the Chinese Academy of Sciences, Tsinghua University, Fudan University, The University of Hong Kong, Southeast University, Dalian University of Technology, Hebei University of Technology, The University of Sheffield, Duke Kunshan University, Nanjing University of Posts and Telecommunications, Jiangsu University, Hefei University of Technology, Nanjing University of Aeronautics and Astronautics, and Jiangsu University of Science and Technology, etc. Mrs. Lehui Ma of Xidian University Press proofread most of the chapters of the book as the executive editor and provided many valuable opinions and suggestions on the content arrangement of the book. The book was solicited, compiled, and edited by Prof. Shijie Xu of The University of Hong Kong and Fudan University (currently). During the early suggestion and conception stage of this book, Academician Yue Hao of Xidian University and Prof. Jinfeng Zhang provided guidance and help, for which the authors express their heartfelt thanks. Due to our limited level, there exist inevitably inaccuracies and even errors in the book, and we sincerely invite readers to criticize and correct them.

Shanghai, China
April 2022

Shijie Xu

Introduction

This book, with the main line of optical characterization of microstructures and optoelectronic devices based on wide bandgap semiconductors, is aimed at senior undergraduates and graduate students. It serves mainly senior undergraduates, graduate students, and researchers in semiconductor optoelectronics and related majors in higher education institutions. The book is arranged according to the principle of "focusing on the frontier topics of wide bandgap semiconductors, emphasizing the basic properties and processes of advanced optoelectronic devices and material microstructures for optical characterization, and aiming to improve the performance of wide bandgap semiconductor optoelectronic devices." From the perspective of applied basic research and the development of advanced optoelectronic devices, scholars who are at the forefront of this field nationwide are organized to write in simple and understandable language, introducing systematically and in detail the frontier research including the growth, doping, stimulated radiation, quantum efficiency measurement, ultraviolet (UV) detection, microsized light-emitting diodes (LEDs), high-efficiency solar cells, and other working principles, optical characterization, performance improvement, etc.

Contents

Chapter 1
Wide Bandgap Semiconductor Laser Diodes: Technology and Characterization

Degang Zhao

1.1 Introduction

Gallium nitride (GaN)-based materials (including indium nitride [InN], aluminum nitride [AlN], and their alloys) are important wide bandgap semiconductor materials with forbidden bandgap widths ranging from 0.7 to 6.2 eV. Their spectral range covers from near-infrared to deep-ultraviolet waveband, and they have significant application value in the field of optoelectronics. The invention of GaN-based blue light-emitting diodes (LEDs) has sparked a revolution in lighting. Three pioneering Japanese scientists, Isamu Akasaki, Hiroshi Amano, and Shuji Nakamura, were awarded the Nobel Prize in Physics in 2014 for the invention of blue LEDs. Compared to LEDs, GaN-based semiconductor laser diodes (LDs) are more challenging optoelectronic devices, with important applications in laser display, laser lighting, laser processing, and other fields. GaN-based blue-green laser diodes are the core light source for laser displays, enabling high-definition (color, geometry), wide color gamut true color display. GaN-based blue laser diodes are small in size and high in brightness, enabling higher brightness laser lighting. GaN-based ultraviolet laser diodes have short wavelengths and high photon energy, enabling high-precision cold processing. GaN-based LDs may even trigger significant technological changes in the fields of display, lighting, and processing.

In 1996, Dr. S. Nakamura who was from Nichia Corporation in Japan reported the world's first GaN-based laser diode [1], with a wavelength of 417 nm, achieving a major breakthrough. Subsequently, GaN-based laser diodes with multiple wavelengths such as ultraviolet and green light were realized. Currently, there still exist challenges in the fabrication of GaN-based LDs. High-quality epitaxial materials are

---

D. Zhao (✉)
Institute of Semiconductors, Chinese Academy of Sciences, Beijing, China
e-mail: dgzhao@semi.ac.cn

S. Xu (ed.), *Optical Characterization of Microstructures and Optoelectronic Devices Based on Wide Band Gap Semiconductors*, Wide Bandgap Semiconductors,
https://doi.org/10.1007/978-981-95-1928-6_1

the foundation, while material epitaxy with metal–organic chemical vapor deposition (MOCVD) is the core, and device technology is also critical. The fundamental architecture of a GaN-based LDs typically comprises a p-GaN layer, a quantum well layer, and an n-GaN layer. This chapter focuses on the fundamental device structure, commencing with an introduction to MOCVD technology. It then proceeds to detail the epitaxial growth of the GaN layer, p-type doping, indium gallium nitride (InGaN) quantum well formation, and associated device fabrication techniques. Additionally, it encompasses the characterization of physical properties, culminating in the fabrication of a high-power GaN-based blue LDs. It should be emphasized here that although wide bandgap semiconductors include GaN, SiC, ZnO, and other materials, only GaN materials have been successful in the field of laser diodes. Therefore, this chapter introduces the technology and characteristics of GaN-based laser diodes.

1.2 Introduction to MOCVD Technology

Common epitaxial growth technologies for GaN materials include molecular beam epitaxy (MBE), hydride vapor-phase epitaxy (HVPE), and metal–organic chemical vapor deposition (MOCVD), etc. MOCVD is also known as metal–organic vapor-phase epitaxy (MOVPE), which was first proposed by H. M. Manasevit in 1968 as a technique for depositing common compound semiconductors using organic metal materials [2]. It is a vapor-phase epitaxy technique that continuously transports metal–organic compounds and hydrides to a heated substrate in the reaction chamber using hydrogen or nitrogen and carries out a series of chemical reactions on the substrate surface to grow the epitaxial layer. MOCVD technology has become the predominant method for epitaxial growth in contemporary semiconductor manufacturing. It possesses numerous advantages, including precise control over the composition, growth rate, interface, and doping concentration of the epitaxial layer; the ability to realize sharp heterojunctions or grow multicomponent alloy compounds with varying semiconductor materials; high purity in the epitaxially grown semiconductor materials; rapid growth rates; excellent reproducibility; ease of achieving multichip and large-area growth; high production capacity; and relatively low costs.

MOCVD growth is a vapor-phase epitaxy process under nonequilibrium conditions using metal–organic compounds for metal transport. Although the MOCVD growth chamber is a single temperature zone, its growth mechanism is quite complex, and the growth process involves transport and multicomponent, multiphase chemical reactions, as shown in Fig. 1.1 [3]. Hydrogen gas carries MO sources, hydrides, and other reaction precursors into the reaction chamber as a carrier gas, and its temperature gradually increases with the gas flow to the heated substrate. The reactions that may occur in the gas-phase include: (1) MO source reacts with hydrides or organic compounds to form adducts. (2) As the temperature further increases, MO sources, hydrides, and adducts will thermally decompose, forming gas-phase nuclei for epitaxial growth preparation. (3) The reaction species in the gas-phase (gas-phase nuclei) diffuse to the heated substrate surface, undergo surface adsorption, surface migration, and surface reaction, and finally enter the lattice to form the epitaxial layer. (4) Some reaction species also undergo substrate surface desorption and diffusion and return to the main gas stream, being carried out of the reaction chamber by the carrier gas. In addition, some gas-phase reaction products are directly carried out of the reaction chamber by the gas flow. It can be seen that the growth reaction process of MOCVD is extremely complex. Although various reaction models have been attempted to be established in many literature reports, they still cannot accurately predict the characteristics of the MOCVD growth system. Therefore, the experience accumulated by researchers over a long period of time is still needed, and a large number of experiments are attempted by continuously adjusting various growth parameters to obtain high-quality epitaxial layers.

The MOCVD growth system generally includes a gas transport subsystem, a reaction chamber subsystem, a tail gas treatment subsystem, a control subsystem, and an in situ monitoring subsystem [4]. Figure 1.2 is a schematic diagram of a low-pressure MOCVD growth system. Among the various subsystems, the in situ monitoring system (also known as the in-place monitoring system) is very important for studying the epitaxial mechanism of materials, mainly used for real-time monitoring of material growth rate, growth morphology, and growth mode, etc., and it is also used to monitor the growth process, providing essential data that enhances the crystal quality. In MOCVD, an optical system is generally used as the in situ detection system, including reflectometer (used to analyze the thickness of the epitaxial layer), emissivity-corrected pyrometer (can obtain the true temperature of the substrate), etc.

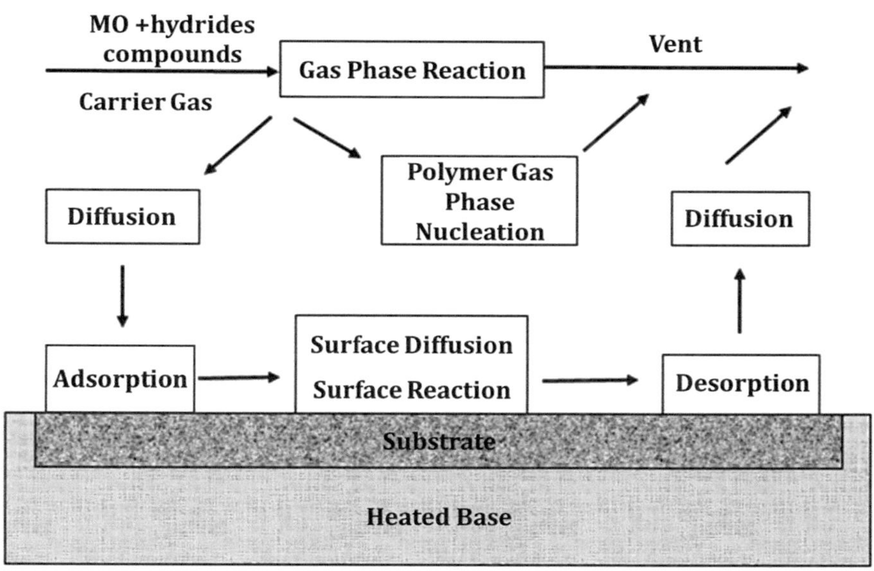

Fig. 1.1 Schematic diagram of MOCVD growth process

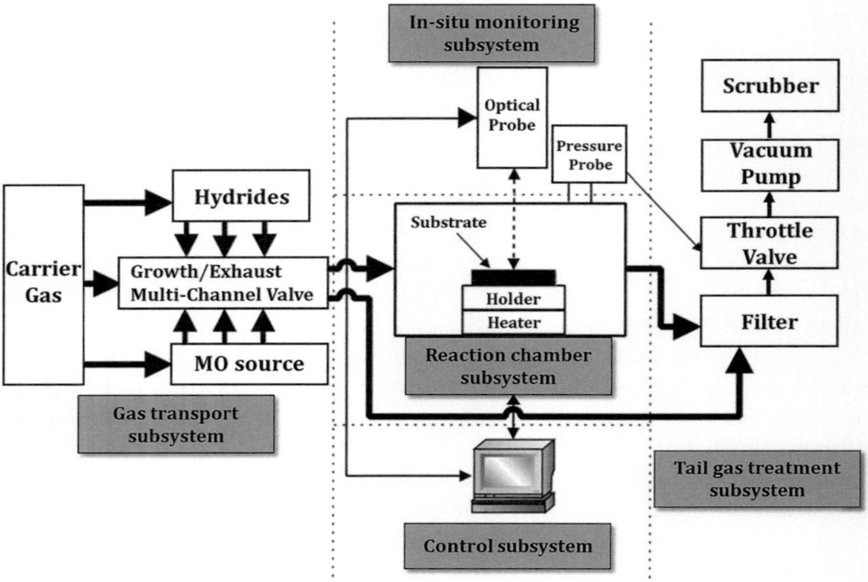

Fig. 1.2 Schematic diagram of low-pressure MOCVD growth system

The MOCVD growth process of GaN materials generally includes high-temperature treatment of sapphire substrates, substrate nitridation, growth of low-temperature buffer layers, and subsequent growth of GaN epitaxial layers. It is a very complex kinetic process. Real-time monitoring of material growth process is very necessary for studying material growth mechanism and improving material quality. Dr. S. Nakamura first introduced in situ monitoring, a useful tool, in the GaN MOCVD growth process [5], which played a significant role in the development of GaN material epitaxy technology. The researchers from Institute of Semiconductors, Chinese Academy of Sciences, have independently developed an in situ normal incidence laser reflectance (LR) monitoring system on low-pressure MOCVD equipment, as shown in Fig. 1.3. The light source for in situ monitoring uses a laser with a power of <1 mW and a wavelength of ~650 nm. The laser light passes through a gap in the high-frequency heating coil and then through a window in the inner tube to illuminate the sample surface at an angle of $\sim 1^{\circ}$. The light reflected from the sample surface is received by a Si detector and transmitted to a lock-in amplifier, and finally, the reflected signal received by the program is collected to obtain the relative change in the reflectance intensity of the sample surface.

Figure 1.4 is a typical in situ reflectance monitoring curve during the MOCVD growth of GaN material [6]. According to the curve, the epitaxial growth process of GaN material can be studied. Generally speaking, the growth process of GaN material is a transition from three-dimensional growth to two-dimensional growth. In the early stage of growth, the surface is relatively rough, which is three-dimensional

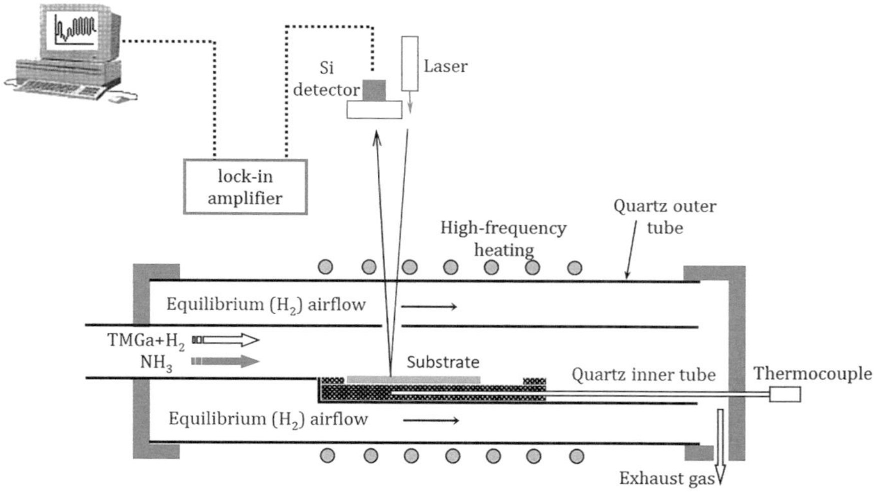

Fig. 1.3 Schematic diagram of the in situ reflectance monitoring system for GaN material MOCVD growth

island growth. As the epitaxial process continues, these islands begin to merge, and dislocations are redirected, resulting in higher-quality GaN material. The in situ monitoring curve in Fig. 1.4 clearly shows this growth process. In the early stage of growth, the reflectance is very low, with no interference fringes, and the surface is very rough. Then oscillating interference fringes gradually form, indicating that the islands begin to merge, and the growth process transitions from three-dimensional growth to two-dimensional growth, resulting in better-quality GaN samples. The advancement of in situ monitoring tools has significantly facilitated the progression of MOCVD epitaxy technology.

Fig. 1.4 Typical in situ reflectance monitoring curve during the MOCVD growth process of GaN material

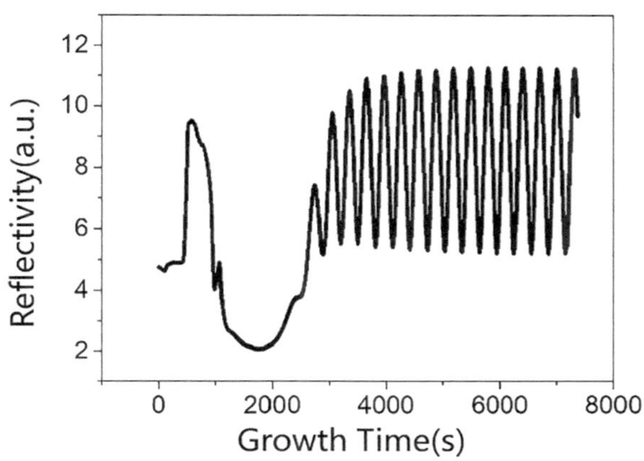

1.3 MOCVD Growth and Characterization of GaN Materials

Due to the difficulty of preparing large-size GaN single crystals and the lack of suitable substrate materials, the growth of GaN epitaxial films is generally carried out on substrates with large mismatches, and heteroepitaxy with large mismatches is very difficult. The early growth process of GaN not only had difficulty in obtaining smooth, crack-free thin films, but the thin films usually had a high background concentration. In 1986, H. Amano and I. Akasaki significantly enhanced the quality of the thin film through a two-step growth process, which involved initially depositing an AlN buffer layer at low temperature followed by the growth of GaN at high temperature [7]. Further research found that when GaN is directly grown on the (0001) surface of sapphire, the growth of GaN is a typical three-dimensional growth mechanism, going through the processes of island formation, island enlargement, three-dimensional growth, and formation of uneven surfaces; for those with a low-temperature buffer layer, processes such as island formation, lateral growth, island merging, and quasi-two-dimensional growth occur. Subsequently, S. Nakamura reported a two-step growth technique of first growing a GaN buffer layer at low temperature and then growing GaN at high temperature [8]. It was also found that the optimal thickness of the low-temperature GaN buffer layer is about 20–25 nm.

In the two-step growth process, the technology of the low-temperature buffer layer is very critical and is crucial to the crystal quality of the subsequent high-temperature GaN epitaxial film. The optimization of the growth conditions of the low-temperature buffer layer can effectively reduce the dislocation density in the epitaxial layer. The introduction of the buffer layer may play the following three roles: (1) The buffer layer provides a nucleation center with the same orientation as the substrate. (2) The buffer layer releases the stress generated by the lattice mismatch between GaN and the substrate, as well as the thermal stress generated by the mismatch of the thermal expansion coefficient. (3) The buffer layer provides a smooth nucleation surface for further growth, reduces the contact angle of its nucleation growth, allows the island-like growth of GaN grains to connect into a surface at a smaller thickness, and transitions to two-dimensional epitaxial growth. It is because of these effects that the low-temperature GaN buffer layer has greatly improved the quality of GaN materials.

Due to the invention and application of the buffer layer and the two-step technology, the technology of GaN epitaxial layer has also made breakthroughs. Blue light-emitting diodes, laser diodes, ultraviolet detectors, high mobility transistors, and other GaN devices have been realized, and many devices have been commercialized. The two-step method has now become the standard process for GaN material epitaxy. The following focuses on the growth of GaN materials on a low-temperature AlN buffer layer [9]. The growth process is: first grow the AlN buffer layer at a low temperature of 540 °C, then raise the temperature to 1040 °C to grow the GaN epitaxial layer. The substrate is (0001) sapphire; ammonia (NH_3), trimethylaluminum (TMAl), and trimethylgallium (TMGa) are the N source, Al source, and Ga source, respectively, and hydrogen (H_2) is the carrier gas. This study primarily investigated the effects of annealing time and thickness of low-temperature AlN layers on the epitaxial growth of GaN materials.

Figure 1.5a and b shows the in situ reflectance monitoring curves of two samples A and B grown by MOCVD, respectively. The thickness of the AlN buffer layer of the two samples is both 20 nm, but the annealing time of the AlN buffer layer is different. The annealing time for the buffer layer of sample A is 1000 s, while that of sample B is 300 s. The in situ monitoring curve can be divided into three parts: (i) the growth of the low-temperature AlN buffer layer; (ii) the annealing of the AlN buffer layer; and (iii) the growth of the GaN epitaxial layer at high temperature. The in situ monitoring curve is very sensitive to changes in surface roughness, which can be used to study its growth process and mechanism. From the growth curves of the two samples, different annealing times have a significant impact on the growth process of the samples. During the annealing process (ii) of the AlN buffer layer of sample A, there is no significant change in the curve, but at the beginning of the growth of the GaN epitaxial layer, the reflectivity is very low. Then, the reflectivity of the curve gradually increases, and some weak oscillation interference fringes begin to appear, and finally, the reflectivity reaches a stable value, and obvious interference fringes can be seen. The growth process of sample B is completely different. Although there is no significant difference in the curve during the annealing process (ii) compared to sample A, the process of GaN epitaxial growth (iii) differs completely. Currently, the GaN epitaxial growth starts with a relatively high reflectivity and obvious interference fringes. During the subsequent growth process, the reflectivity quickly reaches a stable value, and the interference fringes are obvious. The growth of the GaN epitaxial layer generally involves three processes: three-dimensional growth, island coalescing, and quasi-two-dimensional growth. In the early stage of growth, the vertical growth rate is greater than the lateral growth rate, showing three-dimensional growth. At this stage, the epitaxial layer presents an island-like structure, and the surface is very rough. Then, as the growth progresses, the lateral growth rate begins to dominate, the islands begin to gradually coalesce, and the surface begins to gradually flatten. Finally, the islands are basically completely coalesced, and the epitaxial layer begins to show quasi-two-dimensional growth, at which point the surface is very flat. In the growth process of the epitaxial layer of sample A, the initial stage of reflectivity is very low, indicating that the surface is very rough at this time, showing three-dimensional island growth. Then, as the growth process progresses, the reflectivity still increases, and weak interference fringes appear, indicating that the surface gradually flattens, the lateral growth begins to dominate, and the islands begin to coalesce. Finally, the reflectivity of the curve reaches a stable value, and significant interference fringes appear. At this time, the island merging process is basically completed, and quasi-two-dimensional growth begins, and the surface becomes very flat. However, in the epitaxial growth process of sample B, the initial stage of the curve reflectivity is not very low, and there are obvious interference fringes, indicating that the time for three-dimensional island growth is very short. Lateral growth quickly dominates, and the islands coalesce very quickly, showing quasi-two-dimensional growth. From the experimental results, a long annealing time of the AlN buffer layer enhances the lateral epitaxial growth process of the GaN epitaxial layer. The growth processes of the two samples are very different, and the crystal quality of the samples also varies greatly. As shown in Table 1.1, the full width at half maximum (FWHM) of the (002) plane symmetric diffraction and the (102) plane asymmetric diffraction of sample A are narrower than those of sample B. Generally speaking, the FWHM of the (002) plane symmetric diffraction and the (102) plane asymmetric diffraction of GaN material reflects the density of screw dislocations and edge dislocations, respectively. The larger the FWHM, the higher the dislocation density [10, 11]. The X-ray diffraction (XRD) results indicate that the quality of sample A is higher than that of sample B. On the other hand, the carrier mobility of sample A is also higher than that of sample B, further indicating that the quality of sample A is better than that of sample B.

Not only the annealing time of the AlN buffer layer but also the thickness of the low-temperature AlN buffer layer has a significant impact on the growth of the GaN epitaxial layer. In the experiment, the influence mechanisms of AlN buffer layer thicknesses of 45, 30, 20, and 16 nm on the GaN epitaxial layer were studied, and the

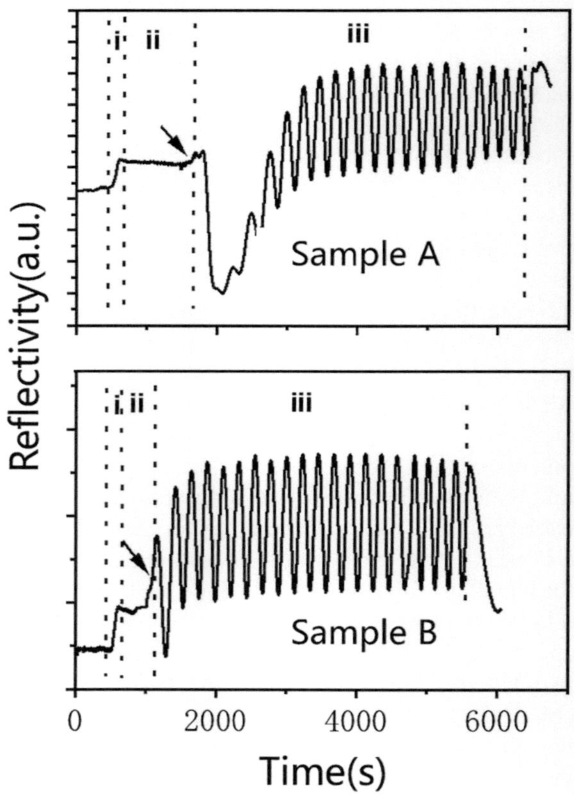

Fig. 1.5 **a** and **b**, respectively, show the in situ monitoring curves of samples A and B grown by MOCVD

Table 1.1 Growth conditions and test results of several GaN layers grown on AlN buffer layers

Sample	AlN buffer layers		XRD FWHM (arcmin)		Mobility (cm^2/Vs)
	Annealing time (s)	Thickness (nm)	(0002)	(10–12)	
A, E	1000	20	6.9	11.2	360
B	300	20	8.1	19.1	142
C	1000	45	10.2	28.6	73
D	1000	30	7.2	13.9	217
F	1000	16	…	…	…

sample numbers were C, D, E (same as sample A but with different codes), and F, respectively. Figure 1.6a, b, c, and d, respectively, represents the in situ monitoring curves of samples C, D, E, and F, and the annealing time of the AlN buffer layer of all these samples is 1000 s. From the curves, the growth process of the four samples is quite different. During the growth process of sample C, the surface reflectance quickly reaches a stable value, and significant interference fringes appear, indicating that the islands coalesce quickly during the growth process, resulting in two-dimensional growth. During the growth process of sample D, the surface reflectance is small at the beginning of growth; as the growth process progresses, the surface reflectance quickly reaches a stable value, and significant interference fringes appear. This indicates that three-dimensional island growth occurs at the beginning of growth, followed by island coalescence, and then quasi-two-dimensional growth appears quickly. During the growth process of sample E, the surface reflectance is very low, and only very weak interference fringes appear; then the reflectance gradually increases with time, and significant interference fringes appear, indicating that sample E has transitioned from three-dimensional growth to two-dimensional growth. During the growth process of sample F, the surface reflectance progressively decreases while surface roughness increases. There is no indication of planarization, and the growth eventually ceases. Specifically, the growth of sample F consistently follows a three-dimensional island pattern. Since the growth processes are so different, the quality of the samples also varies greatly. As shown in Table 1.1, among samples C, D, E, and F, sample E has the best quality, not only having the lowest dislocation density but also the highest carrier mobility. Both too thick and too thin will cause the quality of the GaN epitaxial layer to decrease, and the appropriate buffer layer thickness is conducive to the growth of high-quality GaN epitaxial layers.

In order to further understand the influence mechanism of AlN buffer layer on GaN epitaxial layer, the surface morphology of AlN buffer layer after annealing was also studied. Figure 1.7a, b, and c shows the surface morphology of samples A, B, and C, respectively, after annealing of the low-temperature AlN buffer layer (only before the growth of the epitaxial layer, as shown by the arrows in Figs. 1.5 and 1.6). The sample numbers are represented by a (1000 s annealing, 20 nm thick), b (300 s annealing, 20 nm thick), and c (1000 s annealing, 45 nm thick). As can be seen from the figure, sample A has the largest grain size and the smallest nucleation density,

Fig. 1.6 **a**, **b**, **c**, and **d**, respectively, show the in situ monitoring curves of samples C, D, E, and F

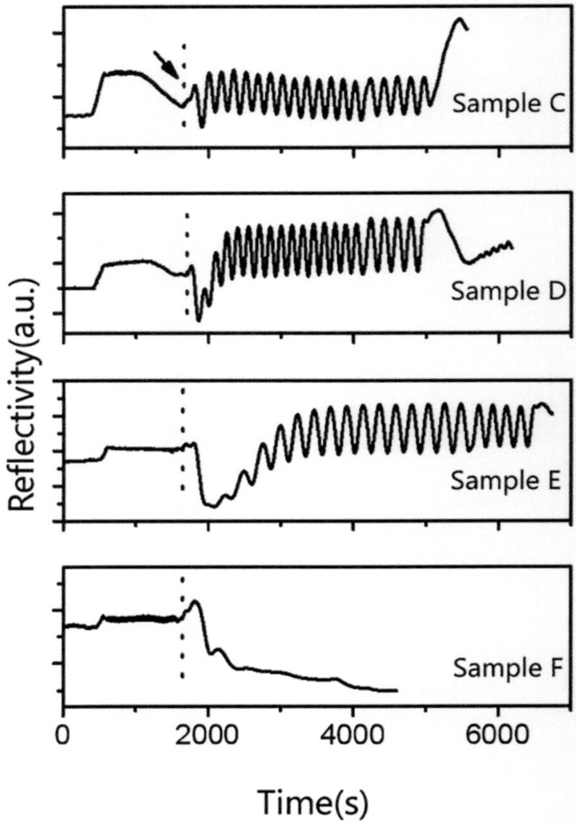

while sample C has the smallest grain size and the largest nucleation density. When sample A grows on AlN buffer layer a, there will be a process of three-dimensional growth, island coalescence, and quasi-two-dimensional growth. During the process of island coalescence, a large number of dislocations will turn and annihilate, thus reducing the dislocation density of the subsequent epitaxial layer and improving the quality. However, when sample C grows on AlN buffer layer c, the small nucleation density causes the islands in the early stage of epitaxial layer growth to quickly coalesce and transition to quasi-two-dimensional growth, allowing a large number of dislocations to penetrate the entire epitaxial layer. If the nucleation density is too small, the islands in the early stage of GaN epitaxial layer growth cannot coalesce and will always maintain a three-dimensional growth process; the quality of the sample will also be poor. Therefore, it can be seen that larger grain size and smaller nucleation density in the low-temperature AlN buffer layer are very important for the growth of high-temperature GaN epitaxial layer. The appropriate thickness of the low-temperature AlN buffer layer after a long time of annealing can achieve this. When growing high-temperature GaN epitaxial layer on low-temperature GaN buffer layer, the surface roughness of GaN buffer layer is very important, and only on the rough GaN buffer layer can a higher-quality GaN epitaxial layer be grown [12]. However, from the results of Fig. 1.7a and c, after the low-temperature AlN buffer layer of sample C is annealed, the reflectivity decreases and the surface becomes rougher. Meanwhile, after the low-temperature AlN buffer layer of sample E is annealed, the reflectivity does not change much, and the surface morphology does not become very rough. Despite this, the quality of the epitaxial layer is very different. Therefore, it can be seen that it is not necessarily the rough low-temperature AlN buffer layer that can grow a high-quality GaN epitaxial layer, which may also be a major difference in the influence mechanism of the low-temperature AlN buffer layer and the low-temperature GaN buffer layer on the epitaxial layer. Of course, more mechanisms need to be further studied through various aspects of work.

The experiment found that not only the growth thickness and annealing time of the buffer layer have an important impact on the quality of the GaN epitaxial layer, but also the nitridation process of the substrate, the V/III control of the epitaxial layer, and other parameters have an important impact on the material quality [13, 14]. Based on extensive research on the epitaxial mechanism and characteristic representation of GaN materials, we have deeply investigated the MOCVD growth mechanism of GaN materials. By adopting unique epitaxial technology, we have grown high-quality GaN epitaxial materials on ordinary flat sapphire substrates. Figure 1.8a shows the variable temperature Hall curve of the high-quality GaN sample. The test results in the figure indicate that the room-temperature carrier mobility of the sample is about 1005 cm^2/Vs [15], and the background carrier concentration is about 1.1×10^{16} cm^{-3}. This is also the best result reported for MOCVD-grown GaN materials on sapphire substrate in the world [16]. Figure 1.8b shows the X-ray diffraction (002) and (102) plane rocking curves. It can be seen that the XRD diffraction half-widths of the (002) and (102) planes of the GaN sample are very narrow, both only 180 arcsec. High-quality materials play an important role in device development.

1.4 p-Type Doping and Characterization of GaN Materials

One of the major problems affecting the development of GaN-based devices is the p-type doping issue. So far, the p-type doping technology of GaN is still a significant factor hindering the development of GaN-based lasers. Mg is the most common acceptor in p-GaN, but the high ionization energy of Mg acceptor (~200 meV) determines that only about 1% of Mg is ionized near room temperature [17]. Additionally, Mg easily forms Mg–H complexes with H during the MOCVD growth process [18, 19], further limiting the ionization of Mg acceptors. Therefore, p-type doping was not achieved in the early development of GaN. It was not until the late 1980s that H. Amano and I. Akasaki from Nagoya University in Japan [20], using the low-energy electron beam irradiation (LEEBI) method, first achieved p-GaN: The hole concentration was 2×10^{16} cm^{-3}, the mobility was 8 $cm^2/V{\cdot}s$, and the resistivity was 35 $\Omega{\cdot}cm$. Subsequently, in 1991, S. Nakamura from Nichia Corporation in Japan made

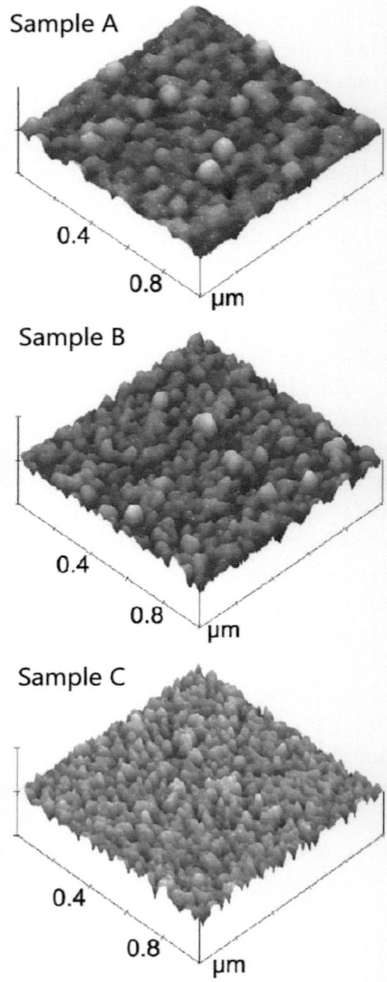

Fig. 1.7 **a**, **b**, and **c** show the surface morphology of samples A, B, and C, respectively, after annealing of the low-temperature AlN buffer layer

an important breakthrough in p-type doping of GaN [21], successfully reducing the resistivity of p-GaN materials from 320 Ω·cm to 0.2 Ω·cm, where the hole concentration was 3×10^{18} cm^{-3} and the mobility was 9 cm^2/Vs. Furthermore, in 1992, S. Nakamura and colleagues from Nichia Corporation discovered the reason why it was difficult to achieve p-GaN: Hydrogen impurities and magnesium acceptors easily form Mg_{Ga}–H complexes, thereby passivating magnesium acceptors and reducing their ionization rate. Then, they first used rapid thermal annealing to destroy Mg_{Ga}–H complexes, release hydrogen impurities, reduce the passivation of hydrogen impurities on magnesium acceptors, increase the ionization rate of magnesium acceptors, and thus obtain p-GaN materials with a hole concentration of 3×10^{17} cm^{-3}, a

Fig. 1.8 Variable temperature Hall test results (**a**) and X-ray diffraction test curves (**b**) of high-quality GaN samples

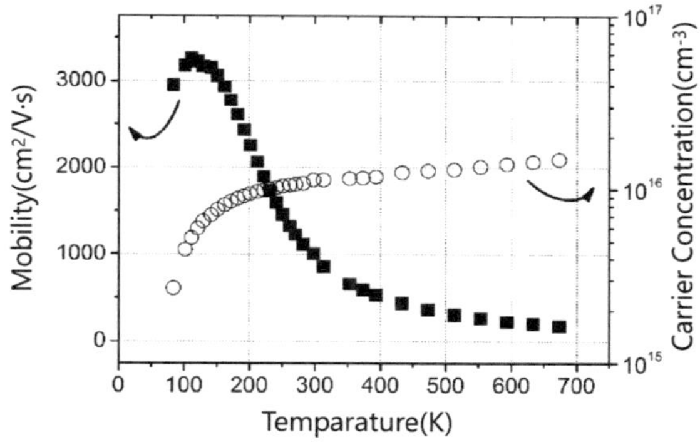

(a) Variable temperature Hall test results

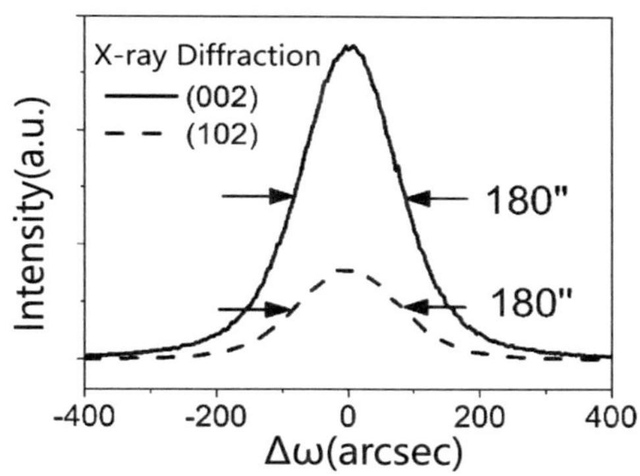

(b) X-ray diffraction test curves

mobility of 10 cm^2/V·s, and a resistivity of 2 Ω·cm [22, 23]. In addition to the influence of Mg–H complexes, factors limiting the Mg doping level in p-type GaN also include defects and impurity compensation. When the Mg doping concentration is increased to a certain level, nitrogen vacancies will also form complexes with some Mg acceptors, showing donor characteristics and causing self-compensation [24, 25]. Suppressing compensation is also crucial for p-type doping.

Generally speaking, GaN-based blue-green laser diodes require high luminescence characteristics of InGaN quantum wells, which also requires p-GaN to be grown at relatively low temperatures (<1000 °C) to protect the quantum wells. However, low-temperature growth of p-GaN can lead to a decline in material quality and electrical performance. We found that impurities such as carbon play a key role. Carbon impurities can form deep-level donors in GaN and compensate with Mg, and the carbon concentration is often high when p-GaN is grown at low temperatures, thereby reducing the carrier concentration of p-GaN [26]. The compensation mechanism and characterization methods of carbon impurities for p-type are briefly introduced below [27].

To study the role of carbon impurities in low-temperature-grown p-GaN, we grew a series of samples using MOCVD, including a high-temperature (T = 1050 °C) grown sample H1 and three low-temperature (T = 950 °C) grown samples L1, L2, and L3. Their growth parameters, surface morphology, and electrical data obtained through Hall testing are all listed in Table 1.2, where the Mg doping concentration of each sample is 1×10^{19} cm^{-3}. It can be seen that although the Mg doping concentration of each sample is the same, the resistivity of the low-temperature-grown p-GaN samples (L1, L2, and L3) is significantly higher than that of the high-temperature-grown sample (H1). Observing the three low-temperature-grown p-GaN samples, it can be seen that as the growth rate decreases, its resistivity shows a downward trend. However, compared with sample H1, the growth rates of L1, L2, and L3 are all relatively low, indicating that the high resistivity of low-temperature-grown p-GaN samples is not caused by a fast growth rate. Furthermore, it was found that the root mean square roughness of the surface of sample L3 is basically the same as that of sample H1, but its resistivity is significantly higher than that of sample H1, indicating that excessive surface roughness is also not the main reason for the high resistivity of low-temperature-grown p-GaN.

In order to determine the cause of the high resistivity of low-temperature-grown p-GaN thin films, the photoluminescence (PL) spectra of high-temperature and low-temperature-grown samples are shown in Fig. 1.9. There are clear differences in the PL spectra of low-resistivity samples (high-temperature growth) and high-resistivity samples (low-temperature growth). In the high-resistivity samples (low-temperature growth), there is only one yellow luminescence band with a central wavelength of 2.27 eV, while in the low-resistivity samples (high-temperature growth), the yellow luminescence band disappears and a relatively strong band-edge emission peak appears. Since in GaN materials, the yellow luminescence band has always been considered to be related to deep-level impurities, the difference in the intensity of the yellow luminescence band may be related to the concentration of deep-level impurities and compensation mechanisms in the samples under study. In order to find possible compensating impurities, a secondary ion mass spectrometer (SIMS) was used to test the depth distribution of carbon and oxygen impurities in samples H1, L1, L2, and L3, as shown in Fig. 1.10. It can be seen that, except for a very thin surface layer, the oxygen impurity concentration in the high-temperature-grown sample H1

Table 1.2 Growth temperature (GT), growth rate (GR), carbon impurity concentration, and resistivity of samples H1, L1, L2, and L3

Sample	GT (°C)	GR (μm/h)	RMS (nm)	Carbon (cm^{-3})	Resistivity (Ω·cm)
H1	1050	1.6	1.25	5.3×10^{16}	1.3
L1	950	1.4	5.68	5.1×10^{17}	1650
L2	950	0.8	1.36	2.6×10^{17}	528
L3	950	0.65	1.29	2.2×10^{17}	515

is basically the same as that in the low-temperature-grown samples. However, it is worth noting that the concentration of carbon impurities in the high-temperature sample (H1) is much lower than that in the low-temperature-grown samples (L1, L2, and L3). We all know that MOCVD uses carbon-containing organic sources as precursors, so carbon impurities are a very important unintentional doping impurity in GaN. Carbon impurities in GaN can take different forms [28, 29], such as substitutional (C_N, C_{Ga}), interstitial (C_i), or complexes. They can act as donors or acceptors, and their formation energy depends on the position of the Fermi level and the stoichiometric conditions of the material growth process. Therefore, defects related to unintentionally doped carbon impurities may be the source of compensating impurities in low-temperature-grown Mg-doped GaN thin films.

According to previous literature [30], the concentration of unintentionally doped carbon impurities is very sensitive to growth conditions, and it varies with growth

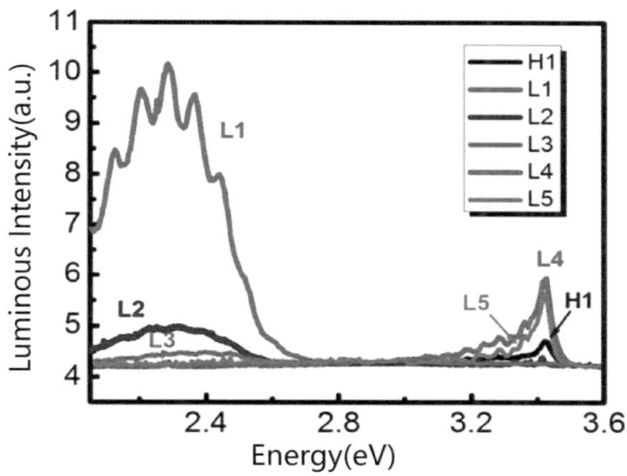

Fig. 1.9 Room-temperature PL spectra of samples H1, L1, L2, L3, L4, and L5

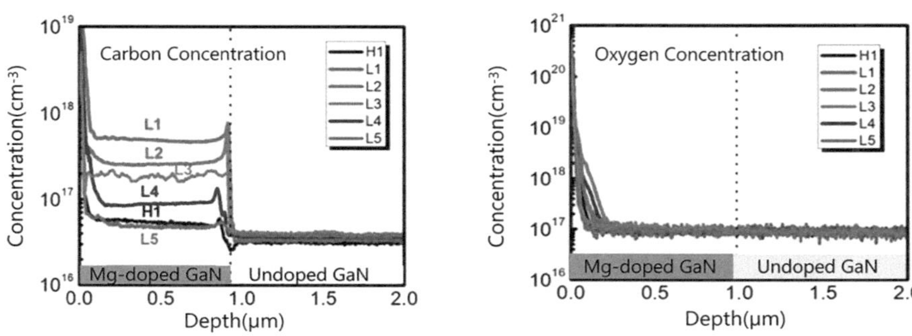

Fig. 1.10 SIMS depth analysis of carbon and oxygen impurities in samples H1, L1, L2, L3, L4, and L5

temperature, pressure, and NH_3/TMGa ratio decreases. Therefore, in order to verify the effect of carbon impurities on resistivity and to obtain low-resistivity low-temperature-grown p-GaN layers, two low-temperature-grown Mg-doped GaN thin films with low carbon impurity concentrations, namely L4 (grown with a higher V/III ratio, NH_3 flow rate increased to 6000 sccm) and L5 (grown with a higher growth pressure, 200 torr) were prepared by MOCVD. The growth parameters and electrical performance parameters of samples L2, L4, and L5 are listed in Table 1.3. Compared with sample L2, the carbon impurity concentration of sample L4 is reduced to 9×10^{16} cm^{-3}, and the resistivity is reduced to 2.2 Ω·cm. For sample L5, both the carbon impurity concentration and resistivity are reduced. Figure 1.11 shows the relationship between p-GaN resistivity and carbon impurity concentration. These results indicate that to achieve low-resistivity p-GaN, the doping concentration of carbon impurities in the p-GaN film must be reduced. On the other hand, as can be seen from Fig. 1.9, samples L4 and L5 with low carbon impurity concentration and low resistivity also do not have a yellow light emission peak. This indicates that the reduced yellow light emission is related to the low carbon impurity concentration and low resistivity.

Studies have shown that carbon impurities are closely related to the yellow luminescence band, and carbon impurities can introduce deep energy levels [31, 32]. Theoretical calculations show that the deep donor complex (C_N–O_N) formed by

Table 1.3 Growth pressure, growth rate, NH_3 flow rate, and carbon impurity concentration of low-temperature-grown samples (L2, L4, and L5)

Sample	GP (torr)	NH_3 (sccm)	GR (μm/h)	Carbon (cm^{-3})	Resistivity (Ω·cm)	Hole concentration (cm^{-3})
L2	100	3000	0.8	2.6×10^{17}	528	–
L4	100	6000	0.8	9×10^{16}	2.2	2.1×10^{17}
L5	200	3000	0.8	4.7×10^{16}	3.0	2×10^{17}

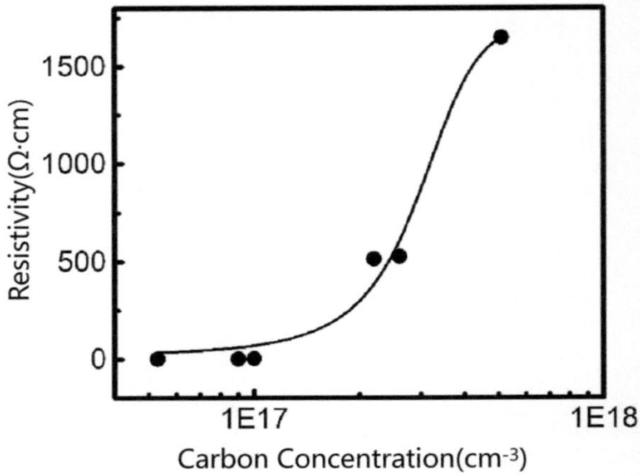

Fig. 1.11 Relationship between p-GaN resistivity and carbon impurity concentration

carbon impurities and oxygen impurities also produces yellow luminescence band [33]. First, the deep donor C_N–O_N complex is more likely to cause low Mg doping levels and high resistivity, which can compensate for the formation of Mg_{Ga} shallow donors, making it difficult for Mg acceptors to ionize into free carriers. Second, the C_N–O_N complex has a relatively low formation energy and has a higher concentration in p-GaN compared to other forms of C. Moreover, in the measured high-resistivity samples, the peak position (2.27 eV) of the PL spectrum is basically consistent with the yellow luminescence band caused by the C_N–O_N complex reported in the literature. Therefore, carbon impurities in low-temperature Mg-doped GaN may preferentially form C_N–O_N complexes and play a donor role in p-GaN, thereby disrupting p-type doping. Methods to reduce the concentration of carbon impurities by changing growth conditions, such as increasing the V/III ratio, increasing growth pressure, increasing growth temperature, and reducing growth rate, can all reduce the resistivity of p-GaN. In addition, we found that the role of H atoms in p-GaN is complex. On the one hand, H atoms can passivate Mg impurities and disrupt p-type; on the other hand, they can also passivate carbon impurities in p-GaN [34], suppressing the compensating effect of carbon impurities and improving p-type. Further in-depth research is needed on p-type doping of GaN materials.

1.5 Growth and Characterization of InGaN Quantum Wells

InGaN quantum wells are the core luminescent area of GaN-based light-emitting devices. The first high-brightness GaN-based LEDs used double heterojunctions [35], followed by quantum wells [36], further improving device performance. Currently, InGaN quantum wells are a basic component of GaN-based blue-green LEDs and laser diodes [37]. Studying the MOCVD growth technology of InGaN quantum wells is crucial for the development of GaN-based laser diodes. Many parameters need to be studied about the growth of InGaN quantum wells, including well layer thickness [38, 39], barrier layer In doping effects, etc. [40]. Here, we mainly introduce the characterization of quantum well growth and luminescent performance after the growth of the low-temperature GaN cap layer on the InGaN well layer [41, 42], which is of reference significance for the growth of high-quality InGaN quantum wells.

Using MOCVD, four InGaN/GaN quantum well samples with different cap layer thicknesses were grown on c-plane sapphire substrates. During growth, trimethylgallium (TMGa), trimethylindium (TMIn), and ammonia (NH_3) were used as the Ga source, In source, and N source, respectively. Silane (SiH_4) and bis(cyclopentadienyl)magnesium (Cp_2 Mg) were used as the dopants for n-type GaN and p-type GaN, respectively. Hydrogen and nitrogen were used as carrier gases for GaN and InGaN growth, respectively. The InGaN/GaN quantum well samples have two quantum well periods. During the growth process of the well layer, the In flow rate remained constant, and the GaN cap layer was grown at the same temperature

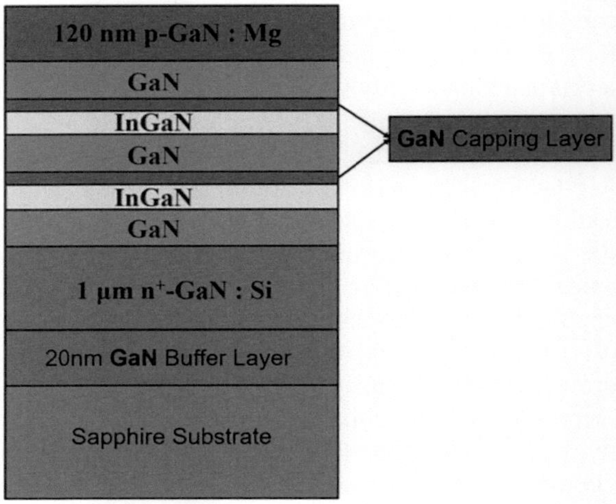

Fig. 1.12 Schematic diagram of InGaN/GaN multiquantum well structure

of 710 °C (Cap layer). Then the growth temperature was raised to 830 °C and maintained for a few seconds, after which the GaN barrier layer began to grow. The only difference in the growth process of the two samples was the difference in the growth time of the cap layer, with the cap layer growth times of samples A and B being 30 s and 200 s, respectively. The schematic diagram of the grown structure parameters is shown in Fig. 1.12.

In order to determine the average In composition, period thickness, and material quality of the InGaN/GaN multiquantum wells, we tested the high-resolution X-ray scanning curves of the two samples. Figure 1.13a shows the X-ray diffraction ω-2θ scanning curve of the (0002) plane of the sample. The −4 level satellite peaks of the two samples can be clearly seen, indicating that the quantum well structure has good periodicity. In addition, the average In composition and period thickness of the sample can be obtained by fitting, as shown in Table 1.4. The results show that as the cap layer thickness increases, the periodic thickness and average In composition of the InGaN/GaN quantum wells slightly increase. In fact, since the growth rate of the GaN cap layer is very low, about 0.006 nm/s, the change in barrier layer thickness is relatively small. However, we noticed that the growth conditions of the GaN cap layer not only affect the barrier layer thickness, but also affect the diffusion, evaporation, and redistribution processes of In atoms during the growth of the InGaN well layer. In addition, to characterize the stress state of the InGaN/GaN multiquantum well structure, we tested the reciprocal space mapping of the GaN (10–15) plane. The test results of sample A are shown in Fig. 1.13b, where the satellite peaks of the quantum wells and the GaN peaks are distributed on the same vertical line, indicating that neither of the two samples has undergone relaxation [43].

We also studied the optical properties. Figure 1.14 shows the variable power photoluminescence spectra of the two samples at 10 K, with an excitation wavelength of 405 nm. The two samples have distinctly different luminescence behaviors. For sample A, there is another luminescence peak A_1 at the low-energy end of the main

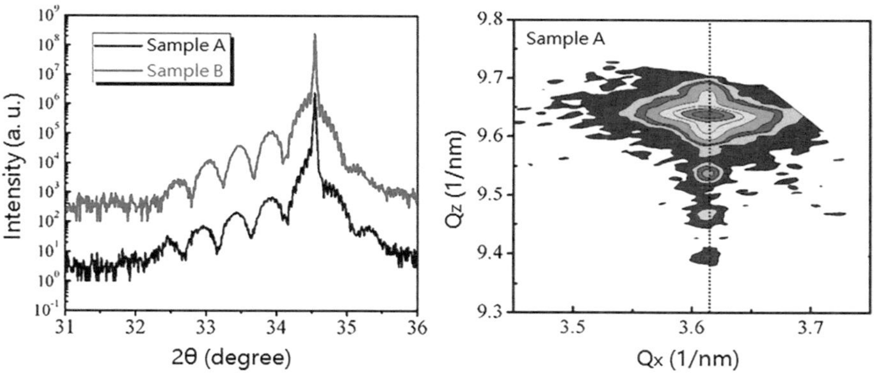

Fig. 1.13 **a** and **b** Show the X-ray diffraction test results of samples A and B

Table 1.4 Fitting results of ω-2θ scanning curves of two samples by XRD test

Sample	Cap layer growth time (s)	Period thickness (nm)	Barrier layer thickness (nm)	Well layer thickness (nm)	In composition of InGaN (%)
A	30	19.10	14.85	4.25	10.0
B	200	19.95	15.50	4.45	10.3

peak A_2. The energy difference between the luminescence peak A_1 and the main peak A_2 is 92 meV; therefore, it can be determined that the luminescence peak A_1 is a phonon sideband. The luminescence spectrum of sample B also accompanies the phonon sideband, defined as peak B_1. From the figure, it can also be seen that when the excitation light power is less than 5 mW, sample B only has one main luminescence peak B_2. However, when the excitation light power is greater than 10 mW, another luminescence peak B_3 appears at the high-energy end of the main peak B_2. As the excitation light power increases, peak B_3 gradually strengthens and replaces peak B_2 as the main peak. Again, we generally believe that at low temperatures, most of the optically excited carriers are distributed on the first excited state related to In clusters, where radiative recombination occurs, producing the luminescence peaks A_2 and B_2 [44].

In order to study the special luminescence peak B_3 in sample B, we tested the variable temperature and variable power photoluminescence spectra of sample B, as shown in Fig. 1.15. Figure 1.15 shows the variable temperature photoluminescence spectra under excitation powers of 5 mW and 40 mW, respectively. It is noted that the double-peak phenomenon in the variable temperature PL spectra under 40 mW excitation power is obvious below 200 K, but it becomes gradually blurred as the temperature continues to rise. Overall, within a very narrow range of excitation power, the main luminescence peak of sample B gradually shifts from the low-energy end luminescence peak to the high-energy end luminescence peak. Under low-power

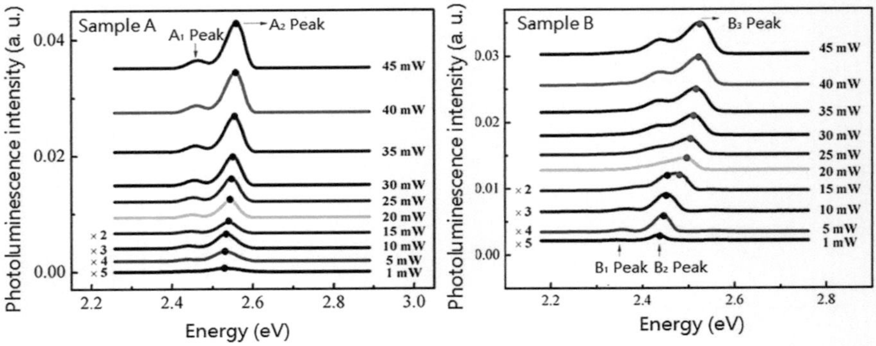

Fig. 1.14 Photoluminescence spectra of sample A (**a**) and sample B (**b**) at 10 K under variable power

conditions, the luminescence peak B_2 dominates; under high-power conditions, the luminescence peak B_3 dominates.

Furthermore, by carefully studying the main luminescence peak energy variation with temperature for the two samples, we can observe some special phenomena. As shown in Fig. 1.16a, when the excitation power of sample A increases from 5 to 40 mW, the PL peak energy variation curve with increasing temperature (hereinafter referred to as the E-T curve) is an "inverted V" curve, which is different from the conventional "S" curve. With the increase of excitation power, apart from the overall blueshift of the peak energy, the inverted V-shaped E-T curve hardly changes. Generally speaking, the inverted "V" temperature-dependent curve can be explained by the combined effect of carrier filling in the luminescent center and bandgap shrinkage at high temperatures [45, 46]. On the other hand, as shown in Fig. 1.16b, when the excitation power was below 5 mW, the E-T curve of sample B shows an inverted V shape, similar to sample A. However, when the excitation power gradually increases

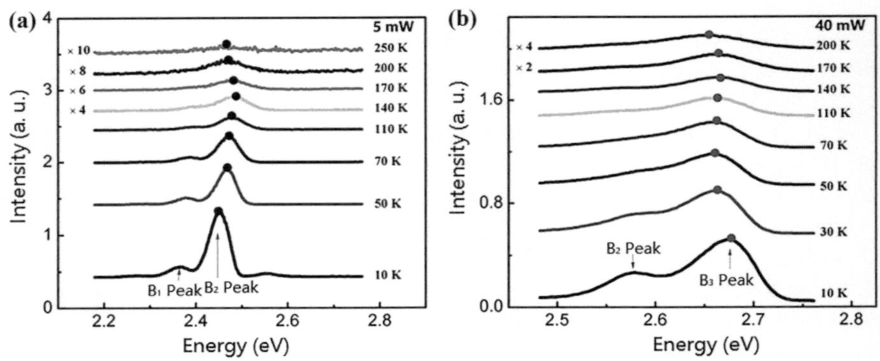

Fig. 1.15 Variable temperature photoluminescence spectra under excitation powers of 5 mW (**a**) and 40 mW (**b**), respectively

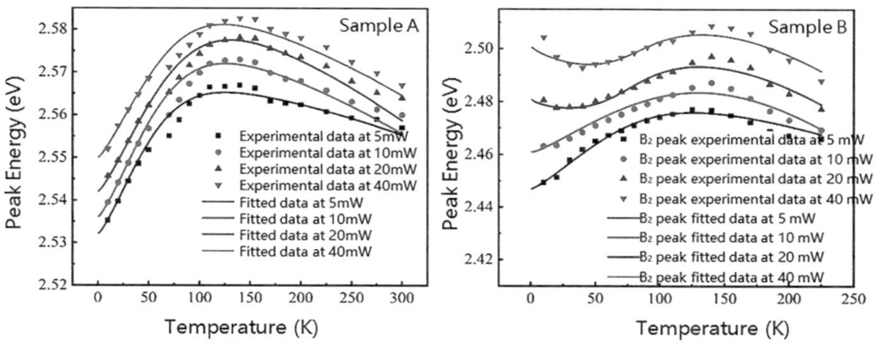

Fig. 1.16 Variation of photoluminescence peak energy with temperature for sample A (**a**) and sample B (**b**)

to 40 mW, a redshift segment first appeared in the lower temperature range, and the E-T curve showed a regular S shape at that time. Obviously, this phenomenon contradicts the traditional view that when the excitation power is large enough, the local effect completely disappears, and the dependence of the luminescence peak energy on temperature strictly follows the Varshni rule [47].

Therefore, in order to quantitatively explain the observed abnormal dependence of the E-T curve on the excitation power, we use the localized-state ensemble (LSE) luminescence model proposed by Professor Xu Shijie and his team members to fit the E-T curve [48]. This model is universal and can be used in all temperature ranges. It can not only fit the "S"-shaped E-T curve, but also fit the "V" or inverted "V"-shaped E-T curve. Furthermore, it can be proven that the LSE model can be simplified to the tail model of Eliseev et al. at high temperatures. In this model, the relationship between peak energy and temperature can be described as:

$$E(T) = E_0 - \frac{\alpha T^2}{\theta + T} - xk_B T$$

where θ is the Debye temperature of the material, α is the Varshni parameter, k_B is the Boltzmann constant, and x can be obtained by numerically solving the following transcendental equation:

$$xe^x = \left[\left(\frac{\sigma}{k_B T}\right)^2 - x\right]\left(\frac{\tau_r}{\tau_{nr}}\right)e^{(E_0 - E_a)/k_B T}$$

where σ is the standard deviation of the localized-state distribution, that is, the width of the Gaussian state density distribution. τ_r and τ_{nr} represent the radiative recombination lifetime and escape lifetime of localized carriers, respectively. E_0 is the central level of the localized states. All energy states below E_a are occupied at

0 K. Thus, E_a is similar to the Fermi level in the Fermi–Dirac distribution. Clearly, E_0 and E_a jointly determine the characteristics of the localized states [49].

Through a large amount of numerical calculation, the fitting parameter results of the two samples are shown in Table 1.5. For sample A, when the excitation light power increases from 5 to 40 mW, E_0 and E_a increase by 19 meV and 18 meV, respectively, while $E_0 - E_a$ and σ remain almost unchanged. This indicates that as the excitation light power increases, more and more carriers are excited. First, some photogenerated carriers will be used to shield the stronger polarization electric field of the InGaN layer, thus leading to an increase in the central energy level E_0. In addition, due to the carrier filling effect, more and more carriers occupy higher-energy levels, thus leading to an increase in the quasi-Fermi energy level E_a of the localized carrier distribution. Overall, $E_0 - E_a$ may represent the combined effect of carrier polarization electric field screening and carrier filling. Therefore, as the excitation light power increases, the luminescence peak of sample A undergoes an overall blueshift. Unlike sample A, for sample B, when the excitation light power increases from 5 to 40 mW, both E_0 and E_a significantly increase by 73 meV and 57 meV, respectively. $E_0 - E_a$ increases by 16 meV, τ_r/τ_{nr} also undergoes changes of several orders of magnitude, and σ slightly decreases. We believe these changes, which are very different from sample A, are due to the different sources of luminescence centers in sample B under 5 mW and 40 mW excitation light power. Therefore, for sample B, we believe that there are two different kinds of localized states. Due to the uneven distribution of In atoms in the InGaN layer, they exhibit different potential energy heights. That is, local regions with higher In composition become deep localized centers, while regions with lower In composition become shallow localized centers. To explain the abnormal luminescence behavior of sample B, we analyzed the redistribution behavior of carriers between the two kinds of localized states with changes in excitation light power and temperature, as shown in Fig. 1.17. At 10 K, under lower excitation light power conditions (5 mW), as shown in Fig. 1.17a, a large number of photogenerated carriers are distributed in the deep localized state, and the low-energy end luminescence peak dominates. But under the higher excitation light power (40 mW), as shown in Fig. 1.17b, more and more photogenerated carriers fill higher-energy levels, and the higher-energy levels in the previous localized state are also filled. As the light power increases, a luminescence peak appears at the high-energy end and gradually dominates. Combined with the fitting results, E_0 and E_a have greatly increased, and the value representing the escape ability of localized carriers has increased by several orders of magnitude. As the temperature increased to 30 K, when the laser power was 5 mW, as shown in Fig. 1.17c, the photogenerated carriers gained a certain amount of thermal energy and were then captured by the shallow localized centers, causing the E-T curve to first undergo a blueshift. However, when the excitation light power increased to 40 mW, as shown in Fig. 1.17d, more photogenerated carriers distributed in the shallow localized centers gain thermal energy and became more easily captured by the deep localized state with strong binding ability. Therefore, under the higher excitation light power conditions, the E-T curve first underwent a redshift. In other words, the abnormal change of the E-T curve with the excitation light power is related to the deep and shallow localized states produced by the uneven distribution of In components in the InGaN well layer of sample B, and the uneven distribution of In components is mainly attributed to the fluctuation of alloy atoms on the atomic scale during the growth process [50].

Upon further analysis, the abnormal luminescence peak that appears in sample B under high excitation power also leads to the abnormal change in the normalized luminescence intensity with temperature. Figure 1.18 shows the change in normalized intensity with temperature for samples A and B at excitation powers of 5 mW and 20 mW, respectively. We notice that the decay of the normalized luminescence intensity

Table 1.5 LSE model fitting results of E-T curves for samples A and B

Sample	Excitation light power (mW)	E_0 (eV)	E_a (eV)	$E_0 - E_a$ (meV)	$\frac{\tau_r}{\tau_{tr}}$	σ (meV)
A	5	2.57	2.532	38	0.004	14
	10	2.579	2.538	41	0.004	14
	20	2.585	2.545	40	0.004	15
	40	2.589	2.55	39	0.003	15
B	5	2.486	2.447	39	0.003	24
	10	2.516	2.461	55	0.009	30
	20	2.541	2.481	60	1.99	19
	40	2.559	2.504	55	15.01	20

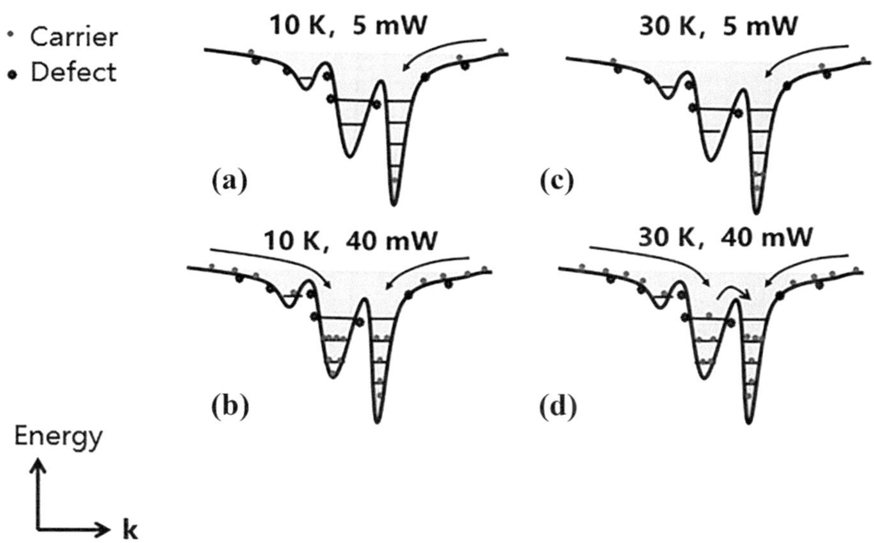

Fig. 1.17 Schematic diagram of carrier distribution in localized states of sample B at 10 K under the excitation power of 5 mW (**a**) and 40 mW (**b**) and at 30 K under the excitation power of 5 mW (**c**) and 40 mW (**d**)

with temperature in sample B is significantly faster than that in sample A. Generally, the thermal decay of the luminescence intensity of InGaN multiquantum well structures is controlled by nonradiative recombination centers and can be described by the Arrhenius equation. Therefore, the faster thermal decay behavior implies that sample B has relatively poor thermal stability. Furthermore, when the excitation power is high enough, a small number of nonradiative recombination centers are activated at relatively low temperatures and are easily filled by a large number of excess carriers. Therefore, under high excitation power, the thermal decay effect of nonradiative recombination centers on luminescence is not very significant. This can successfully explain why the luminescence intensity of sample A at high excitation power hardly changes with temperature in the low-temperature range, and the decay temperature is lower than that at low power. However, the situation with sample B is quite abnormal. When the temperature is below 125 K, the normalized luminescence intensity at an excitation power of 5 mW is surprisingly higher than that at 20 mW. When the temperature is above 125 K, the phenomenon is reversed. At 5 mW, sample B only shows a luminescence peak at the low-energy end, which we believe originates from the deep localized-state luminescence in the quantum well. At 20 mW, another luminescence peak appears at the high-energy end of sample B and dominates the luminescence spectrum, which we believe originates from the shallow localized-state luminescence in the quantum well. Therefore, we can conclude that the deep localized luminescence centers have a higher luminescence efficiency than the shallow localized luminescence centers. We believe that at an excitation power of 20 mW, sample B has two types of localized states involved in luminescence.

Based on the above analyses, we have confirmed that the luminescence peak B_3 of sample B originates from the deep localized state caused by the uneven distribution of indium components in the quantum well. In fact, indium atoms tend to accumulate on the upper surface of the InGaN quantum well layer and form an In-rich layer [51]. The growth of a thicker GaN cap layer at low temperatures is not conducive to the evaporation of In atoms in the In-rich layer, and In atoms can be incorporated into

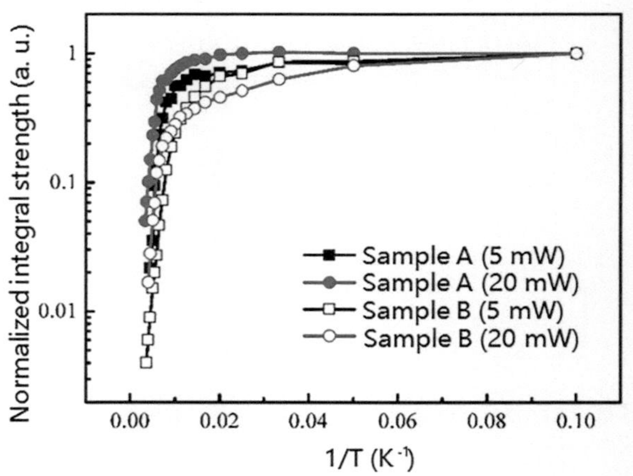

Fig. 1.18 Change in normalized photoluminescence intensity with temperature for samples A and B at 5 mW and 20 mW, respectively

the GaN cap layer and barrier layer [52], which will lead to an increase in the well layer thickness, thereby enhancing the quantum-confined Stark effect (QCSE). The stronger polarization field in the active region of the quantum well will cause more local relaxation, thereby forming different localized states and higher barriers. At the same time, more dislocations and defects are introduced into the subsequent growth of the GaN barrier layer, and In atoms tend to accumulate near the dislocations, thereby forming an uneven distribution [53, 54]. This implies that as the thickness of the GaN cap layer increases, the distribution fluctuations of In atoms become more pronounced. It can be seen that the distribution of In components in the thicker cap layer sample B is very uneven, and appropriately reducing the cap layer thickness is beneficial to improve uniformity. Quantum wells with high uniformity are beneficial for the preparation of laser diodes.

1.6 Device Process and Characterization

The entire development process of GaN-based laser diodes is very complex, including material growth, structural design, and process fabrication. The design of laser diodes must take into account both the distribution characteristics of the electric field and the light field [55, 56]. In other words, structural design also plays a crucial role in their performance [57]. For example, laser diodes are high current injection devices, and their carrier leakage becomes very serious. In traditional GaN-based blue-violet laser diodes, due to the shallow quantum wells, the restriction on electrons is weak, which often causes serious electron leakage current. We found that using InGaN insertion layers effectively reduces the electron leakage current of blue-violet laser diodes [58, 59]. Many literature studies have found that electron injection and leakage have a significant impact on the device [60–62]. We proposed that the polarization-reversed quaternary alloy AlInGaN as an electron barrier layer, GaN/InGaN composite barrier layer structure, etc., can significantly reduce electron leakage [63, 64] and found that the hole barrier layer structure improves the performance of ultraviolet laser diodes [65]. The structural design must also take into account the actual material loss [66], in order to more effectively guide device development. Based on the material growth and structural design, we further studied process technology, and the following mainly introduces the process and preparation process of GaN-based laser diodes [42].

The basic structure of the GaN-based laser diode includes the following in the growth direction: n-electrode, substrate, n-type optical confinement layer, lower waveguide (LWG), active region, upper waveguide (UWG), electron barrier layer, p-type optical confinement layer, contact layer, and p-electrode. The general growth process of the laser structure is as follows, as shown in Fig. 1.19.

Among them, in the epitaxial direction, they are: (1) N-face electrode: Ti/Al/Ti/Au; (2) Substrate: GaN; (3) Optical confinement layer (n-CL): n-AlGaN; (4) Lower waveguide (LWG): GaN or InGaN; (5) Active region: InGaN/GaN quantum well;

Fig. 1.19 Schematic diagram of the three-dimensional structure of the laser

(6) Upper waveguide (UWG): GaN or InGaN; (7) Electron blocking layer (EBL): p-AlGaN; (8) Optical confinement layer (p-CL): p-AlGaN; (9) Contact layer: p-GaN; (10) p-electrode.

The fabrication process of the GaN laser diode is shown in Fig. 1.20.

(1) Sputtering: Clean the laser epitaxial wafer to remove surface contamination and oxide layer, then sputter p-type ohmic contact metal Pd/Pt/Au, and anneal to form a good ohmic contact. (2) Photolithography: Apply photoresist, expose, and pattern the ridge of the laser. (3) Etching: Use an ion beam etching machine (IBE) or inductively coupled plasma etching machine (ICP) to etch and form the ridge of the laser. (4) Deposition: Deposit SiO_2 insulating film on the etched laser epitaxial wafer. (5) Stripping: Soak in acetone, strip the photoresist and SiO_2 on the ridge, exposing the p-type electrode. (6) Photolithography: Pattern the thickened electrode of the laser. (7) Sputtering: Sputter the thickened electrode of the laser. (8) Stripping: Strip the photoresist on the laser epitaxial wafer, at this point the p-side process of the

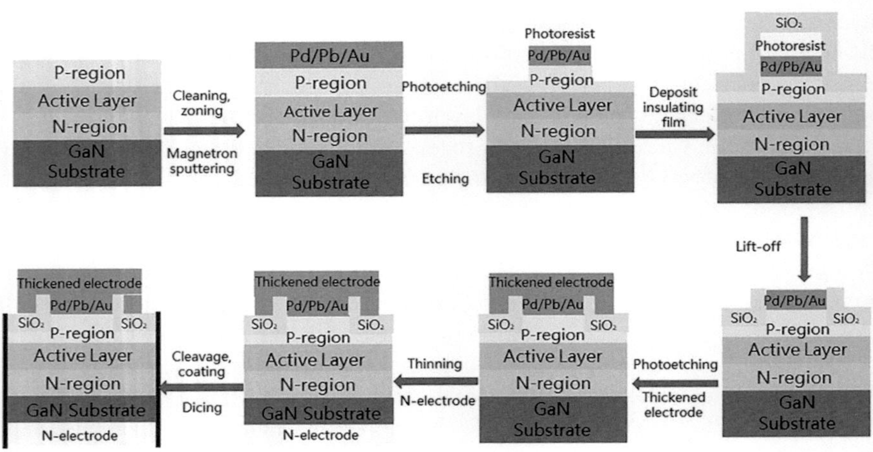

Fig. 1.20 GaN laser fabrication process flowchart

laser is completed. (9) Thinning and polishing: Thin the substrate of the laser; after thinning to about 120 μm, polish to reduce roughness. (10) Etching: Etch the polished laser substrate to form a better ohmic contact with the n-type metal electrode. (11) Sputtering: Sputter the n-electrode Ti/Pt/Au on the back of the laser. (12) Cleaving: Use a laser dicing machine to cleave the laser into bars. (13) Coating: Coat the cavity of the laser. The front and back cavities are coated with different reflectivity films of SiO_2/TiO_2 multilayer dielectric film. (14) Dividing: Divide the laser bars into single laser chips. (15) Packaging: Form a complete device.

There are many technical details in the process of GaN-based laser diodes, including dicing, cleaning, etching, coating, etc., which also affect the performance of the device. Here we briefly introduce the impact of dicing on the process. After obtaining the epitaxial wafer, a laser dicing machine was first used to cut the epitaxial wafer and divide it into four areas as shown in Fig. 1.21. Since the (11–20) direction is the cleavage edge of the GaN material, the slicing perpendicular to the short edge direction can be cleaved without using a laser dicing machine, while the cutting along the (1–100) direction requires a laser dicing machine.

In the experiment, a laser scribing machine was used for scribing. The laser scribing machine is used because the spot diameter obtained after the laser is focused through the optical lens is small, the energy density is high, the speed is fast, the accuracy is high, and the damage to the epitaxial wafer is relatively small. Moreover, it is a noncontact form and does not apply any external force to the epitaxial wafer itself. Generally speaking, we can control the depth of the scribe by controlling parameters such as laser power, scribing speed, and number of scribes to avoid causing too deep penetration and damaging the epitaxial wafer.

At the end of the entire process, when separating the bars into individual cores, we also choose to use a laser scribing machine to scribe on the back of the bar and then split the wafer with a splitting machine. As shown in Fig. 1.22a, the laser scribe (black dot) is very obvious at the front and back cavity of the core, indicating that

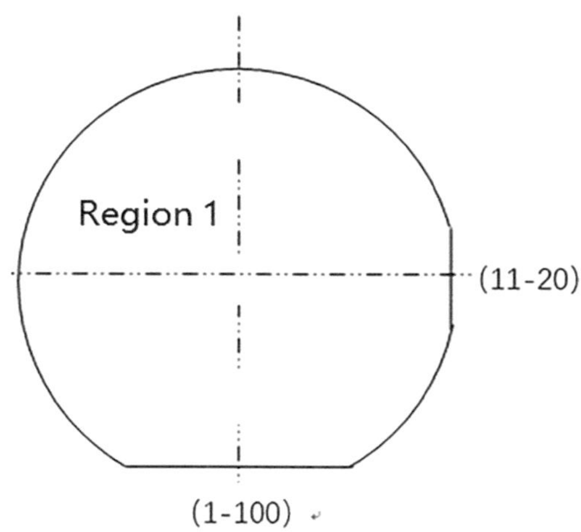

Fig. 1.21 Schematic diagram of GaN-based laser epitaxial wafer

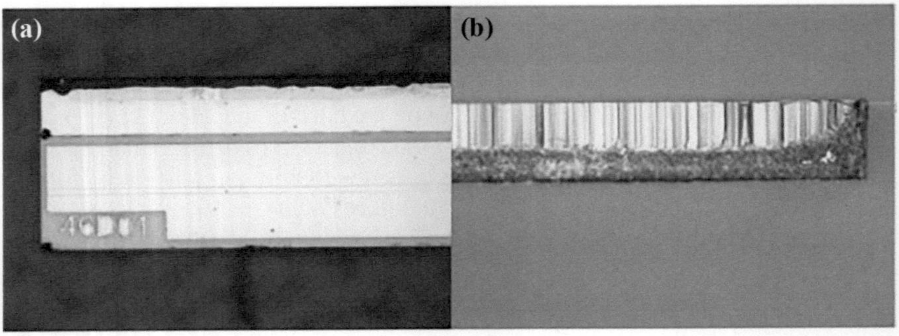

Fig. 1.22 Laser scribe on the surface of the laser from the front (**a**) and cross-sectional view of the laser scribe (**b**)

the laser has penetrated the entire epitaxial wafer from the back at the black dot position, causing damage to the front electrode. During the device testing process, it is very easy to cause a short circuit of the device. At the same time, we extracted the cross-sectional view of the black dot, as shown in Fig. 1.22b. We found that near the front and back cavity, that is, when the laser scribes, the starting and ending positions of the laser contacting the epitaxial wafer penetrate the entire epitaxial layer, while the laser scribes at other positions in the middle penetrate only about one-third of the thickness of the epitaxial wafer. After we changed to another laser scribing machine, no similar phenomenon occurred. Hence, this phenomenon is due to the laser scribing machine equipment. Therefore, it is necessary to carefully adjust the scribing speed, laser power, and number of scribes of the scribing machine.

Under the condition that other parameters remain unchanged, Fig. 1.23a and b, respectively, show the cross-sectional views of the scribes when the scribing speed is 25 and 30 mm/s. It can be clearly seen that when the scribing speed is 30 mm/s, the scribe depth is 34 microns, and the cross section is relatively flat; when the scribing speed decreases to 25 mm/s, the scribe depth can reach 68 microns, and the cross-sectional cracks significantly increase. It can be seen that the scribing speed has a great impact on the scribe depth and cross-sectional structure. Therefore, we must ensure that while the epitaxial wafer can be split, the damage to the epitaxial wafer by the laser is minimized. In the subsequent process steps of separating the core with a laser scribing machine and a splitting machine, this point is particularly important, directly affecting the flatness of the cross section of the separated core and the generation of cracks in the core, thereby affecting the performance of the laser diodes.

In addition to process technology, the physics of the process also needs to be studied in depth. P-type ohmic contact is an important technology, which has an important impact on the working voltage of the laser diodes. We studied the carrier transport mechanism when the Ni/Au metal system forms an ohmic contact with p-GaN and found that the existence of NiO plays an important role in the formation of ohmic contact with p-GaN [67]. After rapid annealing, the direct contact of p-NiO

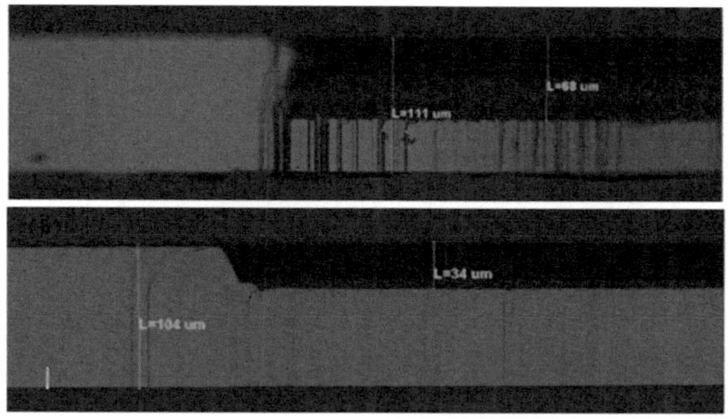

Fig. 1.23 Cross-sectional views of the scribes when the scribing speed is 25 mm/s (**a**) and 30 mm/s (**b**)

with p-GaN reduces the barrier height of the Schottky contact, which is beneficial for the carrier transport process. In addition, the experimental results show that the thickness of the metal Ni has an important impact on the ohmic contact performance of p-GaN, and the optimal thickness of the metal Ni is 15 nm. In addition, we also found that, at the interface between the metal and p-GaN, the appropriate introduction of carbon impurities [68, 69] is beneficial for hole transport and achieves good ohmic contact characteristics.

Based on the solutions for epitaxy, structure, and process, the researchers from the Institute of Semiconductors, Chinese Academy of Sciences, have successfully fabricated GaN-based blue, green, violet, and ultraviolet lasers. Figure 1.24 shows the power–current–voltage curve (PIV curve) and lasing spectrum of a high-power GaN-based blue laser diode. The laser has a stripe width of 30 μm, a cavity length of 1200 μm, a threshold current of 400 mA, a corresponding threshold current density of 1.1 kA/cm^2, a lasing wavelength of 442 nm, and a continuous output power of 6.0 W at room temperature [70]. The successful fabrication of a GaN-based deep-ultraviolet laser diode by us Zhao et al. [71], with a lasing wavelength of 271.8 nm, represents a milestone breakthrough in the field of semiconductor lasers. The realization of GaN-based ultraviolet laser diodes has also attracted international attention [72, 73].

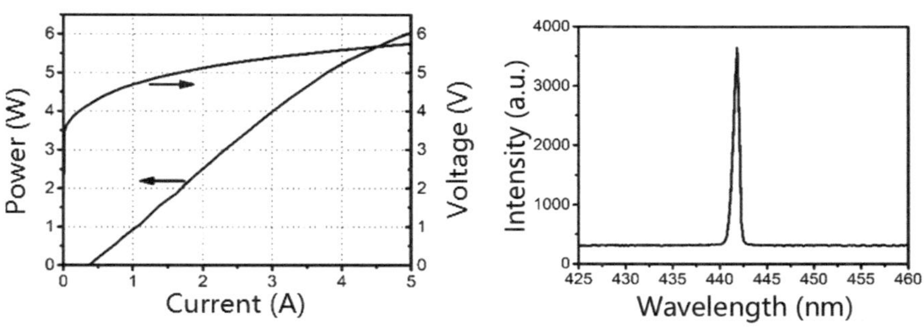

Fig. 1.24 PIV curve and lasing spectrum of GaN-based violet laser

1.7 Summary and Outlook

Since the advent of GaN-based laser diodes, a range of laser diodes with different wavelengths, including blue, ultraviolet, and green, have been successfully realized. From the development trend, GaN-based lasers will move toward multiwavelength and high power. In terms of long wavelengths, GaN-based laser diodes will achieve red or even infrared lasing. In terms of short wavelengths, GaN-based laser diodes will develop toward deep ultraviolet. In 2019, H. Amano and colleagues developed a 271.8 nm GaN-based deep UV laser diodes which represent an important breakthrough in the semiconductor laser diodes [73]. In addition, vertical-cavity surface-emitting lasers are also an important research direction for GaN-based laser diodes. GaN-based laser diodes are high-difficulty devices that require a perfect combination of materials, structures, and processes, and packaging technology also has a significant impact on the reliability [74]. GaN-based laser diodes will ultimately change human life.

References

1. Nakamura S, Senoh M, Nagahama S, Iwasa N, Yamada T, Matsushita T, Kiyoku H, Sugimoto Y (1996) InGaN-based multi-quantum-well-structure laser diodes. Jpn J Appl Phys 35:L74–L76
2. Manasevit HM (1968) Single-crystal gallium arsenide on insulating substrates. Appl Phys Lett 12:156–159
3. Lu D, Duan S (2009) Fundamentals and applications of metal organic chemical vapor deposition. Science Press, Beijing
4. Liu W (2016) Study on the luminescent characteristics of green light InGaN/GaN multi-quantum wells. Doctoral dissertation, Institute of Semiconductors, Chinese Academy of Sciences
5. Nakamura S, Mukai T, Senoh M (1992) In situ monitoring and Hall measurements of GaN grown with GaN buffer layers. J Appl Phys 71:5543–5549
6. Zhao DG, Liu ZS, Zhu JJ, Zhang SM, Jiang DS, Yang H, Liang JW, Li XY, Gong HM (2006) Effect of Al incorporation on the AlGaN growth by metalorganic chemical vapor deposition. Appl Surf Sci 253:2452–2455
7. Amano H, Sawaki N, Akasaki I, Toyoda Y (1986) Metalorganic vapor phase epitaxial growth of a high quality GaN film using an AlN buffer layer. Appl Phys Lett 48:353–355
8. Nakamura S (1991) GaN growth using GaN buffer layer. Jpn J Appl Phys 30:L1705-1707
9. Zhao DG, Zhu JJ, Liu ZS, Zhang SM, Yang H, Jiang DS (2004) Surface morphology of AlN buffer layer and its effect on GaN growth by metalorganic chemical vapor deposition. Appl Phys Lett 85:1499–1501
10. Heinke H, Kirchner V, Einfeldt S, Hommel D (2000) X-ray diffraction analysis of the defect structure in epitaxial GaN. Appl Phys Lett 77:2145–2147
11. Chierchia R, Bottcher T, Heinke H, Einfeldt S, Figge S, Hommel D (2003) Microstructure of heteroepitaxial GaN revealed by x-ray diffraction. J Appl Phys 93:8918–8925
12. Zhao DG, Jiang DS, Zhu JJ, Guo X, Liu ZS, Zhang SM, Wang YT, Yang H (2009) Defect evolution and accompanied change of electrical properties during the GaN growth by metalorganic chemical vapor deposition. J Alloy Compd 487:400–403
13. Le LC, Zhao DG, Wu LL, Deng Y, Jiang DS, Zhu JJ, Liu ZS, Wang H, Zhang SM, Zhang BS, Yang H (2011) Effects of sapphire nitridation on GaN growth by metalorganic chemical vapour deposition. Chin Phys B 20:127306

14. Zhao DG, Jiang DS, Zhu JJ, Liu ZS, Zhang SM, Yang H, Liang JW (2007) Influence of V/III ratio in the initial growth stage on the properties of GaN epilayer deposited on low temperature AlN buffer layer. J Cryst Growth 303:414–418
15. Zhao DG, Yang H, Zhu JJ, Jiang DS, Liu ZS, Zhang SM, Wang YT, Liang JW (2006) Effects of edge dislocations and intentional Si doping on the electron mobility of n-type GaN films. Appl Phys Lett 89:112106
16. Kyle ECH, Kaun SW, Burke PG, Wu F, Wu Y-R, Speck JS (2014) High-electron-mobility GaN grown on free-standing GaN templates by ammonia-based molecular beam epitaxy. J Appl Phys 115:193702
17. Götz W, Johnson NM, Walker J, Bour DP, Street RA (1996) Activation of acceptors in Mg-doped GaN grown by metalorganic chemical vapor deposition. Appl Phys Lett 68:667–669
18. Götz W, Johnson NM, Bour DP, McCluskey MD, Haller EE (1996) Local vibrational modes of the Mg–H acceptor complex in GaN. Appl Phys Lett 69:3725–3727
19. Reboredo FA, Pantelides ST (1999) Novel defect complexes and their role in the p-type doping of GaN. Phys Rev Lett 82:1887–1890
20. Amano H, Kito M, Hiramatsu K, Akasaki I (1989) P-Type conduction in Mg-doped GaN treated with low-energy electron-beam irradiation (LEEBI). Jpn J Appl Phys 28:L2112–L2114
21. Nakamura S, Senoh M, Mukai T (1991) Highly p-Typed Mg-doped GaN films grown with GaN buffer layers. Jpn J Appl Phys 30:L1708–L1711
22. Nakamura S, Iwasa N, Senoh M, Mukai T (1992) Hole compensation mechanism of p-Type GaN films. Jpn J Appl Phys 31:1258–1266
23. Nakamura S, Mukai T, Senoh M, Iwasa N (1992) Thermal annealing effects on p-Type Mg-doped GaN films. Jpn J Appl Phys 31:L139–L142
24. Hautakangas S, Oila J, Alatalo M, Saarinen K, Liszkay L, Seghier D, Gislason HP (2003) Vacancy defects as compensating centers in Mg-doped GaN. Phys Rev Lett 90:137402
25. Obloh H, Bachem KH, Kaufmann U, Kunzer M, Maier M, Ramakrishnan A, Schlotter P (1998) Self-compensation in Mg doped p-Type GaN grown by MOCVD. J Cryst Growth 195:270–273
26. Yang J, Zhao DG, Jiang DS, Chen P, Liu ZS, Le LC, Li XJ, He XG, Liu JP, Zhang SM, Wang H, Zhu JJ, Yang H (2014) Investigation on the compensation effect of residual carbon impurities in low temperature grown Mg doped GaN films. J Appl Phys 115:163704
27. Yang J (2015) Study on the growth of InGaN multi-quantum well materials and solar cell devices. Ph.D. thesis, Institute of Semiconductors, Chinese Academy of Sciences
28. Wright AF (2002) Substitutional and interstitial carbon in wurtzite GaN. J Appl Phys 92:2575–2585
29. Lyons JL, Janotti A, Van de Walle CG (2010) Carbon impurities and the yellow luminescence in GaN. Appl Phys Lett 97:152108
30. Koleske DD, Wickenden AE, Henry RL, Twigg ME (2002) Influence of MOVPE growth conditions on carbon and silicon concentrations in GaN. J Cryst Growth 242:55–69
31. Liang F, Zhao DG, Jiang DS, Liu ZS, Zhu JJ, Chen P, Yang J, Liu ST, Xing Y, Zhang LQ, Li M, Zhang YT, Du GT (2018) Carbon-related defects as a source for the enhancement of yellow luminescence of unintentionally doped GaN. Nanomaterials 8:744
32. Liang F, Zhao DG, Jiang DS, Liu ZS, Zhu JJ, Chen P, Yang J, Liu ST, Xing Y, Zhang LQ (2018) Role of Si and C impurities in yellow and blue luminescence of unintentionally and Si-doped GaN. Nanomaterials 8:1026
33. Demchenko DO, Diallo IC, Reshchikov MA (2013) Yellow luminescence of gallium nitride generated by carbon defect complexes. Phys Rev Lett 110:087404
34. Zhang YH, Liang F, Zhao DG, Jiang DS, Liu ZS, Zhu JJ, Yang J, Liu ST (2020) Hydrogen can passivate carbon impurities in Mg-doped GaN. Nanoscale Res Lett 15:38
35. Nakamura S, Senoh M, Mukai T (1993) High-power InGaN/GaN double-heterostructure violet light emitting diodes. Appl Phys Lett 62:2390–2392
36. Nakamura S, Senoh M, Iwasa N, Nagahama S (1995) High-brightness InGaN blue, green and yellow light-emitting diodes with quantum well structures. Jpn J Appl Phys 34:L797–L799
37. Nakamura S (1998) The roles of structural imperfections in InGaN-based blue light-emitting diodes and laser diodes. Science 281:956–961

38. Peng LY, Zhao DG, Jiang DS, Zhu JJ, Liu ZS, Chen P, Yang J, Liu W, Liang F, Xing Y, Liu ST, Zhang LQ, Wang WJ, Li M, Zhang YT, Du GT (2018) Anomalous electroluminescent blue-shift behavior induced by well widths variance and localization effect in InGaN/GaN multi-quantum wells. Opt Express 26:21736
39. Liu W, Zhao DG, Jiang DS, Chen P, Liu ZS, Zhu JJ, Shi M, Zhao DM, Li X, Liu JP, Zhang SM, Wang H, Yang H (2015) Localization effect in green light emitting InGaN/GaN multiple quantum wells with varying well thickness. J Alloy Compd 625:266–270
40. Wang XW, Yang J, Zhao DG, Jiang DS, Liu ZS, Liu W, Liu ST, Xing Y, Wang WJ, Li M (2018) Influence of In doping in GaN barriers on luminescence properties of InGaN/GaN multiple quantum well LEDs. Superlattices Microstruct 114:32–36
41. Xing Y, Zhao DG, Jiang DS, Liu ZS, Zhu JJ, Chen P, Yang J, Liang F, Liu ST, Zhang LQ (2019) Carrier redistribution between two kinds of localized states in the InGaN/GaN quantum wells studied by photoluminescence. Nanoscale Res Lett 14:88
42. Yao X (2020) Research on key technologies of GaN-based lasers. Ph.D. thesis, Institute of Semiconductors, Chinese Academy of Sciences
43. Rhode SL, Fu WY, Moram MA, Massabuau FC-P, Kappers MJ, McAleese C, Oehler F, Humphreys CJ, Dusane RO, Sahonta S-L (2014) Structure and strain relaxation effects of defects in $In_xGa_{1-x}N$ epilayers. J Appl Phys 116:103513
44. Pozina G, Bergman JP, Monemar B, Takeuchi T, Amano H, Akasaki I (2000) Origin of multiple peak photoluminescence in InGaN/GaN multiple quantum wells. J Appl Phys 88:2677–2681
45. Cho Y-H, Gainer GH, Fischer AJ, Song JJ, Keller S, Mishra UK, DenBaars SP (1998) S-shaped temperature-dependent emission shift and carrier dynamics in InGaN/GaN multiple quantum wells. Appl Phys Lett 73:1370–1372
46. Eliseev PG, Perlin P, Lee J, Osiński M (1997) Blue temperature-induced shift and band-tail emission in InGaN-based light sources. Appl Phys Lett 71:569–571
47. Wang HN, Ji ZW, Qu S, Wang G, Jiang YZ, Liu BL, Xu XG, Mino HF (2012) Influence of excitation power and temperature on photoluminescence in InGaN/GaN multiple quantum wells. Opt Express 20:3392–3940
48. Li Q, Xu SJ, Xie MH, Tong SY (2005) A model for steady-state luminescence of localized-state ensemble. Europhys Lett 71:994–1000
49. Wang ZL, Wang L, Xing YC, Yang D, Yu JD, Hao ZB, Sun CZ, Xiong B, Han YJ, Wang J, Li HT, Luo Y (2017) Consistency on two kinds of localized centers examined from temperature-dependent and time-resolved photoluminescence in InGaN/GaN multiple quantum wells. ACS Photonics 4:2078–2084
50. Watson-Paris D, Godfrey MJ, Dawson P, Oliver RA, Galtrey MJ, Kappers MJ, Humphreys CJ (2011) Carrier localization mechanisms in $In_xGa_{1-x}N$/GaN quantum wells. Phys Rev B 83:115321
51. Van Daele B, Van Tendeloo G, Jacobs K, Moerman I, Leys MR (2004) Formation of metallic In in InGaN/GaN multiquantum wells. Appl Phys Lett 85:4379–4381
52. Hoffmann L, Bremers H, Jönen H, Rossow U, Schowalter M, Mehrtens T, Rosenauer A, Hangleiter A (2013) Atomic scale investigations of ultra-thin GaInN/GaN quantum wells with high indium content. Appl Phys Lett 102:102110
53. Liu JP, Wang YT, Yang H, Jiang DS, Jahn U, Ploog KH (2004) Investigations on V-defects in quaternary AlInGaN epilayers. Appl Phys Lett 84:5449–5451
54. Han D, Ma SF, Jia ZG, Jia W, Liu PZ, Dong HL, Shang L, Zhang AQ, Zhai GM, Li XM, Liu XG, Xu BS (2017) Photoluminescence close to V-shaped pits in the quantum wells and enhanced output power for InGaN light emitting diode. J Phys D Appl Phys 50:475103
55. Liang F, Zhao DG, Jiang DS, Liu ZS, Zhu JJ, Chen P, Yang J, Liu W, Li X, Liu ST, Xing Y, Zhang LQ, Long H, Li M (2017) New design of upper waveguide with unintentionally doped InGaN layer for InGaN-based laser diode. Opt Laser Technol 97:284–289
56. Yang J, Zhao DG, Liu ZS, Jiang DS, Zhu JJ, Chen P, Liang F, Liu ST, Liu W, Xing Y, Li M (2018) Suppression the leakage of optical field and carriers in GaN-based laser diodes by using InGaN barrier layers. IEEE Photonics J 10:1503107

57. Yang J, Zhao DG, Zhu JJ, Liu ZS, Liang F, Jiang DS, Chen P, Liu ST, Liu W, Li M (2019) Enhancing the performance of GaN based LDs by using low In content InGaN instead of GaN as lower waveguide layer. Opt Laser Technol 111:810–813
58. Le LC, Zhao DG, Jiang DS, Chen P, Liu ZS, Yang J, He XG, Li XJ, Liu JP, Zhu JJ, Zhang SM, Yang H (2014) Suppression of electron leakage by inserting a thin undoped InGaN layer prior to electron blocking layer in InGaN-based blue-violet laser diodes. Opt Express 22:11392
59. Le L (2015) Research on the basic problems of GaN-based material growth and light emitting devices. Doctoral dissertation, Institute of Semiconductors, Chinese Academy of Sciences
60. Piprek J, Sink RK, Hansen MA, Bowers JE, DenBaars SP (2000) Simulation and optimization of 420 nm InGaN/GaN laser diodes. Proc SPIE 3944:28–39
61. Hansen M, Piprek J, Pattison PM, Speck JS, Nakamura S, DenBaars SP (2002) Higher efficiency InGaN laser diodes with an improved quantum well capping configuration. Appl Phys Lett 81:4275–4277
62. Kuo YK, Chang YA (2004) Effects of electronic current overflow and inhomogeneous carrier distribution on InGaN quantum-well laser performance. IEEE J Quantum Electron 40:437–444
63. Le LC, Zhao DG, Jiang DS, Chen P, Liu ZS, Zhu JJ, Yang J, Li XJ, He XG, Liu JP, Zhang SM, Yang H (2015) Utilization of polarization-inverted AlInGaN or relatively thinner AlGaN electron blocking layer in InGaN-based blue–violet laser diodes. J Vac Sci Technol, B 33:011209
64. Chen P, Zhao DG, Jiang DS, Zhu JJ, Liu ZS, Le LC, Yang J, Li X, Zhang LQ, Liu JP, Zhang SM, Yang H (2015) The effect of composite GaN/InGaN last barrier layer on electron leakage current and modal gain of InGaN-based multiple quantum well laser diodes. Phys Status Solidi A 212:2936–2943
65. Xing Y, Zhao DG, Jiang DS, Liang F, Chen P, Zhu JJ, Liu ZS, Yang J, Liu W, Liu ST, Zhang LQ, Li M (2017) Suppression of hole leakage by adding a hole blocking layer prior to the first quantum barrier in GaN-based near-ultraviolet laser diodes. Phys Status Solidi A 214:1700320
66. Yang J, Zhao DG, Zhu JJ, Liu ZS, Jiang DS, Chen P, Liang F, Liu ST, Xing Y (2019) Effect of Mg doping concentration of electron blocking layer on the performance of GaN-based laser diodes. Appl Phys B Lasers Opt 125:235
67. Li XJ, Zhao DG, Jiang DS, Liu ZS, Chen P, Zhu JJ, Le LC, Yang J, He XG, Zhang SM, Zhang BS, Liu JP, Yang H (2014) The significant effect of the thickness of Ni film on the performance of the Ni/Au Ohmic contact to p-GaN. J Appl Phys 116:163708
68. Liang F, Zhao DG, Jiang DS, Liu ZS, Zhu JJ, Chen P, Yang J, Liu W, Li X, Liu ST, Xing Y, Zhang LQ, Li M, Zhang J (2017) Influence of residual carbon impurities in a heavily Mg-doped GaN contact layer on an Ohmic contact. Appl Opt 56:4197–4200
69. Liang F, Zhao DG, Jiang DS, Liu ZS, Zhu JJ, Chen P, Yang J, Liu W, Li X, Liu ST, Xing Y, Zhang LQ, Yang H, Long H, Li M (2017) Improvement of Ohmic contact to p-GaN by controlling the residual carbon concentration in p++-GaN layer. J Cryst Growth 467:1–5
70. Liang F, Zhao DG, Liu ZS, Chen P, Yang J, Duan LH, Shi YS, Wang H (2021) GaN-based blue laser diode with 6.0 W of output power under continuous-wave operation at room temperature. J Semicond 42:112801
71. Zhao DG, Yang J, Liu ZS, Chen P, Zhu JJ, Jiang DS, Shi YS, Wang H, Duan LH, Zhang LQ, Yang H (2017) Fabrication of room temperature continuous-wave operation GaN-based ultraviolet laser diodes. J Semicond 38:051001
72. Yasue S, Sato K, Kawase Y, Ikeda J, Sakuragi Y, Iwayama S, Iwaya M, Kamiyama S, Takeuchi T, Akasaki I (2019) The dependence of AlN molar fraction of AlGaN in wet etching by using tetramethylammonium hydroxide aqueous solution. Jpn J Appl Phys 58:SCCC30
73. Zhang Z, Kushimoto M, Sakai T, Sugiyama N, Schowalter LJ, Sasaoka C, Amano H (2019) A 271.8 nm deep-ultraviolet laser diode for room temperature operation. Appl Phys Express 12:124003
74. Wang XW, Liu ZS, Zhao DG, Chen P, Liang F, Yang J (2021) New mechanisms of cavity facet degradation for GaN-based laser diodes. J Appl Phys 129:223106

Chapter 2
Growth, Fabrication, and Characterization of GaN-Based Lasers on Silicon Substrates

Meixin Feng, Jianxun Liu, Qian Sun, and Hui Yang

2.1 Introduction

Group III-nitride semiconductor materials, with their emission wavelengths covering from deep ultraviolet to near-infrared, are widely used in the fabrication of light-emitting diodes (LEDs) and laser diodes (LDs). Compared to LEDs, lasers operate at two orders of magnitude higher current density, making them very challenging. Thanks to the joint efforts of scientists, gallium nitride (GaN)-based lasers have made significant progress, with a threshold current density as low as 0.6 kA/cm^2 and a wall-plug efficiency exceeding 50%. They are not only widely used in fields such as laser display and laser lighting, but also show great application prospects in monolithic integration, visible light communication, atomic clocks, atomic radar, medical instruments, etc. As a result, GaN-based lasers have been included in the long-term development plans of countries around the world and have become an important research topic in the national key R&D program of China.

So far, most GaN-based lasers are grown on expensive 2-inch freestanding GaN substrates (over 10,000 RMB per piece), resulting in a high device cost, which is 2–3 orders of magnitude higher than GaN-based LEDs. In addition, freestanding GaN substrates face issues such as inconsistent tilt angle, dislocation, and residual stress distribution, which severely affect the yield and further increase the device cost. Compared to freestanding GaN substrates, silicon substrates have the following advantages: (1) high quality, larger size, and lower cost together with automated process lines, which can significantly reduce device costs; (2) GaN-based lasers grown on Si substrates are expected to be used as on-chip light sources in silicon

M. Feng (✉) · J. Liu · Q. Sun (✉) · H. Yang
Suzhou Institute of Nano-Tech and Nano-Bionics, Chinese Academy of Sciences, Beijing, China
e-mail: mxfeng2011@sinano.ac.cn

Q. Sun
e-mail: qsun2011@sinano.ac.cn

S. Xu (ed.), *Optical Characterization of Microstructures and Optoelectronic Devices Based on Wide Band Gap Semiconductors*, Wide Bandgap Semiconductors,
https://doi.org/10.1007/978-981-95-1928-6_2

photonics and other fields; (3) GaN-on-Si technology is an innovative technology route with China's independent intellectual property rights. Therefore, GaN-based lasers grown on silicon have become a hot topic and have received widespread attention.

This chapter will first introduce the growth and characterization of GaN-based laser materials on silicon substrates, then elucidate the device fabrication and characterization of GaN-based laser diodes grown on silicon. Finally, the content of this chapter is summarized.

2.2 Growth and Characterization of GaN Lasers on Silicon Substrates

Unlike the homoepitaxial growth of GaN-based lasers on freestanding GaN substrates, the growth of GaN-based lasers on silicon substrates is heteroepitaxial, with a 17% lattice mismatch and a 54% thermal expansion coefficient mismatch between silicon and GaN. This leads to high threading dislocation density (TDD), large tensile stress, and severe wafer warpage in the material of GaN-based lasers on silicon substrates. Moreover, as shown in Fig. 2.1, the structure of a GaN-based laser grown on silicon substrate is complex, including silicon substrate, aluminum nitride/aluminum gallium nitride (AlN/AlGaN) buffer layer, n-GaN contact layer, n-AlGaN cladding layer, n-GaN waveguide layer, indium gallium nitride (InGaN)/GaN multiple-quantum well (MQW) active region, GaN waveguide layer, p-AlGaN electron blocking layer (EBL), p-AlGaN cladding layer, and p-GaN contact layer, etc. The total thickness of the epitaxial layers is about 5–6 μm, which further increases the tensile stress in the laser material, making it prone to cracks and rendering the laser inoperable. Therefore, to develop high-performance GaN-based lasers grown on silicon substrate, it is necessary to solve key scientific issues such as large stress and high defect density in the laser material. This section will provide a detailed introduction to the epitaxial growth and characterization of GaN-based laser materials grown on silicon substrate.

2.2.1 Key Challenges in the Epitaxial Growth of GaN-Based Laser Materials on Silicon Substrates

1. Stress Control

The primary issue faced in growing GaN-based lasers on silicon substrates is the large tensile stress, making it more challenging than epitaxy on sapphire substrates. As shown in Fig. 2.2, when the temperature rises to the GaN growth temperature (such as 1050 °C), the wafer exhibits slight concave warping due to the temperature gradient along the growth direction. For sapphire substrates, since the coefficient of

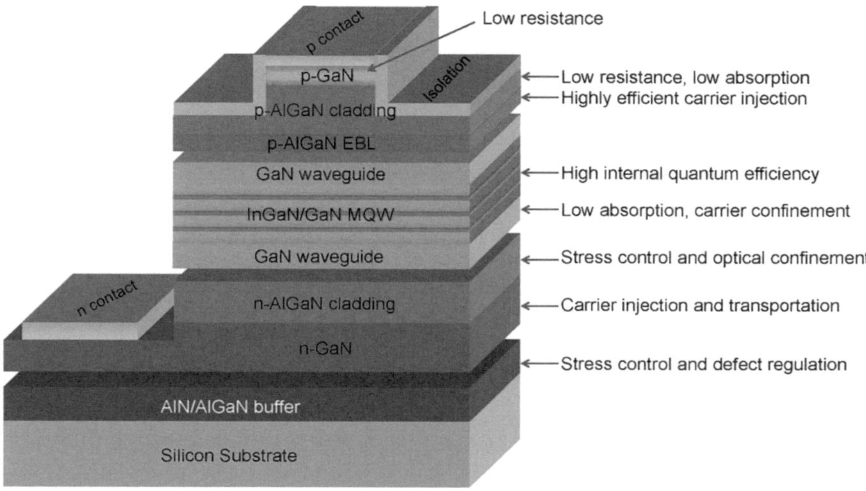

Fig. 2.1 Schematic architecture of GaN-based laser grown on silicon substrate [1]

thermal expansion (CTE) is much larger than that of GaN, the lattice of sapphire substrate shrinks faster during the cooling process after epitaxial growth, resulting in a huge compressive stress within the epilayer, and therefore the epitaxial wafer exhibits convex warping. For silicon substrates, if a stress control layer is not used, due to the 54% thermal expansion coefficient mismatch between silicon and GaN, the shrinkage rate of the GaN lattice is twice that of the silicon substrate during the cooling process after growth. This will induce large tensile stress in the GaN epilayer and huge concave warping, which is prone to cracks. Therefore, it is usually necessary to first grow Al(Ga)N stress control layers on silicon substrate to establish compressive stress to offset the tensile stress during cooling and thus avoid the generation of cracks.

In addition to tensile stress caused by thermal mismatch, the stress control of the epitaxial growth of GaN-based laser materials on silicon substrate is more difficult for several reasons: Due to the limited refractive index difference between GaN and AlGaN, it is usually necessary to epitaxially grow about 2-μm-thick AlGaN cladding layer (with a lattice constant smaller than GaN) to confine the optical field, which will introduce additional tensile stress during the growth process, further increasing the difficulty of stress control. Moreover, considering the material quality and resonant cavity optical design, the total thickness of the GaN-based laser epitaxial layer grown on the silicon substrate usually reaches about 6 μm. The increase in thickness further increases the thermal mismatch tensile stress. Therefore, excellent stress control is crucial for the epitaxial growth of GaN-based laser materials grown on silicon substrate.

To suppress the generation of cracks in GaN materials on silicon substrates, a common method is to use a prestress layer named "handshake buffer layer." That is, first, an AlN nucleation layer with a smaller lattice constant is grown, followed

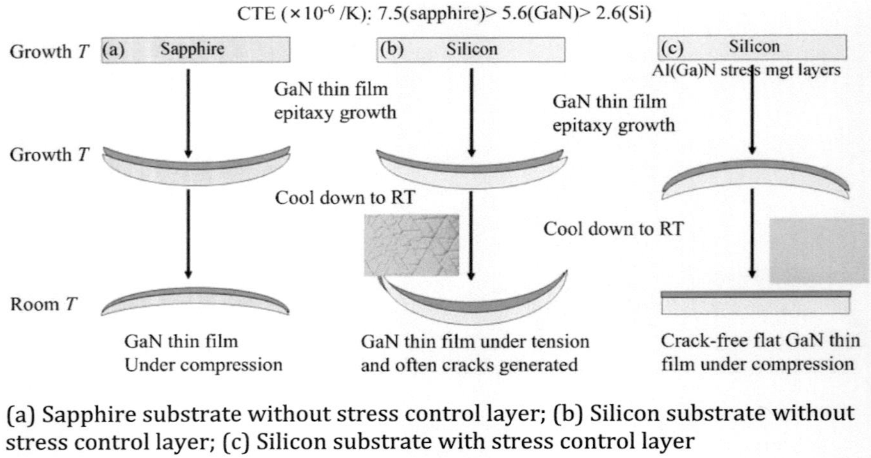

Fig. 2.2 Schematic diagram of stress and warping changes in the film before and after epitaxial growth of GaN on sapphire and silicon substrates [2]

by an Al component gradient AlGaN stress control layer, gradually transitioning to GaN with a larger lattice constant. Benefiting from the positive lattice mismatch between AlN and (Al)GaN, an appropriate amount of compressive stress was established during the high-temperature growth process to offset the tensile stress caused by the difference in thermal expansion coefficients during cooling. There are three common types of stress control layers for GaN on silicon substrates:

(1) Al-component-graded AlN/AlGaN buffer layer [3, 4]: By reducing the Al component in the buffer layer, the lattice constant of AlGaN material increases, thereby gradually establishing compressive stress during the growth process to offset the tensile stress caused by thermal mismatch during the cooling process. This type of buffer layer can not only effectively accumulate compressive stress, but also greatly promote dislocation bending and annihilation, thereby effectively reducing the density of threading dislocations. However, such buffer layers are prone to cause plastic deformation of the silicon substrate and large wafer warping due to excessive accumulated compressive stress.
(2) AlN/AlGaN superlattice [5]: The advantage of this buffer layer is that it is expected to grow thicker GaN films while maintaining good wafer warping. However, because the establishment of compressive stress is relatively slow, the tilting and bending of threading dislocations driven by the compressive stress is limited, resulting in a high dislocation density of the epilayer. In fact, because of the good suppression of vertical leakage and anti-punch-through performance of superlattice, this type of buffer layer is often used to prepare high-voltage, high-power GaN power electronic devices on silicon substrates.
(3) AlN insertion layer [6]: By periodically inserting a certain thickness of AlN in the GaN epitaxial layer, the compressive stress accumulated at the AlN/GaN

interface is greater than the thermal tensile stress. The advantage of this buffer layer is that the residual stress and warping of the GaN epitaxial layer on the silicon substrate can be adjusted by controlling the thickness and number of the AlN insertion layers. However, the high-resistance AlN insertion layer will prevent the vertical transport of current, which is not conducive to the preparation of vertical devices.

2. Suppression of Threading Dislocation Defects

There is a huge lattice mismatch of 17% between the Si(111) and GaN(0002) crystal planes, resulting in a TDD as high as 10^9–10^{10} cm^{-2} in the GaN epitaxial material grown on silicon substrate, which seriously affects the performance and reliability of the laser. Specifically manifested in:

(1) Threading dislocations usually act as nonradiative recombination centers, affecting the internal quantum efficiency of multiple quantum wells.
In comparison with the traditional III–V semiconductor materials, group III-nitride semiconductor materials have a strong exciton effect, so their internal quantum efficiency is less sensitive to threading dislocations. Despite this, such a high density of threading dislocations in GaN materials on silicon substrates will degrade the thermal stability of excitons, thereby causing serious nonradiative recombination, leading to a decrease in the optical gain of the laser active region and an increase in the threshold current density.
(2) Threading dislocations can induce V-shaped pit defects in the quantum well active area, increasing the photon scattering and optical absorption loss.
For InGaN quantum wells, under the influence of the local stress field of dislocations, metal indium atoms easily migrate to the dislocation core to form Cottrell atmospheres, leading to a significant decrease in the growth rate above the In-rich dislocation core, forming V-shaped pit defects. The V-shaped pits will cause a decrease in the flatness of the quantum well interface, leading to increased photon scattering and internal absorption at the interface, significantly reducing the laser gain.
(3) The local stress field of threading dislocations induces impurities and vacancy defects, worsening the optical and electrical performance of the device.
On the one hand, the deep energy level caused by point defects such as impurities and vacancies in the waveguide and quantum well will induce significant absorption of photons, increasing the internal loss of the laser. On the other hand, carbon and oxygen donor impurities in the p-type AlGaN cladding layer will cause a decrease in the hole concentration and an increase in resistivity, leading to an increase in the voltage and Joule heat, and therefore a significant decrease in the reliability of the device.
(4) The dangling bonds (defect states) at the core of the threading dislocation provide leakage channels for carriers, decreasing the carrier injection efficiency and increasing the leakage of the active region of GaN-based laser grown on the silicon substrate, leading to a significant increase in the threshold current density and junction temperature. The increase in junction temperature, in turn,

promotes the diffusion of point defects along the core of the threading dislocation into the active region, causing a rapid decrease in the quantum efficiency of the laser during operation, posing a serious challenge to reliability.

Therefore, reducing the density of threading dislocation defects in silicon substrate GaN materials and improving material quality are crucial for achieving high-performance silicon substrate GaN-based lasers.

2.2.2 Characterization of High-Quality GaN Thin Film on Silicon Substrates

On Si(111) substrate, a 300-nm-thick AlN nucleation layer, a 300-nm-thick $Al_{0.35}Ga_{0.65}N$ stress control layer, a 450-nm-thick $Al_{0.17}Ga_{0.83}N$ stress control layer, and a 3.6-μm-thick GaN template were successively grown. After the growth, the epitaxial wafer surface was observed with the naked eye and no obvious cracks were found. Further characterization analysis is as follows:

1. Observation with an Optical Microscope

As shown in Fig. 2.3a and b, the surface of the epitaxial wafer is smooth and flat, without cracks. Near the edge of the epitaxial wafer, cracks about 0.5 mm long can be observed. The cracks present a 60° angle and extend along the GaN *a*-axis $[11\bar{2}0]$ orientation. The *a*-axis direction of wurtzite GaN material is the easiest direction to cleave, and cleaving along this direction is an important step in preparing the Fabry–Perot cavity facets of GaN ridge waveguide lasers.

2. Atomic Force Microscopy and Cathodoluminescence Measurements

Using an atomic force microscope (AFM) in tapping mode, the microscopic morphology of the GaN thin film grown on silicon substrate is observed. As shown

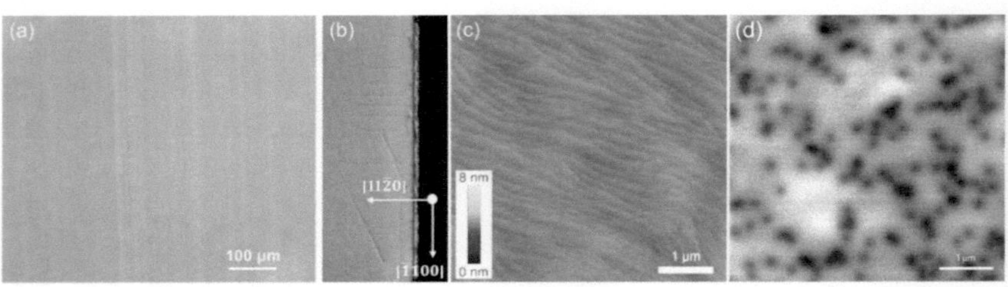

(a) Center of the epitaxial wafer; (b) Edge of the epitaxial wafer; (c) Surface AFM image; (d) Surface CL imaging

Fig. 2.3 Optical microscope image, Atomic force microscope (AFM) image, and Cathodoluminescence (CL) image of GaN epitaxial wafer on a silicon substrate [7]

in Fig. 2.3c, the surface of the GaN sample shows a step-flow morphology with alternating single and double atomic steps, with a root-mean-square roughness of only 0.2 nm.

In GaN materials, threading dislocations are common nonradiative recombination centers. Under the action of cathode electron beams, these areas often appear as dark spots due to their lack of luminescence, forming a stark contrast with areas without dislocations. Therefore, cathodoluminescence (CL) testing is often used to estimate the dislocation density in GaN materials. As shown in Fig. 2.3d, panchromatic CL imaging was performed on the GaN thin films grown on silicon substrate. The dark spots correspond to nonradiative recombination centers formed by threading dislocations. From the density of the dark spots, it can be estimated that the threading dislocation density of the GaN material grown on silicon substrate is about 6×10^8 cm^{-2}.

3. FWHM-χ Test Characterizes the Crystal Quality of GaN Material on Silicon Substrate

According to the mosaic structure theory, due to the large number of threading dislocations in the GaN epitaxial layer, the epitaxial layer presents a mosaic structure [8], with grains being cylinders parallel to the growth direction and their height approximately equal to the film thickness. Usually, the Burgers vector of a screw dislocation is along the <0001> direction, which has a pure shear strain field, thus causing the base plane (0001) to tilt and rotate, with the tilt angle defined as β_t. The Burgers vector of an edge dislocation is within the (0001) base plane, which has a plane strain field, causing the cylindrical surface $\left(10\overline{1}0\right)$ to twist around the *c*-axis, with the twist angle defined as β_e.

In the wurtzite GaN material, the tilt angle β_t and twist angle β_e can be measured by the full width at half maximum (FWHM) of the rocking curve of double- or triple-crystal diffraction in high-resolution X-ray diffraction (XRD). For the tilt angle β_t, the FWHM of the symmetric diffraction of the GaN (0002) plane is commonly used for characterization; however, for the twist angle β_e, due to the thickness limitation of the GaN epitaxial layer, it is almost impossible to test the rocking curve of the $\left(10\overline{1}0\right)$ plane. Therefore, a series of crystal planes with the same crystallographic axis but different χ angles are often scanned obliquely, and then the FWHM-χ relationship curve, that is, FWHM-χ plot, is performed. Finally, the data is fitted according to the formula (2.1), and the FWHM of the XRD rocking curve of the GaN $\left(10\overline{1}0\right)$ plane at a χ angle of 90° can be extrapolated.

$$\beta^2 = (\beta_t \cdot \cos\chi)^2 + (\beta_e \cdot \sin\chi)^2 \tag{2.1}$$

where β corresponds to the half-width of the XRD rocking curve at angle χ.

After obtaining the tilt angle β_t and the twist angle β_e, the densities of screw and edge dislocations can be calculated according to formula (2.2), respectively.

$$D = \frac{\beta^2}{4.35 \cdot |\mathbf{b}|^2} \tag{2.2}$$

where: **b**—Burgers vector of the dislocation

β—tilt or twist angle of the dislocation.

D—density of threading dislocations.

In wurtzite GaN material, the length of the Burgers vector of screw threading dislocations is $|\mathbf{b}| = 0.5185$ nm, and the length of the Burgers vector of edge threading dislocations is $|\mathbf{b}| = 0.3189$ nm.

For the above 3.6-μm-thick GaN epitaxial layer grown on silicon substrate, (0002), $(10\overline{1}1)$, $(10\overline{1}2)$, $(10\overline{1}3)$, $(20\overline{2}1)$, and other crystal faces were tested with XRD double-crystal rocking curve, where (0002) and $(10\overline{1}2)$ crystal face rocking curve half-peak width is about 260 arcsec. The FWHM of the rocking curve of each crystal face is below 300 arcsec. According to formulas (2.1) and (2.2), the screw dislocation density in the thin film can be estimated to be about 1.2×10^8 cm^{-2}, the edge dislocation density is about 4.7×10^8 cm^{-2}, and the total dislocation density is about 5.9×10^8 cm^{-2}, which is very close to the result estimated from the panchromatic CL imaging in Fig. 2.3c.

2.2.3 *Characterization of Stress State and Structural Defect of GaN Lasers Grown on Silicon Substrate*

1. XRD Reciprocal Space Mapping Test

During the XRD reciprocal space mapping (RSM) test, for each step of the corresponding ω value, an ω/2θ scan is performed to obtain the diffraction angle and diffraction intensity information of ω/2θ. The ω/2θ scan corresponds to the size of the crystal plane spacing, and the ω scan corresponds to the direction of the crystal plane in reciprocal space. In this way, with the three variables of ω angle, 2θ angle, and diffraction intensity, the spatial distribution of the crystal plane spacing and diffraction direction within a given scan range can be obtained, thereby more intuitively reflecting the composition, strain relaxation degree, and defects of each epitaxial layer [9].

The RSM test of the nonsymmetric $(10\overline{1}5)$ crystal plane of the epitaxial material of the GaN laser on the silicon substrate was carried out, as shown in Fig. 2.4b. The GaN layer in the epitaxial structure (including the GaN thick layer and the upper and lower GaN waveguide layers), the AlGaN cladding layers, and the InGaN multiple quantum structure all have obvious diffraction peaks, indicating that the crystal quality of these epitaxial layer structures is good. In addition, these signal peaks have the same abscissa Q_x in the RSM diagram, indicating that these epitaxial layers have the same in-plane lattice constant *a*, that is, they are in a co-lattice state. Each coordinate (Q_x, Q_z) in the RSM diagram corresponds to a set of lattice constants *a* and *c*. According to formulas (2.3) and (2.4), the lattice constants at the corresponding peak values are calculated and compared with theoretical values to determine the stress state of the epitaxial layer.

$$Q_x = \frac{2}{3} \cdot \frac{2\pi}{a^2} \tag{2.3}$$

$$Q_z = 5 \cdot 2\pi / c \tag{2.4}$$

During epitaxial growth, a large strain can lead to the generation of mismatch dislocations, and the epitaxial layers gradually become a semicoherent state, that is, lattice relaxation occurs in the epitaxial thin film. The relaxation degree f_r can be obtained according to formula (2.5).

$$f_r = \frac{a - a_s}{a_0 - a_s} \tag{2.5}$$

where a—in-plane lattice constant under stress

a_0—in-plane lattice constant in a fully relaxed state

a_s—lattice constant of the substrate.

According to the coordinate data of the reciprocal space in Fig. 2.4b, the lattice constant and strain relaxation degree of the GaN thin film on the silicon substrate can be obtained using formulas (2.3)–(2.5), as shown in Table 2.1. It can be seen that the AlN nucleation layer is basically in a fully relaxed state, which is also caused by

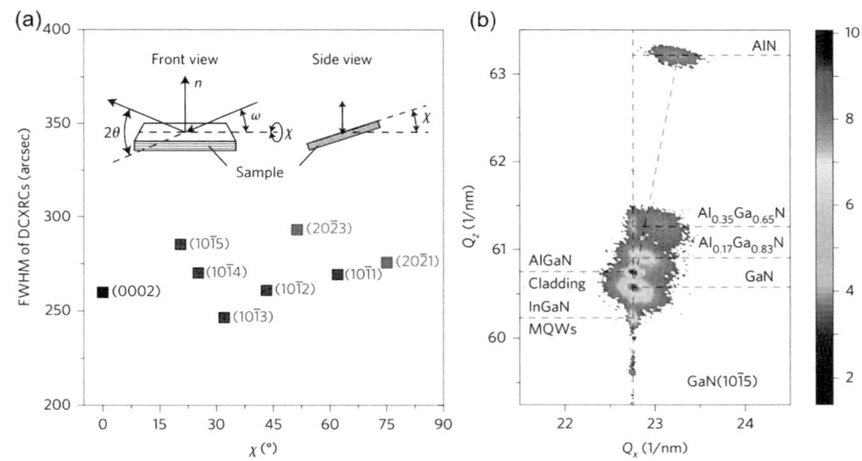

(a) Relationship between the FWHM of the rocking curves of different crystal planes on the <1$\bar{2}$10> zone axis and the χ angle of the GaN grown on Si; (b) Reciprocal space mapping of an InGaN-based LD structure grown on Si measured around (10$\bar{1}$5) asymmetric reflection

Fig. 2.4 XRD double-crystal diffraction results and (10$\bar{1}$5) RSM results of GaN epitaxial layer grown on a silicon substrate [7]

the huge lattice mismatch (−19%) between the Si(111) substrate and AlN material. The actual lattice constants *a* of the GaN template thick layer, $Al_{0.35}Ga_{0.65}N$ and $Al_{0.17}Ga_{0.83}N$ stress control layers are all less than their in-plane lattice constants a_0 in a fully relaxed state. This indicates that they are all under in-plane compressive stress. The in-plane compressive stress induced by the $Al_{0.35}Ga_{0.65}N$ layer is the largest, with a relaxation degree of 0.67, while the relaxation degrees of both $Al_{0.17}Ga_{0.83}N$ layer and GaN layer are 0.86.

It is worth noting that in Fig. 2.4b, the diffraction peaks of the AlGaN cladding layer, InGaN multiple quantum wells, and (In)GaN waveguide layer in the RSM have the same horizontal coordinate Q_x as the GaN thick layer, indicating complete co-epitaxial growth. This lays the key material foundation for realizing room-temperature (RT) electric injection lasing of GaN-based lasers on silicon substrates.

2. Cross-sectional transmission electron microscopy test

To further analyze the types of dislocations in the epitaxial layer and their evolution mechanisms, it is necessary to perform cross-sectional transmission electron microscopy (TEM) tests on the GaN-based laser grown on silicon substrates. Usually, TEM two-beam diffraction imaging is performed on the *m*-plane of GaN, which is an atomic close-packed plane with good dislocation diffraction contrast. Figure 2.5a and b shows TEM images of the same area of the sample under different diffraction vector conditions. The invisibility criterion for dislocations in TEM two-beam diffraction is Lu et al. [10]:

$$\mathbf{g} \cdot \mathbf{b} = 0 \tag{2.6}$$

where: **g**—Diffraction vector under TEM two-beam conditions

b—Burgers vector of the dislocation.

From formula (2.6), for screw dislocations (**b** = <0001>), **g** can be taken as [*h k i* 0] to make screw dislocations invisible; for edge dislocations ($\mathbf{b} = 1/3\langle 11\bar{2}0\rangle$), **g** can be taken as [0 0 0 *l*] which can make the edge dislocation invisible; mixed dislocations will appear in both cases because they contain both screw and edge

Table 2.1 Peak coordinates, lattice constants, and relaxation degrees of each layer of material in the RSM results of GaN-based laser materials on silicon substrates

Epilayer	Q_x (nm^{-1})	Q_z (nm^{-1})	***a*** (nm)	***c*** (nm)	$\boldsymbol{a_0}$ (nm)	f_r
GaN	22.768	60.573	0.31866	0.51865	0.3189	0.86
$Al_{0.17}Ga_{0.83}N$	22.871	60.905	0.31722	0.51582	0.3176	0.86
$Al_{0.35}Ga_{0.65}N$	23.044	61.291	0.31484	0.51257	0.3162	0.67
AlN	23.252	63.203	0.31203	0.49706	0.3112	0.99

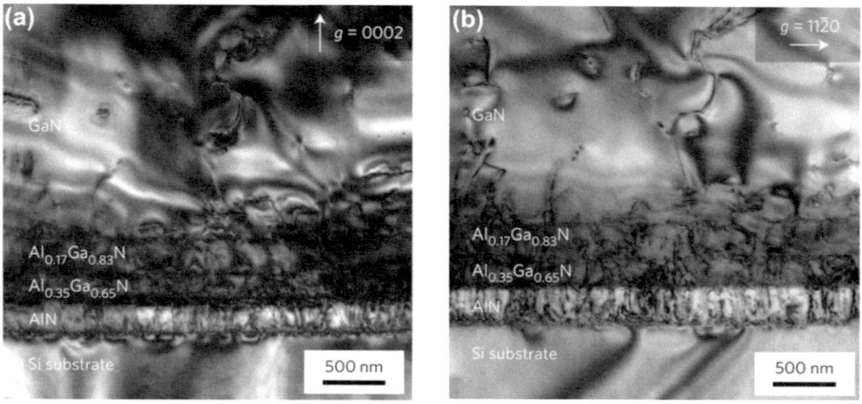

(a) Diffraction vector g= 0002; (b) Diffraction vector g= 11$\bar{2}$0

Fig. 2.5 Two-beam TEM imaging of GaN-based laser materials grown on silicon substrates [7]

dislocation vectors. According to the above criteria, it can be seen from Fig. 2.5 that the number of mixed dislocations in this area is the most, the number of pure edge dislocations is the second, and the number of pure screw dislocations is the least, which is consistent with the distribution of dislocations estimated by high-resolution XRD rocking curves.

2.2.4 Evolution of Stress and Threading Dislocations in High-Quality GaN Materials Grown on Silicon Substrates

Taking a 10-μm-thick GaN film on silicon substrate without cracks as an example, TEM samples were quickly prepared by focused ion beam, and the evolution of threading dislocations was studied using an FEI Tecnai F20 S-TWIN scanning transmission electron microscope [11]. As shown in Fig. 2.6a, a large number of threading dislocations generated in AlN are gradually "filtered" in the AlGaN stress control layer. Within the initial 4 μm thickness of GaN, some threading dislocations even undergo large-angle tilting. As the film thickness increases, the dislocation density gradually decreases, and the tilt angle gradually decreases. Panchromatic CL imaging shown in Fig. 2.6b indicates that the threading dislocation density in the 10-μm-thick GaN film on the silicon substrate is about 5.8×10^7 cm^{-2}. Optical microscope (Fig. 2.6c) and atomic force microscope results (Fig. 2.6d) show that the surface of

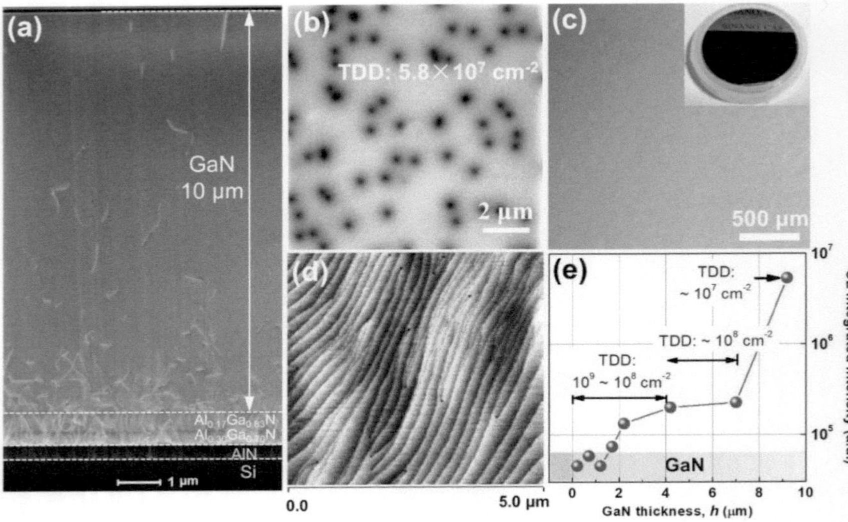

(a) Cross-sectional TEM image; (b) Surface full-spectrum CL image; (c) Surface optical microscope image and epi-wafer image; (d) Surface AFM image; (e) Evolution of the integrated intensity of CL band edge luminescence

Fig. 2.6 Cross-sectional and surface characterization of the GaN materials grown on silicon substrates [11]

the GaN film on the silicon substrate is smooth and flat, without cracks, and exhibits a typical step-flow morphology, with a root-mean-square roughness of only 0.2 nm.

The Gatan MonoCL3+ high-resolution cathodoluminescence spectrometer provides beneficial assistance for studying the optical quality evolution of GaN thick-layer materials on silicon substrates. As shown in Fig. 2.6e, as the GaN film thickness increases and the TDD decreases, the integrated intensity of band-edge luminescence increases by nearly two orders of magnitude. High-quality GaN thick-layer materials on silicon substrates will lay an excellent material foundation for constructing high-performance GaN devices on silicon substrates, such as lasers, detectors, power electronics, and microwave radio frequency devices.

Using the Horiba JY LabRAM high-resolution confocal Raman spectrometer, the frequency shift of the E_2 (high) phonon peak in GaN materials can be characterized by testing in the backscattering mode. The cross-sectional micro-area Raman test was performed on the sample, and the results are shown in Fig. 2.7a. As the thickness of the GaN film on the silicon substrate increases, the peak position of the cross-sectional Raman spectrum gradually blueshifts. For GaN materials, the E_2 (high) phonon peak is at 567.5 cm^{-1} under stress-free conditions. According to the frequency shift $\Delta\omega$ of the thin-film Raman peak relative to the stress-free peak position, combined with formula (2.7):

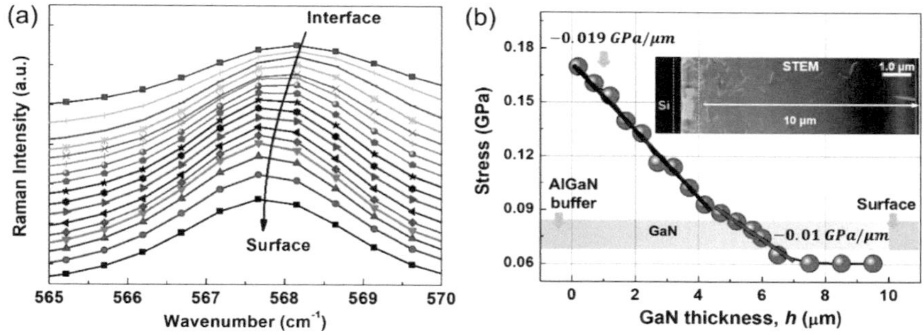

(a) Cross-sectional Raman spectrum along the growth direction; (b) Stress evolution pattern along the growth direction

Fig. 2.7 Cross-sectional Raman test analysis of GaN material on silicon substrate [11]

$$\Delta\omega = K \cdot \sigma \tag{2.7}$$

The residual stress σ of the films at different thicknesses can be calculated.

In the formula:

$\Delta\omega$—wavenumber of Raman shift.

K—Raman coefficient. For GaN material grown on silicon substrate, the Raman coefficient $K = 4.3\ \text{cm}^{-1}\cdot\text{GP}^{-1}$.

According to formula (2.7), the residual stress of GaN thin film at different thickness positions can be calculated, and the relationship between stress and thickness can be obtained, as shown in Fig. 2.7b. Obviously, the change of residual stress in 10-μm-thick GaN on silicon substrate can be divided into three stages: (1) Within the range of 4 μm thickness, the stress decreases rapidly at a rate of −0.019 GPa/μm. (2) Within the range of 4–7 μm thickness, the stress relaxation gradually slows down, with a rate of −0.01 GPa/μm. (3) Within the range of 7–10 μm thickness, the stress in GaN thin film only undergoes slight changes.

2.2.5 *Interaction Mechanism Between Stress and Threading Dislocation*

Based on TEM and Raman test analysis, it can be found that there is a close interaction relationship between threading dislocation and stress [11]. As shown in Fig. 2.8a, under compressive stress, the dislocation moves along the m-axis $\langle 10\bar{1}0\rangle$ tilts, and produces a mismatch dislocation component of the edge type, with its Burgers vector along a-axis $\langle 11\bar{2}0\rangle$ direction. Figure 2.8b shows the typical planar TEM imaging

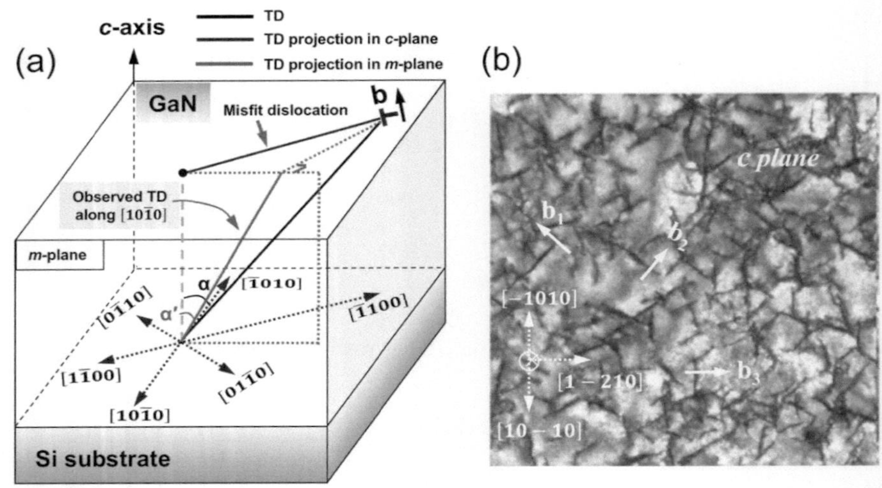

(a) Schematics showing the inclination of TDs in GaN epilayer, which constitutes effective MDs in projection onto the c-plane, consequently relaxing some compressive stress; (b) The observed misfit dislocation by plan-view TEM

Fig. 2.8 Physical model of stress relaxation by threading dislocations [11, 12]

results of the GaN sample, where the crack-like black lines are the projections of the threading dislocations in the *c*-plane. It can be seen that the projection of the Burgers vector of the dislocation is along *a*-axis $\langle 11\bar{2}0\rangle$ direction, showing the characteristics of edge dislocations, further corroborating the dislocation evolution model given in Fig. 2.8a.

However, since edge dislocations can relax compressive stress, the tilting and bending of dislocations will cause the compressive stress accumulated in the thin film to drop rapidly. The larger the tilt angle α, the thicker the extension h, and the longer the projection in the *c*-plane ($L = h \times \tan\alpha$), the greater the consumption of compressive stress. During the heteroepitaxial growth of GaN on silicon substrate, if the compressive stress established by the buffer layer is consumed too much, the thin film is likely to crack due to excessive tensile stress during cooling. Therefore, it is necessary to carefully control the growth conditions of the stress control layer, to make the threading dislocations tilt at a large angle and interact within a thinner thickness range, thereby ensuring sufficient accumulation of compressive stress while improving the thickness and crystal quality of the GaN thin film on the silicon substrate.

2.2.6 Epitaxial Growth and Characterization of the Active Region of GaN-Based Lasers on Silicon Substrates

1.Suppression of interface defects in the quantum well active region
(1) In-rich cluster defects

Under certain growth conditions, V-shaped pits with inverted pyramid can be easily formed at the top of a threading dislocation. Sometimes, the V-shaped pits are filled with In-rich clusters forming a closed loop structure (Fig. 2.9a), often referred to as "trench defects" [13]. Trench defects can decompose during the subsequent high-temperature growth process of p-type GaN, producing voids and metal In precipitation (Fig. 2.9b). Voids and In precipitates often serve as effective channels for nonradiative recombination of carriers, appearing as dark spots in micro-photoluminescence (μ-PL) (Fig. 2.9c), resulting in extremely low internal quantum efficiency of the quantum well [14]. Such problems of trench defects are more pronounced in GaN-based blue-green laser quantum wells with higher In composition. By in situ H_2 etching and heat treatment at the quantum well interface, trench defects can be effectively suppressed, greatly improving the quality of the quantum well interface and optical quality (Fig. 2.9d–f).

(2) V-shaped pit defects

As mentioned in Sect. 2.2.1, metal In atoms migrate to the dislocation core under the action of local stress fields at dislocations, resulting in the formation of Cottrell atmospheres, which significantly slows down the growth rate at the dislocation core, leading to the formation of V-shaped pit defects. The presence of V-shaped pits distorts the quantum well interface, significantly increasing photon scattering and internal absorption losses. Typically, (In)GaN materials are prone to V-shaped pit defects under low temperature, low pressure, and high growth rates. Experimental results from the team of Zhao et al. from the Institute of Semiconductors, Chinese Academy of Sciences, show [17] that the V-shaped pits in InGaN/GaN quantum wells originate from the low-temperature GaN barrier layer. By introducing a small amount of indium during the growth process of the GaN quantum barrier, V-shaped pit defects can be effectively suppressed, reducing the threshold current density of GaN-based lasers by more than half.

2. Control of point defects in the quantum well active region

Unlike light-emitting diodes (LEDs), GaN-based lasers grown on silicon substrate have optical resonator cavity composed of waveguides, cladding layers, and quantum wells. Spontaneous radiation light oscillates and feeds back in the resonator to form

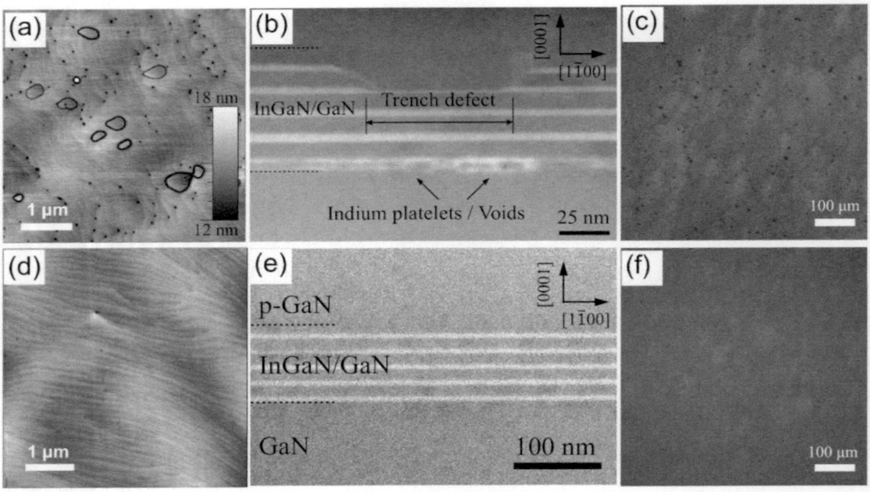

(a) AFM image of a surface with trench defects; (b) Cross-sectional TEM image of a surface with trench defects; (c) μ-PL image of a surface with trench defects; (d) AFM morphology of a surface without trench defects; (e) Cross-sectional TEM image of a surface without trench defects; (f) μ-PL image of a surface without trench defects

Fig. 2.9 Morphology, structure, and luminescent properties of the InGaN quantum well active region [13, 15, 16]

light amplification, which requires high material quality. In addition to threading dislocations, suppressing point defects that cause absorption losses is particularly important for reducing the threshold current density of GaN-based lasers grown on silicon substrate. The GaN waveguide layer, due to its large refractive index and thickness, carries a large part of the light field transport in the resonator. However, since metal–organic chemical vapor deposition (MOCVD) growth of GaN materials is a nonthermal equilibrium growth process, there are high concentrations of point defects (10^{16} cm^{-3} level) in the GaN waveguide material, including vacancy defects, carbon, oxygen impurities, etc. Among them, vacancy defects are an important point defect. They can capture carriers, causing nonradiative recombination, severely affecting the performance of GaN-based lasers grown on silicon. Studies have found that with In doping of GaN materials, the nonradiative recombination coefficient of (In)GaN materials can decrease by nearly three orders of magnitude. In addition, C. Haller et al. from École Polytechnique Fédérale de Lausanne [18] found that there are a large number of vacancy defects on the surface of GaN grown at high temperatures. During the subsequent quantum well growth process, these vacancy defects can diffuse into the quantum well, leading to an increase in nonradiative recombination, which severely affects the internal quantum efficiency. The use of trace amounts of In atoms for isoelectronic doping of the GaN waveguide layer can effectively eliminate the vacancy defects on the GaN surface, greatly enhancing the optical gain of the active area of the laser.

3. Suppression of point defects in the p-type AlGaN cladding layer

For the epitaxial growth of GaN-based laser on silicon substrate, in order to avoid thermal degradation of the InGaN quantum well, the p-AlGaN cladding layer above the quantum well has to be grown at a lower temperature. This will lead to a decrease in material quality, together with an increase in carbon impurities as compensation donors, resulting in a decrease in the hole concentration and an increase in resistivity of p-AlGaN, and therefore a significant increase in the operating voltage of the device. Reducing the growth rate can suppress the incorporation of carbon impurities. Therefore, in order to balance the quality of the quantum well and the resistivity of p-AlGaN, a combination of low temperature and low growth rate is adopted, that is, reducing the growth temperature of p-AlGaN from 950 to 920 °C and the growth rate of p-AlGaN from 17.4 to 8.7 nm/min. Secondary ion mass spectrometry (SIMS) results show that by reducing the growth rate, the carbon impurity concentration in the p-AlGaN cladding layer is reduced by an order of magnitude from 1.4×10^{18} to 1.3×10^{17} cm^{-3}, as shown in Fig. 2.10. Experimental results prove that the reduction in carbon impurity concentration significantly improves the threshold current and threshold voltage of the GaN-based laser on silicon substrate, as detailed in Sect. 2.3.2.

2.3 Device Testing and Characterization of GaN-Based Lasers Grown on Silicon Substrates

In this section, we mainly introduce the device testing and characterization of GaN-based lasers grown on silicon substrates. Based on the research in Sect. 2.2, high-quality GaN-based laser materials were grown on silicon substrates, including a silicon substrate, AlN/AlGaN buffer layer, 3-μm-thick n-GaN contact layer, 1.2-μm-thick n-$Al_{0.05}Ga_{0.95}N$ cladding layer, 80-nm n-GaN waveguide layer, three pairs of 2.5-nm undoped $In_{0.1}Ga_{0.9}N$ multiple quantum wells (MQWs) and four pairs of 7.5-nm undoped $In_{0.02}Ga_{0.98}N$ barrier layers, 60-nm undoped u-GaN waveguide layer, 20-nm p-$Al_{0.2}Ga_{0.8}N$ electron blocking layer (EBL), 500-nm p-$Al_{0.11}Ga_{0.89}N$/GaN superlattice cladding layer and 30-nm p-GaN contact layer, as shown in Fig. 2.11a, with a total epitaxial layer thickness of 5.8 μm. Subsequently, through lithography, etching, deposition, and other micro/nanoprocessing techniques, combined with cleavage and other processes, a GaN-based laser device grown on silicon substrate was fabricated. The device cavity length is 800 μm, and the ridge width is 4 μm, as shown in Fig. 2.11b and c. Afterward, the electroluminescence characteristics of the laser device were tested and characterized.

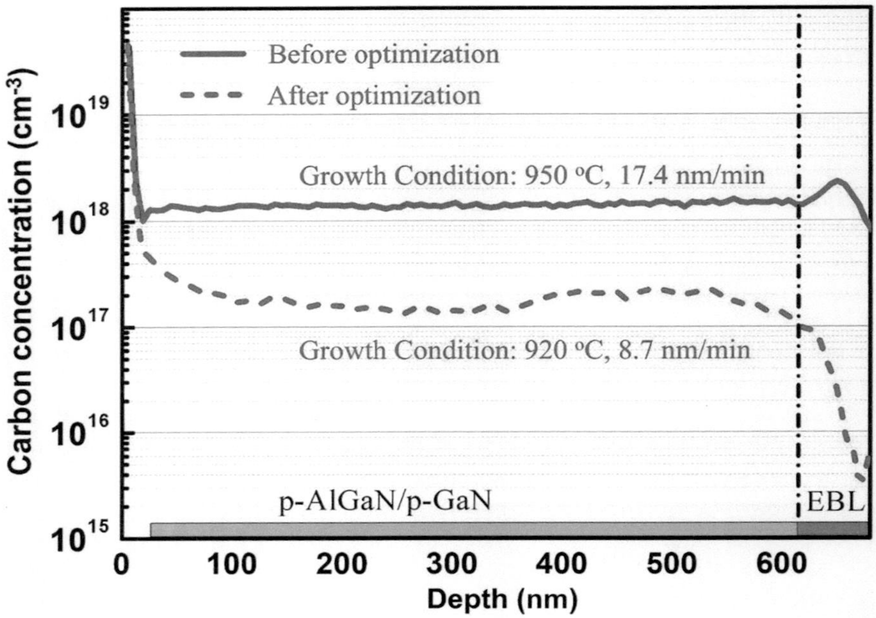

Fig. 2.10 Distribution of carbon impurity concentration in the p-AlGaN cladding layer of the GaN-based laser on silicon substrate before and after optimization [19]

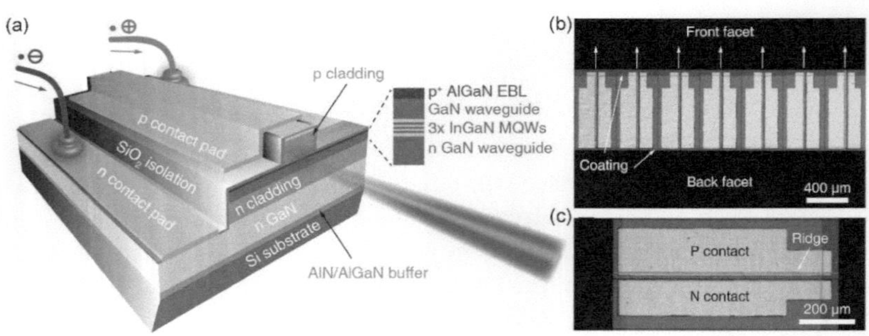

(a) Schematic diagram of GaN-based laser grown on silicon substrate; Optical microscope images of (b) GaN-based laser bar and (c) laser diode grown on silicon substrate

Fig. 2.11 Schematic diagram and optical microscope image of GaN-based laser grown on silicon substrate [7]

2.3.1 Testing and Characterization of GaN-Based Lasers Grown on Silicon Substrate

1. Electroluminescence Spectrum Test

The output spectrum of the GaN-based laser grown on silicon substrate under room-temperature electrical injection was measured, as shown in Fig. 2.12a. When the injection current slowly increased from 50 to 160 mA, due to the shielding effect of the injected carriers, the quantum-confined Stark effect of the multiple quantum wells weakened, and the peak wavelength of the electroluminescence spectrum shifted from 415.9 to 413.4 nm, as shown in Fig. 2.12b. At the same time, the full width at half maximum (FWHM) of the electroluminescence spectrum rapidly narrowed. When the injection current increased to 150 mA, the FWHM of the electroluminescence spectrum quickly reduced to 0.64 nm and remained almost unchanged (less than 1 nm) as the injection current continued to increase.

2. Light Power–Current–Voltage Test

Figure 2.12c shows the light power–current–voltage (L–I–V) curve of the GaN-based laser grown on silicon substrate. There is a clear inflection point at the injection current of 150 mA. When the injection current is less than 150 mA, the light output power of the laser slowly increases with the increase of the injection current. When

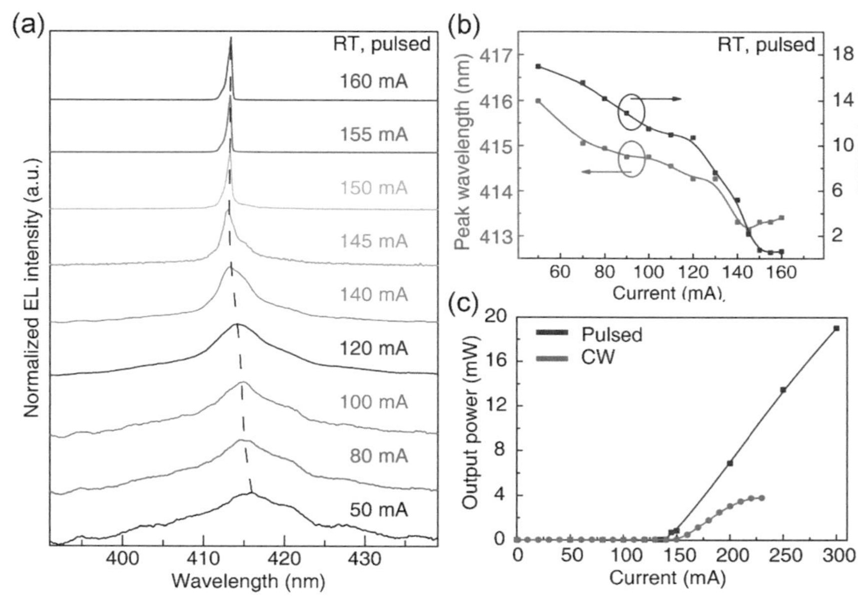

Fig. 2.12 Electroluminescence characteristics of the GaN-based laser grown on silicon substrate [7]

the injection current is greater than 150 mA, the output power of the laser rapidly increases and shows a linear relationship with the injection current, showing obvious lasing characteristics. The figure also shows the I–V curve of the laser. The turn-on voltage of the laser is about 3 V. When the voltage is greater than 3 V, due to the presence of many heterojunction interfaces inside the laser, these heterojunction interfaces form a Schottky barrier, resulting in a large series resistance of the device; therefore, the injection current slowly increases with the voltage. When the injection current is greater than about 30 mA, the Schottky barrier inside the laser is leveled. The series resistance is greatly reduced, and the injection current of the laser rapidly increases with the operation voltage.

3. Far-Field Test and Near-Field Test

The far-field and near-field patterns of the laser were also tested and characterized. Figure 2.13a and b shows the far-field images of the GaN-based laser grown on silicon substrate under 0.8 and 1.07 times the threshold current (Ith = 150 mA). At 0.8 I_{th}, the laser is in the amplified spontaneous emission stage, with no far-field spot. When the injection current is 1.07 Ith, the laser outputs an elliptical far-field spot, with the transverse and longitudinal divergence angles being 6° and 21°, respectively (Fig. 2.13c). The transverse divergence angle is much smaller than the longitudinal divergence angle mainly because the transverse light field is weakly confined by the 4-μm-wide ridge waveguide, while the longitudinal light field is mainly strongly limited by the cladding layers, which has obvious diffraction effects, and results in a larger divergence angle of the laser in the longitudinal direction.

Figure 2.13d–f shows the near-field patterns (NFPs) of the GaN-based laser grown on silicon substrate under 0.2 Ith, 0.8 Ith, and 1.07 Ith, respectively. Under 0.2 Ith, the laser shows spontaneous emission without mode characteristics. When the injection current is 0.8 Ith, the laser shows amplified spontaneous emission, showing the waveguide effect. When the injection current is further increased to 1.07 Ith, the near-field image of the laser shows obvious mode characteristics, and the laser works in single transverse mode. All these characteristics indicate that the GaN-based laser grown on silicon substrate has achieved room-temperature continuous-wave electric injection lasing.

4. Junction temperature and thermal resistance measurement

In the experiment, the junction temperature and thermal resistance of the GaN-based laser grown on silicon substrate were also characterized by the forward-voltage method. As shown in Fig. 2.14a, the voltage of the laser under small current injection is linearly related to the junction temperature. By using a switch, the injection current of the laser can be quickly switched from the operating current to a small current, and then the change in the voltage of the device under small current due to the decrease in junction temperature measured by the oscilloscope is used. Combined with the linear

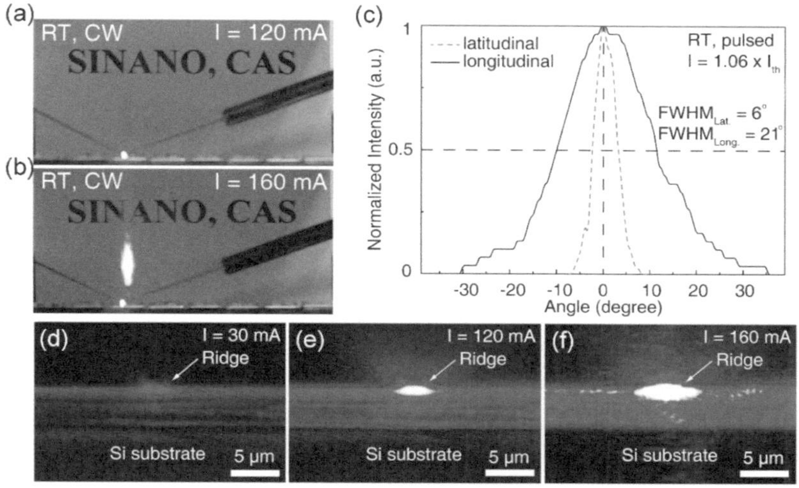

Far-field patterns (FFPs) observed (a) below the threshold current and (b) above the threshold current; (c) Typical FFP divergence angles of the GaN-on-Si LDs measured along the latitudinal and the longitudinal directions; Cross-sectional NFPs of the GaN-on-Si LD observed (d, e) below the threshold (30 and 120 mA, respectively) and (f) above the threshold.

Fig. 2.13 Far-field and near-field results of the GaN-based laser grown on silicon substrate [7]

relationship between the voltage and the junction temperature, the operating junction temperature of the laser can be calculated by assuming that the junction temperature of the laser under small current is approximately equal to room temperature, and then the thermal resistance can be obtained. In the experimental test, the operating current of the laser is 170 mA, the corresponding voltage is 6.1 V, and the output power of the laser is about 1 mW; therefore, the thermal power of the device is 1036 mW. Using the switch in the circuit, the injection current is quickly (300 ns) switched from 170 to 0.3 mA, and then the change in the voltage of the laser over time measured by the oscilloscope is shown in Fig. 2.14b. As the junction temperature drops rapidly, under a constant current of 0.3 mA, the voltage of the laser slowly rises and finally reaches equilibrium. The voltage change is about 85 mV. Combined with the linear coefficient of 3.1 mV/K between the voltage under 0.3 mA injection current and the junction temperature (approximately equal to the ambient temperature), it can be concluded that the temperature rise under 170 mA is 27.4 °C, and the corresponding thermal resistance of the laser is 26.5 K/W.

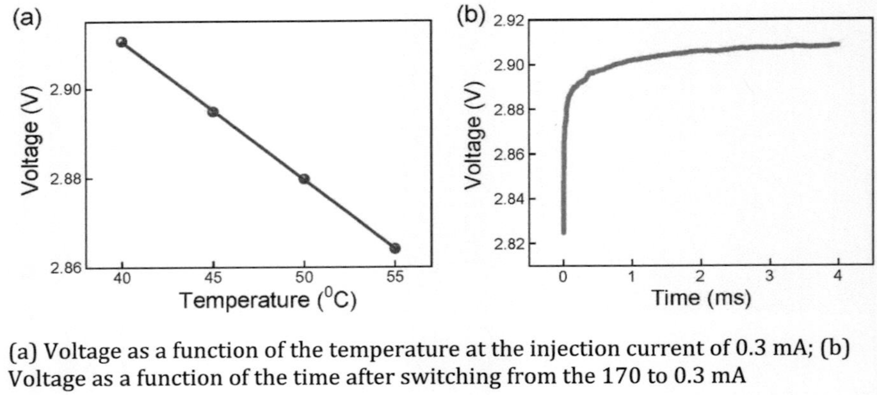

Fig. 2.14 Junction temperature and thermal resistance measurement of the GaN-based laser grown on silicon substrate

2.3.2 Performance Improvement and Characterization of GaN-Based Lasers Grown on Silicon Substrates

1. Improvement of material quality of GaN-based lasers on silicon substrates

As described in Sect. 2.2.6, carbon impurities are easily incorporated into p-AlGaN materials, which not only reduces the electrical conductivity but also increases the optical absorption, severely affecting the performance of the device. The experiment used a combined method of decreasing the growth rate and temperature to reduce the carbon impurities in the p-AlGaN barrier layer by an order of magnitude. Subsequently, the GaN-based lasers grown on silicon substrate before and after optimization were tested and characterized. Figure 2.15c and b shows the micro-photoluminescence images of the active area of the GaN-based lasers on silicon substrate before and after optimization. The luminescence uniformity of the active area after optimization has been significantly improved. Figure 2.15c shows the time-resolved photoluminescence curve at the peak wavelength of the GaN-based laser on silicon substrate under low excitation intensity. After optimization, the nonradiative recombination lifetime τ of the laser is significantly extended, and the internal quantum efficiency has been greatly improved. Figure 2.16 shows the L–I–V curve of the GaN-based laser on silicon substrate before and after optimization. After optimization, the threshold current density and threshold voltage of the laser are reduced to 2.25 kA/cm^2 and 4.7 V, respectively. Under a 4‰ duty cycle pulse current, the lifetime of the laser has been extended from a few minutes to 2.5 h [20].

2. GaN-based laser with an inverted n-side ridge

The hole concentration in group III-nitride semiconductor materials is severely lower than the electron concentration, and the hole mobility is much smaller than the

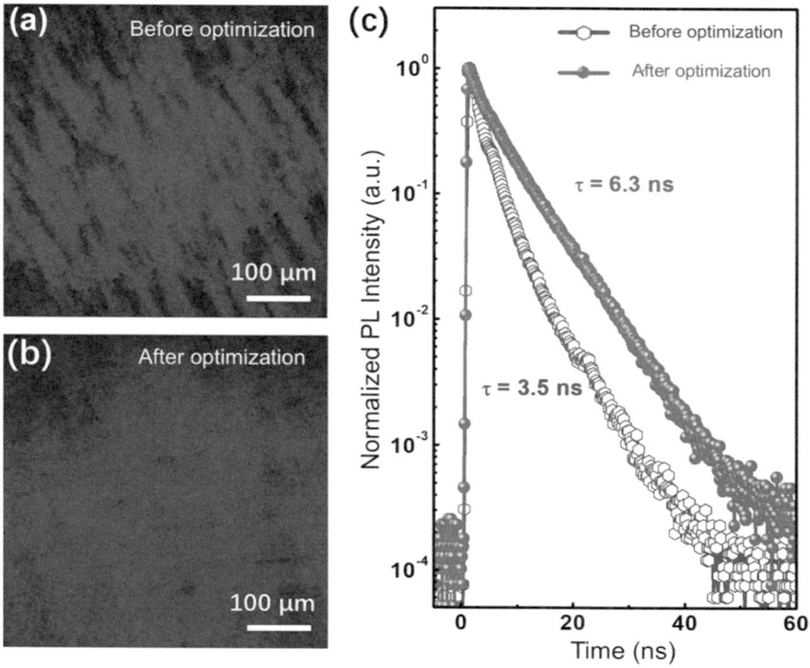

The micro-PL image of GaN-based LDs grown on Si (a) before optimization and (b) after optimization; (c) Time-resolved PL results.

Fig. 2.15 Micro-photoluminescence images and time-resolved photoluminescence curves of GaN-based lasers grown on silicon substrates before and after optimization [19]

Fig. 2.16 L–I–V curves of the GaN-based violet laser grown on silicon substrate before and after optimization [7, 19]

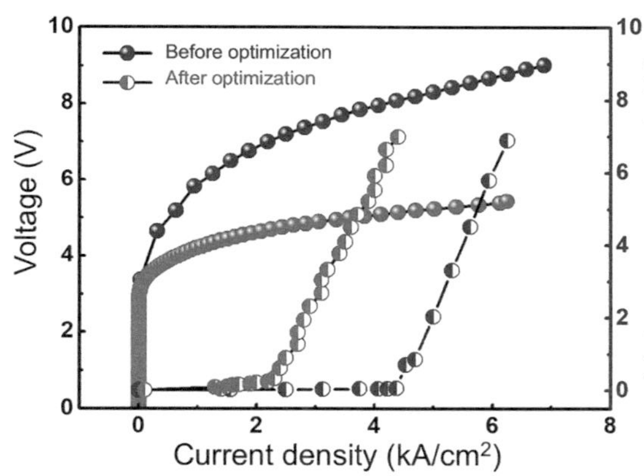

electron mobility, resulting in a much higher resistivity of the p-type layer than the n-type layer. In conventional GaN-based lasers, the ridge is on the p-side, and holes can only be injected into the active area through a narrow ridge, resulting in a small injection area, causing a large series resistance and contact resistance of the p-type layer, high operating voltage of the laser, and large thermal power. When the laser is inverted and packaged, the active area is close to the p-type contact, which can easily cause cavity facet contamination, short circuits, and other problems that cause device failure. Therefore, conventional GaN-based lasers usually use a positive packaging mode, and the heat source is far from the heat sink with a high thermal conductivity, resulting in a large thermal resistance of the device. The high thermal power and thermal resistance cause the device to operate at a high junction temperature, severely affecting the performance and reliability of the device.

In the experiment, a GaN-based laser with an inverted n-side ridge waveguide (nRW) was proposed and fabricated [21]. That is, the laser ridge is moved from the p-side to the n-side, as shown in Fig. 2.17a and b, which can greatly increase the injection area of the holes and significantly reduce the series resistance and ohmic contact resistance on the p-side. Since the series resistivity and ohmic contact resistivity of the n-type material are both small, the n-side ridge has little effect on the series resistance and contact resistance on the n-side; therefore, the total series resistance and contact resistance of the device have been greatly reduced. We tested and characterized the GaN-based laser with an inverted n-side ridge. The test results show that compared with the conventional p-side ridge waveguide (pRW) laser, the differential resistance and operating voltage of the n-side ridge waveguide laser are reduced by 48% and 1.4 V, respectively, as shown in Fig. 2.17c; the thermal resistance and operating junction temperature are reduced by 8 K/W and 25 °C, respectively, as shown in Fig. 2.17d. It is worth noting that the n-side ridge waveguide laser can be directly fabricated on the Si(100) wafer without a bevel angle, which is fully compatible with the silicon-based microelectronics and optoelectronics platform, providing a feasible technical route for the wafer-level heterogeneous integration of optoelectronics such as GaN lasers with Si(100) microelectronics.

3. New type of cladding layer

The AlGaN cladding layer grown on the n-GaN contact layer is subject to tensile stress and is prone to cracking. To avoid cracking, most existing GaN-based lasers use AlGaN cladding layers with low Al content. Due to the small refractive index difference between the low-Al-content AlGaN cladding layer and the GaN waveguide layer, the laser's optical confinement is weak, the confinement factor is small (<3%), and the device threshold current density is high. Using a new type of cladding layer with a low refractive index, such as nanoporous GaN [22], indium tin oxide (ITO) [23], zinc oxide (ZnO) [24], silver (Ag), etc., can significantly increase the refractive index difference between the cladding layer and the waveguide layer without introducing additional tensile stress, significantly enhancing the laser's optical confinement, thereby greatly reducing the threshold current and improving the wall-plug efficiency. In addition, due to the low deposition temperature (<300 °C) of the new

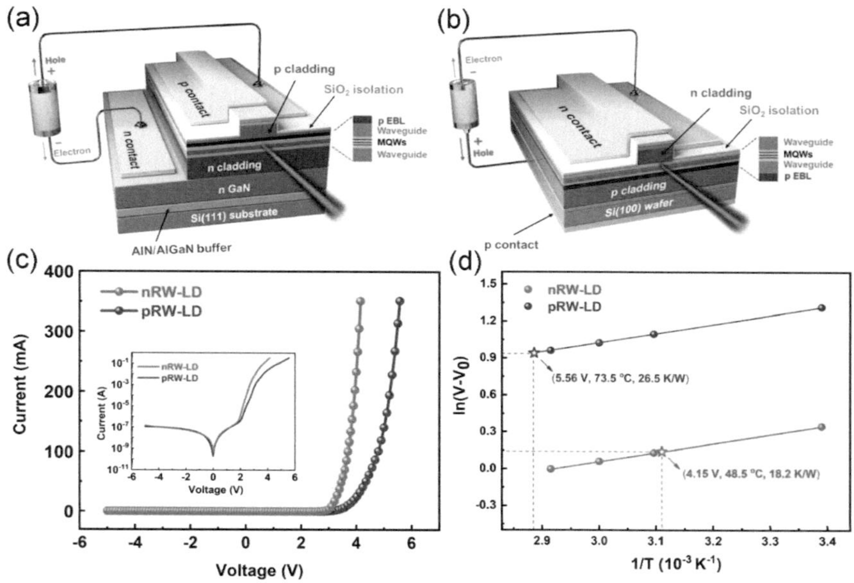

Schematic architecture of GaN-based (a) pRW-LD on Si(111) and (b) nRW-LD on Si(100); (c) I-V characteristics of nRW and pRW LDs measured under various CW injection currents at RT. The inset shows the I-V curves are plotted in the logarithmic scale. (d) Plot of ln(V - V_0) as a function of reciprocal of absolute temperature (1/T) at an injection current of 350 mA.

Fig. 2.17 Schematic diagram of the structure, I–V curve, and thermal resistance measurement curve of the GaN-based laser with an inverted n-side ridge on silicon substrate [21]

type of cladding layer, such as ITO, ZnO, etc., replacing the conventionally high-temperature-grown p-AlGaN cladding layer with the new type of cladding layer can significantly shorten the high-temperature time experienced by the InGaN/GaN multiple quantum wells, effectively reducing the thermal degradation of the multiple quantum wells and improving the material quality and internal quantum efficiency of the active area. Moreover, the resistance of the new type of cladding layer is very small, and replacing the p-AlGaN cladding layer with the new type of cladding layer can also reduce the device series resistance and operating voltage.

2.3.3 *Extension of the Operating Wavelength of GaN-Based Lasers on Silicon Substrates*

Group III-nitride blue lasers and ultraviolet lasers have important application prospects in fields such as laser display, ultraviolet curing, sterilization, and disinfection; therefore, the following will test and characterize the GaN-based blue and ultraviolet lasers on silicon substrates.

1. GaN-based blue laser on silicon substrate

Unlike freestanding GaN substrate lasers, GaN-based lasers on silicon substrates have greater stress and higher defect density. When growing InGaN multiple quantum wells, especially blue quantum wells with higher In components, rich In cluster defects are prone to occur. These rich In cluster defects undergo thermal degradation during the subsequent high-temperature growth of the p-AlGaN cladding layer, resulting in black spots, severely affecting the internal quantum efficiency of the multiple quantum wells, and the device cannot lase. Research has found that by introducing a small amount of hydrogen during the growth of the GaN barrier layer and reducing the growth temperature of the p-AlGaN layer, the rich In cluster defects in the multiple quantum wells can be effectively eliminated, the thermal degradation of the blue multiple quantum wells can be suppressed, and the room-temperature electric injection continuous lasing of the GaN-based blue laser on the silicon substrate is achieved [15]. The lasing spectrum is shown in Fig. 2.18a, with a lasing wavelength of 450 nm. Figure 2.18b shows the L–I–V curve of the blue laser, with a threshold current density of 7.8 kA/cm^2. Recently, by optimizing the Si doping of multiple quantum wells and suppressing the carbon impurities in the p-AlGaN cladding layer, the threshold current density and threshold voltage of the blue laser have been reduced to 4.9 kA/cm^2 and 4.9 V, respectively.

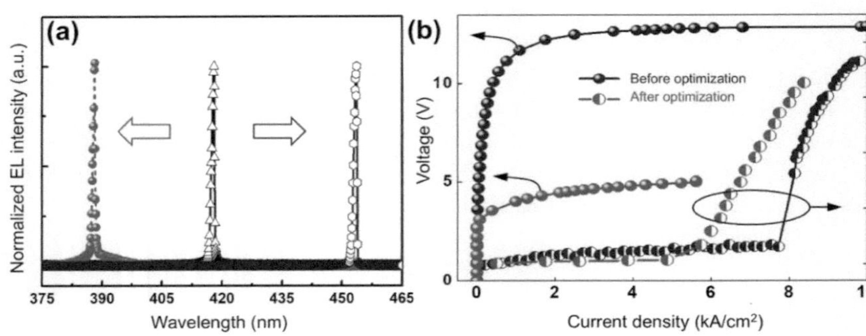

(a) The expansion of lasing wavelength; (b) L-I-V curves of the blue laser before and after optimizations

Fig. 2.18 Extension of lasing wavelength and performance improvement of GaN-based lasers grown on silicon substrates [15, 25]

2. AlGaN-based near-ultraviolet lasers grown on silicon substrates

Unlike InGaN blue multiple quantum wells with higher In components, the In component in near-ultraviolet multiple quantum wells is very low. There is no localized-state effect caused by In component fluctuations, so it is impossible to use the localized-state effect to shield nonradiative recombination caused by dislocation defects, making near-ultraviolet multiple quantum wells very sensitive to dislocation defects. On the other hand, to reduce the threshold current, near-ultraviolet lasers usually use AlGaN cladding layers with higher Al components to enhance optical field confinement. When growing AlGaN layers with higher Al components on GaN templates, AlGaN is subject to the tensile stress of the underlying GaN material and is very prone to cracking. In the experiment, AlGaN templates were used instead of GaN templates to grow AlGaN-based near-ultraviolet lasers, significantly reducing the tensile stress in the laser material and effectively avoiding cracking. Literature reports that compared to screw dislocations and mixed dislocations, edge dislocations have a greater impact on the internal quantum efficiency as nonradiative recombination centers. For this reason, the experiment optimized the AlN/AlGaN buffer layer structure to effectively reduce the edge dislocation density in the AlGaN-on-Si template, improved the growth conditions of the near-ultraviolet multiple quantum wells to greatly reduce the point defect concentration, significantly improved the internal quantum efficiency, and finally achieved room-temperature electric injection lasing of the AlGaN-based near-ultraviolet laser on the silicon substrate [25], with a lasing wavelength of 389 nm, as shown in Fig. 2.18a.

3. Deep-ultraviolet lasers based on AlGaN on silicon substrates

In 2019, the H. Amano team from Nagoya University in Japan achieved the world's first room-temperature electric injection lasing of an AlGaN-based deep-ultraviolet laser, with a lasing wavelength of 271.8 nm, a threshold current of 400 mA, and a threshold voltage of 13.8 V [26]. The high threshold voltage is mainly due to the large series resistance of the p-AlGaN cladding layer and the current crowding effect existing in the coplanar electrode structure of the laser. Using silicon substrate, it is easy to prepare a vertical structure laser, and combined with the n-side ridge waveguide structure, it can greatly reduce the device series resistance and operation voltage, thereby reducing the threshold current. However, due to the huge lattice mismatch and thermal mismatch between silicon and AlGaN, high-Al component AlGaN materials for deep-ultraviolet optoelectronic devices grown on silicon substrates are very prone to cracks and high-density dislocation defects. Therefore, growing high-quality high-Al component AlGaN materials on silicon substrate is very challenging. In the experiment, an AlN/AlGaN superlattice buffer layer was used to grow a crack-free 2-μm $Al_{0.5}Ga_{0.5}N$ thin film on a flat silicon substrate. The full width at half maximum (FWHM) of the XRD double-crystal rocking curves of the (0002) and $(10\bar{1}2)$ planes are 499 and 648 arcsec, respectively, and the corresponding dislocation density is about 3.8×10^9 cm^{-2} [27]. Recently, by optimizing the AlN nucleation layer and AlN/AlGaN superlattice buffer layer, the dislocation density in the $Al_{0.5}Ga_{0.5}N$ thin film grown on silicon substrate has been further reduced to 1×10^9 cm^{-2}. However, for deep-ultraviolet lasers, the dislocation density in the abovementioned AlGaN material on silicon substrate is still high and needs to be further improved.

2.3.4 GaN-Based Lasers on Si(100) Substrates

As mentioned in Sect. 2.2.1, Si(111) substrates are usually used for growing GaN-based materials. However, they are not very compatible with the Si(100) wafers used for complementary metal-oxide semiconductors (CMOS). Therefore, the preparation of GaN-based lasers on Si(100) substrates has also become a development trend. The team of J.E. Bowers at the University of California, Santa Barbara, achieved the heterogeneous integration of InGaN-based lasers and Si(100) wafers through chip bonding, but the efficiency of chip bonding is low and it is not compatible with large-scale silicon CMOS platforms [28]. The team of Professor H. Amano at Nagoya University in Japan grew a full structure of semipolar GaN lasers on a graphical Si(100) substrate, but only achieved optically pumped lasing [29]. In 2019, the team of Professor Shen Bo at Peking University used single-crystal graphene as a buffer layer to epitaxially grow single-crystal GaN thin films on a Si(100) substrate, demonstrating the great prospects for growing group III-nitrides on a Si(100) substrate [30]. However, the dislocation density of all reported GaN materials grown on Si(100) substrate is still very high, which is not suitable for practical device applications, and further improvement of material quality is needed. Another technical route uses silicon-on-insulator (SOI) substrate Si(111)-on-SiO_2-on-Si(100) to epitaxially grow GaN-based lasers, which is expected to achieve monolithic integration of GaN-based optoelectronic devices and Si(100) CMOS.

2.4 Chapter Summary

GaN-based lasers grown on silicon substrates have received extensive interest in recent years and provide highly promising applications in laser display, laser lighting, and silicon-based photonics. Unlike lasers homoepitaxially grown on freestanding GaN substrates, GaN-based lasers heteroepitaxially grown on silicon substrates show huge lattice mismatch and thermal mismatch between silicon and GaN, resulting in high dislocation density and large residual stress in the epilayer, severely affecting device performance and reliability. In recent years, significant progress has been made in GaN-based lasers grown on silicon, but the output power and reliability of the devices need to be further improved. This chapter focused on the epitaxial growth, device process preparation, and characterization of GaN-based lasers on silicon substrates, specifically as follows:

1. Introduced the characterization methods of stress and defects in GaN materials grown on silicon substrates and elucidated the interaction mechanism of stress and threading dislocations.
2. Using the AlN/AlGaN buffer layer technology that combines stress control and dislocation filtering, a high-quality GaN thin film grown on silicon substrate with a threading dislocation density as low as 5.8×10^7 cm^{-2} has been realized.
3. Introduced the characterization methods for defects such as In-rich clusters, V-pit defects, and point defects in the quantum wells of GaN-based lasers on silicon, discussed their impacts on laser performance, and finally proposed corresponding solutions.
4. Performed testing characterization of GaN-based laser devices on silicon substrates, including electroluminescence spectra, L–I–V curves, near-field images, far-field images, junction temperature tests, etc., confirming the lasing of the laser.
5. Introduced the development of GaN-based lasers grown on silicon substrates. The device threshold current and operation voltage have been reduced to 2.25 kA/cm^2 and 4.7 V, respectively, close to the level of freestanding GaN substrate lasers; however, the reliability of GaN-based lasers grown on silicon substrates needs to be further improved. New technologies, including n-side ridge waveguide laser structures and new cladding layers, are expected to significantly improve the performance of GaN-based laser devices on silicon substrates.
6. By further carrying out high-quality GaN material heteroepitaxial growth on silicon substrates, point defect engineering, quantum dot active area technology, novel cladding layers, and laser structure, it is expected to significantly improve the performance and reliability of GaN-based lasers grown on silicon, realizing the high-efficiency, long-lifetime GaN-based lasers on silicon and also their on-chip integrated applications in the near future.

References

1. Feng M, Liu J, Sun Q et al (2021) III-Nitride semiconductor lasers grown on Si. Prog Quantum Electron 77:100323
2. Sun Q, Yan W, Feng M et al (2016) GaN-on-Si blue/white LEDs: epitaxy, chip, and package. J Semicond 37(4):044006
3. Cheng K, Leys M, Degroote S et al (2006) Flat GaN epitaxial layers grown on Si(111) by metalorganic vapor phase epitaxy using step-graded AlGaN intermediate layers. J Electron Mater 35(4):592–598
4. Leung B, Han J, Sun Q (2014) Strain relaxation and dislocation reduction in AlGaN step-graded buffer for crack-free GaN on Si (111). Phys Status Solidi C 11(3):437–441
5. Wang H, Liang H, Wang Y et al (2010) Effects of AlGaN/AlN stacked interlayers on GaN growth on Si (111). Chin Phys Lett 27(3):038103
6. Dadgar A, Hempel T, Bläsing J et al (2011) Improving GaN-on-silicon properties for GaN device epitaxy. Phys Status Solidi C 8(5):1503–1508
7. Sun Y, Zhou K, Sun Q et al (2016) Room-temperature continuous-wave electrically injected InGaN-based laser directly grown on Si. Nat Photonics 10(9):595–599

8. Xu Z (2007) Detection and analysis of semiconductors, 2nd edn. Science Press, Beijing
9. Mai Z (2015) X-ray characterization of thin film structures, 2nd edn. Science Press, Beijing
10. Lu L, Shen B, Xu FJ et al (2007) Morphology of threading dislocations in high-resistivity GaN films observed by transmission electron microscopy. J Appl Phys 102(3):033510
11. Liu J, Huang Y, Sun X et al (2019) Wafer-scale crack-free 10 μm-thick GaN with a dislocation density of 5.8×10^7 cm^{-2} grown on Si. J Phys D Appl Phys 52(42):425102
12. Cantu P, Wu F, Waltereit P et al (2005) Role of inclined threading dislocations in stress relaxation in mismatched layers. J Appl Phys 97(10):103534
13. Massabuau FCP, Davies MJ, Oehler F et al (2014) The impact of trench defects in InGaN/GaN light emitting diodes and implications for the "green gap" problem. Appl Phys Lett 105(11):112110
14. Li Z, Liu J, Feng M et al (2013) Suppression of thermal degradation of InGaN/GaN quantum wells in green laser diode structures during the epitaxial growth. Appl Phys Lett 103(15):152109
15. Sun Y, Zhou K, Feng M et al (2018) Room-temperature continuous-wave electrically pumped InGaN/GaN quantum well blue laser diode directly grown on Si. Light Sci & Appl 7:13
16. Jia C, Yu T, Lu H et al (2013) Performance improvement of GaN-based LEDs with step stage InGaN/GaN strain relief layers in GaN-based blue LEDs. Opt Express 21(7):8444–8449
17. Yang J, Zhao DG, Jiang DS et al (2020) Suppression of the formation of V-pits in InGaN/GaN multi-quantum well growth and its effect on the performance of GaN based laser diodes. J Alloy Compd 822:153571
18. Haller C, Carlin JF, Jacopin G et al (2018) GaN surface as the source of non-radiative defects in InGaN/GaN quantum wells. Appl Phys Lett 113(11):111106
19. Liu J, Wang J, Sun X et al (2019) Performance improvement of InGaN-based laser grown on Si by suppressing point defects. Opt Express 27(18):25943–25952
20. Tang Y, Feng M, Wen P et al (2020) Degradation study of InGaN-based laser diodes grown on Si. J Phys D Appl Phys 53(39):395103
21. Zhou R, Feng M, Wang J et al (2020) InGaN-based lasers with an inverted ridge waveguide heterogeneously integrated on Si(100). ACS Photonics 7(10):2636–2642
22. Yuan G, Xiong K, Zhang C et al (2016) Optical engineering of modal gain in a III-nitride laser with nanoporous GaN. ACS Photonics 3(9):1604–1610
23. Hardy MT, Holder CO, Feezell DF et al (2013) Indium-tin-oxide clad blue and true green semipolar InGaN/GaN laser diodes. Appl Phys Lett 103(8):081103
24. Myzaferi A, Mughal AJ, Cohen DA et al (2018) Zinc oxide clad limited area epitaxy semipolar III-nitride laser diodes. Opt Express 26(10):12490–12498
25. Feng M, Li Z, Wang J et al (2018) Room-temperature electrically injected AlGaN-based near-ultraviolet laser grown on Si. ACS Photonics 5(3):699–704
26. Zhang Z, Kushimoto M, Sakai T et al (2019) A 271.8 nm deep-ultraviolet laser diode for room temperature operation. Appl Phys Express 12(12):124003
27. Huang Y, Liu J, Sun X et al (2020) Crack-free high quality 2 μm-thick $Al_{0.5}Ga_{0.5}N$ grown on a Si substrate with a superlattice transition layer. CrystEngComm 22(7):1160–1165
28. Kamei T, Kamikawa T, Araki M et al (2020) Research toward a heterogeneously integrated InGaN laser on silicon. Phys Status Solidi A 217(7):1900770
29. Murase T, Tanikawa T, Honda Y et al (2011) Optical properties of (1–101) InGaN/GaN MQW stripe laser structure on Si substrate. Phys Status Solidi C 8:2160–2162
30. Feng Y, Yang X, Zhang Z et al (2019) Epitaxy of single-crystalline GaN film on CMOS-compatible Si(100) substrate buffered by graphene. Adv Func Mater 29(42):1905056

Chapter 3
Efficiency Characterization of GaN-Based LEDs

Lai Wang, Boyang Lu, and Yi Luo

3.1 Overview

Thanks to the successful development of gallium nitride (GaN)-based blue light-emitting diodes (LEDs), energy-efficient and environmentally friendly semiconductor solid-state white light sources have been realized, which garnered the 2014 Nobel Prize in Physics. Following years of technological advancement, LEDs have been extensively adopted in solid-state lighting (SSL) applications. Moreover, they hold significant potential for application in the fields of display technology and visible light communication [1].

The widespread attention garnered by light-emitting diodes (LEDs) is attributed to their numerous advantages over traditional light-emitting devices, including compact size, high brightness, long lifespan, and superior energy efficiency. Notably, their high energy efficiency stands out as a primary characteristic. LEDs can achieve approximately 80% energy savings when compared to incandescent lamps of equivalent brightness, a factor of particular importance in light of the increasingly depleted nonrenewable energy sources. According to data from the US Department of Energy in 2017, lighting accounts for about 16% of the total electricity consumption in the United States, a figure that also surpasses 10% in China. Consequently, governments have implemented policies to actively promote research and development in solid-state lighting (SSL) technology. For instance, the US Department of Energy initiated the SSL program as early as 2000, and China proposed its semiconductor lighting development plan in 2003, thereby launching the National Semiconductor Lighting Project. Thanks to the collaborative efforts of nations worldwide, the semiconductor lighting industry, with LEDs as the core device, has reached a high level of maturity, with the internal quantum efficiency of GaN-based blue LEDs exceeding 90%, and

L. Wang (✉) · B. Lu · Y. Luo
Department of Electronic Engineering, Tsinghua University, Beijing, China
e-mail: wanglai@tsinghua.edu.cn

S. Xu (ed.), *Optical Characterization of Microstructures and Optoelectronic Devices Based on Wide Band Gap Semiconductors*, Wide Bandgap Semiconductors,
https://doi.org/10.1007/978-981-95-1928-6_3

the external quantum efficiency also surpassing 70%. Currently, nearly 50% of traditional lighting sources in China have been replaced by LED products, resulting in an annual electricity savings of about 280 billion kWh, equivalent to the generating capacity of the Three Gorges Hydropower Projects. Does this imply that research into LED efficiency is no longer significant? The answer is a resounding no. This issue can be examined from both application and physical perspectives.

From an application perspective, there remains potential for improving the efficiency of light-emitting diodes (LEDs) used in solid-state lighting (hereinafter referred to as "lighting LEDs") [2–4]. Experimental findings indicate that while the brightness of GaN-based LEDs increases with the injection current density, the efficiency initially rises but subsequently decreases as the injection current density continues to increase, as illustrated in Fig. 3.1. The peak efficiency is observed at a current density ranging from a few A/cm^2 to tens of A/cm^2. Lighting LEDs are designed to achieve high brightness with a minimal number of chips and a small chip area, necessitating operation at a higher current density, typically around 35 A/cm^2. Evidently, there is a loss in efficiency when LEDs are operated under these conditions compared to their peak efficiency, a phenomenon known as "efficiency droop." To elucidate the physical mechanisms underlying this phenomenon, extensive research has been conducted in recent years, with main findings attributing the cause to Auger recombination, carrier leakage, and thermal effects, among others [5, 6]. Although a consensus has not yet been reached within the academic community, efforts to mitigate efficiency droop based on current understanding are already underway. These endeavors can be broadly categorized into two approaches: (1) enhancing the existing LED device structure, which includes employing multi-quantum well (MQW) structures to reduce carrier concentration, growing electron-blocking layers to minimize carrier leakage, and improving chip heat dissipation design, and (2) exploring large-size LED chip epitaxy methods to fabricate large-area light-emitting surface light sources, which can lower the operating current density and obviate the need for additional optical systems for light uniformity, as depicted in Fig. 3.2. Clearly, regardless of the approach adopted, a more precise efficiency characterization method is essential to guide further advancements in the manufacturing process.

From a physical perspective, the efficiency characterization method must be continually updated in alignment with the advancement of device physics. While considerable experience has been accumulated with regard to lighting LEDs, these insights are not directly transferable to other LED devices. For instance, consider micro-LEDs used for display purposes. Micro-LEDs are miniature LED chips, often less than a hundred micrometers in size, with each chip serving as an independent light-emitting unit within a display screen. Current mainstream display technologies include liquid crystal display (LCD) and organic light-emitting diode (OLED) technologies. LCD technology primarily consists of a backlight source, polarizing plates, and a liquid crystal layer. Light emitted by the backlight source is transformed into polarized light through the lower polarizing plate. The arrangement of liquid crystal molecules is altered by an external electric field, thereby controlling the polarization direction of the incident light. The upper polarizing plate then blocks light with

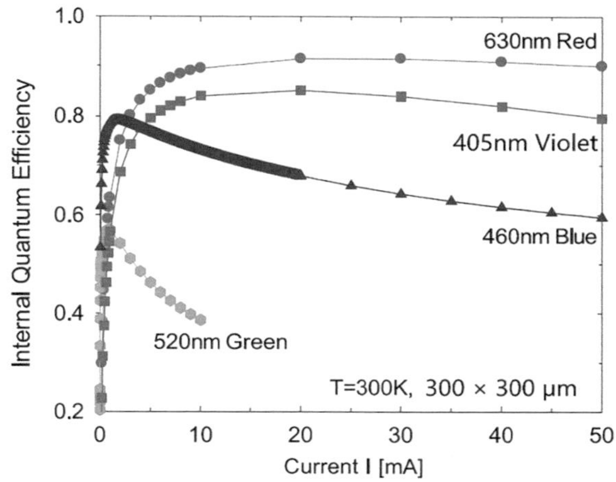

Fig. 3.1 LED efficiency drop phenomenon, cited from reference [3]

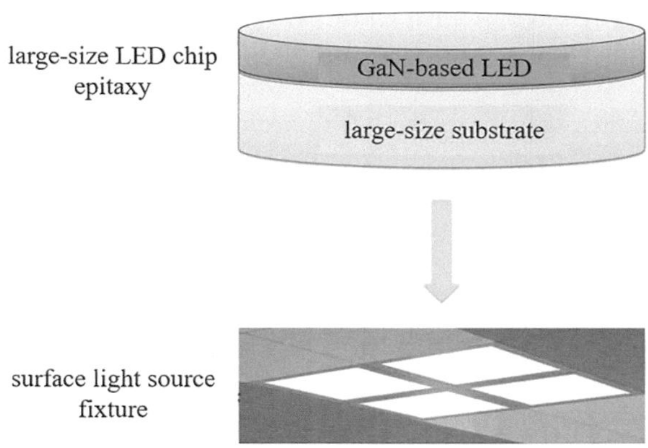

Fig. 3.2 Using large-size LED chips to solve efficiency drop

mismatched polarization directions to control the light's on/off state. OLED technology utilizes amorphous organic materials to create active light-emitting devices, which can bypass the substrate material lattice matching requirements of epitaxial growth and employ techniques such as spin coating, evaporation, and inkjet printing to deposit films onto substrates like glass or plastic. The main drawback of LCDs is their low efficiency, necessitating continuous backlighting and passage through multiple optical layers, such as the liquid crystal layer and polarizing plates. Despite this, LCDs remain widely used in mid-to-low-end and large-screen consumer electronics due to their mature manufacturing process. OLEDs, with their active light-emitting properties, have improved light-emitting efficiency and can achieve flexible displays. However, the instability of organic materials leads to a shorter lifespan for OLEDs, and their high manufacturing cost and difficulty in scaling to large sizes limit their use to mid-to-high-end consumer electronics products. Indeed, even with

OLED technology, there are complaints about the insufficient battery life of electronic products such as smartphones, largely due to the high power consumption of display devices (often exceeding 50% of a smartphone's screen power consumption). To address the power consumption and other limitations of display technologies, micro-LEDs have been introduced, with the hope that they will inherit the high efficiency and energy-saving attributes of lighting LEDs. However, micro-LEDs exhibit a significant difference compared to lighting LEDs: The chip size is very small, the number of chips is large, and for regular display applications, each chip requires lower brightness and operates at a smaller current density, usually less than 1 A/cm^2. This difference in injection level leads to variations in carrier transport and recombination processes, necessitating the use of different models when analyzing the impact of various physical processes on efficiency, as shown in Fig. 3.3. For example, it is commonly assumed that the carrier injection efficiency of lighting LEDs can be approximated as 100% at moderate current densities, but this assumption does not hold for micro-LEDs due to the significant impact of etching damage and the pronounced sidewall leakage phenomenon, which results in very low injection efficiency for micro-LEDs. Therefore, it is essential to develop more rational efficiency characterization methods tailored to different devices.

In summary, efficiency remains a pivotal technical indicator for LEDs that necessitates ongoing research and development. The subsequent sections will delineate the definition of LED efficiency and the commonly employed characterization methods. By integrating typical experimental outcomes, this work will illustrate the application of efficiency characterization in assessing LED performance and carrier dynamics. Finally, the study will examine the applicability and limitations of various characterization methods and will provide insights into the future development trajectory of efficiency characterization.

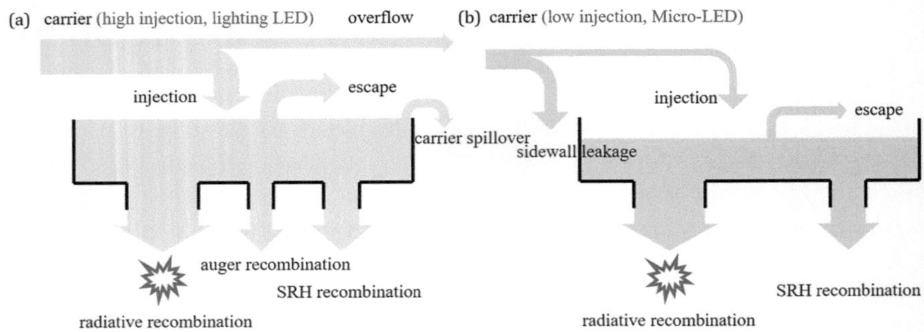

Fig. 3.3 Schematic diagram of LED carrier transport and recombination process: **a** lighting LED and **b** micro-LED

3.2 Definition of LED Efficiency

LEDs are semiconductor devices that convert electrical energy into light energy. Similar to ordinary diodes, which are typically made from materials such as silicon (Si) and germanium (Ge), LEDs consist of a PN junction structure. However, LEDs are fabricated from direct bandgap materials that have a higher probability of radiative transitions. Under forward bias conditions, where the P and N regions of the LED are connected to the positive and negative terminals of a power supply, respectively, the barrier height at the PN junction is reduced. Electrons and holes diffuse toward each other, meet, and recombine, emitting light in the process. This recombination, where electrons return from an excited state to the ground state and release energy in the form of photons, is known as electroluminescence (EL). The energy of the emitted photons is determined by the width of the material's bandgap, as illustrated in Fig. 3.4.

To increase the likelihood of electron–hole encounters, a low-dimensional structure, such as quantum wells or quantum dots, typically composed of low-doped intrinsic semiconductor material, is inserted between the P and N regions of the LED. This structure confines the injected carriers and is referred to as the "active region." Furthermore, when the LED is exposed to light with photon energy exceeding the bandgap width, electrons in the valence band absorb the incident photon energy, transition to the conduction band, and subsequently radiate back to the ground state. This process is known as photoluminescence (PL).

The overall efficiency of an LED is quantified by the wall-plug efficiency (WPE, η_{WPE}), which is defined as the ratio of the optical power ($P_{\mathrm{p,out}}$) emitted by the LED into the external environment to the electrical power ($P_{\mathrm{e,in}}$) supplied to the LED, expressed as follows:

$$\eta_{\mathrm{WPE}} = \frac{P_{\mathrm{p,out}}}{P_{\mathrm{e,in}}} \tag{3.1}$$

Fig. 3.4 Schematic diagram of LED energy bands

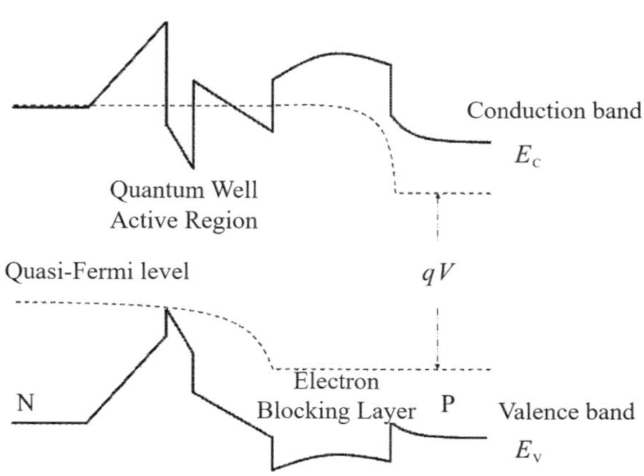

where the subscript "E" denotes physical quantities related to electricity and the subscript "P" denotes physical quantities related to light. η_{WPE} reflects the LED device's capability to convert excitation energy into radiant energy per unit time. $P_{e,in}$ can be determined by measuring the forward voltage (V) and current (I), such that $P_{e,in} = V \times I$. $P_{p,out}$ is typically measured using an integrating sphere, which involves placing the LED chip at the center of a hollow sphere coated with white diffuse material on the inner wall. The light emitted by the LED undergoes multiple reflections off the inner wall, resulting in uniform illuminance. The light power at the sphere's wall aperture can then be measured to calculate $P_{p,out}$, making η_{WPE} a measurable physical quantity. This measurement method is known as the "4π geometric structure." For light sources that do not emit light backward, the "2π geometric structure" can also be employed.

For light-emitting devices, the human eye's response to LED radiation is of particular interest. The human retina contains two types of photoreceptor cells: cone cells and rod cells. Cone cells are responsible for responding to bright light and color, whereas rod cells are highly sensitive to light intensity but not to color. During daylight or in brightly lit environments, cone cells are primarily active, resulting in vision that perceives both brightness and color, known as "photopic vision." In contrast, during nighttime or in dark environments, rod cells are primarily active, leading to a monochromatic, gray-black perception without color, referred to as "scotopic vision." The degree of visual response induced by radiation of different wavelengths is termed the "luminosity function," as depicted in Fig. 3.5. Under photopic conditions, the human eye is most sensitive to yellow-green light (approximately 555 nm), while under scotopic conditions, the eye is most sensitive to light at 507 nm. As the wavelength increases or decreases, the response diminishes; beyond the visible light spectrum (380–780 nm), there is no response. Assuming the light power at various wavelengths emitted by the LED is $P_{p,out}(\lambda)$, the luminous flux of the LED radiation can be calculated as shown in Eq. (3.2).

$$\Phi = 683\mathrm{lm/W} \times \int_{380\mathrm{nm}}^{780\mathrm{nm}} \left[P_{p,out}(\lambda) \times \bar{y}(\lambda)\right] d\lambda \tag{3.2}$$

Here, 683 lm/W represents the maximum luminous flux producible per unit of light power; the depicted image corresponds to the visual response function. The luminous efficacy (η_Φ) of the LED is defined as the ratio of the luminous flux (Φ) emitted by the LED into the external environment to the electrical power ($P_{e,in}$) supplied to the LED, as presented in Eq. (3.3).

$$\eta_\Phi = \frac{\Phi}{P_{e,in}} \tag{3.3}$$

η_Φ is a measurable physical quantity, assessed using the same method as for η_{WPE}. Indeed, η_Φ and η_{WPE} are closely related; the distinction lies in the use of the human eye as a detector for η_Φ, whereas η_{WPE} measures light power using a detector with a flat spectral response, reflecting the difference between photometry and radiometry.

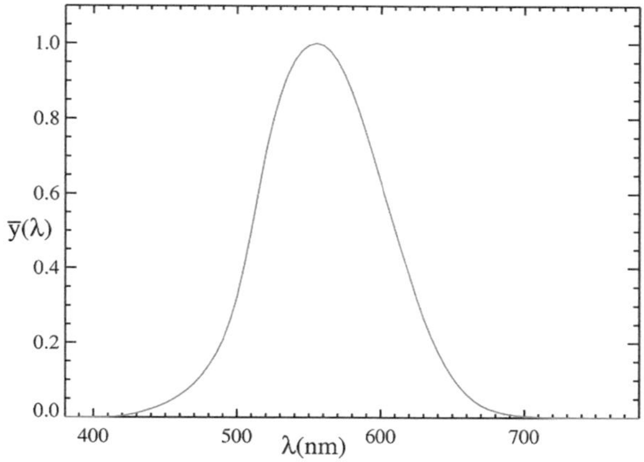

Fig. 3.5 Visual response function under bright visual conditions

Specifically, photometric quantities such as luminous flux (unit: lm), luminous intensity (unit: cd), and luminance (unit: cd/m^2) represent the total visual stimulation to the human eye and are applicable only within the visible light spectrum. Radiometric quantities, including radiant flux (unit: W), radiant intensity (unit: W/sr), and radiance (unit: W/sr/m^2), reflect objective electromagnetic radiation measurements and are applicable across the entire electromagnetic spectrum. It is important to note that for efficiencies that cannot be directly measured, relative values are commonly used. Therefore, the luminous intensity discussed subsequently employs arbitrary units (a.u.) and does not correspond to the photometric quantity of luminous intensity.

The wall-plug efficiency η_{WPE} and the internal quantum efficiency η_{Φ} are metrics that characterize the macroscopic energy conversion efficiency, which is a critical indicator for LEDs. However, we cannot optimize η_{WPE} and η_{Φ} directly, as the photoelectric conversion process encompasses multiple physical processes, including excitation, energy transport, and radiation emission, with each stage having factors that contribute to efficiency loss. Consequently, it is essential to adhere to the principle of controlling variables, to decompose these physical processes with a focus on carrier dynamics, and to analyze them individually.

First, due to the voltage drop across the series resistance of the LED, which is primarily composed of the ohmic contact resistance of the metal electrode, the resistance of the semiconductor material, and the resistance of the semiconductor bulk, as well as the energy loss associated with carrier relaxation to the lowest energy state after injection into the active region, the potential energy of the carriers supplied by the power source cannot be fully converted into photon energy. The voltage efficiency (η_{VTG}) is defined as the ratio of the average photon energy emitted by the LED to the external world to the potential energy of the carriers injected into the LED ($E_{\mathrm{e,in}}$), that is:

$$\eta_{\mathrm{VTG}} = \frac{\overline{E}_{\mathrm{p,out}}}{E_{\mathrm{e,in}}} = \frac{P_{\mathrm{p,out}} / n_{\mathrm{p,out}}}{qV} \tag{3.4}$$

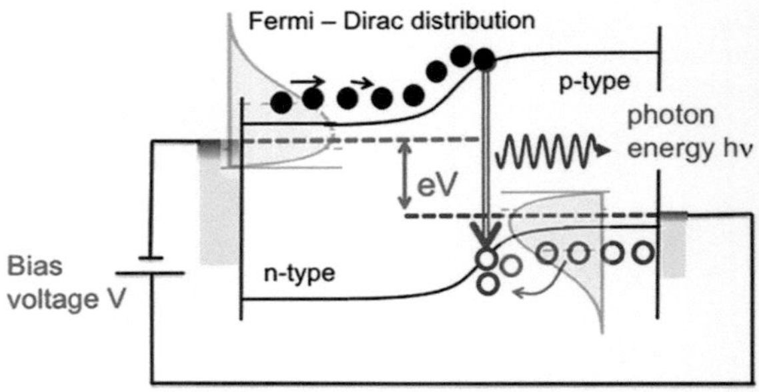

Fig. 3.6 Subthreshold radiation phenomenon of blue LED, illustration quoted from reference [4]

where $q \approx 1.6 \times 10^{-19}$ C is the basic charge and $n_{\mathrm{p,out}}$ is the number of photons radiated to the outside world per unit time by the LED, which can be obtained by measuring the output power at different wavelengths $P_{\mathrm{p,out}}(\lambda)$, that is:

$$n_{\mathrm{p,out}} = \int_0^{\infty} \frac{P_{\mathrm{p,out}}(\lambda)}{hc/\lambda} \mathrm{d}\lambda \tag{3.5}$$

where $h \approx 6.63 \times 10^{-34}$ J · s is the Planck constant, $c \approx 3 \times 10^8$ m/s is the speed of light in a vacuum, and hc/λ is the photon energy. η_{VTG} is a measurable physical quantity, which reflects the ability of carrier energy to be converted into photon energy. Assuming that the total number of carriers injected into the LED per unit time $n_{\mathrm{e,in}} = I/q$ is equal to $n_{\mathrm{p,out}}$, that is, each carrier injected into the LED can be converted into a photon and radiated out, then $\eta_{\mathrm{WPE}} = \eta_{\mathrm{VTG}}$. η_{VTG} usually decreases with the increase in forward voltage V, so it is necessary to reduce the working voltage of the LED as much as possible and reduce the series resistance to improve η_{VTG}. It is worth noting that the LED can emit significant light when V is less than the potential difference corresponding to the bandgap width; that is, the injected electron energy is less than the radiated photon energy; at this time, $\eta_{\mathrm{VTG}} > 100\%$; this phenomenon is called "subthreshold radiation"; there is still no reasonable explanation, as shown in Fig. 3.6.

The carriers injected into the LED from the power source are not all directed into the active region. There are mechanisms that lead to efficiency reduction, such as surface recombination, defect-related tunneling, and carrier overflow during transport. The injection efficiency (η_{INJ}) is defined as the ratio of the number of carriers injected into the active region per unit time ($n_{\mathrm{e,active}}$) to the total number of carriers injected into the LED per unit time ($n_{\mathrm{e,in}}$), that is:

$$\eta_{\mathrm{INJ}} = \frac{n_{\mathrm{e,active}}}{n_{\mathrm{e,in}}} \tag{3.6}$$

η_{INJ} cannot be directly measured, and the mechanism that reduces the injection efficiency is still not fully understood, so it is very difficult to optimize the injection efficiency.

Once carriers are successfully injected into the active region, the carrier concentration within this area increases, facilitating recombination. While some carriers undergo radiative recombination, converting their energy into light, there are also physical processes such as nonradiative recombination, carrier leakage, and carrier escape. The radiative efficiency (η_{RAD}) is defined as the ratio of the number of photons emitted from the active region per unit time ($n_{\mathrm{p,active}}$) to the number of carriers injected into the active region per unit time ($n_{\mathrm{e,active}}$), that is:

$$\eta_{\mathrm{RAD}} = \frac{n_{\mathrm{p,active}}}{n_{\mathrm{e,active}}} \tag{3.7}$$

η_{RAD} cannot be directly measured; rather, it reflects the competition between radiative and nonradiative processes. Ensuring that radiative recombination prevails is a critical aspect of designing the LED epitaxial structure. Moreover, η_{INJ} and η_{RAD} are not entirely independent of each other, as η_{INJ} influences the carrier concentration within the active region, which in turn is closely related to η_{RAD}.

The photons emitted from the active region must traverse the bulk material of the LED before they can emerge into the external environment and be detected by observers. This process is known as "light extraction." The light extraction efficiency (η_{LEE}) is defined as the ratio of the number of photons $n_{\mathrm{p,out}}$ emitted into the external world per unit time by the LED to the number of photons $n_{\mathrm{p,active}}$ emitted from the active region per unit time, that is:

$$\eta_{\mathrm{LEE}} = \frac{n_{\mathrm{p,out}}}{n_{\mathrm{p,active}}} \tag{3.8}$$

Since $n_{\mathrm{p,active}}$ cannot be directly measured, η_{LEE} remains an unmeasurable physical quantity. For conventional LED devices, η_{LEE} is a significant factor that limits overall efficiency. The refractive index of GaN material in the blue wavelength range at room temperature is approximately 2.4, which is substantially different from that of the external environment (refractive index of about 1) and the sapphire substrate (refractive index of about 1.7). This mismatch leads to total internal reflection, trapping most of the emitted photons within the LED. During the process of multiple reflections, these photons are absorbed by the material and converted into heat energy through mechanisms such as self-absorption, impurity absorption, and free-carrier absorption, which can affect the stability of the LED device. Consequently, extensive efforts have been directed toward enhancing the light extraction from LED chips, and various methods for simulating the calculation of η_{LEE} have been developed.

In summary, η_{WPE} can be decomposed into four components: η_{VTG}, η_{INJ}, η_{RAD}, and η_{LEE}, that is:

$$\eta_{\text{WPE}} = \eta_{\text{VTG}} \times \eta_{\text{INJ}} \times \eta_{\text{RAD}} \times \eta_{\text{LEE}} \tag{3.9}$$

We are more accustomed to the concepts of internal quantum efficiency (IQE) and external quantum efficiency (EQE). In these contexts, "quantum efficiency" refers to the ratio of the number of radiated photons to the number of carriers generated by excitation, which is a comparison of the input and output from a quantum perspective. The internal quantum efficiency (IQE) is defined as the ratio of the number of photons, $n_{\text{p,active}}$, radiated from the active region per unit time to the total number of carriers, $n_{\text{e,in}}$, injected into the LED per unit time. Mathematically, this can be expressed as follows:

$$\text{IQE} = \frac{n_{\text{p,active}}}{n_{\text{e,in}}} = \eta_{\text{INJ}} \times \eta_{\text{RAD}} \tag{3.10}$$

Under photoexcitation conditions, nonequilibrium carriers are generated by the absorption of incident photons by the active region material, a process that does not involve carrier injection. Consequently, IQE as defined in this context, often referred to as IQE_{P}, includes only the radiative efficiency (η_{RAD}). This is expressed as the ratio of the number of photons, $n_{\text{p,active}}$, radiated from the active region per unit time to the total number of photons, $n_{\text{p,in}}$, incident on the LED per unit time. EQE is defined as the ratio of the number of photons, $n_{\text{p,out}}$, radiated to the outside world by the LED per unit time to the total number of carriers, $n_{\text{e,in}}$, injected into the LED per unit time. This can be expressed as follows:

$$\text{EQE} = \frac{n_{\text{p,out}}}{n_{\text{e,in}}} = \eta_{\text{INJ}} \times \eta_{\text{RAD}} \times \eta_{\text{LEE}} = \text{IQE} \times \eta_{\text{LEE}} \tag{3.11}$$

Under photoexcitation conditions, EQE equals the ratio of the number of photons radiated to the outside world by the LED per unit time $n_{\text{p,out}}$ to the total number of photons incident on the LED per unit time $n_{\text{p,in}}$, denoted as EQE_{P}. Obviously, IQE cannot be directly measured, but EQE is a measurable physical quantity. For the design of LED epitaxial structures, IQE, which reflects the carrier recombination process, is the most critical indicator. The characterization method of IQE will be discussed in detail below.

3.3 Characterization Methods of Internal Quantum Efficiency of LED

The characterization methods for IQE can be categorized into the following three types:

(1) Electrical Injection: This method is frequently employed for analyzing LED chips. Energy input is achieved through electrical injection, and IQE is determined by measuring the EL characteristics and the EL intensity–current–voltage (I-V) characteristics under various conditions. The electrical injection testing approach is more aligned with the actual operating conditions of the LED; however, the IQE derived from this test is influenced by both carrier injection and carrier recombination, which can complicate the analysis.
(2) Photoexcitation: This method is commonly used for analyzing LED epitaxial wafers. Energy input is supplied through photoexcitation, and IQE is calculated by measuring PL characteristics and the PL intensity–excitation power characteristics under different conditions. The photoexcitation testing method allows for a direct observation of carrier recombination behavior, but the testing conditions for IQE differ significantly from the actual operating conditions of the LED, which may not accurately characterize the efficiency of the LED during operation.
(3) Calculation of η_{LEE}: Given that EQE is the product of IQE and η_{LEE}, if η_{LEE} can be obtained through simulation calculations, IQE can be derived in conjunction with the test results of EQE. Since η_{LEE} is highly sensitive to the LED's chip structure and environmental parameters, any alteration in the device structure necessitates reestablishing the model and recalculating, leading to the consideration that the IQE obtained by this method may not be sufficiently accurate and lacks universality. Therefore, this method will not be discussed further in this context.

Since IQE cannot be measured directly, it is imperative to establish a reasonable physical model and extract the necessary parameters from experimental data. The most widely utilized model to date is the ABC model [7], which postulates the existence of three recombination channels within the active region: Shockley–Read–Hall (SRH) nonradiative recombination, bimolecular radiative recombination, and Auger nonradiative recombination. Under the assumption that the concentrations of electrons and holes are equal and that carriers are solely subject to recombination processes, we can derive the following relationships:

$$G = An + Bn^2 + Cn^3 \tag{3.12}$$

where G is the carrier generation rate and A, B, and C are the SRH recombination coefficient, bimolecular recombination coefficient, and Auger recombination coefficient, respectively. For GaN-based materials, their magnitudes are approximately $10^7\ \mathrm{s}^{-1}$, $10^{-11}\ \mathrm{cm}^3 \cdot \mathrm{s}^{-1}$, and $10^{-30}\ \mathrm{cm}^6 \cdot \mathrm{s}^{-1}$, respectively. The exact values depend on factors such as the crystal quality of the material, the structure of the active region, temperature, and injection level. The number of photons radiated from the active region $n_{\mathrm{p,active}}$ is proportional to the bimolecular recombination rate, so IQE can be represented as the ratio of the bimolecular recombination rate Bn^2 to the carrier generation rate G, that is

$$\mathrm{IQE} = \frac{Bn^2}{An + Bn^2 + Cn^3} \tag{3.13}$$

Obviously, if we can get the accurate values or the relationship between A, B, and C, we can calculate IQE.

The ABC model was initially employed to elucidate the phenomenon of efficiency degradation under high injection conditions and has been successful in this regard. However, the ABC model is not without its limitations. For instance, the model primarily considers band-to-band transitions involving conduction band electrons and valence band holes, while neglecting the recombination process associated with excitons. Excitons, unlike free electrons and holes, are bound electron–hole pairs characterized by a Coulomb interaction. Due to the shielding of the Coulomb field by free carriers, stable excitons are typically observed in wide bandgap semiconductor materials with high crystal quality or at low temperatures. Nevertheless, in quantum-confined structures, the increased overlap of the wave functions of the electrons and holes that constitute the exciton allows the exciton effect to be discerned even at room temperature. Additionally, the ABC model assumes that carriers within the active region are solely involved in recombination processes, overlooking other physical phenomena such as carrier overflow and escape, as depicted in Fig. 3.7. Consequently, the temporal evolution of carrier concentration may not align with the model's predictions. To obtain a more precise IQE, it is necessary to make rational corrections to the physical model based on the sample's characteristics. For samples with similar structures but with localized differences, it can be postulated that their physical parameters are largely identical, with only the parameters influenced by these local discrepancies requiring distinction. Results obtained through this approach are expected to be more accurate and facilitate easier comparison.

Below, we will delineate five commonly employed methods for characterizing IQE, along with the corresponding experimental outcomes. These examples will illustrate the utility of efficiency characterization in assessing LED performance and elucidating carrier dynamics.

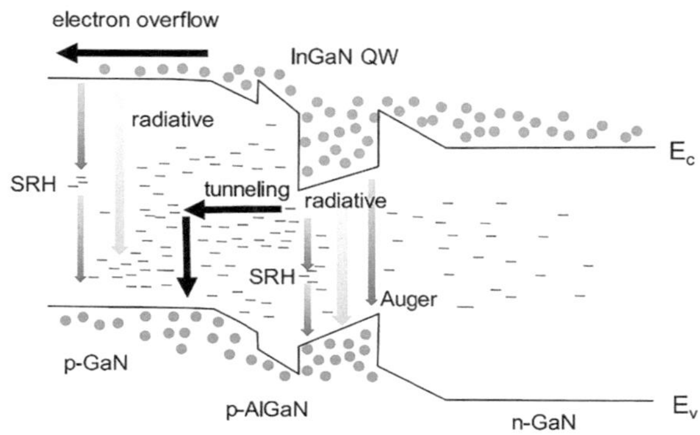

Fig. 3.7 Schematic diagram of GaN-based LED current components, quoted from reference [3]

3.3.1 Efficiency Characterization Based on Electrical Injection

1. Temperature-Dependent Electroluminescence

Through the extraction of experimental data, it is very difficult to calculate the absolute value of IQE accurately by extracting the accurate values of parameters *A*, *B*, and *C*, but it is very easy to calculate the relative value of IQE. The temperature-dependent electroluminescence (TDEL) method is based on this idea and is built on two assumptions [8–10]:

(1) The light extraction efficiency η_{LEE} is constant, so the ratio of IQE to EQE remains unchanged under all current *I* and temperature *T* conditions. This allows us to use the absolute value of IQE under a certain test condition as a reference (denoted as IQE_{EF}, the corresponding EQE is EQE_{REF}), and calculate the absolute value of IQE under all test conditions through the ratio of EQE to EQE_{REF} under other test conditions (i.e., the relative value of IQE).

$$\text{IQE}(I, T) = \frac{\text{EQE}(I, T)}{\text{EQE}_{\text{REF}}} \times \text{IQE}_{\text{REF}} \tag{3.14}$$

(2) At low temperatures (denoted as T_{REF}), the radiation efficiency η_{RAD} is 100%, and the injection efficiency η_{INJ} is 100% when EQE reaches the maximum value EQE_{REF} (the corresponding current is denoted as I_{REF}), so $\text{IQE}(I_{\text{REF}}, T_{\text{REF}}) = 100\%$, and this is used as IQE_{REF}.

In actual operation, we measure the EQE under different current and temperature conditions in the same optical path configuration, usually taking the maximum EQE under all test conditions, EQE_{MAX}, as the EQE_{REF} when $\text{IQE} = 100\%$. Figure 3.8 shows the normalized EQE (i.e., IQE obtained using the TDEL method) of the blue LED under different current and temperature conditions. From an experimental perspective, as the temperature decreases, the maximum EQE gradually stabilizes at EQE_{REF} and EQE_{REF} remains stable near I_{REF}, so the assumption that $\eta_{\text{INJ}} = \eta_{\text{RAD}}$

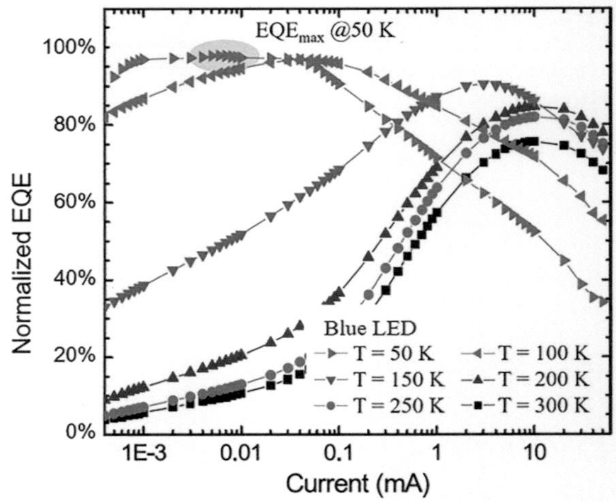

Fig. 3.8 TDEL method, cited from reference [3]

= 100% at EQE_{REF} seems reasonable; from a physical perspective, the number of material defects N_{defect} satisfied

$$N_{defect}(T) \propto \exp\left(-\frac{E_a}{k_B T}\right) \quad (3.15)$$

where $k_B \approx 1.38 \times 10^{-23}$ J/K is the Boltzmann constant and E_a is the activation energy of the defect. At low temperatures, the carriers at the defect sites do not have enough energy to become free carriers. These states, which usually play the role of nonradiative recombination centers (NRCs), are occupied, so the carriers will not be captured by them; that is, it can be assumed that there is no loss in the carrier transport and recombination process at low temperatures.

However, experimental observations have indicated that EQE does not necessarily increase monotonically with decreasing temperature; that is, EQE_{REF} may be achieved at higher temperatures, as depicted in Fig. 3.9. In such cases, the TDEL method no longer applies, and the cause of this phenomenon may be attributed to suboptimal P-type doping. Magnesium (Mg) atoms are commonly used as the P-type dopant for gallium nitride (GaN), yet the ionization energy of Mg atoms is substantial (greater than 140 meV). This high ionization energy prevents the holes generated by Mg atoms at low temperatures from acquiring sufficient energy to transition to the valence band. Consequently, the concentration of holes injected into the active region is exceedingly low, precluding significant recombination. As the temperature increases, the acceptor impurities begin to ionize, thereby injecting holes into the active region, leading to higher luminescence efficiency at elevated temperatures compared to that at lower temperatures.

In addition, even when the current and temperature conditions for achieving EQE_{REF} are appropriate, the TDEL method may still be inadequate. Figure 3.10 presents EL spectra of near-ultraviolet LEDs at varying temperatures, along with the

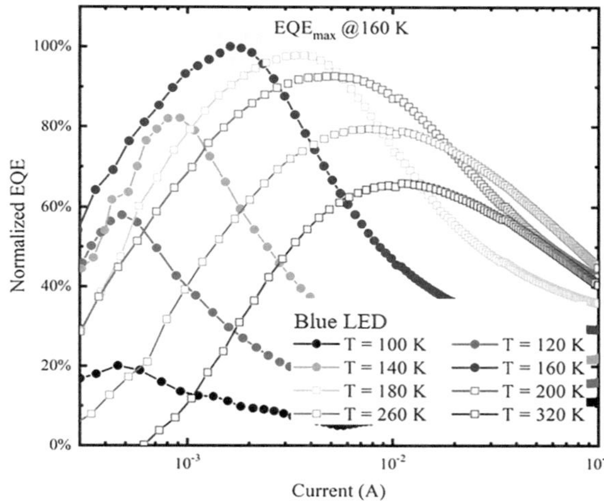

Fig. 3.9 Maximum EQE achieved at higher temperatures

IQE values obtained through the TDEL method. To streamline the testing process, a current of 100 mA is selected for this analysis. It is observed that IQE decreases with increasing temperature, which appears to be a plausible calculation outcome. However, at lower temperatures, the LED emits visible light, a phenomenon attributed to the electron leakage current. That is, the high ionization energy of Mg atoms results in an imbalance of electron and hole concentrations in the active region at low temperatures. Electrons accumulate in the active region, generating an electric field that impedes further injection of electrons into the active region. Furthermore, the quantum well barrier height of near-ultraviolet LEDs is relatively low, allowing electrons injected into the quantum well to readily escape the barrier's confinement and recombine with un-ionized acceptor impurities in the P-region, emitting longer-wavelength visible light. As the temperature increases, this electron and hole concentration imbalance is mitigated, leading to more recombination occurring in the active region. Consequently, the visible light emission diminishes, while the near-ultraviolet radiation intensifies. In this experiment, the assumption that η_{INJ} is 100% at low temperatures is clearly invalid, rendering the calculation of IQE using the TDEL method meaningless.

The TDEL method is simple in principle, is easy to operate, and reflects the efficiency of LEDs under actual working conditions. It is one of the most commonly used methods for characterizing IQE. Because it requires variable temperature experiments, it takes a long time to adjust and stabilize the temperature, and it is necessary to pay attention to the impact of the change in η_{INJ} with temperature on the experimental results.

2. Ideality Factor Analysis [11]

According to the Shockley diode equation, the I-V characteristic curve of an LED can be expressed as follows:

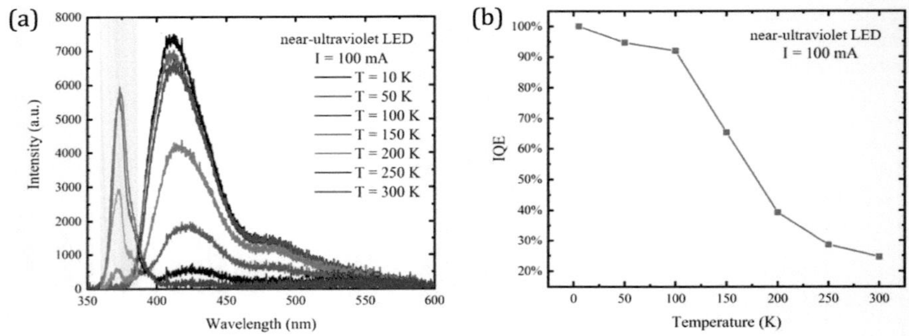

Fig. 3.10 Variable temperature EL spectra and IQE of near-ultraviolet LEDs: **a** EL spectrum and **b** IQE calculated using the TDEL method

$$I = I_s\left[\exp\left(\frac{qV}{n_{\text{ideal}}k_BT}\right) - 1\right] \approx I_s\exp\left(\frac{qV}{n_{\text{ideal}}k_BT}\right) \tag{3.16}$$

where I_S is the reverse saturation current and n_{ideal} is known as the "ideality factor," which is a correction to the original equation, reflecting the components of the diode current: When the diffusion current dominates, $n_{\text{ideal}} = 1$; when the recombination current dominates, $n_{\text{ideal}} = 2$.

Shockley derived this equation based on the relatively simple structure of silicon homojunction diodes. However, GaN-based LEDs typically employ MQW structures, leading to a more complex relationship between n_{ideal} and the current. As depicted in Fig. 3.11b, n_{ideal} exhibits a non-monotonic trend with increasing current, initially decreasing and then increasing. It is commonly accepted that the higher values of n_ideal in the low current range are attributed to tunneling current contributions. In the high current range, the larger n_{ideal} values are believed to be due to voltage division caused by series resistance. In the current range where n_{ideal} reaches its minimum, the influence of these two interfering factors is less pronounced. Therefore, by analyzing the current components within this specific range, insights into the carrier recombination behavior can be garnered.

Specifically, in this current range, the total current I_{TOTAL} may have four current components: SRH recombination current I_{SRH}, radiative recombination current I_{RAD}, the recombination current I_{AUG}, and the leakage current I_{LEAK}. Since the voltage corresponding to the vicinity of the minimum value of n_{ideal} is less than the turn-on voltage of the LED, the current component I_{SRH} should dominate, and the effect of I_{AUG} can be ignored, that is:

$$\begin{aligned} I_{\text{TOTAL}} &= I_{\text{SRH}} + I_{\text{RAD}} + I_{\text{LEAK}} \\ &= I_{s1}\exp\left(\frac{qV}{2k_BT}\right) + I_{s2}\exp\left(\frac{qV}{k_BT}\right) + I_{s3}\exp\left(\frac{qV}{n_{\text{LEAK}}k_BT}\right) \end{aligned} \tag{3.17}$$

IQE can be represented as the ratio of I_{RAD} to I_{TOTAL}, that is:

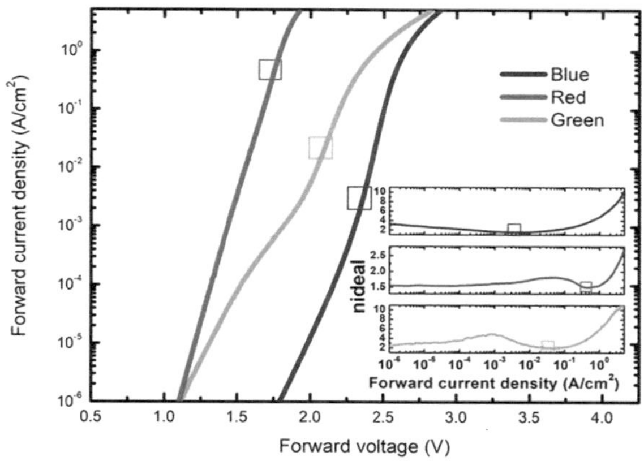

Fig. 3.11 Blue, red, and green LED's *I*-*V* curve and n_{ideal}

$$\mathrm{IQE}=\frac{I_{\mathrm{RAD}}}{I_{\mathrm{TOTAL}}}=\frac{I_{s2}\exp\left(\frac{qV}{k_B T}\right)}{I_{\mathrm{TOTAL}}} \tag{3.18}$$

This expression only contains one unknown parameter I_{S2}, which can be obtained by fitting the *I*-*V* curve in this current range. Assuming the light extraction efficiency η_{LEE} is constant, and the IQE at the minimum value of n_{ideal} (corresponding current is I_{nMIN}) is IQE(I_{nMIN}), the IQE in the entire current range can be obtained by combining the EQE measurement results in the entire current range, that is:

$$\mathrm{IQE}(I)=\frac{1}{\eta_{\mathrm{LEE}}}\times \mathrm{EQE}(I)=\frac{\mathrm{IQE}(I_n)}{\mathrm{EQE}(I_n)}\times \mathrm{EQE}(I) \tag{3.19}$$

Experimental findings indicate that for blue and (gallium arsenide-based) red LEDs, the first two terms are sufficient to accurately fit the I-V curve, implying that the effect of I_{LEAK} can be neglected. In contrast, for green LEDs, I_{LEAK} must be taken into account. This is due to the higher indium (In) content in green LEDs and the generally lower material quality compared to blue and red LEDs. Within this current range, nonlinear effects can be disregarded, and the parallel conductance model can be employed to approximate I_{LEAK}, assuming that it is directly proportional to V. Figure 3.12 displays the IQE of blue and green LEDs across the entire current range, as obtained using the ideal factor analysis method. It is observed that the peak IQE of blue LEDs is higher than that of green LEDs, and the calculated ηLEE value (approximately 60%) aligns with our expectations.

To encapsulate the ideal factor analysis method, initially, measure the I-V curve of the LED device and, then, calculate the ideal factor, n_{ideal}, and fit the I-V curve within the current range where n_{ideal} assumes its minimum value to derive the ratio of the radiative recombination current, I_{RAD}, to the total current, I_{TOTAL}, which corresponds to the IQE in the vicinity of the minimum n_{ideal}. Subsequently, measure the EQE of the LED device across the entire current range. Utilize the IQE at the minimum n_{ideal} to

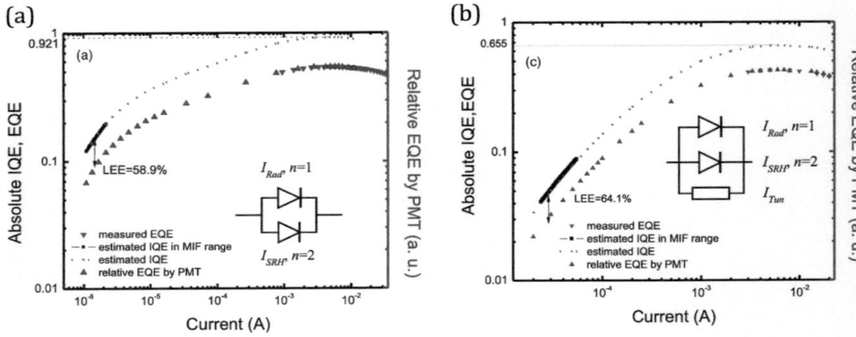

Fig. 3.12 Ideal factor analysis method: **a** blue LED and **b** green LED

calculate the light extraction efficiency, η_{LEE}, and assume it to be constant throughout the current range. This enables the calculation of IQE across the entire current range in conjunction with the EQE measurement outcomes. The ideal factor analysis method necessitates only routine electrical and optical testing of LEDs at room temperature. However, it demands high precision in current and voltage measurements to ensure the accuracy of the fitting. This method is applicable primarily to LEDs with superior material quality (it is essential to disregard other current components when n_{ideal} is at its minimum); thus, it has certain limitations.

3.3.2 Efficiency Characterization Based on Light Injection

1. Temperature-Dependent Photoluminescence

The temperature-dependent photoluminescence (TDPL) method is consistent with the TDEL method in concept, measuring the external quantum efficiency $\mathrm{EQE_P}(P, T)$ under different excitation power P and temperature T conditions, which is proportional to the ratio of PL intensity $I_{\mathrm{PL}}(P, T)$ to P, that is:

$$\mathrm{EQE_P}(P, T) \propto \frac{I_{\mathrm{PL}}(P, T)}{P} \tag{3.20}$$

Normalize by the maximum value of $\mathrm{EQE_P}(P, T)$, $\mathrm{EQE_{P,REF}}$, to obtain the $\mathrm{IQE_P}$ under all test conditions, that is:

$$\mathrm{IQE_P}(P, T) = \frac{\mathrm{EQE_P}(P, T)}{\mathrm{EQE_{P,REF}}} \tag{3.21}$$

Figure 3.13 depicts the PL spectra of the green LED at various temperatures, along with the $\mathrm{IQE_P}$ obtained through TDPL method. To streamline the testing process, an excitation power of 250 mW is chosen. The data reveal that IQE_P initially increases and then decreases as the temperature rises, indicating that the temperature at which IQE_P reaches its maximum is not the lowest temperature. This behavior parallels the phenomenon observed in the TDEL method, suggesting a correlation between carrier recombination processes and this observation. That is, the radiative efficiency, η_{RAD}, may not always increase monotonically with decreasing temperature under certain circumstances, although the precise reason remains unclear. One hypothesis is that at lower temperatures, carriers predominantly exist as excitons, and as the temperature increases, excitons gradually transition to free carriers, with the recombination mechanism shifting from exciton recombination to bimolecular recombination. During this transition, the maximum IQE_P is observed. As the temperature continues to rise, nonradiative recombination processes intensify, and the radiative recombination coefficient, B, may decrease due to alterations in carrier distribution, leading to a decline in IQE_P. However, this lacks compelling experimental validation. Furthermore, temperature variations affect the refractive index of the sample. Based on the Kramers–Kronig relationship, it can be inferred that the sample absorbs the excitation light to varying extents at different temperatures. Figure 3.14 presents the correlation between the absorption rate and temperature for the blue LED at an excitation power of 100 mW. The absorption rate, α, is defined as follows:

$$\alpha = \frac{P_A}{P} = \frac{P - P_R - P_T}{P} \tag{3.22}$$

where P_A is the absorption power of the sample, P_R is the power reflected by the sample surface, and P_T is the transmission power through the sample. Obviously, the change in PL intensity with temperature caused by the dependence of the absorption rate on temperature is not caused by the carrier recombination process, so its impact should be excluded in the efficiency calculation; that is, the denominator in Formula (3.20) should be corrected to P_A.

In addition, as the temperature T rises, NRC is gradually activated and causes the fluorescence intensity I_{PL} to gradually decrease, as shown in Formula (3.15), so by

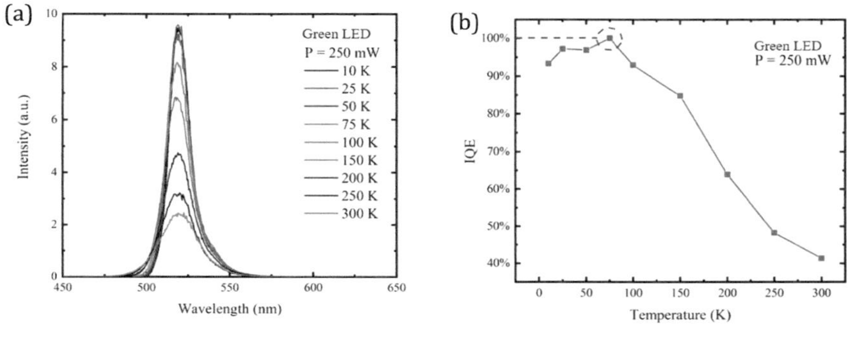

Fig. 3.13 TDPL method: **a** PL spectrum and **b** IQE_P calculated using the TDPL method

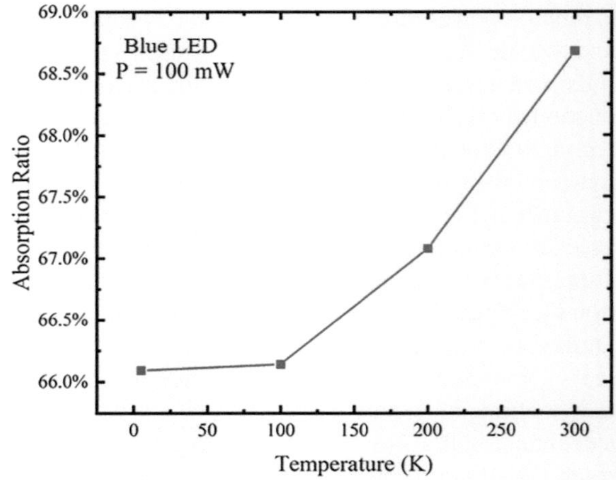

Fig. 3.14 The effect of temperature on absorption rate

analyzing I_{PL}–T, information about NRC can be obtained from the curve. As shown in Fig. 3.15, the I_{PL}–T curve is fitted with Arrhenius, that is:

$$I_{PL}(T) = \frac{I_{PL,0}}{1 + \sum_i A_i \exp\left(-\frac{E_i}{k_B T}\right)} \tag{3.23}$$

where $I_{PL,0}$ is a constant, E_i is the activation energy of the ith type of NRC, and A_i is the pre-factor, which is proportional to the number of the ith type of NRC. For GaN-based materials, it is generally believed that there are two types of NRCs. The NRC with smaller activation energy is believed to be related to the roughness of the

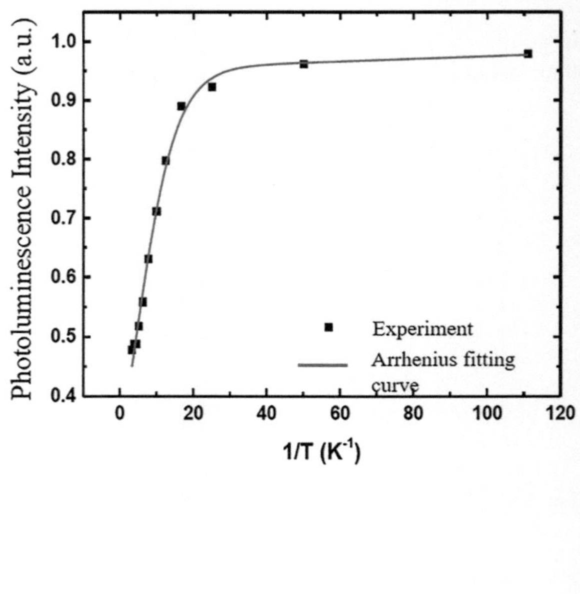

Fig. 3.15 Fitting the I_{PL}–T curve with Arrhenius

quantum well interface; the NRC with larger activation energy (usually 30 meV ~ 60 meV) is believed to be related to dislocations.

In addition, when the test sample possesses a fully formed PN junction structure, the forward electric field generated by carrier escape will influence the PL spectrum. Carrier escape is a mechanism by which carriers, initially bound within quantum confinement structures, can become free carriers through certain processes and migrate to the P and N regions under the influence of the built-in electric field of the PN junction. This results in the formation of an open-circuit voltage under open-circuit conditions and a short-circuit current under short-circuit conditions. The exact mechanism of carrier escape is not yet fully understood, but potential explanations include thermal excitation, direct tunneling, and defect-assisted tunneling. However, the experimental evidence for carrier escape is clear. In steady-state PL testing, as depicted in Fig. 3.16a, the conditions of open and short circuits have a pronounced impact on the intensity and shape of the PL spectrum: the PL intensity under short-circuit conditions is lower than that under open-circuit conditions, suggesting that carriers can effectively escape under short-circuit conditions, thereby reducing the number of carriers available for recombination. The peak wavelength of the PL spectrum under short-circuit conditions is blue-shifted compared to that under open-circuit conditions, because the open-circuit voltage generated by carrier escape enhances the polarization field within the quantum well, which in turn lengthens the emission wavelength. In transient PL testing, as shown in Fig. 3.16b, following excitation, it is observed that the fluorescence decay under short-circuit conditions is more rapid than under open-circuit conditions, indicating that carrier escape is a swift process. Some research suggests that when testing samples with a complete PN junction structure, the sample should be short-circuited to mitigate the influence of carrier escape under open-circuit conditions on the experimental outcomes. Moreover, the phenomenon of carrier escape enables the use of LEDs as detectors, which is anticipated to broaden the application scenarios of display devices, such as enabling under-screen fingerprint recognition without the necessity for additional detectors.

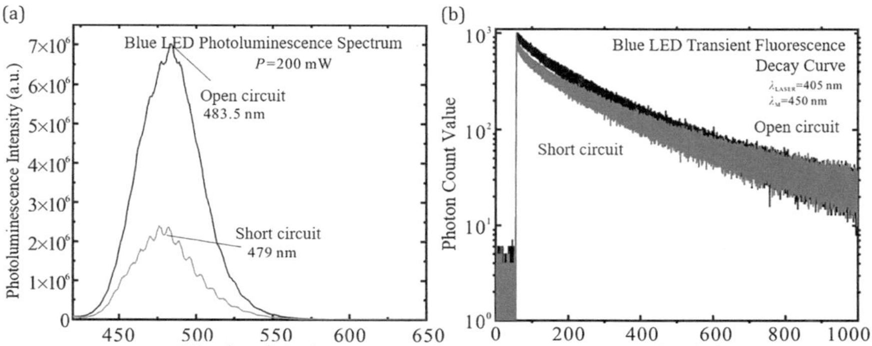

Fig. 3.16 Experimental results of carrier escape: **a** steady-state PL test and **b** transient PL test

2. Power-Dependent Photoluminescence

The power-dependent photoluminescence (PDPL) method involves the utilization of a power-adjustable continuous-wave laser to irradiate the LED sample, while recording the photoluminescence intensity, I_{PL}, at various excitation power levels, P_{laser}. The recombination coefficient is then determined by resolving the nonlinear relationship between the excitation intensity and the fluorescence intensity. The relationship between the carrier generation rate, G, and P_{laser} can be articulated as follows:

$$G = P_{laser} \times \frac{(1 - R)\alpha}{h\nu_{laser} A_{spot}} \tag{3.24}$$

where R is the reflectance of the sample surface to the laser, α is the absorption rate of the material to the laser, $h\nu_{laser}$ is the energy of the laser photon, and A $_{SPOT}$ is the spot area of the laser irradiated on the sample. According to the ABC model, under steady-state conditions, $G = An + Bn^2 + Cn^3$, and I_{PL} is proportional to the radiative recombination term Bn^2, denoted as I $_{PL} = \beta Bn^2$. Combining the above equations, we can get the relationship between P_{laser} and I_{PL}, that is:

$$\begin{aligned} P_{laser} = mG &= \frac{mA}{(B\beta)^{\frac{1}{2}}} I_{PL}^{\frac{1}{2}} + \frac{m}{\beta} I_{PL} + \frac{mC}{(B\beta)^{\frac{3}{2}}} I_{PL}^{\frac{3}{2}} \\ &= Q_1 I_{PL}^{\frac{1}{2}} + Q_2 I_{PL} + Q_3 I_{PL}^{\frac{3}{2}} \end{aligned} \tag{3.25}$$

where $m = (h\nu_{laser} A_{SPOT})/[(1\text{-}R)\alpha]$ is a constant and Q_1, Q_2, and Q_3 are unknown parameters, which can be obtained by measuring P_{laser} and I_{PL} The relationship curve is fitted as shown in Fig. 3.17a. These parameters reflect the proportional relationship between the composite coefficients, that is:

$$\frac{Q_1}{(Q_2)^{\frac{1}{2}}} = \frac{mA}{(mB)^{\frac{1}{2}}}, \frac{Q_3}{(Q_2)^{\frac{3}{2}}} = \frac{mC}{(mB)^{\frac{3}{2}}}$$

Rewrite Eq. (3.25) in the form of an equation with $(mB)^{1/2}$ n as the unknown, that is:

$$\begin{aligned} P_{laser} &= mG \\ &= \frac{mA}{(mB)^{\frac{1}{2}}}\left[(mB)^{\frac{1}{2}} n\right] + \left[(mB)^{\frac{1}{2}} n\right]^2 + \frac{mC}{(mB)^{\frac{3}{2}}}\left[(mB)^{\frac{1}{2}} n\right]^3 \\ &= \frac{Q_1}{(Q_2)^{\frac{1}{2}}}\left[(mB)^{\frac{1}{2}} n\right] + \left[(mB)^{\frac{1}{2}} n\right]^2 + \frac{Q_3}{(Q_2)^{\frac{3}{2}}}\left[(mB)^{\frac{1}{2}} n\right]^3 \end{aligned} \tag{3.26}$$

Solving this cubic equation, we can obtain the IQE_P under different excitation powers.

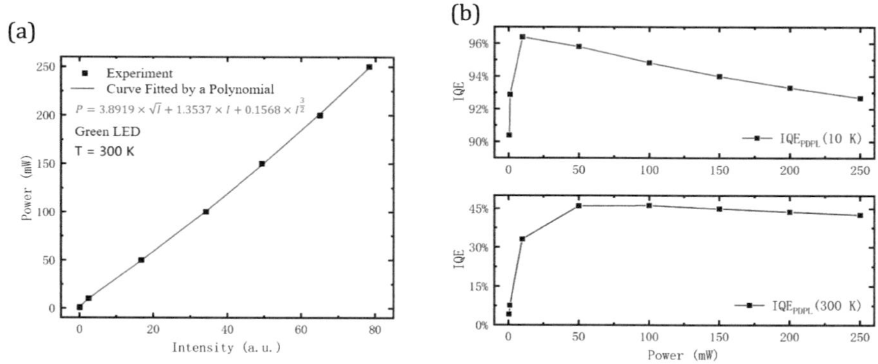

Fig. 3.17 PDPL method: **a** parameter fitting and **b** calculating IQE$_P$ using the PDPL method

$$\mathrm{IQE_P}(P_{\mathrm{laser}}) = \frac{Bn^2}{G} = \frac{mBn^2}{mG} = \frac{\left[(mB)^{\frac{1}{2}}n\right]^2}{P_{\mathrm{laser}}} \tag{3.27}$$

Figure 3.17b shows the IQE$_P$ of the green LED obtained by the PDPL method at temperatures of 10 K and 300 K, respectively. It can be seen that the efficiency of the LED at low temperatures is higher than at room temperature, but the efficiency decline is more severe. This phenomenon can be explained by the longer carrier lifetime τ at low temperatures leading to a higher carrier concentration $n = G \times \tau$ at the same excitation power P_{laser} (i.e., generation rate G) than at high temperatures. In addition, the low-temperature IQE$_P$ obtained by the PDPL method is not equal to 100%, indicating that there is an error in the assumption that the IQE$_P$ of the TDEL and TDPL methods can reach 100% at low temperatures. In fact, each IQE testing method has a certain error, so it is necessary to use multiple methods for characterization. Specifically, the TDPL method and the PDPL method can be conveniently combined due to their similar operation methods. Specifically, first measure the EQE$_P$ at different excitation powers at low temperature (denoted as T_{REF}); according to the TDPL method and take the maximum value as a reference, the corresponding excitation power is denoted as P_{REF}, satisfying IQE$_{\mathrm{P,TDPL}}$ (T_{REF}, P_{REF}) = 100%; measure the EQE $_\mathrm{P}$ at different excitation powers at high temperature (denoted as T_{M}); normalize it to obtain the IQE$_{\mathrm{P,TDPL}}$ (T_{M}, P) at different excitation powers at high temperature; then calculate the IQE $_{\mathrm{P,PDPL}}$ (T_{REF}, P) and IQE $_{\mathrm{P,PDPL}}$ (T_{M}, P); the IQE$_{\mathrm{P,PDPL}}$ obtained using the PDPL method (T_{REF}, P_{REF}) is often not equal to 100%, and the results obtained by the TDPL method need to be corrected, that is:

$$\mathrm{IQE_{P,ADJ}}(T_{\mathrm{M}}, P) = \mathrm{IQE_{P,TDPL}}(T_{\mathrm{M}}, P) \times \mathrm{IQE_{P,PDPL}}(T_{\mathrm{REF}}, P_{\mathrm{REF}}) \tag{3.28}$$

Then, compare the corrected TDPL result $IQE_{P,ADJ}$ (T_M, P) with the PDPL result $IQE_{P,PDPL}$ at high temperature (T_M, P). If the two match well, it can be considered that the measured IQE_P is relatively accurate.

The PDPL method necessitates testing only at room temperature and requires a broad range of excitation power variations to obtain more precise fitting parameters; thus, it also demands a considerable amount of time for the experimental procedure. Additionally, the PDPL method assumes that A, B, C, and m are independent of the injection level; this assumption does not entirely conform to the actual physical processes, which may be a source of error.

Both the TDPL and PDPL methods are classified as steady-state testing procedures. The TDPL method yields IQE_P corresponding to various temperatures at a specific excitation power; conversely, the PDPL method yields IQE_P corresponding to different excitation powers at a constant temperature. Therefore, under ideal conditions, these two methods should yield equivalent outcomes.

3. Time-Resolved Photoluminescence

Time-resolved photoluminescence (TRPL) is a transient testing technique that typically employs ultrashort pulse lasers to excite LED samples, recording the decay curve of fluorescence intensity within the pulse duration. This method directly reflects the temporal variation of carrier concentration and can provide a more intuitive analysis of the carrier recombination mechanisms [12–14].

Figure 3.18a presents the fluorescence decay curves for blue LEDs at various wavelengths, which exhibit non-exponential decay characteristics. This indicates the presence of multiple recombination channels with distinct decay rates. If the radiative recombination lifetime, τ_R, and the nonradiative recombination lifetime, τ_{NR}, can be extracted from the fluorescence decay curve, the IQE_P can be calculated as follows:

$$IQE_P = \frac{\frac{1}{\tau_R}}{\frac{1}{\tau_{NR}} + \frac{1}{\tau_R}} \tag{3.29}$$

Numerous models have been proposed to elucidate the non-exponential nature of the decay curves, including the stretched exponential model and the double exponential model; however, these models often lack explicit physical interpretations. In this context, we present the bimolecular recombination model derived from the ABC model, wherein the temporal evolution of the carrier concentration can be expressed as follows:

$$\frac{\partial n}{\partial t} = G - An - Bn^2 - Cn^3 \tag{3.30}$$

Under pulse excitation conditions, the effect of Auger recombination can be disregarded. The time point at which the laser pulse irradiates the sample is taken as the starting point, with the initial condition being $n(0) = n_0$. Solving the differential equation yields:

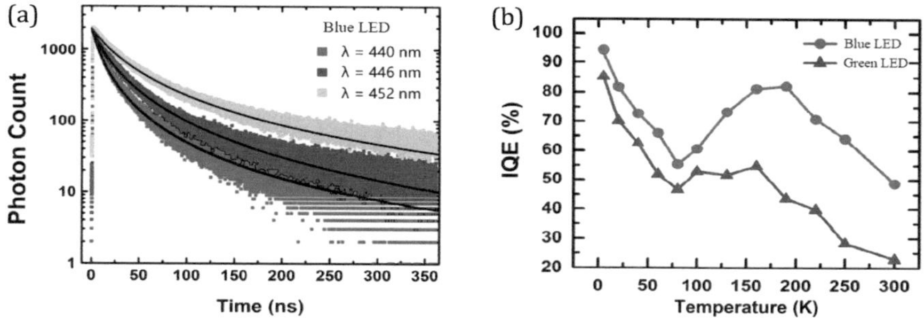

Fig. 3.18 TRPL method: **a** fluorescence decay curve and **b** calculation of IQE_P using the TRPL method

$$n = \frac{A}{B\left[e^{(At+k)} - 1\right]} \tag{3.31}$$

where $k = \ln(A/(Bn_0) + 1)$ is a constant. For bimolecular radiative recombination, the PL intensity is proportional to the radiative recombination rate, so

$$I = \beta n^2 = \frac{\beta\left(\frac{A}{B}\right)^2}{\left[e^{(At+k)} - 1\right]^2} = \frac{M_1}{\left(M_2 e^{M_3 t} - 1\right)^2} \tag{3.32}$$

where $M_1 = \beta(A/B)^2 = (A^2 I_0)/(Bn_0)^2$, $M_2 = e^k = A/(Bn_0) + 1$, and $M_3 = A$ are unknown parameters, which can be obtained by fitting the measured fluorescence decay curve.

The number of photons generated per unit volume per cycle T is as follows:

$$\begin{aligned} N_{\text{photon}} &= \int_0^T Bn^2 \mathrm{d}t \approx \frac{A^2}{B} \int_0^\infty \frac{1}{\left[e^{(At+k)} - 1\right]^2} \mathrm{d}t \\ &= n_0 + \frac{A}{B}\left[\ln\left(\frac{A}{Bn_0}\right) - \ln\left(\frac{A}{Bn_0} + 1\right)\right] \end{aligned} \tag{3.33}$$

The number of photo-generated carriers injected per unit volume per cycle is $n_{p,in} = n_0$, so IQE_P can be represented as follows:

$$\begin{aligned} IQE_P &= \frac{N_{\text{photon}}}{N_{\text{carrier}}} = 1 + \frac{A}{Bn_0}\left[\ln\left(\frac{A}{Bn_0}\right) - \ln\left(\frac{A}{Bn_0} + 1\right)\right] \\ &= 1 + (M_2 - 1)[\ln(M_2 - 1) - \ln(M_2)] \end{aligned} \tag{3.34}$$

Figure 3.18b displays the IQE_P values for blue and green LEDs as obtained via the TRPL method. It is observed that the IQE_P exhibits a trend of first decreasing, then increasing, and subsequently decreasing again as the temperature rises. This behavior can be accounted for by the redistribution of localized carriers. In the temperature range where redistribution is not predominant (5–60 and 190–300 K), the nonradiative recombination rate increases continuously with temperature, while the radiative recombination rate decreases correspondingly, resulting in a decline of $\mathrm{IQE_P}$ with increasing temperature. Within the temperature range where redistribution is dominant (80–160 K), carriers with weaker localization gain sufficient energy to migrate into regions with stronger localization. The likelihood of these carriers being captured by nonradiative centers (NRCs) diminishes, leading to a reduction in the nonradiative recombination rate and consequently an increase in $\mathrm{IQE_P}$ with temperature.

The quantum-confined Stark effect is present in polar GaN-based materials, which diminishes the overlap between electron and hole wave functions within the quantum well. Consequently, the bimolecular recombination model, which neglects the exciton recombination process, can be applied. Nonetheless, there may be issues where the bimolecular recombination model fails to adequately fit general fluorescence decay curves. Additionally, the TRPL method testing system is quite costly, making it challenging for widespread adoption.

The TRPL method is a transient test. Compared with the steady-state test that generates a constant carrier concentration per unit time, the TRPL method uses ultra-short pulse lasers to generate a certain carrier concentration n_0 at the moment ($t = 0$). Assuming that n_0 can be maintained, then the TRPL method can be transformed into a steady-state test; at this time, $\partial n/\partial t = 0$; Eq. (3.30) degenerates into the ABC model expression of Eq. (3.12), and $\mathrm{IQE_P}$ can be expressed as follows:

$$\mathrm{IQE_P} = \frac{Bn_0^2}{An_0 + Bn_0^2} = \frac{1}{\frac{A}{Bn_0} + 1} = \frac{1}{M_2} \tag{3.35}$$

That is, the fitting parameters obtained by the TRPL method can be used to simulate the steady-state test that generates a constant carrier concentration n_0 per unit time.

3.4 Summary and Outlook

With the rapid advancement of GaN-based LED technology, the methods for efficiency characterization and the understanding of the underlying physical processes are continually evolving and deepening. Driven by novel devices and applications, the need for accurate and reliable efficiency characterization methods is becoming increasingly significant. An ideal efficiency characterization method should possess the following attributes:

(1) It should rely solely on experimentally measurable physical quantities (such as current, voltage, and radiant power emitted to the outside world), without the necessity of assuming any physical parameters (such as chip size, epitaxial structure, and carrier recombination rates).
(2) It should, as much as possible, utilize relative values of physical quantities for calculation to minimize errors.
(3) The test conditions should closely mimic the actual operating conditions of the LED.
(4) The test system should be as straightforward and cost-effective as possible, with the required testing time kept to a minimum.

Although a universal efficiency characterization method has not yet been identified, we already possess numerous powerful tools. To ensure the accuracy of efficiency test results, it is advisable to employ a combination of multiple characterization methods, and the results obtained from different methods should be analyzed in conjunction with the sample characteristics.

Moreover, there are still many unresolved issues concerning the efficiency characterization of GaN-based LEDs that warrant further exploration. For instance, electrical injection and photoexcitation are two such characterization methods; however, as previously discussed, the physical processes they entail are similar. If a method could be developed to integrate electrical injection testing with photoexcitation testing, the injection efficiency η_{INJ} could be calculated [15], providing a basis for independent optimization of η_{INJ}. Furthermore, the emergence of micro-LEDs renders the study of carrier transport and recombination behavior under low injection conditions particularly relevant [16, 17]. When the applied bias voltage is reduced, although the number of photons emitted by the LED to the outside world may decrease, the average photon energy remains consistently close to the material's bandgap width and does not decrease substantially. Consequently, the voltage efficiency η_{VTG} in this range may exceed 100%. If manufacturing processes can be enhanced to achieve a sufficiently high EQE, it may be feasible to attain a photoelectric conversion efficiency η_{WPE} greater than 100%, effectively realizing the conversion of thermal energy into light energy [18].

References

1. Qi K (2012) Principles of luminescence and luminescent materials. University of Electronic Science and Technology Press, Chengdu
2. Lai W, Yuchen X, Zhibiao H et al (2015) Evaluation of the internal quantum efficiency of InGaN blue LEDs. Sci China Phys Mech Astron 45:067304
3. Seong TY, Han J, Amano H et al (2013) III-Nitride based light emitting diodes and applications. Springer, Netherlands, Dordrecht
4. Weisbuch C (2019) On the search for efficient solid state light emitters: past, present, future. ECS J Solid State Sci Technol 9(1):016022
5. Wang L, Jin J, Mi C et al (2017) A review on experimental measurements for understanding efficiency droop in InGaN-based light-emitting diodes. Materials 10(11):1233

6. Hangleiter A (2019) Recombination dynamics in GaInN/GaN quantum wells. Semicond Sci Technol 34(7):073002
7. Karpov S (2015) ABC-model for interpretation of internal quantum efficiency and its droop in III-nitride LEDs: a review. Opt Quant Electron 47(6):1293–1303
8. Wang J, Wang L, Zhao W et al (2010) Understanding efficiency droop effect in InGaN/GaN multiple-quantum-well blue light-emitting diodes with different degree of carrier localization. Appl Phys Lett 97(20):201112
9. Wang J, Wang L, Wang L et al (2012) An improved carrier rate model to evaluate internal quantum efficiency and analyze efficiency droop origin of InGaN based light-emitting diodes. J Appl Phys 112(2):023107
10. Lu B, Wang L, Hao Z et al (2019) Study on optical properties and internal quantum efficiency measurement of GaN-based green LEDs. Appl Sci 9(3):383
11. Mi C, Wang L, Jin J et al (2019) Estimating internal quantum efficiency of light-emitting diodes from current–voltage curves. Appl Phys Express 12(3):032002
12. Wang Z, Wang L, Xing Y et al (2017) Consistency on two kinds of localized centers examined from temperature-dependent and time-resolved photoluminescence in InGaN/GaN multiple quantum wells. ACS Photonics 4(8):2078–2084
13. Xing Y, Wang L, Yang D et al (2017) A novel model on time-resolved photoluminescence measurements of polar InGaN/GaN multi-quantum-well structures. Sci Rep 7(1):1–9
14. Xing Y, Wang L, Wang Z et al (2017) A comparative study of photoluminescence internal quantum efficiency determination method in InGaN/GaN multi-quantum-wells. J Appl Phys 122(13):135701
15. Masui H, Nakamura S, DenBaars SP (2008) Experimental technique to correlate optical excitation intensities with electrical excitation intensities for semiconductor optoelectronic device characterization. Semicond Sci Technol 23(8):085018
16. Schubert MF, Xu J, Dai Q et al (2009) On resonant optical excitation and carrier escape in GaInN/GaN quantum wells. Appl Phys Lett 94(8):081114
17. Yang H, Ma Z, Jiang Y et al (2017) The enhanced photo absorption and carrier transportation of InGaN/GaN quantum wells for photodiode detector applications. Sci Rep 7(1):1–6
18. Kuritzky LY, Weisbuch C, Speck JS (2018) Prospects for 100% wall-plug efficient III-nitride LEDs. Opt Express 26(13):16600–16608

Chapter 4
Fabrication and Optoelectronic Properties of GaN-Based Micro-LEDs

Pengfei Tian, Yanzhe Li, Lintao Yan, Xinyi Lu, and Zexing Yuan

In recent years, micro-light-emitting diodes (LEDs) have undergone rapid development in research and application due to their excellent optoelectronic properties. This chapter compares the display characteristics of micro-LEDs with LCDs and OLEDs, and introduces the applications of micro-LEDs in display and communication. Subsequently, the fabrication processes of micro-LED devices are introduced, including substrate stripping, quantum dot integration with micro-LEDs, and other key technologies for preparing full-color micro-LED displays. Finally, the size effect, temperature effect, aging, and modulation bandwidth characteristics of micro-LEDs are analyzed.

4.1 Introduction to Micro-LED

In recent years, with the continuous expansion of TVs, computers, large-area display screens, and various smart device markets, display technology has been constantly developed toward high resolution, large area, and high reliability. Micro-light-emitting diodes (LEDs), as one of the fastest-developing display technologies, are widely considered to be an optimal display technology. Global tech giants are constantly researching micro-LEDs and striving to explore commercialization pathways, such as wearable displays, virtual reality (VR), augmented reality (AR), biomedical applications, and visible light communication (VLC). At the 2012 Consumer Electronics Show (CES) exhibition, Sony displayed a large-sized flat-panel display system based on micro-LEDs, which significantly improved

P. Tian (✉) · Y. Li · L. Yan · X. Lu · Z. Yuan
Institute for Electric Light Sources, School of Information Science and Technology, Fudan University, Shanghai, China
e-mail: pftian@fudan.edu.cn

S. Xu (ed.), *Optical Characterization of Microstructures and Optoelectronic Devices Based on Wide Band Gap Semiconductors*, Wide Bandgap Semiconductors,
https://doi.org/10.1007/978-981-95-1928-6_4

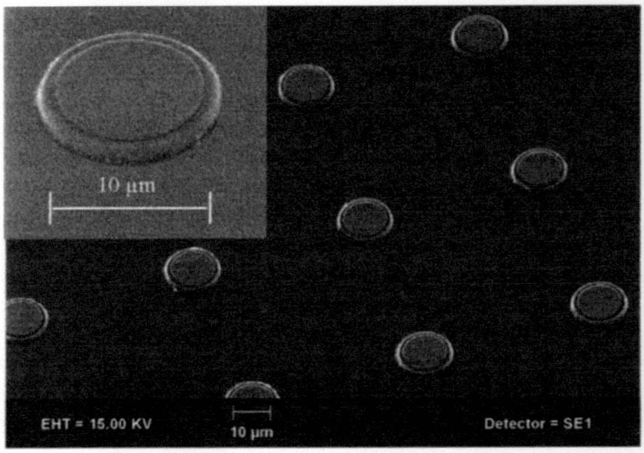

Fig. 4.1 SEM image of a 12 μm diameter cylindrical micro-LED [1]

color, response speed, and contrast ratio compared to traditional display devices; Samsung also released a 110-inch micro-LED TV in 2020. Currently, the industry and academia, both domestically and abroad, are heavily investing in the research and development of micro-LED display technology.

The concept of micro-LED was first proposed by Hongxing Jiang's research group [[1]]. Figure 4.1 shows a scanning electron microscope (SEM) image of a 12-μm-diameter micro-LED. The group used inductively coupled plasma (ICP) etching to prepare indium gallium nitride (InGaN)/gallium nitride (GaN) quantum well-based micro-LEDs, and then tested the device characteristics, including current–voltage characteristics, light power, emission spectrum, and compared them with those of traditional lighting LEDs. The results showed that although the micro-LED only changed in size compared to the traditional LED, the device characteristics of the two were significantly different, which gave researchers new expectations for the application prospects of micro-LEDs.

In 2008, Martin Dawson's research group successfully integrated micro-LEDs with color conversion materials in an inverted micro-LED array [2]. As shown in Fig. 4.2, this work used a 360 nm wavelength UV micro-LED array to excite cadmium selenide (CdSe) quantum dots (QDs) to achieve UV to red and green color conversion. This indicates that after mixing a certain proportion of CdSe quantum dots with UV micro-LEDs, not only can white light be emitted, but also a full-color display can be achieved.

As shown in Table 4.1, compared to liquid–crystal display (LCD) and organic light-emitting diode (OLED) displays, micro-LEDs have unparalleled superiority in display quality and lifespan [3–7]. Compared with LCDs, micro-LED, due to its self-luminous characteristics, does not require color filters or backlight units. Moreover, the thickness of micro-LED displays is smaller than that of LCDs. In addition, the response time of micro-LED devices is at the ns level, which is much faster than that of LCD and OLED displays. Therefore, micro-LED has great application prospects not only in display but also in high-speed visible light communication. Even under higher current density, micro-LEDs still exhibit good heat dissipation, and the brightness,

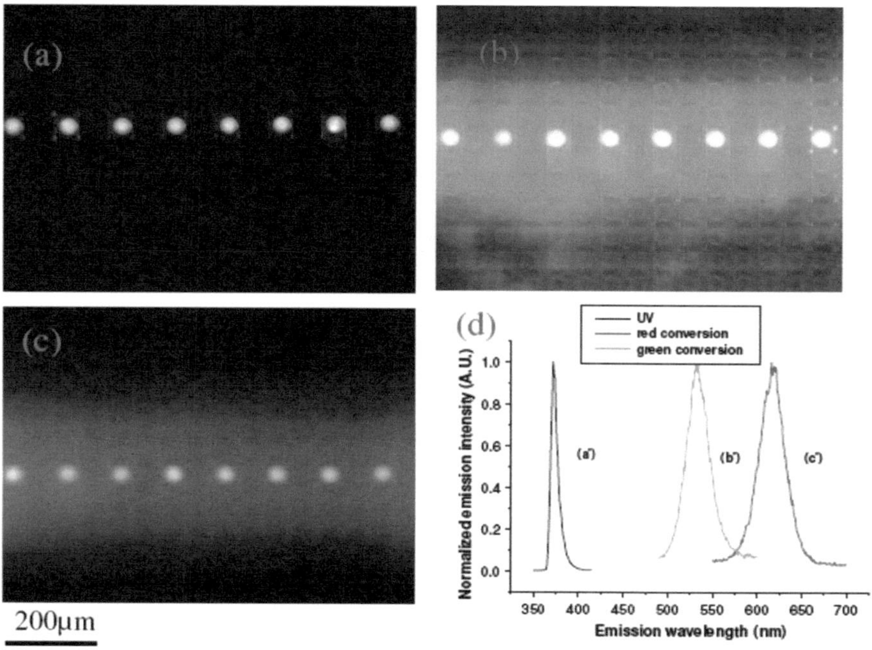

Fig. 4.2 Multicolor display of micro-LED based on quantum dots [2]; **a** Image of 360 nm wavelength UV micro-LED; **b** Green light conversion image; **c** Red light conversion image; **d** Spectrum of the three types of emitted light

contrast, and other display characteristics of micro-LED pixels are superior to those of LCD and OLED displays. The reliability of micro-LED can be inferred from the lifespan of lighting LEDs (which typically exceeds tens of thousands of hours), making micro-LED lifespan more competitive than that of LCD and OLED. Even for large-screen display scenarios, multiple micro-LED displays can be seamlessly spliced, representing another key advantage of micro-LED displays compared with current large-screen LCDs. The color gamut is another critical indicator of display performance. Micro-LED displays exhibit a color gamut exceeding 100% NTSC, enabling vivid color reproduction.

Although the advantages of micro-LEDs are obvious, there are also technical difficulties and challenges. On the one hand, based on existing material growth technology, it is still difficult to grow high-efficiency red, green, and blue (RGB) micro-LED devices on the same epitaxial substrate; on the other hand, due to the limited size of epitaxial wafers, it is difficult to meet the requirements of large-size displays. Therefore, achieving a large-area full-color display is the main challenge in current micro-LED display technology. At present, several solutions have been proposed. One is to transfer micro-LEDs of different light-emitting wavelengths to the same substrate or drive panel through mass transfer technology, and the other is

Table 4.1 Comparison of parameters of LCD, OLED, and micro-LED displays [7]

Parameter	LCD	OLED	Micro-LED
Luminous mode	Backlight module	Self-luminous	Self-luminous
Brightness (cd/m^2)	3000	5000	100000
Material	Inorganic	Organic	Inorganic
Contrast	5000	∞	∞
PPI	>300	>6000	>10000
Color rendering	75% NTSC	>100% NTSC	>100% NTSC
Response time	5 ms	10 μs	0.2 ns
Lifetime	Medium	Medium	Long
Temperature stability	233–373 K	238–358 K	15–500 K

to prepare blue or purple micro-LED arrays, then integrate color conversion materials onto the micro-LEDs to achieve RGB primary colors, which are the two most commonly used methods. Of course, there are other methods, such as growing two quantum wells with different light-emitting wavelengths through material growth technology [8], then adjusting the light color of micro-LED by changing the injected current to achieve a full-color micro-LED display; there is also synthesizing RGB light-emitting micro-LED modules through a three-color prism to achieve full-color micro-LED projection modules; recently, a vertical bonding technology for RGB full-color micro-LED integration has been proposed.

As mentioned earlier, in addition to gaining widespread attention in displays, micro-LEDs also perform well in high-speed visible light communication (VLC). Due to the low capacitance of micro-LEDs and their ability to withstand high current density, they show better performance in VLC than traditional LEDs. In addition, micro-LED serves both as a transmitter and a receiver, which has promoted the rapid development of high-speed visible light communication based on micro-LED. In 2010, Martin Dawson's research group used a 16 × 16 micro-LED array emitting at 450 nm to develop a high-speed VLC system. As shown in Fig. 4.3, the modulation bandwidth of this device reaches over 245 MHz [9]. Figure 4.4 illustrates the bit error ratio (BER) versus received optical power under different pseudorandom sequence lengths. When the length of the pseudorandom sequence is $2^7 - 1$ bits and $2^9 - 1$ bits, the highest error-free transmission rate reaches 1 Gb/s; when the length of the pseudorandom sequence is $2^{31} - 1$ bits, the highest transmission rate without error can reach 622 Mb/s.

In 2017, the research group led by Tian Pengfei built an underwater optical communication system using micro-LEDs as the transmitter. At a transmission distance of 5.4 m, the real-time communication rate reached 200 Mb/s [10]. In 2019, they used a micro-LED array as a photoelectric detector to develop a 2 × 2 multiple-input multiple-output (MIMO) visible light communication system with

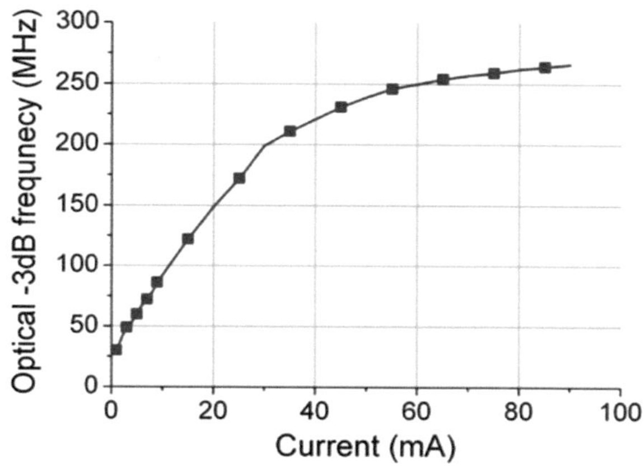

Fig. 4.3 Relationship between modulation bandwidth and current of 450 nm micro-LED [9]

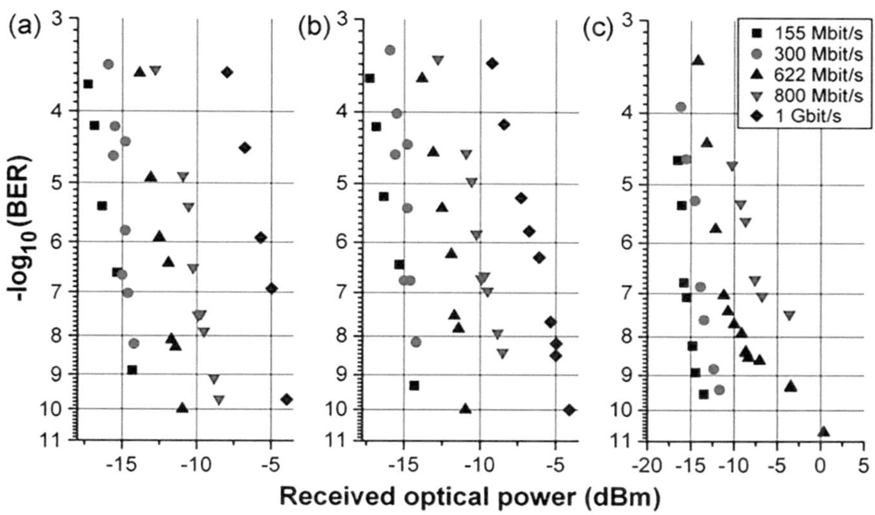

Fig. 4.4 Relationship between error rate and received optical power of 450 nm micro-LED array [9]. The lengths of the pseudorandom sequences are respectively **a** $2^7 - 1$ bits; **b** $2^9 - 1$ bits; **c** $2^{31} - 1$ bits

a data transmission rate of 350 Mb/s [11]. In 2021, the research group also used micro-LEDs as solar cells to collect light energy, to achieve the integration of underwater wireless charging and underwater wireless optical communication [12].

Of course, due to various challenges and difficulties, the commercial mass production of micro-LEDs has not yet been realized, which also limits their widespread adoption in commercial applications. Nevertheless, we believe that with the continuous development of micro-LED technology, it will eventually make its own contribution not only in the fields of display and optical communication but also in emerging fields such as optogenetics, medical detection, and biomedicine.

4.2 GaN-Based Micro-LED Devices

Currently, sapphire is one of the most commonly used substrates for GaN-based LED epitaxial wafers due to its mature preparation process. Silicon substrate LEDs have the advantages of low cost, easy substrate removal, and exhibit unique advantages in the preparation of long-wavelength, high-efficiency red LEDs. In view of the difficulties in realizing full-color micro-LED displays, this section will introduce the preparation methods for micro-LED devices on sapphire and silicon substrates, as well as the approach of integrating quantum dots with micro-LEDs for color conversion.

4.2.1 Preparation of Sapphire Substrate Micro-LED

Figure 4.5 shows the schematic diagram of the process for preparing micro-LED devices on a sapphire substrate [12]. First, the GaN buffer layer, n-type gallium nitride (n-GaN) layer, InGaN/GaN multiple quantum well (MQW), and p-type gallium nitride (p-GaN) layer epitaxial structures are sequentially grown on the sapphire substrate. This epitaxial material is often grown using metal organic chemical vapor deposition (MOCVD) technology. Then, an indium tin oxide (ITO) film is prepared on top of the epitaxial structure by magnetron sputtering as a current spreading layer. Subsequently, mesa structures are etched through spin-coating photoresist, ultraviolet lithography, and ICP etching on the epitaxial wafer, followed by thermal annealing to form p-type ohmic contacts. Then, a SiO_2 insulating layer is deposited by plasma-enhanced chemical vapor deposition (PECVD) technology, and SiO_2 holes are opened. Finally, a certain thickness of titanium/gold (Ti/Au) metal is grown on the surface of the n-GaN layer and ITO layer as electrodes by magnetron sputtering.

Substrate stripping technology is an important process in the preparation of sapphire substrates for micro-LED devices. With the development of wearable devices and biomedical applications, the demand for flexible micro-LED devices is increasing. Due to the poor thermal conductivity of sapphire substrates, they cannot fully meet the heat dissipation requirements of the devices when fabricating high-power devices, making substrate stripping necessary for flexible applications. In addition, transferring the micro-LEDs from the sapphire substrate to the driving substrates is also an important step in achieving a full-color display.

Laser lift-off (LLO) technology is generally used to achieve the stripping of substrates such as sapphire that are transparent to ultraviolet light. Figure 4.6 shows a

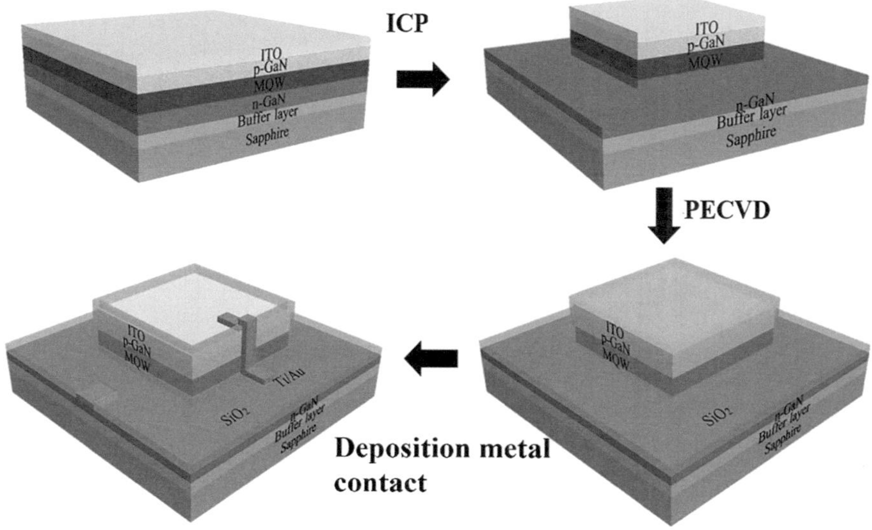

Fig. 4.5 Schematic diagram of the fabrication process of a single GaN-based micro-LED device [13]

schematic diagram of the process of stripping the sapphire substrate and transferring the device using LLO technology [13]. First, a bonding layer is coated on the transition substrate using special methods such as spin coating [14]. Next, the micro-LED device on the sapphire substrate is temporarily bonded to the transition substrate, usually using organic materials, epoxy resins, or metal alloys as the bonding layer. Then, the sapphire substrate is irradiated with a laser, causing the GaN at the junction of the buffer layer and the substrate to absorb energy and decompose, ultimately separating the device from the substrate.

After LLO substrate stripping, micro-LEDs can be transferred onto target substrates, especially flexible ones, via transfer printing technology. Figure 4.7 shows a schematic of fabricating flexible micro-LEDs using a blue InGaN LED epitaxial

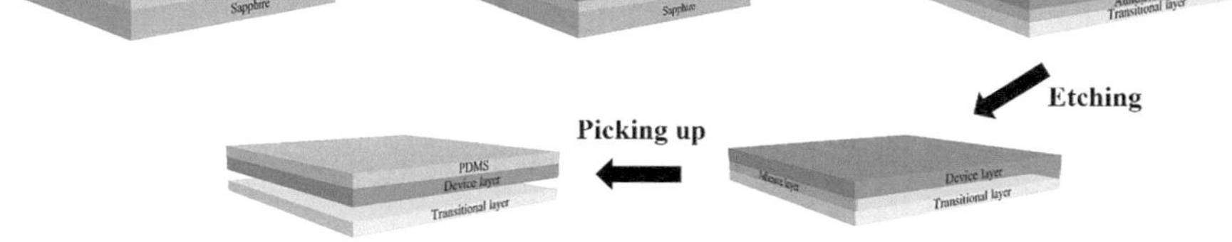

Fig. 4.6 Schematic diagram of the process of laser lift-off (LLO) stripping sapphire substrate and device transfer [13]

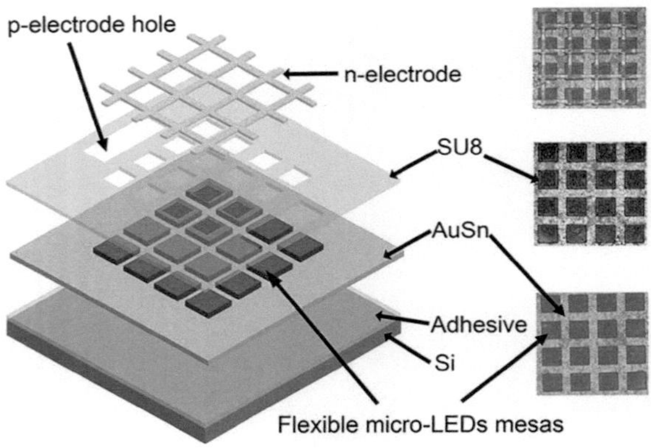

Fig. 4.7 Schematic diagram of preparing flexible micro-LED based on large-area mesa pattern [15]. On the left is a schematic diagram of the processing steps of flexible micro-LED based on a sapphire substrate; on the right are microscope photos during the preparation process. Bottom: flexible micro-LED mesa; middle: SU8 photoresist used as an insulating layer; top: interconnect electrodes for flexible micro-LED array; the size of flexible micro-LED pixels is 140 μm

wafer grown on a sapphire substrate [15]. First, a large-area mesa of 1 × 1 mm is etched into the sapphire substrate by ICP etching. Then, an ITO layer and a titanium/silver/titanium/gold (Ti/Ag/Ti/Au) layer are deposited on the photoresist mask layer to achieve p-type ohmic contact and mirror function, respectively. Then, palladium/titanium/gold (Pd/Ti/Au) is deposited on the silicon wafer as a metal bonding layer to prepare the transfer substrate. A 50-μm-thick flexible gold–tin (AuSn) layer is used as the bonding layer to bond the LED epitaxial wafer of the sapphire substrate to the transfer substrate. Finally, the sapphire substrate is separated from the LED chip via LLO, and the device is heated in hot water to remove the silicon (Si) substrate and AuSn layer, completing the fabrication of flexible micro-LED devices.

The specific experimental results are shown in Fig. 4.8. Among them, (a) shows the I–V characteristics of a single pixel before and after the bending test of a 4 × 4 flexible micro-LED array. (b) shows the L–I characteristics of a flexible micro-LED with a bending radius of 6 mm. To verify the accuracy of the experimental results, 6 pixels were randomly selected for I–V and L–I characteristic tests. As can be seen from the inset, the six-pixel points only show a fluctuation of ±10%. This good uniformity is attributed to the excellent conductivity of the underlying AuSn substrate. The L–I characteristics indicate that due to the high thermal conductivity of the AuSn substrate, the current density of the flexible micro-LED can be maintained at approximately 357 A/cm^2 before thermal saturation, enabling the device to be applied in fields such as flexible high-brightness displays and VLC.

Sapphire substrate is the most commonly used substrate for GaN-based LED epitaxial wafers, owing to its excellent mechanical properties, chemical stability, high strength, high hardness, and low visible light absorption, enabling advantages such as backside light emission or double-sided light emission. However, sapphire

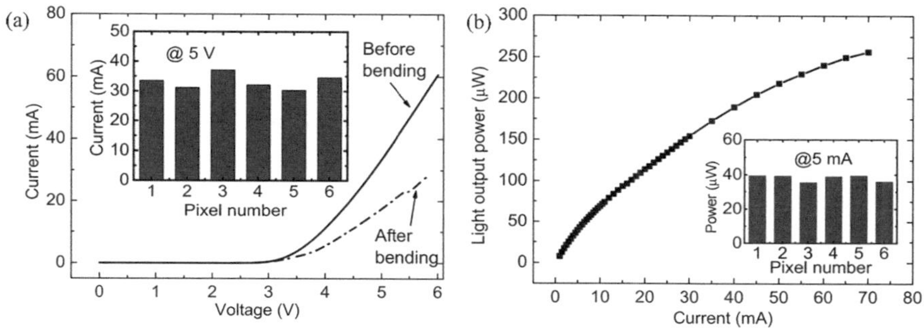

Fig. 4.8 Photoelectric characteristics of flexible micro-LED [15]; **a** Typical I–V characteristics of flexible micro-LED before and after bending 10 times. Inset: The current of 6 randomly selected pixels at 5 V before the bending test; **b** L–I characteristics of flexible micro-LED with a curvature radius of 6 mm. **b** Inset: The output optical power of 6 randomly selected pixels at 5 mA

substrates exhibit poor thermal conductivity and are difficult to remove. To simplify the fabrication process of flexible micro-LED devices, researchers are also exploring Si-based micro-LEDs.

4.2.2 Preparation of Si Substrate Micro-LED

Figure 4.9 shows the schematic diagram of the structure of the Si substrate micro-LED [16]. This epitaxial structure is grown on a 6-inch Si(111) wafer using MOCVD technology. The structure includes an aluminum gallium nitride (AlGaN) buffer layer, n-GaN layer, InGaN/GaN MQW layer, AlGaN electron barrier layer, and p-GaN layer, with a total thickness of 2 μm. Subsequently, a Ni/Au (10/25 nm) metal layer was deposited on the p-GaN to form a current spreading layer. Then, using reactive ion etching (RIE) and ICP etching, the nickel/gold (Ni/Au) metal layer and GaN layer were etched, respectively, to obtain a columnar platform with a diameter of 45 μm, and then annealed in an annealing furnace to achieve good ohmic contact. Then, a Ti/Au (50/200 nm) layer was deposited on the platform as a common n-electrode for the array, and SiO_2 was deposited by PECVD. The passivation layer was then etched using a wet method to create holes in the SiO_2 of each platform. Finally, Ti/Au (50/200 nm) was prepared as the p-electrode. Figure 4.9b shows a photograph of the micro-LED array under an optical microscope.

Figure 4.10 shows the EL spectrum of the Si substrate micro-LED and the luminescent image of the 10 × 10 Si substrate micro-LED array. As shown in Fig. 4.10a, under a driving current of 1.6 mA DC, the peak wavelength of this Si substrate micro-LED is at 470 nm. The multiple peaks in the spectrum are mainly attributed to light reflections occurring at various interfaces within the device. Figure 4.10b is an image of the 10 × 10 Si substrate micro-LED array, individually controlled and emitting light to display the capital letter "I". It can be seen from the figure that the

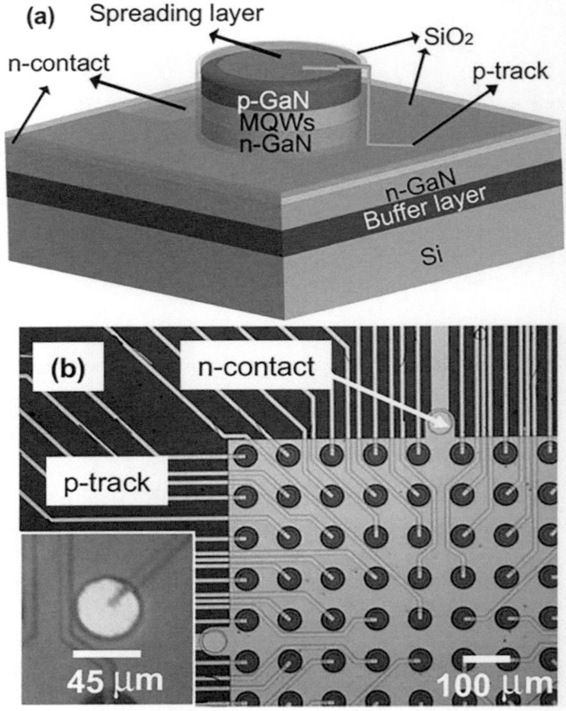

Fig. 4.9 Structure of Si substrate micro-LED [16]; **a** Schematic diagram of the structure of the Si substrate micro-LED; **b** 10 × 10 micro-LED array under an optical microscope

brightness of all 8 light-emitting pixels is very consistent, which indicates that the electrical and optical properties between the pixels are consistent.

Unlike sapphire substrates, micro-LED devices based on Si substrates generally use wet etching to strip the substrate; the most common method is to remove the

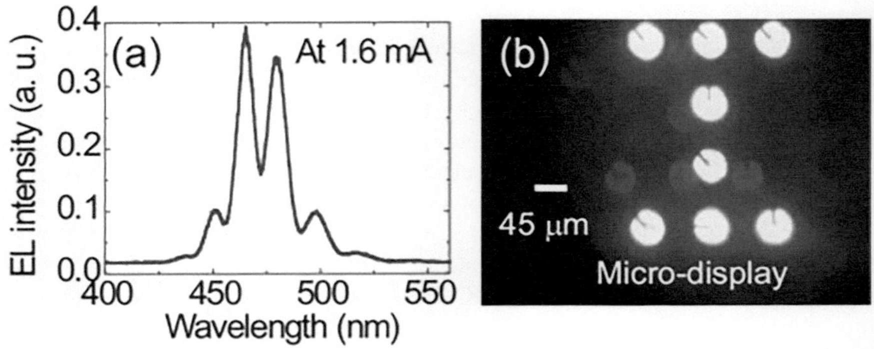

Fig. 4.10 Characteristics of Si substrate micro-LED and its array [16]; **a** Electroluminescence (EL) spectrum under 1.6 mA injection current; **b** Luminescent image of Micro-LED array

Si substrate using a potassium hydroxide (KOH) solution. Specifically, the device structure is deeply etched into the Si substrate by ICP etching, and after fixing the device with an anchor structure, it is immersed in a KOH solution for anisotropic wet etching until the micro-LED is suspended, connected to the substrate only by the anchor structure. After wet etching, the device is only supported by the anchor structure, making it highly fragile; therefore, it must be handled with care during subsequent transfer printing processes.

Figure 4.11 shows a schematic diagram of the process for preparing flexible GaN-based micro-LEDs using Si substrate epitaxial materials, as well as the corresponding SEM images. First, micro-LED devices based on Si substrates are prepared, with a specific process similar to that described above. For the ohmic contact of n-type GaN, ICP etching is first used for 90 s until the surface of the n-type GaN layer is etched, then a chromium/gold (Cr/Au) layer is deposited on the n-type GaN and p-type GaN layers, serving as n-type and p-type electrodes, respectively. The electrodes are annealed at 600 °C in a nitrogen environment for 1 min using a rapid thermal annealing (RTA) device to achieve ohmic contact. To protect the ohmic contact electrodes, an additional mask composed of polyethylene oxide (PEO), Au, and Cr (1 mm/200 nm/10 nm) is deposited on the electrodes. For the PEO layer, ICP etching is used for 25 min, and for Au, a gold etching solution is used for 1 min to etch the surface of the GaN-based micro-LED structure. Then, the substrate of the GaN-based micro-LED is etched away using anisotropic wet etching with a KOH solution. During the wet etching process with a high concentration of alkaline solution, a protective gold mask can be made on the electrode to prevent the electrode from being damaged. After the Si substrate is stripped, the transfer printing step can be carried out: the PDMS stamp is evenly contacted with the Au-protected GaN-based micro-LED. By controlling the transfer rate, the PDMS stamp can pick up and place the GaN-based micro-LED. When transferring the GaN-based micro-LED to the flexible substrate, polyurethane (PU) is coated on the flexible substrate, which is then irradiated with ultraviolet light to optically cure the PU. Finally, the PDMS stamp is removed, and the GaN-based micro-LED can be bonded to the flexible substrate. Finally, UV-sensitive epoxy resin (SU8) is spin-coated on the GaN/PU/LCP substrate. Then, a Cr/Au layer is sputter-deposited to complete the process of transferring the Si-based micro-LED to the flexible substrate. Compared to sapphire, Si growth technology is more mature, making it easier to obtain low-cost, large-size, high-quality substrates. Additionally, Si substrates offer the advantages of high thermal conductivity, good electrical conductivity, and ease of stripping. However, growing GaN on Si is more challenging, and Si's absorption of visible light can both reduce the external quantum efficiency of Si-based micro-LED devices.

4.2.3 Micro-LED and Quantum Dot Integrated Devices

As mentioned earlier, full-color micro-LED displays can also be achieved by integrating quantum dot color conversion materials. For example, metal halide perovskite

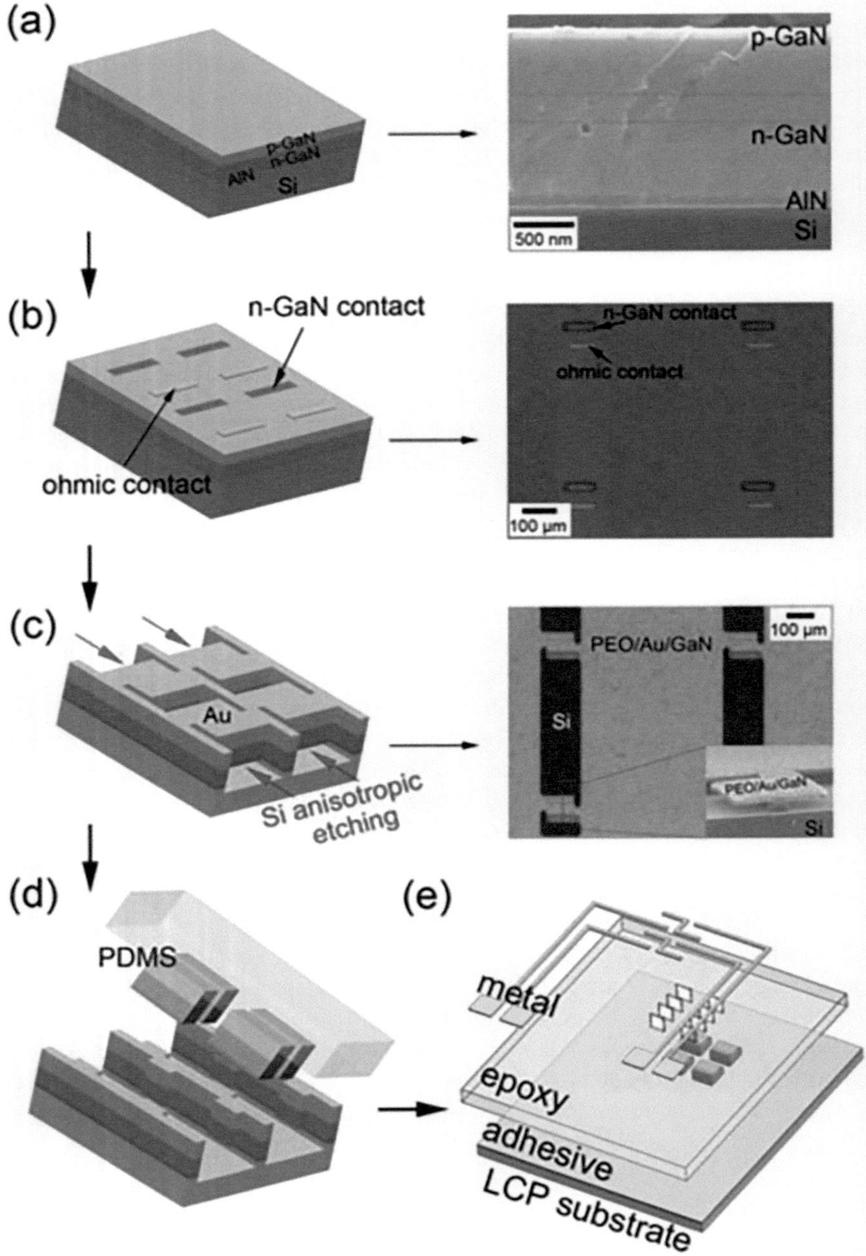

Fig. 4.11 Schematic diagram and corresponding pictures of the process of preparing flexible GaN-based micro-LEDs on a flexible substrate [17]; **a** Micro-LED structure on Si substrate; **b** Ohmic contact on micro-LED; **c** Anisotropic etching of KOH; **d** Process of transferring the patterned micro-LED array with a PDMS stamp; **e** Preparation of flexible substrate micro-LED array device

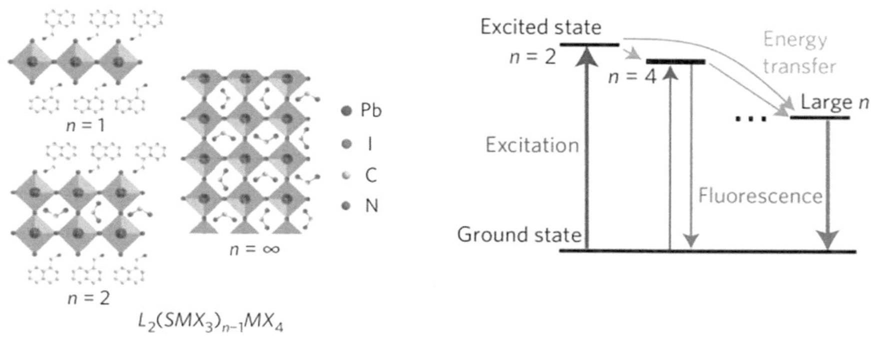

Fig. 4.12 Two-dimensional perovskite structure diagram and the energy transfer process in its multi-quantum well structure [20]

materials are typical quantum dot materials. Perovskite materials are direct bandgap semiconductor luminescent materials, characterized by narrow half-peak width, wide color gamut, low synthesis temperature, photoluminescence quantum yield up to 100%, and the ability to adjust the halogen ratio to control the luminescent spectrum for full-spectrum luminescence [18,19]. Figure 4.12 shows a diagram of the two-dimensional perovskite structure and the energy transfer process in their multi-quantum well structure.

Currently, aerosol jetting technology is the most commonly used method for integrating micro-LEDs and quantum dots. This is mainly because this technology can accurately deposit a thin and uniform QD layer, thereby improving the resolution of the quantum dot-integrated micro-LED full-color display.

The technical process of aerosol jetting QDs is shown in Fig. 4.13 [21]. Firstly, the epitaxial material is etched to the surface of the sapphire substrate to create isolation trenches using ICP etching technology. This process isolates the UV micro-LED array with a luminescent wavelength of 395 nm, facilitating the integration of different quantum dots for full-color display. Then, the QD solution is atomized and rolled into the airflow. Next, by adjusting the carrier gas flow and sheath gas flow, the RGB three-color QD spray is deposited on the surface of the micro-LED device in a certain proportion. To improve the external quantum efficiency of the device, a distributed Bragg reflector (DBR) can be prepared on top of the device to prevent UV photon leakage. Finally, the UV micro-LED array integrated with QDs is packaged, with each sub-pixel maintained at a certain distance, and a full-color display of micro-LED is achieved by adjusting the injected current amplitude.

Figure 4.14 shows the schematic diagram of the experiment of micro-LED used for optical communication with color conversion by perovskite quantum dots [22]. The blue light emitted by the micro-LED is used to excite cesium lead halide ($CsPbBr_{1.8}I_{1.2}$) perovskite quantum dots (YQDs) to produce white light. The output white light is collimated by the transmitter lens, focused by the receiver lens, and finally detected by the photodetector. A filter is used to remove blue light from the

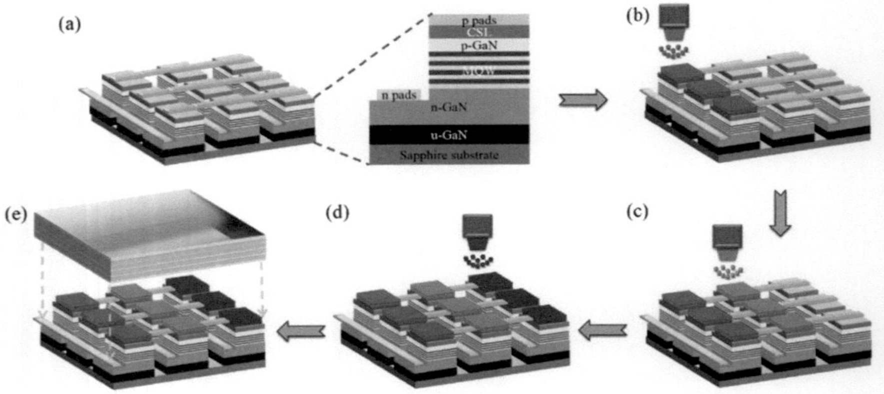

Fig. 4.13 Aerosol jetting technology flowchart [21]

micro-LED emission for measuring the optical communication characteristics of the quantum dots. Figure 4.15 shows the luminescence and color coordinate schematic of the quantum dot integrated micro-LED. The modulation bandwidth of the packaged 80 × 80 μm blue micro-LED is 160 MHz, with a peak emission wavelength of 445 nm. The peak wavelength of the quantum dots is 560 nm, and the corresponding International Commission on Illumination (CIE) color coordinates are (0.27, 0.30). Without filters and equalization techniques, the bandwidth of the white light system can reach 85 MHz. At the same time, the bandwidth of using YQDs for color conversion is still as high as 73 MHz. When non-return-to-zero on–off keying (NRZ-OOK) is used as the modulation scheme, the transmission rate can reach up to 300 Mbps, and the bit error rate is 2.0×10^{-3}, which is far lower than the forward error correction standard of 3.8×10^{-3} required for error-free data transmission. After six months, the white light system and YQDs maintain high bandwidths of 83 MHz and 70 MHz, respectively, indicating the high stability of the YQD-based color conversion system in VLC.

Due to the simple preparation process and excellent optical properties—such as narrow luminescence spectrum, high luminescence efficiency, and tunable luminescence wavelength, perovskite quantum dots exhibit high stability. Therefore, they have a wide range of application scenarios in full-color display and VLC. However, due to the mechanism of color conversion, this will reduce the external quantum efficiency of the integrated micro-LED device after assembly.

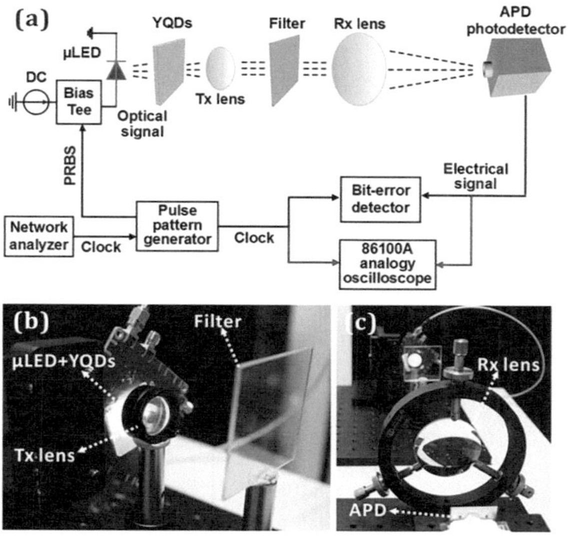

Fig. 4.14 Schematic diagram of micro-LED optical communication experimental link and experiment [22]; **a** Schematic diagram of micro-LED optical communication with color conversion by perovskite quantum dots; **b** Integration of quantum dots and micro-LED at the optical path transmitter end; **c** Optical path receiver end

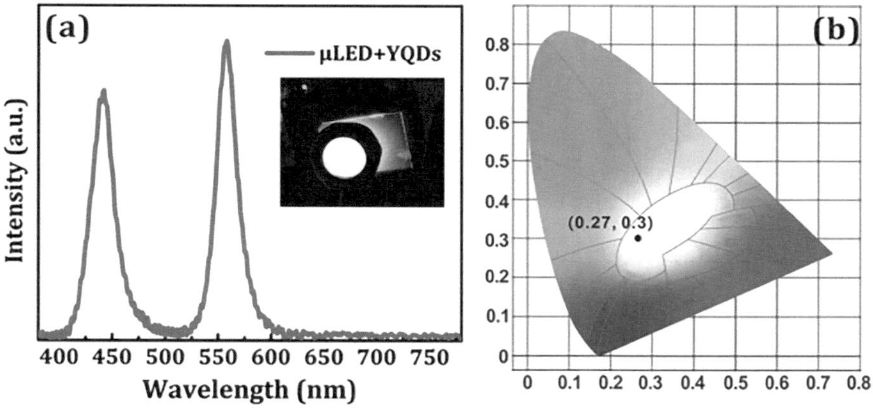

Fig. 4.15 Schematic diagram of quantum dot integrated micro-LED luminescence and its color coordinates [22]; **a** Spectrum of white light produced by blue micro-LED integrated with quantum dots. Inset: Photo of the integrated device emitting white light; **b** Commission Internationale de l'Éclairage (CIE) coordinates corresponding to the emitted white light

4.3 Micro-LED Optoelectronic Characteristics

The unique device structure of micro-LED exhibits distinctive optoelectronic characteristics, which play an important role in display and visible light communication. In this section, we will focus on the relationships between the quantum efficiency and other optoelectronic characteristics of micro-LED and various factors such as size and temperature. We will also study and analyze the aging and bandwidth characteristics of micro-LEDs [23, 37–39].

4.3.1 Micro-LED Size Effect

Micro-LED often leads to a decrease in external quantum efficiency (EQE) under high injection current density, which is efficiency degradation. At present, many mechanisms have been proposed to explain the efficiency degradation. Recent studies have shown that compared with traditional large-sized LEDs, the size reduction of micro-LEDs gives them unique electrical and optical characteristics. Therefore, this section will introduce the impact of micro-LED size changes on current–voltage characteristics and quantum efficiency, and provide a detailed explanation of their underlying mechanisms.

First, GaN-based blue micro-LEDs of different sizes were prepared on a sapphire substrate, with pixel diameters of 6, 25, 50, and 105 μm, and the current–voltage (I–V) curves and current density–voltage (J–V) curves of different sizes micro-LEDs were tested. As shown in Fig. 4.16, the four micro-LEDs exhibit a strong size-dependent voltage characteristic with respect to injection current density. The 6-μm micro-LED achieves an injection current density of 4000 A/cm^2 at 5 V, while the 105-μm micro-LED only reaches 600 A/cm^2 under the same voltage, indicating that smaller micro-LEDs have higher injection current density at 5 V [23].

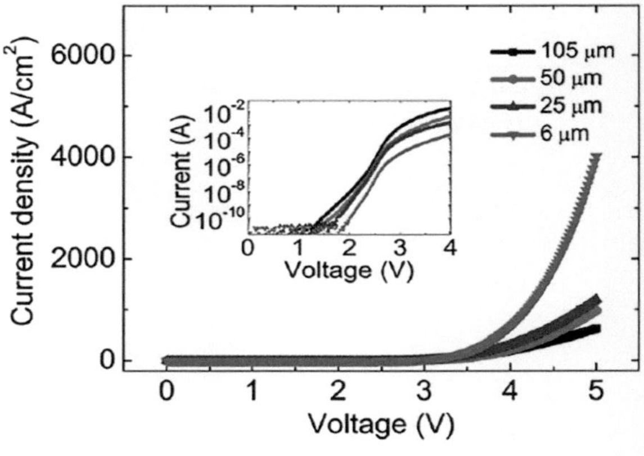

Fig. 4.16 Current density–voltage curves of different-sized micro-LEDs. Inset: Current–voltage curves [23]

To estimate the current spreading effect related to size, the function relationship between the current density distribution in the mesa and the distance from the edge of the mesa to the center was calculated, and then the ratio of the current density at the center and edge of the micro-LED mesa was obtained. The function relationship is as follows:

$$J(x) = J(0)\exp\left(-\frac{x}{L_s}\right) \tag{4.1}$$

$$L_s = \sqrt{\frac{(\rho_c + \rho_p t_p)t_n}{\rho_n}} \tag{4.2}$$

where:

$J(0)$—Current density at the edge of the platform

ρ_c, ρ_p, ρ_n—Resistivities of the p electrode, p-type layer, and n-type layer

t_p, t_n—Thickness of the p-type and n-type layers

L_s—Current spreading length, used to define current crowding in micro-LEDs.

Calculations show that for micro-LEDs with diameters of 105 μm and 6 μm, the ratio of current density at the mesa center to the mesa edge is 0.48 and 0.96, respectively. This indicates that compared with large-sized micro-LEDs, small-sized micro-LEDs have a more uniform current density distribution. Moreover, it has been reported that under higher injection current densities, the phenomenon of current crowding becomes more pronounced [24]. Therefore, with the increase in voltage, smaller-sized micro-LEDs are conducive to more uniform current spreading, alleviating the effect of current crowding, and thus improving their optoelectronic characteristics. As current crowding can lead to local Auger recombination, electron leakage, and other efficiency-impacting factors, especially under very high current densities, it is expected that for smaller-sized micro-LEDs, uniform current spreading will be beneficial for achieving higher efficiency under high injection current densities.

Analyzing the internal and external quantum efficiency of micro-LEDs can effectively reflect their optoelectronic conversion characteristics. In quantum wells, carrier recombination is described by the ABC model, which includes three primary electron–hole recombination mechanisms: Shockley–Read–Hall (SRH) recombination, Auger recombination, and radiative recombination. The calculation formula is as follows:

$$I_{QW} = I_{SRH} + I_{rad} + I_{Auger} = qV_{active}\left(An + Bn^2 + Cn^3\right) \tag{4.3}$$

where in the formula:

q—Electron charge

V_{active}—Effective recombination volume of the active layer

n—Carrier concentration in the quantum well

A, B, C—SRH recombination, radiative recombination, Auger recombination coefficients

I_{QW}—Total current in the quantum well

I_{SRH}—Defect recombination current

I_{rad}—Radiative recombination current

I_{Auger}—Auger recombination current.

By testing the change in output optical power with current for micro-LEDs of different sizes, the EQE curve with current density was calculated, and the peak EQE current density and efficiency drop with diameter for micro-LEDs were shown. As can be seen from the inset of Fig. 4.17a, for all micro-LED sizes, the EQE peaks at low current densities and then decreases with increasing current density. At low current densities, smaller micro-LEDs have lower efficiency, but at high current densities, they have higher efficiency, which can be attributed to better current spreading capability in smaller micro-LEDs. Also, from Fig. 4.17b, it can be seen that as the size of the micro-LED decreases, the current density corresponding to the peak EQE increases, and the percentage of efficiency drop decreases. This study controlled for variables such as the materials used and the device structure of the micro-LEDs, with size being the only changing factor, which is the main reason for the change in efficiency of the micro-LEDs.

To directly prove the impact of sidewall defects, Sample A and Sample B were prepared, where the heat annealing time for Sample A was 2 min and that for Sample B was 3 min. The EQE of Sample B changes with size as shown in Fig. 4.18.

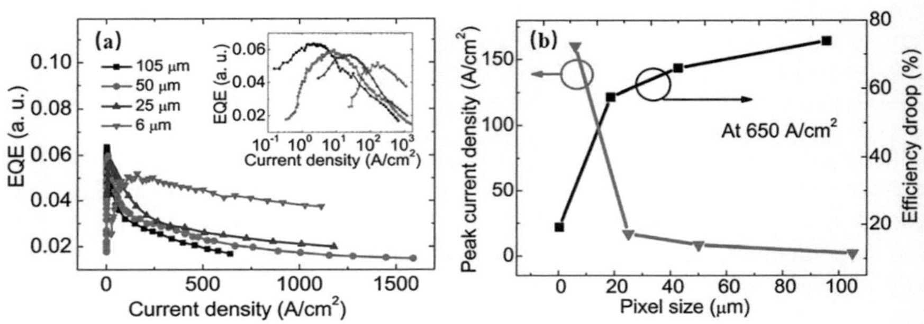

Fig. 4.17 External quantum efficiency of micro-LEDs of different sizes as a function of current density [23]; **a** The variation curve of the external quantum efficiency of different sizes of micro-LEDs in Sample A with current density. Inset: The variation curve of the external quantum efficiency with the logarithmic coordinates of the current density; **b** The curve of the current density and the percentage of efficiency decline corresponding to the peak external quantum efficiency with size. Sample A: Heat annealing time is 2 min

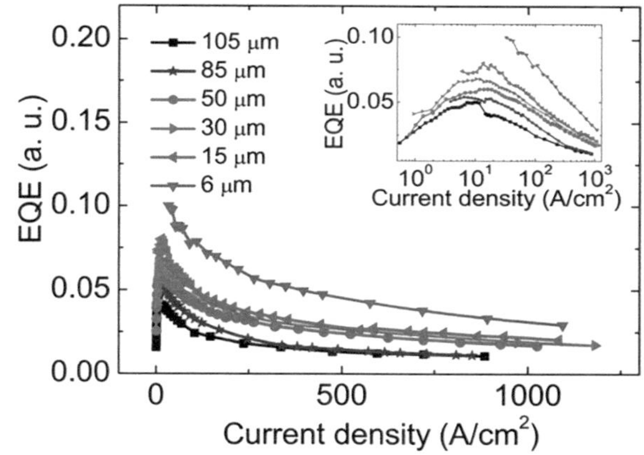

Fig. 4.18 The variation curve of the external quantum efficiency of different sizes of micro-LEDs in Sample B with current density. Inset: The variation curve of the external quantum efficiency with the logarithmic coordinates of the current density. Sample B: Heat annealing time is 3 min [23]

Compared with Sample A, at low current densities, the peak EQE of small-sized micro-LEDs in Sample B is significantly reduced, while the variation at high current densities is less pronounced. These results indicate that the sidewall defects in Sample B can be more effectively repaired by prolonged heat treatment. Therefore, at lower current densities, repairing sidewall defects has a significant impact on enhancing the efficiency of smaller micro-LEDs [25].

At present, a large amount of research has been conducted internationally on the size effect of micro-LEDs, including the impact mechanism of sidewall defects [25–29], the technology to repair sidewall defects [30, 31], and the excellent heat dissipation characteristics of micro-LEDs [32]. These studies not only provide explanations for efficiency and the mechanisms of efficiency decline but also propose methods to improve micro-LED efficiency and alleviate efficiency decline issues. Through these methods, the efficiency of micro-LEDs in actual applications can be improved, enabling them to play a role in a wider range of applications.

4.3.2 Micro-LED Temperature Effects

Due to the junction temperature effect of micro-LEDs, and considering that the working environment temperature will also change in actual applications, it is particularly important to study the impact of temperature on micro-LED performance and clarify the mechanism.

A blue micro-LED with a diameter of 40 μm on a sapphire substrate emitting light from the front was selected, and the variation curve of the micro-LED external quantum efficiency and modulation bandwidth f −3 dB with temperature was tested [33]. Figure 4.19a shows the EQE of the micro-LED as a function of current density at temperatures increasing in 25 K increments from 300 to 500 K. It can be seen from the figure that at low current densities, as the temperature increases, the external quantum

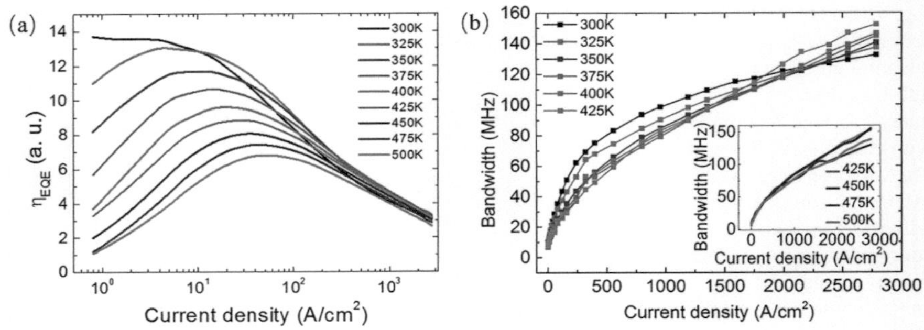

Fig. 4.19 The variation curve of the EQE and modulation bandwidth of the micro-LED with current density [33]; **a** The curve of EQE with current density under the temperature conditions from 300 to 500 K; **b** The curve of modulation bandwidth with current density under the temperature conditions from 300 to 425 K

efficiency of the micro-LED decreases. To analyze this phenomenon, Fig. 4.19b shows the variation in modulation bandwidth f -3 dB of the micro-LED with current density under different temperatures. Results show that at low current densities, the external quantum efficiency decreases with increasing temperature, while the modulation bandwidth decreases at lower currents but increases at higher currents as temperature rises.

The curve of modulation bandwidth in Fig. 4.19b further yields the calculation formula for carrier concentration under the injected current I:

$$n = \frac{1}{qV_{active}} \int_0^I \eta_{inj}(I')\tau(I')dI' \tag{4.4}$$

where: q—electron charge

V_{active}—effective recombination volume of the active layer

I—current

τ—differential carrier lifetime.

Due to the distribution of carriers in more quantum wells, fluctuations in indium (In) content, and other possible factors, V_{active} may vary under different currents/temperatures, and the precise calibration of V_{active} under different currents and temperatures is very challenging. Our analysis follows the general assumption that only the quantum well closest to p-GaN has a large number of carriers and dominates the carrier recombination process. Therefore, in this work, a quantum well with a thickness of 2.8 nm is used to estimate V_{active}. In addition, the carrier injection efficiency is assumed to be 1 for all temperatures. Electron leakage has little effect at low current densities, but may affect the efficiency of micro-LEDs at high current densities. The trend of the carrier recombination coefficient obtained later

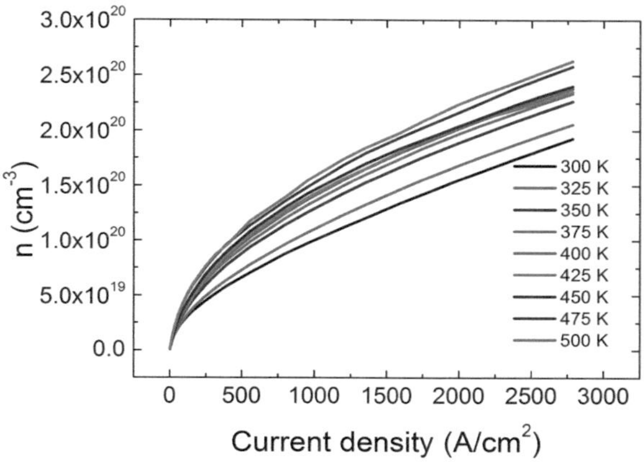

Fig. 4.20 Change curve of carrier concentration in quantum well with current density under temperature conditions from 300 to 500 K [33]

with temperature changes is consistent throughout the current density range, so it can be considered that the impact of electron leakage on the results can be ignored.

Figure 4.20 shows the change curve of carrier concentration with current density under temperature conditions from 300 to 500 K. As shown, the carrier concentration in the quantum well increases with rising temperature.

By fitting the experimental curves related to the Shockley–Read–Hall nonradiative recombination, radiative recombination, Auger recombination (ABC) coefficients at different temperatures using the ABC model, parameters for temperatures ranging from 300 to 500 K are obtained. It is found that as the temperature increases, the SRH recombination coefficient A increases, while the radiative recombination coefficient B decreases with the increase of temperature. As shown in Fig. 4.21, at lower current densities, the temperature dependence of coefficient B is more pronounced, while at higher current densities, its temperature dependence weakens, attributed to the band filling effect. This result is consistent with the results reported in other literature [34]. As shown in Fig. 4.22, similar to the radiative recombination coefficient B, the Auger recombination coefficient C also decreases with the increase of temperature, and as the carrier concentration increases, the temperature dependence of coefficient C weakens, which is also attributed to the band filling effect.

The I–V characteristics and EQE of micro-LEDs change with temperature, and in practical applications, the temperature will constantly change, thereby affecting some important parameters of micro-LEDs in display and communication applications. Therefore, it is very important to analyze the mechanism of temperature effects on carrier recombination and micro-LED performance, which provides a way to solve the problem of efficiency decline and helps to clarify the trend of small signal modulation bandwidth of micro-LEDs in visible light communication.

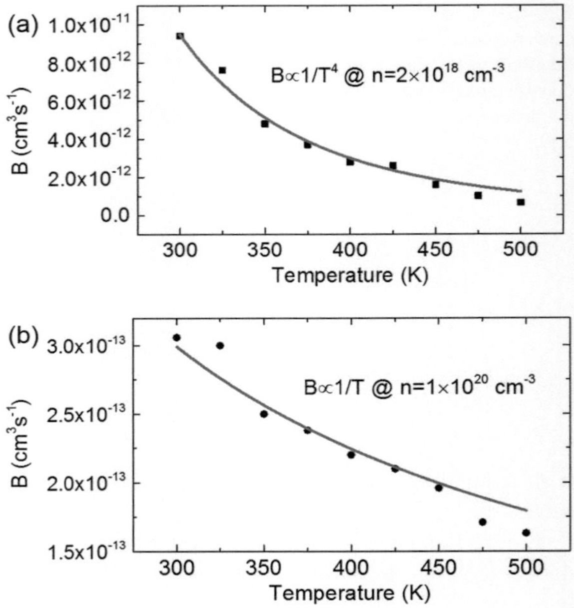

Fig. 4.21 Change curve of radiative recombination coefficient B with temperature [33]; **a** n = 2 × 10^{18} cm^{-3} (B ∝ $1/T^4$); **b** n = 1 × 10^{20} cm^{-3} (B ∝ $1/T^{3/2}$) Change curve of the radiative recombination coefficient B with temperature

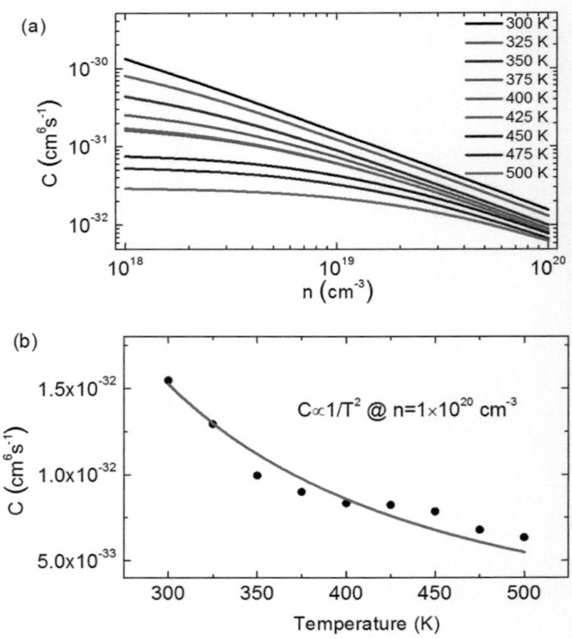

Fig. 4.22 Change curve of Auger recombination coefficient C with temperature [33]; **a** Change curve of Auger recombination coefficient C with carrier concentration under temperature conditions from 300 to 500 K; **b** n = 1 × 10^{20} cm^{-3} Change curve of Auger recombination coefficient C with temperature

4.3.3 Micro-LED Aging Characteristics

In the previous two sections, we discussed the physical mechanisms affecting the optoelectronic performance of micro-LEDs caused by size and temperature. However, given the unique structure of micro-LEDs, it is essential to study their aging characteristics under specific driving current densities.

As mentioned in the previous two sections, due to the size effect of micro-LEDs, their size has a great impact on their lifespan. Compared with large-area LEDs, micro-LEDs can work at extremely high injection current densities to achieve high-density output optical power and high modulation bandwidth in their applications. The characteristics of micro-LEDs under high current density, including electrical characteristics, light output, electroluminescent spectral shift, efficiency droop, and light modulation bandwidth, have been widely studied. However, the aging characteristics and reliability under such high current densities, that is, the key characteristics of how the performance of micro-LEDs changes with the increase of driving time, are relatively less studied. Due to the different driving current densities and sizes of micro-LEDs, their lifespan should be defined in different applications. At an extremely high current density of 3.5 kA/cm^2, the output optical power of blue InGaN micro-LEDs of different sizes was measured until they decayed to 90% of the decay time (L90). The L90 lifespan facilitates comparison of the aging behavior of micro-LEDs of various sizes and can be used to predict their reliability.

First, micro-LEDs of different sizes were selected, and their output optical power was tested under identical temperature and current density conditions [35], as shown in Fig. 4.23. It can be seen that the 20 μm micro-LED shows superior aging characteristics to other sizes of micro-LEDs in the experiment. By testing the light–current (L–I) curves of three micro-LED sizes before and after aging, we found that aging leads to a decrease in output optical power. Moreover, as working time increases, the output optical power of the 20-μm micro-LED under high current density will eventually stabilize or decline. Moreover, as the working time extends, it can be expected that the output optical power of the 20 μm micro-LED under high current density will eventually decrease. Usually, under the same working current density, smaller micro-LEDs will show more stable aging characteristics than larger micro-LEDs.

As shown in Fig. 4.23, the output optical power of micro-LEDs exhibits two stages: first, an increase and then a decrease. Therefore, the aging characteristics of micro-LEDs are likely governed by the competition between two mechanisms, which can be analyzed by testing the I–V characteristic curves at different aging times. The interplay between p-GaN layer degradation and active region defects causes the output optical power of micro-LEDs to first increase and then decrease. Smaller micro-LEDs exhibit less output optical power degradation compared to larger ones, attributed to their lower junction temperature. Without increasing the junction temperature, high current density will not cause significant degradation of micro-LEDs at room temperature, but the combined effect of high current density and high junction temperature will cause micro-LEDs to show strong degradation.

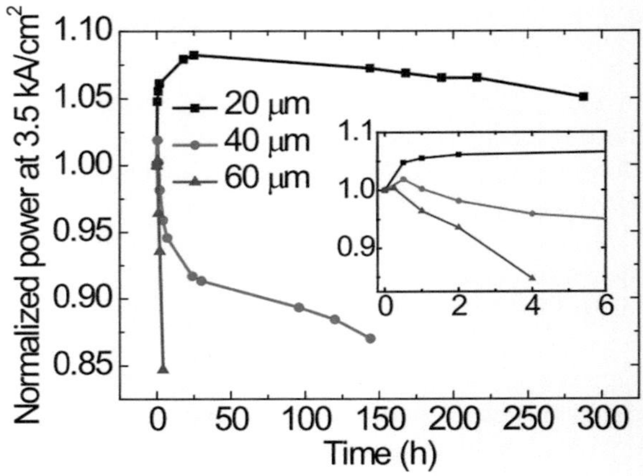

Fig. 4.23 Change curve of normalized output optical power of micro-LEDs of different sizes with working time. Inset: 0–6 h data [35]

Compared to large-size LEDs, the smaller dimensions of micro-LEDs allow for higher driving current density, superior current spreading capability, and thermal diffusion ability. As shown in Fig. 4.24, micro-LED arrays have a higher thermal saturation current density, which gives them a higher modulation bandwidth [36]. Therefore, high-power LEDs constructed from micro-LED arrays have been developed, featuring high output optical power and less light output degradation, making them suitable for solid-state lighting and visible light communication applications.

4.3.4 Micro-LED Optoelectronic Modulation Bandwidth Characteristics

GaN-based LEDs have been widely used in fields such as solid-state lighting and visible light communication, and in visible light communication, the achievable communication rate depends on the limited bandwidth. Therefore, studying the modulation bandwidth changes of micro-LEDs and the underlying dominant mechanisms is of great significance. The modulation bandwidth of GaN-based micro-LEDs is mainly attributed to carrier lifetime and resistance–capacitance (RC) effects. As micro-LEDs for communication typically operate at high injection current densities, the carrier recombination mechanisms under such conditions must also be analyzed. The modulation bandwidth is affected by various factors, such as large-signal modulation depth. Analyzing the relationship between bandwidth and these factors, along with their mechanisms, can effectively enhance the bandwidth in VLC systems.

First, a 40-μm-dameter GaN-based blue micro-LED with an inverted chip structure [38] is selected to enable backside light emission. During the measurement process, since the rise/fall time of the measurement system is much faster than that of the micro-LED, the rise/fall time of the system can be ignored and will not affect the measurement results. By measuring the frequency response of the micro-LED

Fig. 4.24 Normalized light power over time for micro-LED arrays and traditional large-size LEDs [35]; **a** Under current density of 100 A/cm^2; **b** Under current density of 200 A/cm^2 Aging characteristics of normalized output light power over time for micro-LED arrays and traditional large-size LEDs. Inset: Data for 0–3 h

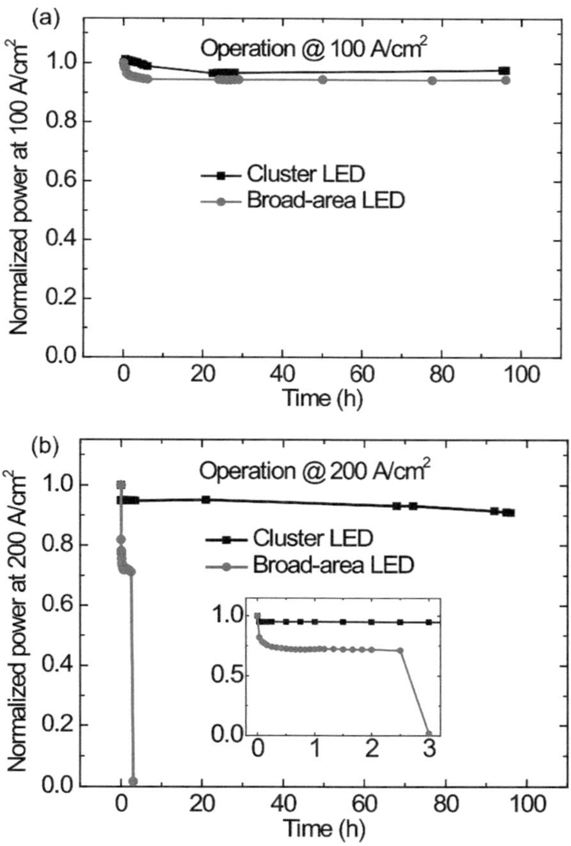

under different currents and extracting the −3 dB modulation bandwidth as well as the modulation bandwidth under different input power, as shown in Fig. 4.25, it can be observed that the modulation bandwidth has a trend of decreasing with the increase of modulation power. The bandwidth under different input modulation voltages is directly compared by measuring the corresponding signal peak voltage on the micro-LED under a specific input power.

To further analyze the relationship between modulation bandwidth and signal rise/fall time, the rise and fall times were measured at a modulation depth of 2 V. The results show that as the modulation depth increases, the sum of the rise and fall times also increases, and the trend of the sum of the rise and fall times is consistent with the trend of the fall time. During the rise time, as the input signal voltage increases, the time constant increases, causing the modulation bandwidth to decrease; during the fall time, as the modulation voltage decreases, the modulation bandwidth increases, which is consistent with the bandwidth change in Fig. 4.25, and as shown in Fig. 4.25, these effects are more pronounced when the modulation depth is large.

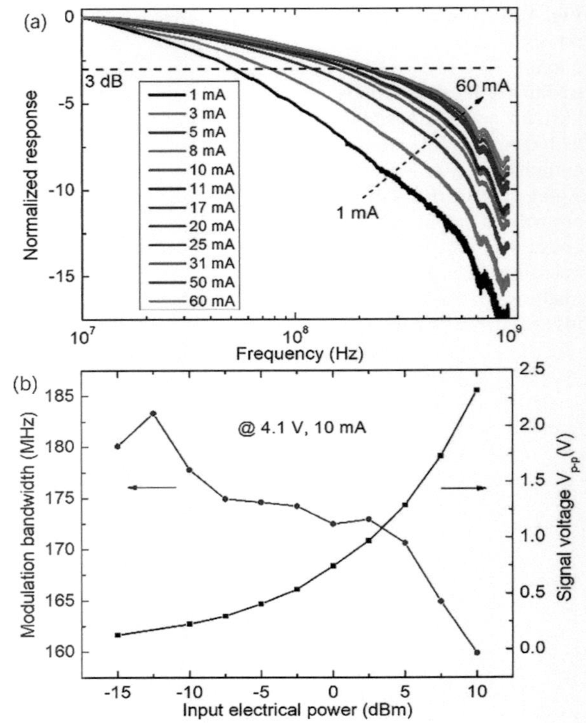

Fig. 4.25 Change in micro-LED bandwidth with current density [38]; **a** Frequency response characteristics of 40 μm micro-LED under 1–60 mA current; **b** Relationship between modulation bandwidth, signal voltage, and input power

At the same time, under an injection current of 35 mA and a transmission distance of 1 m, the relationship between the bit error rate and the modulation depth at different communication rates was tested, as shown in Fig. 4.26. At lower communication rates, the bit error rate decreases with the increase of modulation depth, but at higher communication rates, the bit error rate does not significantly change with the modulation depth.

Since the maximum modulation bandwidth that micro-LED can reach is affected by the carrier lifetime in the quantum well, controlling the recombination process of carriers in GaN-based micro-LED is very important [39]. Micro-LEDs with emission wavelengths of 450 or 520 nm were selected, their differential carrier lifetimes under different currents were measured, and the ABC model was used to analyze the mechanism.

According to the ABC model, the differential carrier lifetime τ is given by the following formula:

$$\tau^{-1} = A + 2Bn + 3Cn^2 \tag{4.5}$$

From the relationship between the differential carrier lifetime and the frequency response bandwidth, the relationship between the carrier lifetime and the carrier

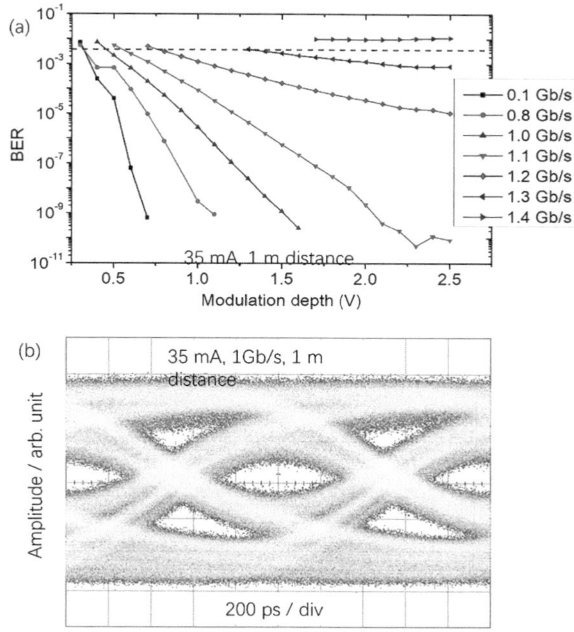

Fig. 4.26 Schematic diagram of micro-LED communication characteristics [38]; **a** Relationship between bit error rate and modulation depth at different communication rates; **b** Eye diagram at 1 Gbps when the injection current is 35 mA and the distance is 1 m

density in the quantum well was derived, as shown in Fig. 4.27. It was observed that when the value of parameter A changed, the derived parameters B and C exhibited minor variations.

Figure 4.28 shows the obtained coefficient B changes with carrier concentration. The B value of the 520 nm micro-LED is usually an order of magnitude lower

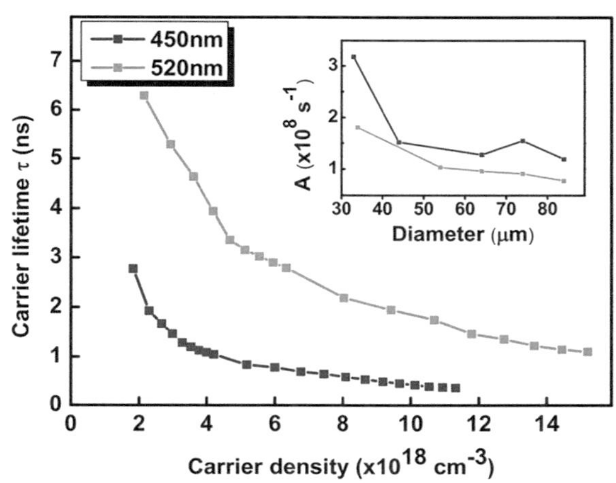

Fig. 4.27 Relationship between differential carrier lifetime and carrier concentration in quantum wells for two devices [39]

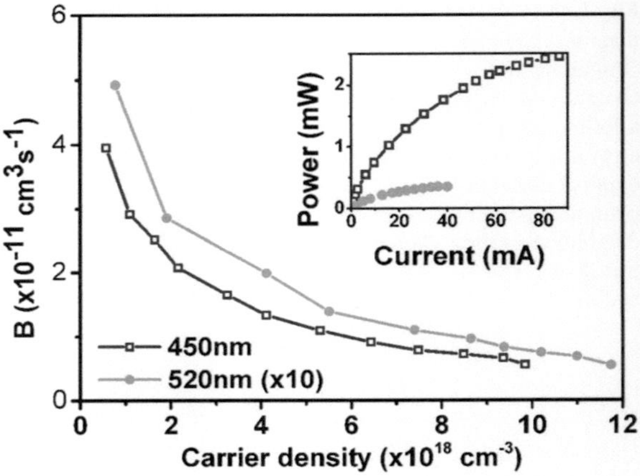

Fig. 4.28 Relationship between the radiation coefficient B values obtained from the L–I curves of the two devices and the carrier concentration. Inset: Relationship between output optical power and current [39]

than that of the 450 nm micro-LED, which is caused by the stronger piezoelectric polarization effect in InGaN devices with higher In content.

For the coefficient C, assuming an injection efficiency of 100%, Fig. 4.29 shows the Auger recombination current I_c values for two wavelengths of micro-LEDs. Apart from the lower carrier density, the data can be fitted into a straight line, indicating that the Auger recombination coefficient does not change significantly within this range. The C value obtained from this process is $C_{450} = (1 \pm 0.3) \times 10^{-29}\text{cm}^6 \cdot \text{s}^{-1}$ and $C_{520} = (3 \pm 1) \times 10^{-30}\text{cm}^6 \cdot \text{s}^{-1}$, and there is no significant change for micro-LEDs of different diameters. Through the above work, SRH recombination and Auger recombination can be considered as nonradiative recombination mechanisms to understand carrier lifetime, and there is no need to explain with any carrier leakage mechanism or changes in injection efficiency.

In addition, for GaN-based LEDs, as the thickness of the buffer layer increases, the dislocation density in the epitaxial material decreases, and the wavelength blue-shifts, leading to higher efficiency for LEDs with thicker GaN buffer layers [40]. However, by analyzing the rate equation, it is found that defect-related nonradiative recombination helps to increase the modulation bandwidth; that is, LEDs with thinner GaN buffer layers have higher modulation bandwidth. By comparing the modulation bandwidth and current characteristics of LEDs with two buffer layer thicknesses, as shown in Fig. 4.30, it is evident that increasing the buffer layer thickness leads to a decrease in bandwidth across different currents.

Using the ABC model, it is known that the SRH recombination related to defects in LEDs with thinner buffer layers increases, leading to a higher modulation bandwidth, providing a method to increase the modulation bandwidth of visible light communication. However, this will inevitably lead to a decrease in LED efficiency. Therefore, in practical applications such as solid-state lighting and visible light communication, the trade-off between these two characteristics should be considered.

Studying the capacitive effect of micro-LEDs is of great significance, as it helps further analyze the switching characteristics, modulation characteristics, and other

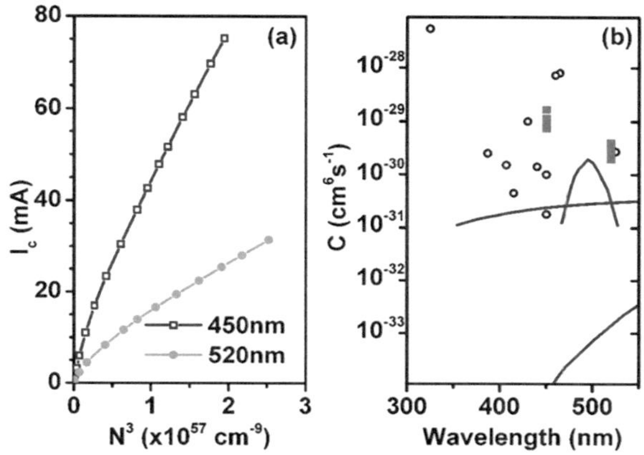

Fig. 4.29 Auger recombination related curves [39]; **a** The Auger recombination current values of the two devices. The Auger coefficient C is given by the slope; **b** The values of the Auger coefficient C found in this work and reported in previous literature. The red squares represent the values of C obtained in this work, the black hollow circles represent the experimental measurements of C in the literature, and the blue curve represents the theoretical values of C in the literature

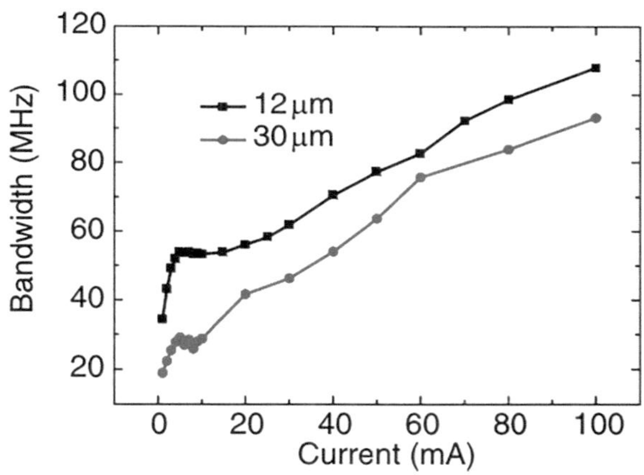

Fig. 4.30 Relationship between modulation bandwidth and current characteristics of LEDs with 12 and 30 μm GaN buffer layer thickness [40]

performance of micro-LED devices [24]. A study on the size-dependent capacitance characteristics of InGaN-based micro-LED arrays under reverse and forward biases was conducted. For these devices, it can be observed that the capacitance values under reverse bias and low forward bias are linearly proportional to the size of the micro-LED. At the same time, a negative capacitance effect was measured under higher forward bias. The results show a small deviation in the linear relationship between negative capacitance and LED size, which can be attributed to variations in injection current densities and heat dissipation capabilities across different micro-LED arrays. This result can be explained by assuming that the negative capacitance effect is governed by carrier density and junction temperature rather than directly by the applied bias voltage, consistent with theoretical analysis. Since changes in size in micro-LED devices can affect their performance in many ways, such as modulation bandwidth, the impact of size on micro-LED capacitance should be studied in detail.

Currently, many international studies have been conducted on the modulation bandwidth effect of micro-LEDs, mainly including the use of high modulation bandwidth micro-LEDs to achieve long-distance high-speed visible light communication, such as achieving high modulation bandwidth through integrated micro-LED arrays [41], integrating micro-LEDs with high electron mobility transistors (HEMTs) or quantum dots [22,42].

This section systematically analyzes the optoelectronic characteristics of micro-LEDs from four aspects—size effect, temperature effect, aging characteristics, and optoelectronic modulation bandwidth—through a combination of experimental and theoretical analyses. It explains the underlying mechanisms of these effects and proposes strategies to improve micro-LED efficiency and mitigate efficiency droop. In practical applications such as displays, lighting, and visible light communication, adjusting parameters like micro-LED size and ambient temperature can efficiently meet the requirements of different scenarios.

4.4 Conclusion

This chapter mainly introduces the applications of micro-LEDs in displays, lighting, and visible light communication, and focuses on the processes of preparing micro-LED devices on different substrates, including substrate stripping, quantum dot-micro-LED integration, and other key technologies for fabricating full-color micro-LED displays. Finally, the size effect, temperature effect, aging characteristics, modulation bandwidth characteristics, and their intrinsic mechanisms in micro-LEDs are analyzed. The integration of micro-LED arrays and efficiency enhancement remains a significant challenge currently. In the future, with the rapid advancement of key technologies, large-area, low-cost, and high-efficiency micro-LED devices are expected to be more effectively applied in display technologies and visible light communication fields.

References

1. Jin SX, Li JY, Lin JZ et al (2000) GaN microdisk light emitting diodes. Appl Phys Lett 76(5):631–633
2. Gong Z, Gu E, Jin SR et al (2008) Efficient flip-chip InGaN micro-pixellated light-emitting diode arrays: promising candidates for micro-displays and colour conversion. J Phys D Appl Phys 41(9):094002
3. Lee HE, Shin JH, Park JH et al (2019) Micro light-emitting diodes for display and flexible biomedical applications. Adv Func Mater 29(24):1808075

4. Lee HE, Choi JH, Lee SH et al (2018) Monolithic flexible vertical GaN light-emitting diodes for a transparent wireless brain optical stimulator. Adv Mater 30(28):1800649
5. Lin CC, Fang YH, Kao MJ et al (2018) 59–2: Invited paper: ultra-fine pitch thin-film micro-LED display for indoor applications. SID Symp Dig Tech Pap 49(1):782–785
6. Yoon JK, Park EM, Son JS et al (2013) The study of picture quality of OLED TV with WRGB OLEDs structure. SID Symp Dig Tech Pap 44(1):326–329
7. Pengfei T (2021) Micro-LED display technology. Shanghai Jiao Tong University Press, Shanghai
8. Wang Z, Zhu S, Shan X et al (2021) Full-color micro-LED display based on a single chip with two types of InGaN/GaN MQWs. Opt Lett 46(17):4358–4361
9. McKendry JJD, Green RP, Kelly AE et al (2010) High-speed visible light communications using individual pixels in a micro light-emitting diode array. IEEE Photonics Technol Lett 22(18):1346–1348
10. Tian P, Liu X, Yi S et al (2017) High-speed underwater optical wireless communication using a blue GaN-based micro-LED. Opt Express 25(2):1193–1201
11. Liu X, Lin R, Chen H et al (2019) High-bandwidth InGaN self-powered detector arrays toward MIMO visible light communication based on micro-LED arrays. ACS Photonics 6(12):3186–3195
12. Lin R, Liu X, Zhou G et al (2021) InGaN micro-LED array enabled advanced underwater wireless optical communication and underwater charging. Adv Opt Mater 9:2002211
13. Zhou X, Tian P, Chen X et al (2020) Growth, transfer printing and colour conversion techniques towards full-colour micro-LED display. Prog Quantum Electron 71:100263
14. Chan CKT, Bibl A (2015) Adhesive wafer bonding with controlled thickness variation. U.S. Patent 9087764
15. Tian P, McKendry JJD, Gu E et al (2016) Fabrication, characterization and applications of flexible vertical InGaN micro-light emitting diode arrays. Opt Express 24(1):699–707
16. Tian P, McKendry JJD, Gong Z et al (2014) Characteristics and applications of micro-pixelated GaN-based light emitting diodes on Si substrates. J Appl Phys 115(3):033112
17. Lee SY, Park KI, Huh C et al (2012) Water-resistant flexible GaN LED on a liquid crystal polymer substrate for implantable biomedical applications. Nano Energy 1(1):145–151
18. Quan LN, García de Arquer FP, Sabatini RP et al (2018) Perovskites for light emission. Adv Mater 30(45):1801996
19. Era M, Morimoto S, Tsutsui T et al (1994) Organic-inorganic heterostructure electroluminescent device using a layered perovskite semiconductor $(C_6H_5C_2H_4NH_3)_2PbI_4$. Appl Phys Lett 65(6):676–678
20. Wang N, Cheng L, Ge R et al (2016) Perovskite light-emitting diodes based on solution-processed self-organized multiple quantum wells. Nat Photonics 10(11):699–704
21. Han HV, Lin HY, Lin CC et al (2015) Resonant-enhanced full-color emission of quantum-dot-based micro-LED display technology. Opt Express 23(25):32504–32515
22. Mei S, Liu X, Zhang W et al (2018) High-bandwidth white-light system combining a micro-LED with perovskite quantum dots for visible light communication. ACS Appl Mater Interfaces 10(6):5641–5648
23. Tian P, McKendry JJD, Gong Z et al (2012) Size-dependent efficiency and efficiency droop of blue InGaN micro-light emitting diodes. Appl Phys Lett 101(23):231110
24. Yang W, Zhang S, McKendry JJD et al (2014) Size-dependent capacitance study on InGaN-based micro-light-emitting diodes. J Appl Phys 116(4):044512
25. Ryu HY, Shim JI (2011) Effect of current spreading on the efficiency droop of InGaN light-emitting diodes. Opt Express 19(4):2886–2894
26. Meyaard DS, Shan Q, Cho J et al (2012) Temperature dependent efficiency droop in GaInN light-emitting diodes with different current densities. Appl Phys Lett 100(8):081106
27. Wierer JJ Jr, Tansu N (2019) III-nitride micro-LEDs for efficient emissive displays. Laser Photonics Rev 13(9):1900141
28. Smith JM, Ley R, Wong MS et al (2020) Comparison of size-dependent characteristics of blue and green InGaN micro-LEDs down to 1 μm in diameter. Appl Phys Lett 116(7):071102

29. Hwang D, Mughal A, Pynn CD et al (2017) Sustained high external quantum efficiency in ultrasmall blue III-nitride micro-LEDs. Appl Phys Express 10(3):032101
30. Olivier F, Daami A, Licitra C et al (2017) Shockley-Read-Hall and Auger non-radiative recombination in GaN based LEDs: a size effect study. Appl Phys Lett 111(2):022104
31. Wong MS, Hwang D, Alhassan AI et al (2018) High efficiency of III-nitride micro-light-emitting diodes by sidewall passivation using atomic layer deposition. Opt Express 26(16):21324–21331
32. Huang SC, Li H, Zhang ZH et al (2017) Superior characteristics of microscale light emitting diodes through tightly lateral oxide-confined scheme. Appl Phys Lett 110(2):021108
33. Tian P, McKendry JJD, Herrnsdorf J et al (2014) Temperature-dependent efficiency droop of blue InGaN micro-light emitting diodes. Appl Phys Lett 105(17):171107
34. Lu S, Liu W, Zhang ZH et al (2014) Low thermal-mass LEDs: size effect and limits. Opt Express 22(26):32200–32207
35. Tian P, Althumali A, Gu E et al (2016) Aging characteristics of blue InGaN micro-light emitting diodes at an extremely high current density of 3.5 kA cm^{-2}. Semicond Sci Technol 31(4):045005
36. Guo X, Schubert EF (2001) Current crowding and optical saturation effects in GaInN/GaN light-emitting diodes grown on insulating substrates. Appl Phys Lett 78(21):3337–3339
37. Meneghini M, Trevisanello LR, Meneghesso G et al (2008) A review on the reliability of GaN-based LEDs. IEEE Trans Device Mater Reliab 8(2):323–331
38. Tian P, Wu Z, Liu X et al (2018) Large-signal modulation characteristics of a GaN-based micro-LED for Gbps visible-light communication. Appl Phys Express 11(4):044101
39. Green RP, McKendry JJD, Massoubre D et al (2013) Modulation bandwidth studies of recombination processes in blue and green InGaN quantum well micro-light-emitting diodes. Appl Phys Lett 102(9):091103
40. Tian P, Edwards PR, Wallace MJ et al (2017) Characteristics of GaN-based light emitting diodes with different thicknesses of buffer layer grown by HVPE and MOCVD. J Phys D Appl Phys 50(7):075101
41. Lan H, Tseng I, Lin Y et al (2020) High-speed integrated micro-LED array for visible light communication. Opt Lett 45(8):2203–2206
42. Cai Y, Haggar JIH, Zhu C et al (2021) Direct epitaxial approach to achieve a monolithic on-chip integration of a HEMT and a single micro-LED with a high-modulation bandwidth. ACS Appl Electron Mater 3(1):445–450

Chapter 5
Key Factors Limiting the External Quantum Efficiency of GaN-Based Micro-LEDs and Remedies

Hang Sheng, Chunshuang Chu, Kangkai Tian, Yonghui Zhang, and Zihui Zhang

5.1 Research Background of Micro-LED

Nowadays, the rapid development of smartphones, tablets, and high-resolution displays has undoubtedly led to an increasing demand for high-performance displays in the entire industry. In the past few decades, the mainstream display systems have been based on liquid crystal display (LCD) and organic light-emitting diode (OLED) technologies, which have almost occupied the entire smart display field. The former has advantages such as long life, low cost, portability, and high brightness. The latter has characteristics of self-luminescence and high contrast [1, 2]. Therefore, display technologies based on LCD and OLED are likely to continue to serve the display field in the next few years [3–9]. However, there are still some imperfections in these two display technologies at the current stage. For example, LCDs have poor color conversion efficiency and low color saturation contrast [10]; OLEDs have problems with brightness and lifetime [11–13]. Display technology based on micro-LED (abbreviated as μLED) has excellent brightness, lifetime, resolution, and efficiency, so it is gradually entering the vision of the community. The first μLED chip was manufactured and reported by a research team from the University of Texas in 2000 [14, 15]. Since then, due to its unique performance characteristics and broad potential application prospects, it has been a research hotspot in recent years. Figure 5.1 shows the major potential application fields of μLED in recent years, which are mainly concentrated in smart watches, portable displays, laptops, micro projection displays, augmented reality (AR) and virtual reality (VR), and other emerging display technologies [16–24]. In recent years, due to the demand of the high-level consumer market, the size and spacing of individual pixels in the μLED display system are

H. Sheng · C. Chu · K. Tian · Y. Zhang · Z. Zhang (✉)
School of Electronic Information Engineering, Hebei University of Technology, Tianjin, China
e-mail: zh.zhang@gdut.edu.cn

S. Xu (ed.), *Optical Characterization of Microstructures and Optoelectronic Devices Based on Wide Band Gap Semiconductors*, Wide Bandgap Semiconductors,
https://doi.org/10.1007/978-981-95-1928-6_5

getting smaller and smaller, i.e., from the earliest 73 PPI (pixel density) in 2007 to the 300,000 PPI announced by VueRea in Canada in 2019. The pixel density of μLEDs used in smart displays will be larger and larger in the future. In addition to having a very good application prospect in the display field, the large frequency bandwidth and fast frequency response characteristics of μLED also make μLED widely used in visible light communication (VLC) [25–30]. At the current stage, although the market prospects of μLED are broad, there are still many bottlenecks to overcome, including high process equipment requirements, full-color display and mass transfer, and other key technical problems, as well as extremely low external quantum efficiency (EQE) under reduced device size.

1. Full-Color Display Technology Based on Micro-LED

The full-colorization of micro-LED is currently a hot research direction. In today's market trend of pursuing colorization and high resolution, high contrast ratio, various research institutes and organizations have proposed a variety of effective solutions. The methods for micro-LED to achieve full-color can be mainly summarized into the following two categories: RGB three-color LED scheme and UV/blue LED excitation color conversion material. The former is based on the color mixing principle of three primary colors (red, green, and blue). As is well known, the three primary RGB colors can achieve most of the color gamut in nature through a certain ratio. Therefore, gallium arsenide/indium phosphide (GaAs/InP)-based red light LED and

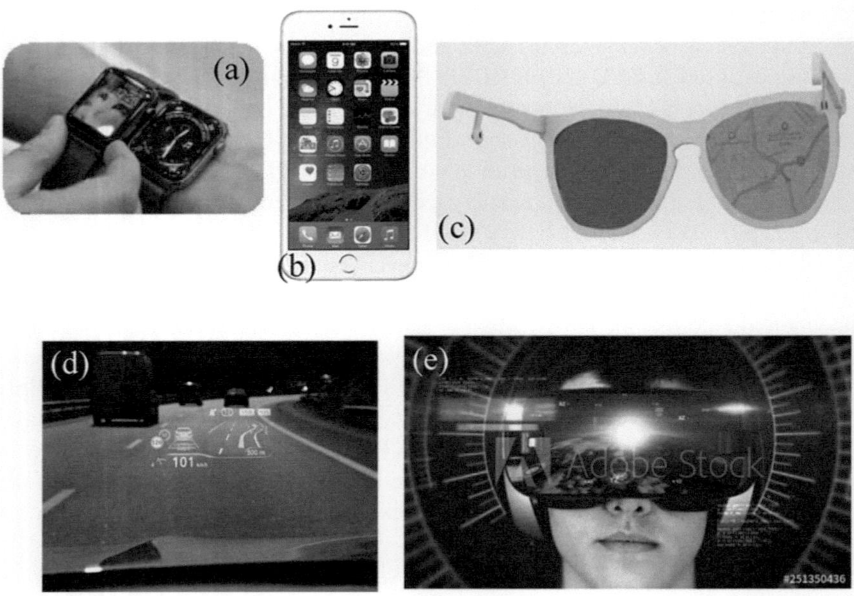

Fig. 5.1 Micro-LED display applications in **a** smart wearables; **b** smart cellphones; **c** smart glasses; **d** vehicle dashboard; **e** AR, VR

GaN-based blue-green light LED can be integrated [31–34], and then the brightness of each LED can be controlled by adjusting the applied bias and the size of the injected current to achieve color matching, achieving a full-color effect [33]. The advantage of this full-color display technology based on the RGB scheme is that it has a wide display color gamut and fast frequency response. However, it is very difficult to accurately transfer millions of μLEDs to complex drive circuit substrates [16]. At the same time, because the actual output current of the drive chip will have an error with the theoretical current, and each LED in a single pixel has a certain half-wave width and light decay phenomenon, which affects the color rendering of the LED. Another widely adopted full-color technology in the industry is to use phosphors or quantum dots and other color conversion materials to achieve full-color displays [35]. GaN-based blue LEDs can pump color conversion materials to produce green and red light, while GaN-based UV LEDs can pump color conversion materials to produce blue, green, and red light [36–43]. Because quantum dot materials can excite light of a specific wavelength by controlling their size, the emitted color can cover the entire visible light spectrum range from blue to red, and they have high color conversion efficiency, narrow half-peak width, and wide absorption spectrum, so they have very high color purity and saturation. Secondly, the quantum dot structure is simple, thin, and can be curled, so it is widely used in micro-LED full-color display and even flexible display [47–49]. In recent years, cadmium selenide (CdSe) and InP-based quantum dot materials have shown excellent performance in making full-color displays and have received widespread attention [44–46]. The commonly used method is to spin-coat uniform and size-controllable quantum dots using spin coating and mist spraying technology, and then further excite the quantum dots with blue light or ultraviolet light to achieve color conversion.

2. **Massive Transfer Technology of Micro-LED Display**

Another technical challenge of μLED display is the massive transfer technology. Massive transfer, also known as thin-film transfer, is the process of transferring μLED devices to a thin-film transistor (TFT) backplane with a specific current drive and assembling them into a two-dimensional periodic array. Since there are many pixels to be transferred, in general, a 500 PPI mobile phone screen requires 8 million pixel particles; the allowable error for transfer is extremely small, so the accuracy needs to be strictly controlled. Secondly, to suppress the crosstalk among individual pixels and achieve high-quality imaging effects, the substrate needs to be further thinned or even peeled off, which again increases the difficulty for the massive transfer scheme [55, 56]. The display market based on μLED urgently needs a low-cost and efficient transfer technology [50]. Today's transfer technologies are mainly divided into two categories, namely physical transfer and chemical transfer, among which representative technologies include electrostatic adsorption transfer, micro transfer printing technology, and wafer-level monolithic hybrid integration [31, 50, 52–54]. However, so far, no technology has been widely accepted by the industry in all indicators, and a detailed description of the existing massive transfer technology can be found in the literature [51].

5.2 Size Effect of Micro-LED

In 2000, Jiang and others first proposed the micro-LED structure and conducted the first study on the impact of LED size on device performance [15]. In addition, over the past few decades, systematic research has been conducted on the size effect of μLEDs. It was found that as the size of the μLED chip decreases, the performance of the device also changes accordingly, which is first reflected in the optical and thermal performance of the device. Firstly, the reduction in μLED size can increase the light output from the sidewalls to a certain extent, thereby improving the light extraction efficiency of the device [57]. However, this sidewall light output can also cause optical crosstalk between pixels [42]. Therefore, measures need to be taken to reduce the lateral propagation of light and improve the light extraction on the side of the flip-chip substrate, which can be achieved by adopting the RCLED structure to control the direction of photon transmission and reduce optical crosstalk [58, 59]. At the same time, the reduction in μLED chip size also causes the device's heat dissipation to increase, so the junction temperature of small-sized devices is significantly lower than that of large-sized devices under the same current density [60–62]. Experimental data show that the thermal resistance of the device decreases linearly with the decrease in LED size [63]. Figures 5.2a and b are the curves of light power versus current density under experimental testing and simulation calculation of indium gallium nitride (InGaN)/GaN LED devices of different sizes. According to Fig. 5.2a, when the LED size is $25 \times 25\ \mu m^2$, the output power significantly decreases at 5000 A/cm^2, and when the LED size is $200 \times 200\ \mu m^2$, the light output power decreases at a current density of 2000 A/cm^2. The simulation calculation results of Fig. 5.2b also explain this issue, and the agreement with the experimental test structure is obtained.

According to previous studies [64–66], the performance of large-sized LED devices will be affected by the current crowding effect, especially when the injection current density is large, such that the current tends to accumulate locally. However, when the chip size decreases, the lateral resistance gradually decreases, so μLEDs

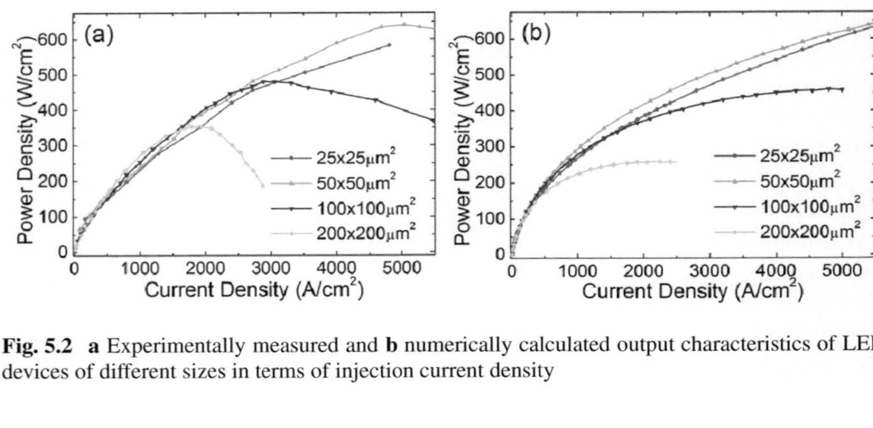

Fig. 5.2 **a** Experimentally measured and **b** numerically calculated output characteristics of LED devices of different sizes in terms of injection current density

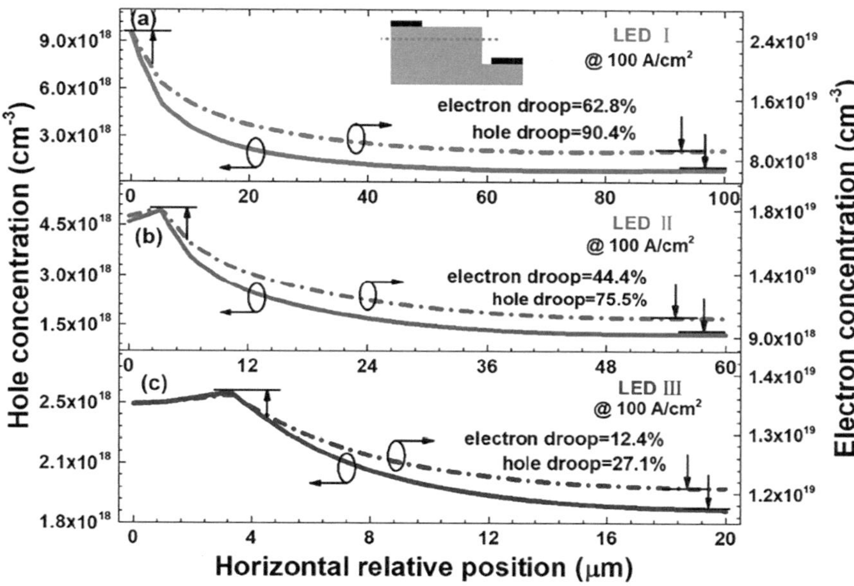

Fig. 5.3 Lateral carrier distribution in the active area of devices of different sizes [69]

have a better current spreading effect [67, 68]. Figure 5.3 shows the lateral carrier distribution in the first quantum well near the p-region side in devices of different sizes [69]. For LEDs I, II, and III, the decreased levels in electrons at the mesa edge are 62.8%, 44.4%, and 12.4%, respectively, and the holes are 90.4%, 75.5%, and 27.1% respectively. This further indicates that as the device size decreases, the lateral resistance decreases, so the lateral spreading effect of the current is greatly improved.

5.3 The Fundamental Reason for the Lower External Quantum Efficiency of Micro-LEDs

The peak EQE of GaN-based μLED devices reported so far is generally smaller than 15% [68, 69, 71, 73, 74]. The μLED reported by Hwang and others is currently the best-performing device, with a chip size of 20 × 20 μm^2, and its peak EQE can reach 40.2% [9]. The driving current of the actual μLED devices used for display is very small, but the efficiency of the device at such a small injection current is very low, and it is not high enough to produce a sufficiently high luminous intensity [72]. Studies have shown that at such a low operating current level, the phenomenon of electron leakage is not obviously observed [75]. Secondly, the carrier concentration in the

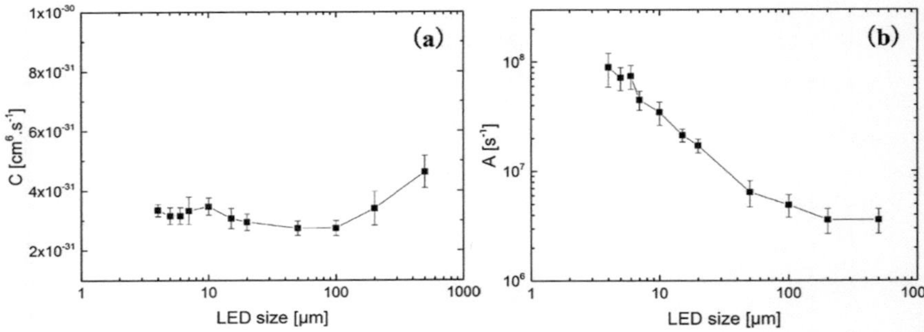

Fig. 5.4 **a** Auger recombination coefficient; **b** Defect-related nonradiative recombination coefficient in terms of different device sizes

quantum well at a small injection current level is also very low, so the Auger recombination rate inside the device is not obvious. The calculation results from Fig. 5.4a show that the Auger recombination rate in the LED does not strongly change with different chip sizes. However, Fig. 5.4b shows that as the size of the LED decreases, the Shockley–Read–Hall (SRH) nonradiative recombination rate will significantly increase [73]. Therefore, for small-sized μLED devices, at such low driving current levels, the internal quantum efficiency (IQE) of small-sized devices mainly depends on defect-induced nonradiative recombination. Therefore, suppressing defects at the sidewall surface is one of the main tasks for improving the performance of μLED devices.

Previous studies have shown that μLED has a good current spreading effect, which means that carriers can easily spread to the edge of the device mesa and combine with defects at the edge to produce nonradiative recombination. For large-sized LED devices, although there is also nonradiative recombination caused by defects at the sidewall edge or surface, the area of edge damage accounts for a small proportion of the entire device, so the impact on device performance is not significant, but for μLED, this impact on performance is hard to ignore [9, 69–71]. In addition, the mass transfer process can also cause chip damage and defects, which will undoubtedly further cause device performance degradation.

As is well known, in the full-color display of micro-LED based on RGB technology, blue and green light can be achieved by GaN-based μLED, and red light can generally be obtained by gallium phosphide (GaP)/GaAs-based μLED. However, these two types of μLED chips have a large difference in thickness and driving voltage, which brings certain difficulties to the design of mass transfer and driving circuit. In addition, GaAs-based μLED is very sensitive to defects, which makes the EQE of the current red-light GaAs-based μLED very low [76, 112]. Therefore, to avoid these issues, the most direct solution is to adopt high In-component GaN-based red light μLED [76]. However, for InGaN/GaN LEDs grown along the [0001] direction, there is a large lattice mismatch at the quantum well/quantum barrier (QW/QB) interface, and a strong polarization field is caused. This will cause separation between the electron and hole wave functions, which is called the quantum-confined Stark effect (QCSE). QCSE will make the radiative recombination efficiency of the device decrease, especially for high In-component InGaN/GaN quantum wells [77]. Therefore, for green and red light GaN-based μLEDs, some technical methods need to be adopted to alleviate the QCSE effect caused by lattice mismatch, thereby improving the EQE of the device.

In addition, although the reduction in device size can improve the light extraction efficiency (LEE) [57], the light extraction efficiency at the device sidewalls will also increase, which will cause optical crosstalk between adjacent pixels [42]. Therefore, it is also necessary to improve the light extraction efficiency of the light-emitting surface by adjusting the directionality of the light emission, thereby improving the EQE of the device.

5.4 Methods and Measures to Improve the EQE of Micro-LED

According to the previous description, sidewall defects have a significant impact on the performance of μLED devices. There have been many research reports on the methods for suppressing nonradiative recombination at surface defects. Most of the technical solutions mainly start from the following aspects: (1) eliminating sidewall defects by conducting thermal annealing process, (2) using plasma-enhanced chemical vapor deposition (PECVD) or atomic layer deposition (ALD) to grow passivation layers to suppress surface defects, (3) removing surface defects through chemical treatment, and (4) adjusting lateral current spreading, and limiting the injected current to the central area of the device far from the edge defect area of the platform.

1. **Thermal Annealing**

Firstly, Tian and others attempted to reduce the density of sidewall defects caused by mesa etching by optimizing the thermal annealing time [74]. As shown in Fig. 5.5, for a $6 \times 6\ \mu m^2$ LED device, when the annealing time is increased from 2 to 3 min, the peak EQE of the device can be increased from 0.05 to 0.10. At the same time, after increasing the annealing time, the current density for the peak EQE is reduced from about 100 A/cm^2 (see Fig. 5.5a) to 40 A/cm^2 (see Fig. 5.5b), which further indicates the reduction of the nonradiative recombination rate caused by defects. In addition, the efficiency droop of the $6 \times 6\ \mu m^2$ LED device in Fig. 5.5b is quite obvious, indicating that the increase in carrier concentration in multiple quantum wells (MQWs) will cause Auger recombination, leading to significant efficiency droop. According to our previous research [57], sidewall defects consume a large number of carriers, resulting in a very low carrier concentration injected into MQWs, and such a low carrier concentration in MQWs does not cause significant Auger recombination. Therefore, for small-sized μLEDs with a high surface defect density, the efficiency droop caused by Auger recombination is not very significant, and the main reason affecting device performance is the nonradiative recombination caused by defects. Further analysis of Fig. 5.5a and b shows that for large-sized

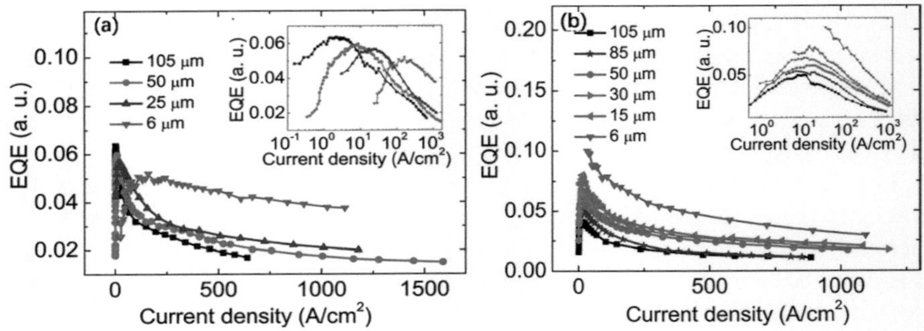

Fig. 5.5 External quantum efficiency of μLEDs of different sizes under different annealing times of **a** 2 min; **b** 3 min [74]

LED devices, changing the thermal annealing time has a smaller impact on EQE, which also shows that sidewall defects have a smaller impact on large-sized devices. Therefore, for small-sized μLEDs, properly optimizing the thermal annealing process can effectively suppress the density of sidewall defects and further reduce nonradiative SRH recombination [78].

2. **Sidewall Passivation and Chemical Treatment**

Although the thermal annealing process can reduce the defect density of the sidewalls to a certain level, it cannot completely suppress the plasma-caused damage on the sidewalls. Therefore, additional device treatments are needed to further reduce the defect density of the sidewalls. Traditional large-sized LEDs and mini-LEDs usually use potassium hydroxide (KOH) or ammonium sulfide to chemically treat the mesa sidewalls [78, 88–90], and this wet treatment can form a relatively smooth sidewall surface [91]. However, for traditional large-sized LEDs, the defective area accounts for a small proportion of the total volume, and thus, this chemical treatment is not significant enough to improve device performance. However, for small-sized μLEDs, appropriate chemical treatment can help improve the EQE of the device [92]. Figure 5.6 shows the electroluminescence (EL) diagram of six different sizes of LED devices that are biased to the current density of 1 A/cm^2 [92]. The sidewalls of the μLED devices on the left have been treated with chemical wet etching, and an ALD-based passivation layer is also deposited on the sidewalls. The reference devices on the right have not been provided with any treatment. Comparative studies have shown that the luminescence intensity of small-sized LEDs that have not experienced chemical treatment is very dim, e.g., for the 10 × 10 μm^2 μLED devices, it is even impossible to collect luminescence images. In addition, as the device size increases, for example, the 100 × 100 μm^2 LED device, the chemical treatment has a smaller impact on the device. The electroluminescence test results on the right side of Fig. 5.6 can also show that the sidewall surface defects can act as a channel for current leakage. The strong leakage current in the edge area causes the luminescence intensity in this area to be much stronger than that in the center of the device. The device on the left that uses KOH chemical treatment on the sidewalls has relatively

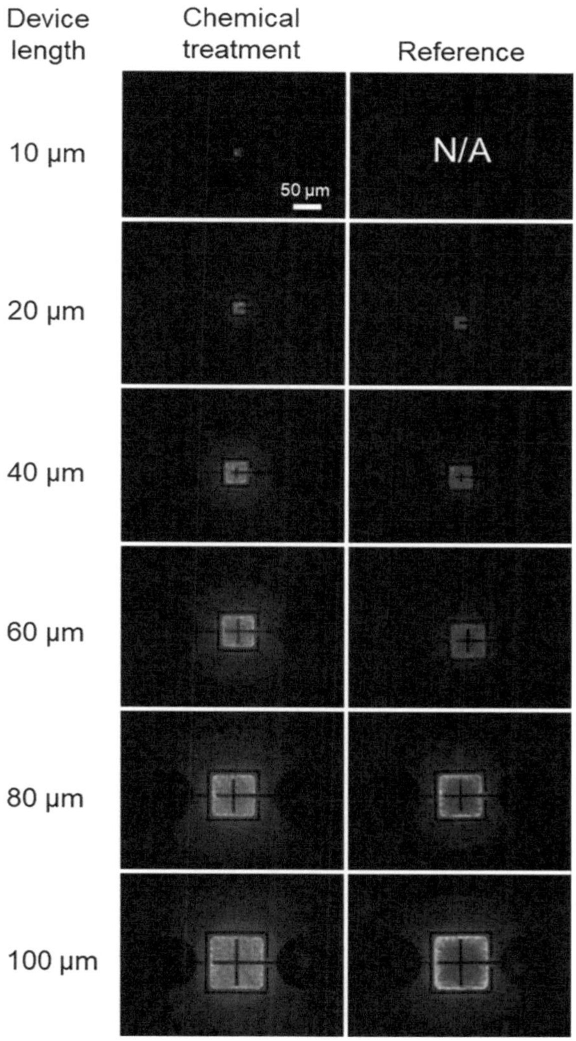

Fig. 5.6 Comparison of electroluminescence (EL) of LED devices with different sizes before/after sidewall defect treatment [92]

uniform luminescence intensity across the entire platform, which also indicates that edge defects are suppressed and sidewall leakage is reduced.

Figure 5.7a and b are the EQE in terms of the injection current for LED devices with different chip sizes in Fig. 5.6. For $20 \times 20\ \mu m^2$ and $10 \times 10\ \mu m^2$ small-sized LED devices, after chemical treatment in KOH solution, the EQE has been significantly improved, while for large-sized LED devices, there is almost no change in EQE before and after KOH solution chemical treatment. In addition, it can be found from Fig. 5.7a that at the current density of 100 A/cm^2, for LED devices with sizes of $10 \times 10\ \mu m^2$, $20 \times 20\ \mu m^2$, $40 \times 40\ \mu m^2$, $60 \times 60\ \mu m^2$, $80 \times 80\ \mu m^2$ and 100×100

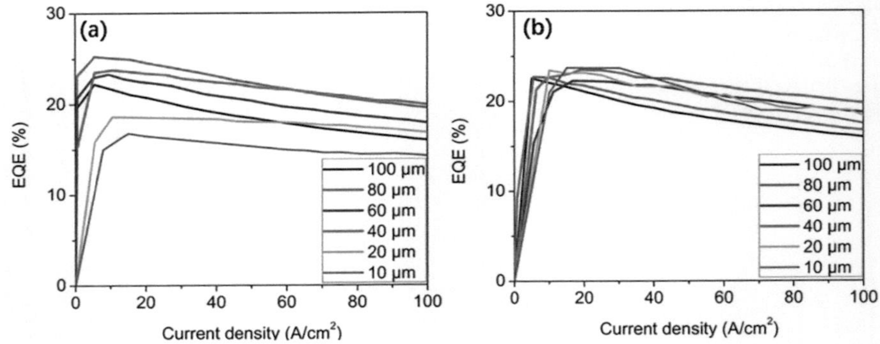

Fig. 5.7 EQE in terms of injection current for LEDs with different sizes **a** before the sidewall defect management; **b** after the sidewall defect management

µm^2, the degrees of efficiency droop are 17.65%, 10.81%, 13.04%, 21.93%, 22% and 31.82%, respectively, while for the devices in Fig. 5.7b that have undergone KOH solution chemical treatment, the degrees of efficiency droop are 29.17%, 23.4%, 14.89%, 17.78%, 30.43%, and 32.61%, respectively. Taking the small-sized LED device of 10 × 10 µm^2 as an example, after KOH solution chemical treatment, the degree of efficiency droop significantly increases, which is due to the reduction in surface defect density, the increase in carrier injection efficiency, and the increase in carrier concentration in the MQW. These factors, in general, lead to an increase in the internal Auger recombination rate of the device, causing a significant efficiency droop [69]. However, for the large-sized LED device of 100 × 100 µm^2, there is not much difference in the degree of efficiency droop before and after chemical treatment, which further indicates that surface defects have little impact on the performance of large-sized LED devices.

Of course, merely using the method of chemically treating the sidewalls of the device is not enough to eliminate surface defects. Usually, it should be combined with sidewall passivation technology to further remove defects and damages [79]. In addition to reducing sidewall defects, sidewall passivation can also improve the light extraction efficiency of the device [80, 81]. The commonly used passivation layers are mainly deposited by using PECVD and ALD methods. The advantage of PECVD is that its deposition rate is very fast. The insulating passivation layer deposited by the PECVD system can improve the electrical and optical performance of the device to a certain level, such as reducing the leakage current and improving the light extraction efficiency [80–83]. However, the quality of the passivation layer deposited by the PECVD system is not dense enough, which makes some surface defects unable to be effectively passivated [111]. Therefore, due to the precise controllable thickness, high quality, and density of the insulating layer, it is now more inclined to use the ALD system to deposit the passivation layer [84]. Figure 5.8a and b are the EQE curves of µLED devices of 100 × 100 µm^2 and 20 × 20 µm^2 in different sidewall treatment conditions. LED-1 is a reference device that has not undergone

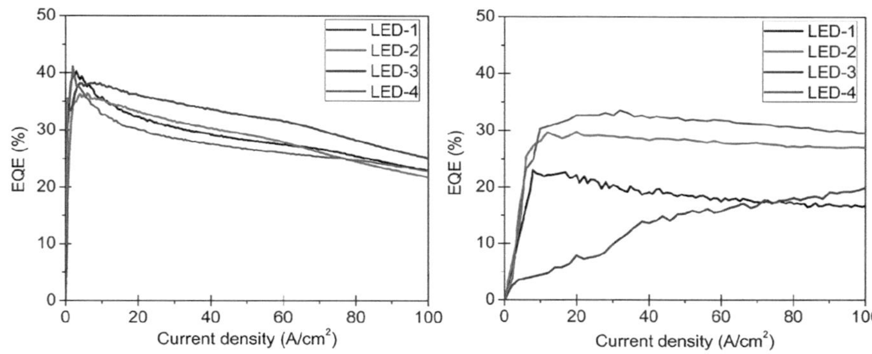

Fig. 5.8 External quantum efficiency in terms of the injection current after conducting different sidewall treatments for **a** 100 × 100 μm^2 LEDs and **b** 20 × 20 μm^2 LEDs

any sidewall treatment. On the sidewall of LED-2, a passivation layer was deposited using an ALD system. The sidewall of LED-3 has the passivation layer grown by the PECVD system, and the wet chemical treatment was subsequently conducted. The sidewall of LED-4 has the passivation layer grown by the ALD system, and the wet chemical treatment was subsequently conducted. According to Fig. 5.8a, we can conclude that whether to use sidewall treatment or not does not have a significant impact on the EQE of large-size LED devices of 100 × 100 μm^2. However, as can be seen from Fig. 5.8b, for the 20 × 20 μm^2 μLED, there is a significant difference in the EQE of the device under different sidewall treatments. At a current density of 100 A/cm^2, the EQE of LED-1 without any treatment is only 16%, while the EQE of LED-2 using ALD-based sidewall passivation technology is as high as 28%. This shows that the method based on ALD sidewall passivation significantly reduces the defect density of the μLED sidewall surface. Secondly, the EQE of LED-4 using ALD passivation technology and hydrofluoric acid (HF) chemical treatment is 31%, while the EQE of LED-3 using PECVD passivation technology and HF chemical treatment is only 20%. This also shows that the ALD passivation layer is more significant in suppressing sidewall damage.

Figure 5.9a shows the electroluminescence images of LED-1, LED-2, LED-3, and LED-4 of different chip sizes. Taking the 10 × 10 μm^2 μLED as an example, the luminous intensity of LED-2 is significantly enhanced when compared to LED-1. For LED-3, especially for small-sized devices, even though its sidewall uses a PECVD passivation layer and wet chemical treatment technology, it still cannot collect luminescent images. This is partly because the sidewall passivation layer based on PECVD weakens the transparency of the ITO layer [85, 86]; on the other hand, it also indicates that the effect of removing surface defects by the passivation layer deposited by PECVD is not ideal. In contrast, LED-4, which has experienced ALD-based passivation technology and HF chemical treatment, shows better luminescence characteristics than other devices. This also reflects the importance of using ALD-deposited passivation layers and chemical treatment for dealing with defects on

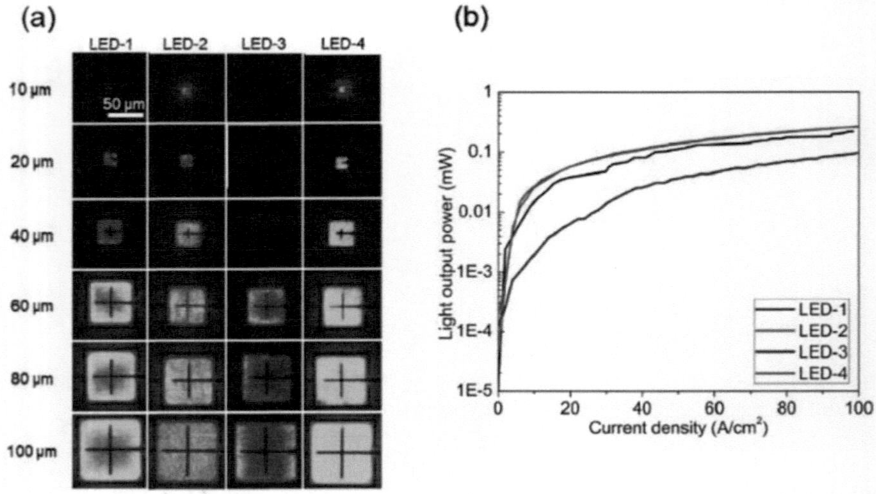

Fig. 5.9 **a** Electroluminescence (EL) images of LED devices with different chip sizes; **b** Output light power for 20 × 20 μm^2 μLEDs

the mesa sidewalls. As has been mentioned earlier, surface passivation and chemical treatment can remove defects on the sidewalls, thereby reducing the paths for leakage, allowing carriers to distribute more evenly on the mesa, which also helps to produce a more uniform light distribution. Figure 5.9b shows the output optical power of LED-1, LED-2, LED-3, and LED-4 with a size of 20 × 20 μm^2. It can be seen that LED-4, which uses an ALD-based sidewall passivation layer and HF chemical treatment, has much better output characteristics. In summary, for small-sized μLED devices, adopting ALD-deposited passivation layers on the device sidewalls and chemical treatment can alleviate the impact of surface defects on device performance to a certain extent [87].

3. **Control of Lateral Current Spreading**

In addition to growing high-quality passivation layers, another effective method to suppress surface recombination is to control the internal current spreading length, i.e., to limit the injected current in the middle of the device as much as possible and keep it away from the edge sidewalls. One method is to design a buried tunnel junction to limit the current in the middle of the device. This helps prevent the carriers from reaching the edge of the device, thereby reducing the nonradiative recombination caused by defects at the edge of the device. The reduced nonradiative recombination at the device edges thus improves the device EQE [70, 93, 94]. Figure 5.10a and b show the current–voltage (I–V) and EQE variation characteristics curves of different LEDs [95]. The standard LED is a reference device without a tunnel junction (TJ), and LED-1, LED-2, and LED-3 are LEDs with TJs. The silicon doping concentrations

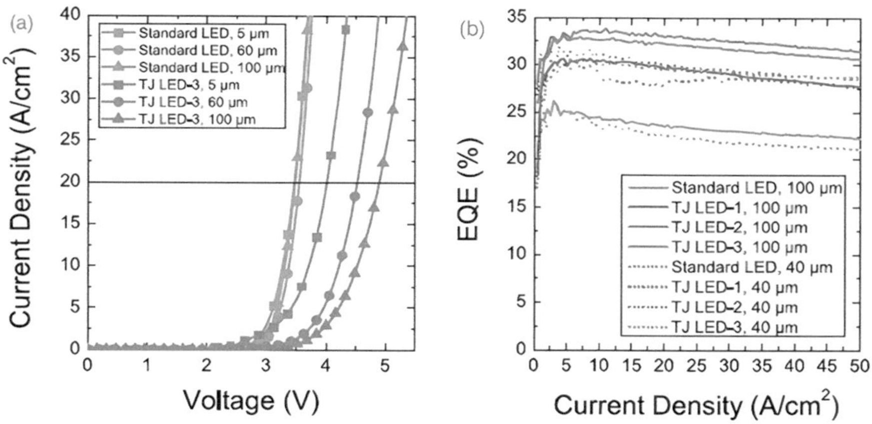

Fig. 5.10 Comparison of **a** I–V characteristics and **b** EQEs for different LED devices with/without buried tunnel junction LED devices of different sizes

in the n + −GaN layer of the TJ in LED-1, LED-2, and LED-3 are 7.0×10^{19} cm^{-3}, 1.1×10^{20} cm^{-3}, and 1.7×10^{20} cm^{-3}, respectively. According to Fig. 5.10(a), we find that LEDs with a buried tunnel junction all have a high forward bias, which is because the ionization rate of the Mg-doped p + −GaN layer is very low [113, 114]. Such low doping efficiency will inevitably reduce the tunneling probability of carriers [115, 116]. In addition, due to the increase in internal spreading resistance and p-type gallium nitride (p-GaN) resistance, although the forward voltage of the 100×100 μm^2 LED is increased compared to the 5×5 μm^2 LED, the EQE of LEDs with a buried tunnel junction has greatly improved when compared to the reference device. As can be seen from Fig. 5.10b, for a 40×40 μm^2 μLED device with a buried tunnel junction, the peak EQE can reach 34%, which is much higher than the 25% of the reference device.

Furthermore, similar to the design concept of the confinement aperture in GaN-based vertical-cavity surface-emitting lasers (VCSELs) [116], researchers have also designed confinement apertures for μLEDs to achieve the same function as buried tunnel junctions [96]. The materials for making apertures generally include silicon dioxide (SiO_2), silicon nitride (SiN_x), etc., which can achieve significant lateral current confinement, thereby greatly alleviating nonradiative recombination at the edge. However, for either buried tunnel junctions or a confinement aperture structure, the lower Mg doping efficiency of the p + −GaN layer and the secondary etching required for making the confinement apertures both increase the difficulty of the process, respectively.

Another effective method to adjust the current spreading effect can be achieved by controlling the vertical resistance of the LED [117]. For large-sized LED devices, it is possible to use positive–negative-positive (PNP) or negative–positive-negative (NPN) junctions inside the device to appropriately increase the vertical resistivity to further improve the current spreading of the device [64, 117–119]. However, for small-sized devices, it is necessary to reduce the vertical resistivity to better limit the current. A simple method to confine the current in the device center is to appropriately reduce the quantum barrier thickness [97]. Another advantage of thinning the quantum barrier is that it can enhance the hole injection efficiency [98]. As shown in Fig. 5.11, both calculated and experimental results show that using thin quantum barriers can improve the EQE of the μLED device. To better illustrate the effect of thinning the quantum barrier on reducing surface recombination, Fig. 5.12 calculates the ratio of nonradiative recombination and radiative recombination (R SRH/R Rad) caused by defects at the edge of the device platform. It can be clearly seen that as the quantum barrier thins, the ratio of R SRH/R Rad also decreases. This further illustrates that the current spreading can be effectively controlled by thinning the quantum barrier, thereby further reducing the impact of nonradiative recombination at the edge on device performance.

Another issue for GaN-based LEDs is the polarization effect of the material itself. As has been mentioned earlier, to reduce the complexity of the display process, using high In-component GaN-based red LEDs to achieve full-color display is a good choice. It is well known that there are serious piezoelectric and spontaneous

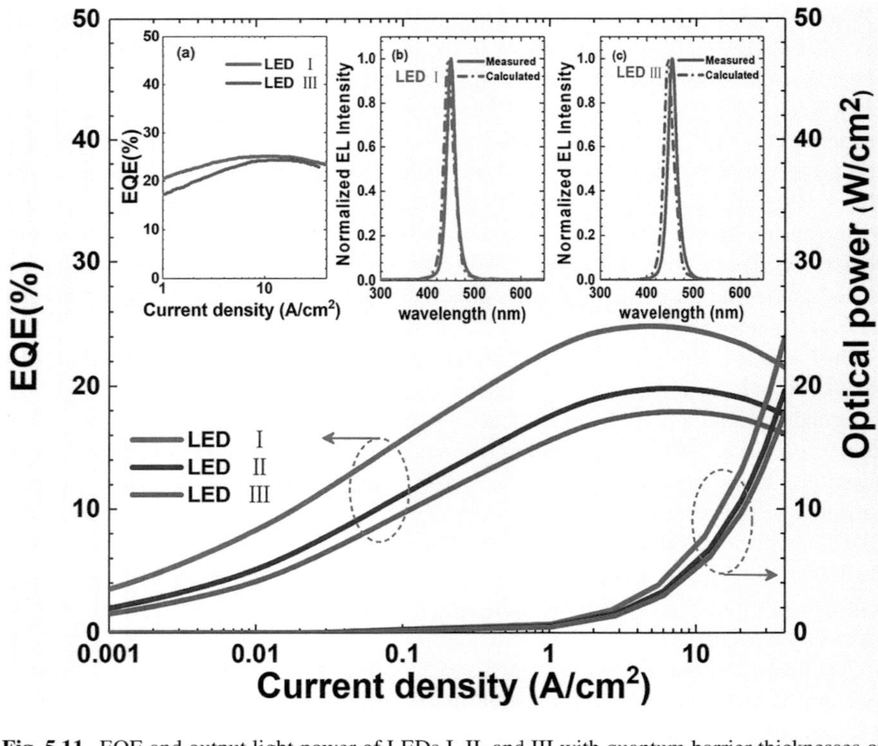

Fig. 5.11 EQE and output light power of LEDs I, II, and III with quantum barrier thicknesses of 6, 9, and 12 nm, respectively. **a** Experimentally measured EQE for LEDs I and III; **b** and **c** are the measured and simulated electroluminescence spectra of LEDs I and III, respectively [97]

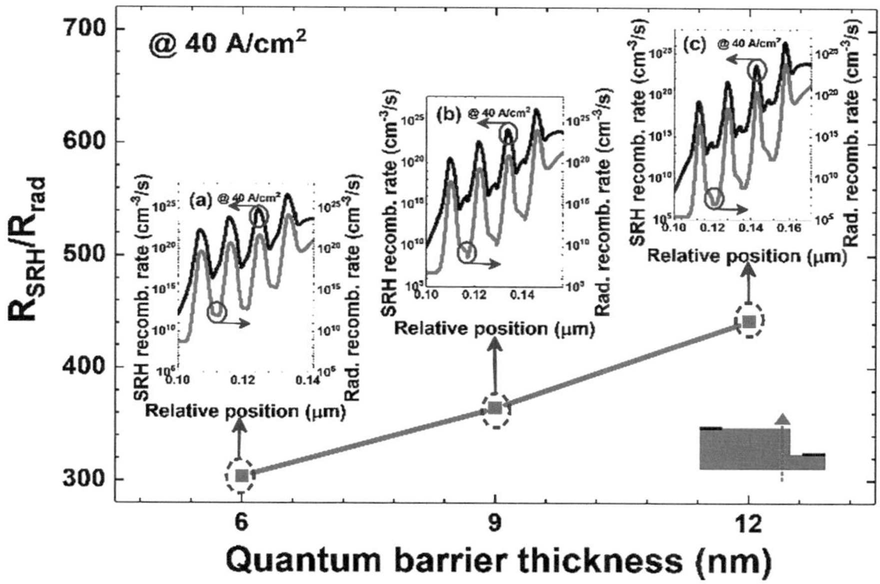

Fig. 5.12 Ratios of SRH recombination rate and radiative recombination rate for LEDs with different quantum barrier thicknesses. **a**, **b**, and **c** are radiative recombination profiles and SRH recombination for μLEDs I, II, and III, respectively [97]

polarization effects at the heterojunction interface for III-nitride semiconductors [120, 121], and this polarization effect will also cause a very strong quantum confined Stark effect (QCSE), reducing the overlap of electron and hole wave functions. This leads to the reduced internal quantum efficiency (IQE) for GaN-based μLED, especially high In-component green and red μLED [76, 122]. The most direct method to eliminate this effect is to grow InGaN/GaN MQWs on nonpolar or semipolar substrates [99, 100]. Nevertheless, these nonpolar or semipolar, high-quality GaN-based LEDs can only be grown on GaN substrates, and the nonnegligible problem with GaN substrates is the high cost [101]. Therefore, most μLEDs currently still use [0001] sapphire substrates and [111] Si substrates [123, 124]. Therefore, the polarization-induced electric field causing QCSE in the quantum wells is naturally a problem that needs to be solved. Studies have found that the stress caused by lattice mismatch at the heterojunction interface of the device can be released by changing the device size. As a result, attempts to use nano-patterned LEDs to achieve stress release have been made [102, 103]. Commonly, nano-column LEDs are used to release most of the stress in InGaN/GaN MQWs, thereby further improving device performance [104, 105]. For InGaN/GaN MQWs with a higher In-component, the IQE of the nano-column μLED device has a significant improvement [105]. Figure 5.13a shows the structure of a nano-ring LED (NRLED) manufactured by nano-sphere lithography [106]. Figure 5.13b, c, d, and e show the stress distribution curves in NRLED devices with nano-ring widths of 300 nm, 120 nm, 80 nm, and 40 nm, respectively. The

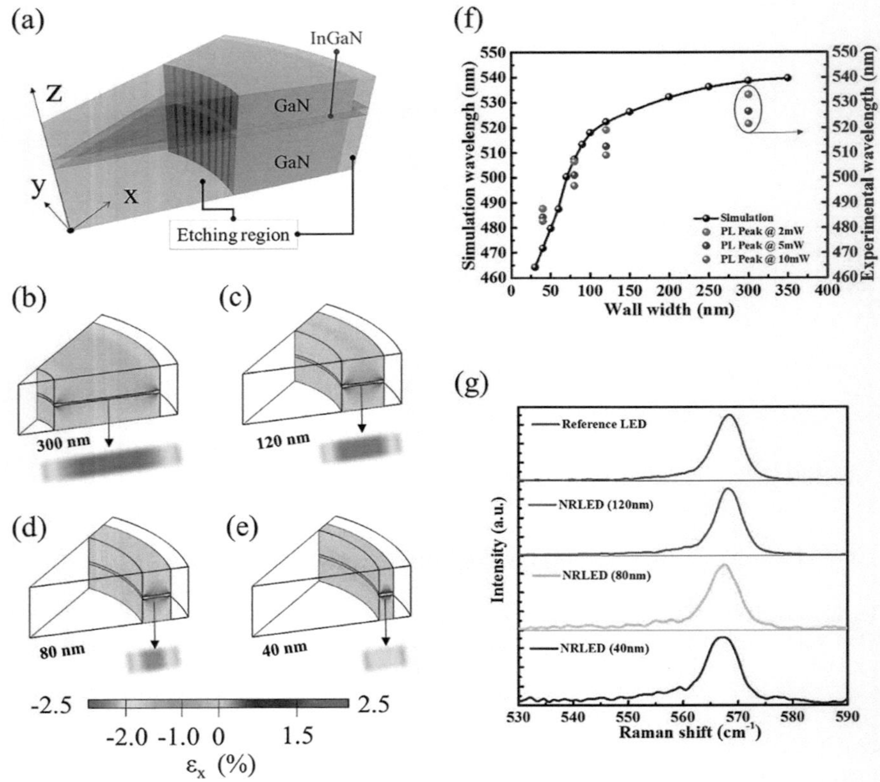

Fig. 5.13 **a** Schematic diagram of the nano-ring LED device; **b**, **c**, **d**, and **e** show the stress distribution calculation results of the nano-rings with widths of 300 nm, 120 nm, 80 nm, 40 nm, respectively; **f** shows the simulated and experimentally measured wavelength shift; and **g** shows the Raman spectroscopy test results of LED devices with different widths of nano-rings [106]

stress distribution results show that the internal stress of the device can be effectively released by reducing the width of the nano-ring. On the one hand, the maximum stress in the device is distributed in the center of the nano-ring, and as the width of the ring decreases, the stress in the central area decreases from $-$ 2.3% to $-$1.7%. On the other hand, as shown in Fig. 5.13f, the reduction in the width of the ring also leads to a blue shift in the photoluminescence emission peak, which further verifies the release of strain inside the device. Figure 5.13g shows the Raman test frequency shifts of NRLEDs with nano-ring widths of 300 nm, 120 nm, 80 nm, and 40 nm, which are 569.49 cm^{-1}, 568.10 cm^{-1}, 567.42 cm^{-1}, and 567.23 cm^{-1}, respectively. The corresponding stress sizes are 0.89 GPa, 0.71 GPa, 0.41 GPa, and 0.32 GPa, respectively. This also indicates that the small width of the nano-ring leads to a greater degree of stress release and an even higher IQE for the device.

Although the proposed nano-LED structure can suppress QCSE, the efficiency of green and red μLEDs in high In-component InGaN quantum wells is still very low. To further reduce the impact of lattice mismatch during the growth of quantum wells, a new type of InGaN substrate has been proposed [107]. Experimental results show that, based on this substrate, photoluminescence spectra covering blue to amber can be well achieved. When the emission wavelength is 536 nm and 566 nm, the IQE values are 31% and 10%, respectively. Therefore, the green and red μLEDs grown epitaxially on this new type of InGaN substrate are also one of the methods to alleviate the internal QCSE of the device.

In addition to IQE, the EQE of μLED is also affected by the light extraction efficiency (LEE). As has been mentioned earlier, sidewall passivation can increase the critical angle of light emission from the sidewalls. This then reduces the reflection of the sidewalls and thus improves the light extraction efficiency of the sidewalls. However, the light emitted from the sidewalls can cause optical crosstalk and affect the display effect. For commonly inverted μLED devices, laser lift-off (LLO) technology is generally used to remove the sapphire substrate and prevent photons from escaping from the sidewalls. This further reduces the optical crosstalk on the substrate side [108]. However, the large refractive index difference between GaN material and air can cause significant interface reflection, which leads to a decrease in LEE. As shown in Fig. 5.14, by calculating the far-field distribution of a $10 \times 10 \mu m^2$ μLED on the side and bottom, it can be seen that when compared to devices with completely removed substrates, μLED devices with sapphire substrates can transmit more light on the left and right peripheries due to the lateral propagation of light within the sapphire substrate, causing serious optical crosstalk. Therefore, for such inverted μLED devices, to reduce or eliminate optical crosstalk, the sapphire substrate is usually removed by laser lift-off [108]. At the same time, as shown in Fig. 5.14a, removing the sapphire substrate will also sacrifice the light extraction efficiency of the bottom of the μLED, which will also reduce the EQE of the device. Currently, there are several common methods to improve the light extraction efficiency of μLED devices. The first one is using a sidewall passivation layer based on an omnidirectional reflector (ODR) to reduce light loss and improve LEE [84, 110]. Secondly, by using a truncated pyramid shape and optimizing the tilt angle of the platform, the bottom LEE can be increased to 58.8% [109]. In addition, studies have shown that using high-reflectivity metal Ag as a p-type electrode can also improve the LEE of μLED [125, 126].

5.5 Summary

In this chapter, we have reviewed the key challenges currently encountered by μLED and the potential bottlenecks in display technology based on μLEDs. In addition to the technical difficulties of achieving full-color display and massive transfer, we have analyzed the main reasons for the low EQE of μLED devices and have summarized the common measures to improve EQE. First, to reduce the impact of μLED defects on device performance, the current main approaches are: (1) optimizing thermal annealing process to reduce the density of sidewall defects, (2) using

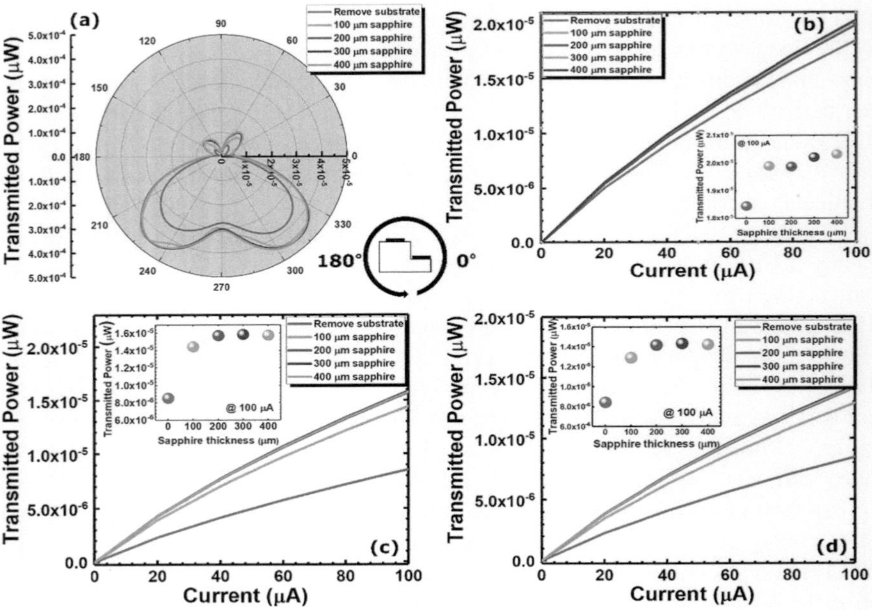

Fig. 5.14 Light intensity distribution at different angles of LED devices with different substrate thicknesses. **a**, **b**, **c**, and **d** are the light distribution on the bottom, left, and right sides, respectively

plasma-enhanced chemical vapor deposition (PECVD) or atomic layer deposition (ALD) systems to grow passivation layers to suppress sidewall defects, (3) wet chemical treatment to remove surface defects, and (4) modulating current diffusion length to confine carriers in the central device away from edge defect areas. In addition, to alleviate the QCSE of high-In-component green and red GaN-μLEDs, there are currently two mainstream methods, namely using nano-LED structures or new InGaN substrates. Finally, we have also summarized the known methods to further improve EQE by improving LEE. Therefore, we firmly believe that the information summarized in this chapter is to allow the community to have a deeper understanding of the device mechanism of μLED, aiming to better assist all sectors of society in manufacturing and researching high-efficiency GaN-based μLED.

References

1. Tan G, Zhu R, Tsai Y-S, Lee K-C, Luo Z, Lee Y-Z, Wu S-T (2016) High ambient contrast ratio OLED and QLED without a circular polarizer. J Phys D: Appl Phys 49(31):31501
2. Chen H, Tan G, Wu S-T (2017) Ambient contrast ratio of LCDs and OLED displays. Opt Express 25(26):33643–33656
3. Vettese D (2010) Liquid crystal on silicon. Nat Photon 4(11):752–754
4. Yang D-K, Wu S-T (2006) Fundamentals-of-liquid. Wiley & Sons Ltd.

5. Chen H, He J, Wu S-T (2017) Recent advances on quantum-dot-enhanced liquid-crystal displays. IEEE J Sel Top Quantum Electron 23(5):1–11
6. Park Y-S, Lee S, Kim K-H, Kim S-Y, Lee J-H, Kim J-J (2013) Exciplex-forming co-host for organic light-emitting diodes with ultimate efficiency. Adv Funct Mater 23(39):4914–4920
7. Tsujimura T (2017) OLED-display-fundamentals-and-applications, 2nd edn. Wiley & Sons, Inc.
8. Tang CW, VanSlyke SA (1987) Organic electroluminescent diodes. Appl Phys Lett 51(12):913–915
9. Hwang D, Mughal A, Pynn CD, Nakamura S, DenBaars SP (2017) Sustained high external quantum efficiency in ultrasmall blue III–nitride micro-LEDs. Appl Phys Express 10(032101):1–4
10. Schadt M (2009) Milestone in the history of field-effect liquid crystal displays and materials. Jpn J Appl Phys 48(3)
11. Féry C, Racine B, Vaufrey D, Doyeux H, Cinà S (2005) Physical mechanism responsible for the stretched exponential decay behavior of aging organic light-emitting diodes. Appl Phys Lett 87(213502):1–3
12. Chen HW, Lee JH, Lin BY, Chen S, Wu ST (2018) Liquid crystal display and organic light-emitting diode display: present status and future perspectives. Light Sci Appl 7:17168
13. Murawski C, Leo K, Gather MC (2013) Efficiency roll-off in organic light-emitting diodes. Adv Mater 25(47):6801–6827
14. Jiang HX, Jin SX, Li J, Shakya J, Lin JY (2001) III-Nitride blue microdisplays. Appl Phys Lett 78(9):1303–1305
15. Jin SX, Li J, Lin JY, Jiang HX (2000) InGaN/GaN quantum well interconnected microdisk light emitting diodes. Appl Phys Lett 77(20):3236–3238
16. Wu T, Sher C-W, Lin Y, Lee C-F, Liang S, Lu Y, Huang Chen S-W, Guo W, Kuo H-C, Chen Z (2018) Mini-LED and micro-LED: promising candidates for the next generation display technology. Appl Sci 8:1557
17. Jiang HX, Lin JY (2013) Nitride micro-LEDs and beyond-a decade progress review. Opt Express 21(3):475–484
18. Yole Développement (2017) MicroLED displays could disrupt LCD and OLED. www.yole.fr/MicroLEDDisplays_Market.aspx.W1lAV_knaEs
19. Griffiths AD, Herrnsdorf J, McKendry JJD, Strain MJ, Dawson MD (2019) GaN micro-LED structured light sources for multi-modal optical wireless communications systems. Phil Trans R Soc A 378:20190185
20. Li N, Han K, Spratt W, Bedell S, Ott J, Hopstaken M, Libsch F, Li Q, Sadana D (2019) Ultra-low-power sub-photon-voltage high-efficiency light-emitting diodes. Nat Photon 13(9):588–592
21. Tan G, Huang Y, Li M-C, Lee S-L, Wu S-T (2018) High dynamic range liquid crystal displays with a mini-LED backlight. Opt Express 26(13):16572–16584
22. Tian P, McKendry JJ, Gu E, Chen Z, Sun Y, Zhang G, Dawson MD, Liu R (2016) Fabrication, characterization and applications of flexible vertical InGaN micro-light emitting diode arrays. Opt Express 24(1):699–707
23. Zhang K, Peng D, Lau KM, Liu Z (2017) Fully-integrated active matrix programmable UV and blue micro-LED display system-on-panel (SoP). J Soc Inf Disp 25(4):240–248
24. Lin JY, Jiang HX (2020) Development of microLED. Appl Phys Lett 116(10)
25. Arvanitakis GN, Bian R, McKendry JJ, Cheng C, Xie E, He X, Yang G, Islim MS, Purwita AA, Gu E, Haas H, Dawson MD (2020) Gb/s underwater wireless optical communications using series-connected GaN micro-LED arrays. IEEE Photon J 12(2)
26. Chen CH, Hargis M, Woodall JM, Melloch MR, Reynolds JS, Yablonovitch E, Wang W (1999) GHz bandwidth GaAs light-emitting diodes. Appl Phys Lett 74(21):3140–3142
27. Monavarian M, Rashidi A, Aragon AA, Oh SH, Rishinaramangalam AK, DenBaars SP, Feezell D (2018) Impact of crystal orientation on the modulation bandwidth of InGaN/GaN light-emitting diodes. Appl Phys Lett 112(4)

28. Rajbhandari S, McKendry JJD, Herrnsdorf J, Chun H, Faulkner G, Haas H, Watson IM, O'Brien D, Dawson MD (2017) A review of gallium nitride LEDs for multi-gigabit-per-second visible light data communications. Semicond Sci Technol 32(2)
29. Rashidi A, Monavarian M, Aragon A, Okur S, Nami M, Rishinaramangalam A, Mishkat-Ul-Masabih S, Feezell D (2017) High-speed nonpolar InGaN/GaN LEDs for visible-light communication. IEEE Photonics Technol Lett 29(4):381–384
30. Rashidi A, Monavarian M, Aragon A, Rishinaramangalam A, Feezell D (2018) Nonpolar m-Plane InGaN/GaN micro-scale light-emitting diode with 1.5 GHz modulation bandwidth. IEEE Electron Device Lett 39(4):520–523
31. Bower CA, Meitl MA, Raymond B, Radauscher E, Cok R, Bonafede S, Gomez D, Moore T, Prevatte C, Fisher B, Rotzoll R, Melnik GA, Fecioru A, Trindade AJ (2017) Emissive displays with transfer-printed assemblies of 8 μm × 15 μm inorganic light-emitting diodes. Photon Res 5(2)
32. Beckers A, Fahle D, Mauder C, Kruecken T, Boyd AR, Heuken M (2018) Enabling the next era of display technologies by micro LED MOCVD processing. SID Symp Dig Tech Pap 49(601)
33. Peng D, Zhang K, Chao VS-D, Mo W, Lau KM, Liu Z (2016) Full-color pixelated-addressable light emitting diode on transparent substrate (LEDoTS) micro-displays by CoB. J Disp Technol 12(7):742–746
34. Peng D, Zhang K, Liu Z (2017) Design and fabrication of fine-pitch pixelated-addressed micro-LED arrays on printed circuit board for display and communication applications. IEEE J Electron Devices Soc 5(1):90–94
35. Liu Z, Lin CH, Hyun BR, Sher CW, Lv Z, Luo B, Jiang F, Wu T, Ho CH, Kuo HC, He JH (2020) Micro-light-emitting diodes with quantum dots in display technology. Light Sci Appl 9(83)
36. He J, Chen H, Chen H, Wang Y, Wu ST, Dong Y (2017) Hybrid downconverters with green perovskite-polymer composite films for wide color gamut displays. Opt Express 25(11):12915–12925
37. Liu ZJ, Wong KM, Chong WC, Lau KM (2011) Active matrix programmable monolithic light emitting diodes on silicon (LEDoS) displays. SID Symp Dig Tech Pap
38. Chen D-C, Liu Z-G, Deng Z-H, Wang C, Cao Y-G, Liu Q-L (2014) Optimization of light efficacy and angular color uniformity by hybrid phosphor particle size for white light-emitting diode. In: The 10th China international forum on solid state lighting (ChinaSSL), vol 33, pp 348–352. Beijing
39. Wei F, Li S, Bai X, Liu Z (2018) Hybrid full color micro-led displays with quantum dots. SID Symp. Dig. Tech. Pap 49:697–699
40. Han HV, Lin HY, Lin CC, Chong WC, Li JR, Chen KJ, Yu P, Chen TM, Chen HM, Lau KM, Kuo HC (2015) Resonant-enhanced full-color emission of quantum-dot-based micro LED display technology. Opt Express 23(25):32504–32515
41. Kim HM, Ryu M, Cha JHJ, Kim HS, Jeong T, Jang J (2019) Ten micrometer pixel, quantum dots color conversion layer for high resolution and full color active matrix micro-LED display. J Soc Inf Disp 27(6):347–353
42. Lin H-Y, Sher C-W, Hsieh D-H, Chen X-Y, Chen H-MP, Chen T-M, Lau K-M, Chen C-H, Lin C-C, Kuo H-C (2017) Optical cross-talk reduction in a quantum-dot-based full-color micro-light-emitting-diode display by a lithographic-fabricated photoresist mold. Photon Res 5(5)
43. Pust P, Weiler V, Hecht C, Tucks A, Wochnik AS, Henss AK, Wiechert D, Scheu C, Schmidt PJ, Schnick W (2014) Narrow-band red-emitting $Sr[LiAl_3N_4]:Eu^{2+}$ as a next-generation LED-phosphor material. Nat Mater 13(9):891–896
44. Dai X, Zhang Z, Jin Y, Niu Y, Cao H, Liang X, Chen L, Wang J, Peng X (2014) Solution-processed, high-performance light-emitting diodes based on quantum dots. Nature 515(7525):96–99
45. Chen Y, Xing W, Liu Y, Zhang X, Xie Y, Shen C, Liu JG, Geng C, Xu S (2020) Efficient and stable CdSe/CdS/ZnS quantum rods-in-matrix assembly for white LED application. Nanomater (Basel) 10(2)

46. Liu X, Liu Y, Xu S, Geng C, Xie Y, Zhang ZH, Zhang Y, Bi W (2017) Formation of "Steady Size" state for accurate size control of CdSe and CdS quantum dots. J Phys Chem Lett 8(15):3576–3580
47. Ding T, Yang X, Ke L, Liu Y, Tan W-Y, Wang N, Zhu X-H, Sun XW (2016) Improved quantum dot light-emitting diodes with a cathode interfacial layer. Org Electron 32:89–93
48. Wu Z, Liu P, Zhang W, Wang K, Sun XW (2020) Development of InP quantum dot-based light-emitting diodes. ACS Energy Lett 5(4):1095–1106
49. Zhu S, Song Y, Wang J, Wan H, Zhang Y, Ning Y, Yang B (2017) Photoluminescence mechanism in graphene quantum dots: quantum confinement effect and surface/edge state. Nano Today 13:10–14
50. Cok RS, Meitl M, Rotzoll R, Melnik G, Fecioru A, Trindade AJ, Raymond B, Bonafede S, Gomez D, Moore T, Prevatte C, Radauscher E, Goodwin S, Hines P, Bower CA (2017) Inorganic light-emitting diode displays using micro-transfer printing. J Soc Inf Disp 25(10):589–609
51. Zhou X, Tian P, Sher C-W, Wu J, Liu H, Liu R, Kuo H-C (2020) Growth, transfer printing and colour conversion techniques towards full-colour micro-LED display. Prog Quantum Electron 71:100263
52. Jung A, Choi JH, Jang SH, Han SJ (2019) Review of micro-light-emitting-diode technology for micro-display applications. SID Symp Dig Tech Pap 50(1)
53. Corbett B, Loi R, Zhou W, Liu D, Ma Z (2017) Transfer print techniques for heterogeneous integration of photonic components. Prog Quantum Electron 52:1–17
54. Zhang L, Ou F, Chong WC, Chen Y, Li Q (2018) Wafer-scale monolithic hybrid integration of Si-based IC and III-V epi-layers-A mass manufacturable approach for active matrix micro-LED micro-displays. J Soc Inf Disp 26(3):137–145
55. Hwang D, Yonkee BP, Addin BS, Farrell RM, Nakamura S, Speck JS, DenBaars S (2016) Photoelectrochemical liftoff of LEDs grown on freestanding c-plane GaN substrates. Opt Express 24(20):22875–22880
56. Yoon J, Jo S, Chun IS, Jung I, Kim HS, Meitl M, Menard E, Li X, Coleman JJ, Paik U, Rogers JA (2010) GaAs photovoltaics and optoelectronics using releasable multilayer epitaxial assemblies. Nature 465(7296):329–333
57. Choi HW, Jeon CW, Dawson MD, Edwards PR, Martin RW, Tripathy S (2003) Mechanism of enhanced light output efficiency in InGaN-based microlight emitting diodes. J Appl Phys 93(10):5978–5982
58. Hu X-L, Liu W-J, Weng G-E, Zhang J-Y, Lv X-Q, Liang M-M, Chen M, Huang H-J, Ying L-Y, Zhang B-P (2012) Fabrication and characterization of high-quality factor GaN-based resonant-cavity blue light-emitting diodes. IEEE Photonics Technol Lett 24(17):1472–1474
59. Tsai C-L, Yen C-T, Huang W-J, Xu Z-F, Ko S-C (2013) InGaN-based resonant-cavity light-emitting diodes fabricated with a TaO SiO distributed Bragg reflector and metal reflector for visible light communications. J Disp Technol 9(5):365–370
60. De Santi C, Meneghini M, La Grassa M, Galler B, Zeisel R, Goano M, Dominici S, Mandurrino M, Bertazzi F, Robidas D, Meneghesso G, Zanoni E (2016) Role of defects in the thermal droop of InGaN-based light emitting diodes. J Appl Phys 119(9)
61. Gong Z, Jin S, Chen Y, McKendry J, Massoubre D, Watson IM, Gu E, Dawson MD (2010) Size-dependent light output, spectral shift, and self-heating of 400 nm InGaN light-emitting diodes. J Appl Phys 107(1)
62. Lu S, Liu W, Zhang ZH, Tan ST, Ju Z, Ji Y, Zhang X, Zhang Y, Zhu B, Kyaw Z, Hasanov N, Sun XW, Demir HV (2014) Low thermal-mass LEDs: size effect and limits. Opt Express 22(26):32200–32207
63. LoboPloch N, Rodriguez H, Stolmacker C, Hoppe M, Lapeyrade M, Stellmach J, Mehnke F, Wernicke T, Knauer A, Kueller V, Weyers M, Einfeldt S, Kneissl M (2013) Effective thermal management in ultraviolet light-emitting diodes with micro-LED arrays. IEEE Trans Electron Devices 60(2):782–786
64. Che J, Chu C, Tian K, Kou J, Shao H, Zhang Y, Bi W, Zhang ZH (2018) On the p-AlGaN/n-AlGaN/p-AlGaN current spreading layer for AlGaN-based deep ultraviolet light-emitting diodes. Nanoscale Res Lett 13(1)

65. Liu Y-J, Huang C-C, Chen T-Y, Hsu C-S, Liou J-K, Liu W-C (2011) Improved performance of an InGaN-based light-emitting diode with a p-GaN/n-GaN barrier junction. IEEE J Quantum Electron 47(6):755–761
66. Park J-S, Kim JK, Cho J, Seong T-Y (2017) Review—group III-nitride-based ultraviolet light-emitting diodes: ways of increasing external quantum efficiency. ECS J. Solid State Sci. Technol 6(4):Q42–Q52
67. Herrnsdorf J, McKendry JJD, Shuailong Z, Enyuan X, Ferreira R, Massoubre D, Zuhdi AM, Henderson RK, Underwood I, Watson S, Kelly AE, Gu E, Dawson MD (2015) Active-matrix GaN micro light-emitting diode display with unprecedented brightness. IEEE Trans Electron Devices 62(6):1918–1925
68. Konoplev SS, Bulashevich KA, Karpov SY (2018) From large-size to micro-LEDs: scaling trends revealed by modeling. Phys Status Solidi A 215(10)
69. Kou J, Shen C-C, Shao H, Che J, Hou X, Chu C, Tian K, Zhang Y, Zhang Z-H, Kuo H-C (2019) Impact of the surface recombination on InGaN/GaN-based blue micro-light emitting diodes. Opt Express 27(12):A643–A653
70. Malinverni M, Martin D, Grandjean N (2015) InGaN based micro light emitting diodes featuring a buried GaN tunnel junction. Appl Phys Lett 107(5)
71. Olivier F, Tirano S, Dupré L, Aventurier B, Largeron C, Templier F (2017) Influence of size-reduction on the performances of GaN-based micro-LEDs for display application. J Lumin 191:112–116
72. Huang Y, Tan G, Gou F, Li M-C, Lee S-L, Wu S-T (2019) Prospects and challenges of mini-LED and micro-LED displays. J Soc Inf Disp 27(7):387–401
73. Olivier F, Daami A, Licitra C, Templier F (2017) Shockley-Read-Hall and Auger non-radiative recombination in GaN based LEDs: a size effect study. Appl Phys Lett 111(2)
74. Tian P, McKendry JJD, Gong Z, Guilhabert B, Watson IM, Gu E, Chen Z, Zhang G, Dawson MD (2012) Size-dependent efficiency and efficiency droop of blue InGaN micro-light emitting diodes. Appl Phys Lett 101(23)
75. Cho J, Schubert EF, Kim JK (2013) Efficiency droop in light-emitting diodes: challenges and countermeasures. Laser Photonics Rev 7(3):408–421
76. Mukal T, Yamada M, Nakamura S (1999) Characteristics of InGaN-based UV/blue/green/amber/red light-emitting diodes. Jpn J Appl Phys 38:3976–3981
77. Hwang J, Hashimoto R, Saito S, Nunoue S (2014) Development of InGaN-based red LED grown on (0001) polar surface. Appl Phys Express 7:071003
78. Yang Y, Cao XA (2009) Removing plasma-induced sidewall damage in GaN-based light-emitting diodes by annealing and wet chemical treatments. J Vac Sci & Technol B Microelectron Nanometer Struct 27(6)
79. Zuo P, Zhao B, Yan S, Yue G, Yang H, Li Y, Wu H, Jiang Y, Jia H, Zhou J, Chen H (2016) Improved optical and electrical performances of GaN-based light emitting diodes with nano truncated cone SiO_2 passivation layer. Opt Quantum Electron 48(5)
80. Yang C-M, Kim D-S, Lee S-G, Lee J-H, Lee YS, Lee J-H (2012) Improvement in electrical and optical performances of GaN-based LED with SiO_2/Al_2O_3 double dielectric stack layer. IEEE Electron Device Lett 33(4):564–566
81. Yiyun Z, Enqing G, Zhi L, Tongbo W, Jing L, Xiaoyan Y, Guohong W (2012) Light extraction efficiency improvement by curved GaN sidewalls in InGaN-based light-emitting diodes. IEEE Photonics Technol Lett 24(4):243–245
82. Chen W, Hu G, Lin J, Jiang J, Liu M, Yang Y, Hu G, Lin Y, Wu Z, Liu Y, Zhang B (2015) High-performance, single-pyramid micro light-emitting diode with leakage current confinement layer. Appl Phys Express 8(3)
83. Ley RT, Smith JM, Wong MS, Margalith T, Nakamura S, DenBaars SP, Gordon MJ (2020) Revealing the importance of light extraction efficiency in InGaN/GaN microLEDs via chemical treatment and dielectric passivation. Appl Phys Lett 116(25)
84. Wong MS, Hwang D, Alhassan AI, Lee C, Ley R, Nakamura S, DenBaars SP (2018) High efficiency of III-nitride micro-light-emitting diodes by sidewall passivation using atomic layer deposition. Opt Express 26(16):21324–21331

85. Son KS, Choi DL, Lee HN, Lee WG (2002) The interfacial reaction between ITO and silicon nitride deposited by PECVD in fringe field switching device. Curr Appl Phys 2(3):229–232
86. Guenther G, Schierning G, Theissmann R, Kruk R, Schmechel R, Baehtz C, Prodi-Schwab A (2008) Formation of metallic indium-tin phase from indium-tin-oxide nanoparticles under reducing conditions and its influence on the electrical properties. J Appl Phys 104(3)
87. Huang Chen S-W, Shen C-C, Wu T, Liao Z-Y, Chen L-F, Zhou J-R, Lee C-F, Lin C-H, Lin C-C, Sher C-W, Lee P-T, Tzou A-J, Chen Z, Kuo H-C (2019) Full-color monolithic hybrid quantum dot nanoring micro light-emitting diodes with improved efficiency using atomic layer deposition and nonradiative resonant energy transfer. Photon Res 7(4)
88. Choi WH, You G, Abraham M, Yu S-Y, Liu J, Wang L, Xu J, Mohney SE (2014) Sidewall passivation for InGaN/GaN nanopillar light emitting diodes. J Appl Phys 116(1)
89. Tang B, Miao J, Liu Y, Wan H, Li N, Zhou S, Gui C (2019) Enhanced light extraction of flip-chip mini-LEDs with prism-structured sidewall. Nanomaterials 9(3)
90. Nedy JG, Young NG, Kelchner KM, Hu Y, Farrell RM, Nakamura S, DenBaars SP, Weisbuch C, Speck JS (2015) Low damage dry etch for III-nitride light emitters. Semicond Sci Technol 30(8)
91. Yue Y, Yan X, Li W, Xing HG, Jena D, Fay P (2014) Faceted sidewall etching of n-GaN on sapphire by photoelectrochemical wet processing. J Vac Sci Technol B 32(6)
92. Wong MS, Lee C, Myers DJ, Hwang D, Kearns JA, Li T, Speck JS, Nakamura S, DenBaars SP (2019) Size-independent peak efficiency of III-nitride micro-light-emitting-diodes using chemical treatment and sidewall passivation. Appl Phys Express 12(9)
93. Lee S, Forman CA, Lee C, Kearns J, Young EC, Leonard JT, Cohen DA, Speck JS, Nakamura S, DenBaars SP (2018)GaN-based vertical-cavity surface-emitting lasers with tunnel junction contacts grown by metal-organic chemical vapor deposition. Appl Phys Express 11(6)
94. Young EC, Yonkee BP, Wu F, Oh SH, DenBaars SP, Nakamura S, Speck JS (2016) Hybrid tunnel junction contacts to III–nitride light-emitting diodes. Appl Phys Express 9(2)
95. Hwang D, Mughal AJ, Wong MS, Alhassan AI, Nakamura S, DenBaars SP (2018) Micro-light-emitting diodes with III–nitride tunnel junction contacts grown by metalorganic chemical vapor deposition. Appl Phys Express 11(1)
96. Huang S-C, Li H, Zhang Z-H, Chen H, Wang S-C, Lu T-C (2017) Superior characteristics of microscale light emitting diodes through tightly lateral oxide-confined scheme. Appl Phys Lett 110(2)
97. Chang L, Yeh Y-W, Hang S, Tian K, Kou J, Bi W, Zhang Y, Zhang Z-H, Liu Z, Kuo H-C (2020) Alternative strategy to reduce surface recombination for InGaN/GaN micro-light- emitting diodes—thinning the quantum barriers to manage the current spreading. Nanoscale Res Lett
98. Chen B-C, Chang C-Y, Fu Y-K, Huang K-F, Lu Y-H, Su Y-K (2011) Improved performance of InGaN/GaN light-emitting diodes with thin intermediate barriers. IEEE Photonics Technol Lett 23(22):1682–1684
99. Chakraborty A, Haskell BA, Keller S, Speck JS, DenBaars SP, Nakamura S, Mishra UK (2004) Nonpolar InGaN/GaN emitters on reduced-defect lateral epitaxially overgrown a-plane GaN with drive-current-independent electroluminescence emission peak. Appl Phys Lett 85(22):5143–5145
100. Sharma R, Pattison PM, Masui H, Farrell RM, Baker TJ, Haskell BA, Wu F, DenBaars SP, Speck JS, Nakamura S (2005) Demonstration of a semipolar (1013) InGaN/GaN green light emitting diode. Appl Phys Lett 87(23)
101. Masui H, Nakamura S, DenBaars SP, Mishra UK (2010) Nonpolar and semipolar III-nitride light-emitting diodes: achievements and challenges. IEEE Trans Electron Devices 57(1):88–100
102. Bai J, Wang Q, Wang T (2012) Characterization of InGaN-based nanorod light emitting diodes with different indium compositions. J Appl Phys 111(11)
103. Koester R, Sager D, Quitsch WA, Pfingsten O, Poloczek A, Blumenthal S, Keller G, Prost W, Bacher G, Tegude FJ (2015) High-speed GaN/GaInN nanowire array light-emitting diode on silicon (111). Nano Lett 15(4):2318–2323

104. Ramesh V, Kikuchi A, Kishino K, Funato M, Kawakami Y (2010) Strain relaxation effect by nanotexturing InGaN/GaN multiple quantum well. J Appl Phys 107(11)
105. Wang Q, Bai J, Gong YP, Wang T (2011) Influence of strain relaxation on the optical properties of InGaN/GaN multiple quantum well nanorods. J Phys D Appl Phys 44(39)
106. Wang SW, Hong KB, Tsai YL, Teng CH, Tzou AJ, Chu YC, Lee PT, Ku PC, Lin CC, Kuo HC (2017) Wavelength tunable InGaN/GaN nano-ring LEDs via nano-sphere lithography. Sci Rep 7
107. Even A, Laval G, Ledoux O, Ferret P, Sotta D, Guiot E, Levy F, Robin IC, Dussaigne A (2017) Enhanced In incorporation in full InGaN heterostructure grown on relaxed InGaN pseudo-substrate. Appl Phys Lett 110(26)
108. Li S-H, Lin C-P, Fang Y-H, Kuo W-H, Wu M-H, Chao C-L, Horng R-H, Su G-DJ (2019) Performance analysis of GaN-based micro light-emitting diodes by laser lift-off process. Solid State Electron Lett 1(2):58–63
109. Bulashevich K, Konoplev S, Karpov S (2018) Effect of die shape and size on performance of III-nitride micro-LEDs: a modeling study. Photonics 5(4)
110. Lin CH, Lai CF, Ko TS, Huang HW, Kuo HC, Hung YY, Leung KM, Yu CC, Tsai RJ, Lee CK, Lu TC, Wang SC (2006) Enhancement of InGaN–GaN indium–tin–oxide flip-chip light-emitting diodes with TiO_2–SiO_2 multilayer stack omnidirectional reflector. IEEE Photonics Technol Lett 18(19):2050–2052
111. Dingemans G, Kessels WMM (2010) Influence of the deposition temperature on the c-Si surface passivation by Al_2O_3 films synthesized by ALD and PECVD. Electrochem Solid-State Lett 13(3):H76–H79
112. Bulashevich KA, Karpov SY (2016) Impact of surface recombination on efficiency of III-nitride light-emitting diodes. Phys Status Solidi RRL 10(6):480–484
113. Smorchkova IP, Haus E, Heying B, Kozodoy P, Fini P, Ibbetson JP, Keller S, DenBaars SP, Speck JS, Mishra UK (2000) Mg doping of GaN layers grown by plasma-assisted molecular-beam epitaxy. Appl Phys Lett 76(718)
114. Zhang ZH, Tan ST, Kyaw Z, Ji Y, Liu W, Ju ZG, Hasanov N, Sun XW, Demir HV (2013) InGaN/GaN light-emitting diode with a polarization tunnel junction. Appl Phys Lett 102:193508
115. Li LP, Shi Q, Tian KK, Chu CS, Fang MQ, Meng RL, Zhang YH, Zhang ZH, Bi WG (2017) A dielectric-constant-controlled tunnel junction for III-nitride light-emitting diodes. Phys Status Solidi A 1600937
116. Lu T-C, Kao C-C, Kuo H-C, Huang G-S, Wang S-C (2008) CW lasing of current injection blue GaN-based vertical cavity surface emitting laser. Appl Phys Lett 92:141102
117. Zhang Z-H, Tan ST, Liu W, Ju ZG, Zheng K, Kyaw Z, Ji Y, Hasanov N, Sun XW, Demir HV (2013) Improved InGaN/GaN light-emitting diodes with a p-GaN/n-GaN/p-GaN/n-GaN/p-GaN current-spreading layer. Opt Express 21(4):4958–4969
118. Che JM, Shao H, Kou JQ, Tian KK, Chu CS, Hou X, Zhang YH, Sun Q, Zhang ZH (2019) Improving the current spreading by locally modulating the doping type in the n-AlGaN layer for AlGaN-based deep ultraviolet light-emitting diodes. Nanoscale Res Lett 14:268
119. Che JM, Shao H, Chang L, Kou JQ, Tian KK, Chu CS, Hou X, Zhang YH, Bi WG, Zhang ZH (2020) Doping-induced energy barriers to improve the current spreading effect for AlGaN-based ultraviolet-B light-emitting diodes. IEEE Electron Device Lett 41(7):1001–1004
120. Zhang ZH, Liu W, Ju ZG, Tan ST, Ji Y, Kyaw Z, Zhang XL, Wang LC, Sun XW, Demir HV (2014) Self-screening of the quantum confined Stark effect by the polarization induced bulk charges in the quantum barriers. Appl Phys Lett 104:243501
121. Tian KK, Chu CS, Shao H, Che JM, Kou JQ, Fang MQ, Zhang YH, Bi WG, Zhang ZH (2018) On the polarization effect of the p-EBL/p-AlGaN/p-GaN structure for AlGaN-based deep-ultraviolet light-emitting diodes. Superlattices Microstruct 122:280–285
122. Ling SC, Lu TC, Chang SP, Chen JR, Kuo H-C, Wang S-C (2010) Low efficiency droop in blue-green m-plane InGaN/GaN light emitting diodes. Appl Phys Lett 96:231101
123. Ma J, Zhu XL, Wong KM, Zou XB, Lau KM (2013) Improved GaN-based LED grown on silicon (111) substrates using stress/dislocation-engineered interlayers. J. Crist. Growth 370:265–268

124. Kim TI, Jung YH, Song JZ, Kim D, Li YH, Kim H-S, Song I-S, Wierer JJ, Pao HA, Huang YG, Rogers JA (2012) High-efficiency, microscale GaN light-emitting diodes and their thermal properties on unusual substrates. Small 8(11):1643–1649
125. Dupré L, Marra M, Verney V, Aventurier B, Henry F, Olivier F, Tirano S, Daami A, Templier F (2017) Processing and characterization of high resolution GaN/InGaN LED arrays at 10 micron pitch for micro display applications. Proc SPIE 10104:22
126. Olivier F, Daami A, Dupré L, Henry F, Aventurier B, Templier F (2017) 25-4: Investigation and improvement of 10μm Pixel-pitch GaN-based micro-LED arrays with very high brightness. SID Symp Dig Tech Pap 253

Chapter 6
Growth and Optical Characterization of Nonpolar and Semipolar GaN LEDs

Yun Zhang, Kun Xing, and Tao Wang

6.1 Nonpolar and Semipolar III-Nitrides

6.1.1 Polarization in III-Nitrides

1. **Spontaneous Polarization and Piezoelectric Polarization**

The crystal axis length ratio of a perfect hexagonal cell is $\frac{c}{a} = \sqrt{\frac{8}{3}} \approx 1.633$. But the length ratios of wurtzite III-nitrides are all smaller than this perfect value, with a lack of inversion symmetry. This leaves a displacement between the center of metal atoms and the center of N atoms, leading to a spontaneous polarization along the $-c$ axis direction. On the other hand, the residual strain in the crystal also produces another polarization in III-nitrides, named the piezoelectric polarization. For example, for the *c*-plane indium gallium nitride/gallium nitride (InGaN/GaN) quantum well, the compressive stress in the InGaN layer produces a piezoelectric polarization along the $-c$ direction. Theoretical calculation shows that the electric field of the piezoelectric polarization is at the order of MV/cm level [1]. Under the combined impact of the spontaneous polarization and the piezoelectric polarization, an internal electric field is formed along the *c*-direction with the appearance of the charges at the heterostructure interface. This internal electric field can severely affect the band structure and carrier distribution in the III-nitride heterostructure.

Y. Zhang (✉)
Jiangsu University, Zhenjiang, China
e-mail: yzhang@ujs.edu.cn

K. Xing
Hefei University of Technology, Hefei, China

T. Wang
University of Sheffield, Sheffield, UK

S. Xu (ed.), *Optical Characterization of Microstructures and Optoelectronic Devices Based on Wide Band Gap Semiconductors*, Wide Bandgap Semiconductors,
https://doi.org/10.1007/978-981-95-1928-6_6

2. Influence of Internal Electric Field on Optoelectronic Devices

For the optoelectronic devices, such as the InGaN-based light-emitting diode (LED), the energy bands of the quantum well will tilt due to the internal electric field, which confines the electrons and holes at the two opposite sides of the quantum well. As a result of that, the overlap of the electron and hole wave functions decreases, leading to a drop in the probability of the transitions from the conduction band to the valence band. This phenomenon is also known as the quantum-confined Stark effect (QCSE). The QCSE is usually considered a major reason for the decrease of internal quantum efficiency (IQE) in InGaN/GaN quantum well structures. For the InGaN/GaN quantum well structures with a high indium (In) component, the lattice mismatch becomes more significant. The resulting compressive strain in the InGaN quantum well layer is consequently large, producing a more pronounced QCSE effect. This effect leads to a sharp drop in the IQE of the LEDs in yellow and green light. This lack of high-efficiency green LEDs is commonly referred to as the "green gap" issue.

The tilted band structure also reduces the radiative energy levels where electrons and holes transport, causing the emission peak of the *c*-plane InGaN/GaN quantum well to shift to longer wavelength. As the number of carriers in the quantum well increases, the Coulomb effect introduces a shielding effect that can weaken the impact of the polarization field, thereby raising the radiative energy levels of the electron and hole. Consequently, the emission peak of the quantum well shifts toward the blue. This color drift phenomenon, which changes with the number of injected carriers, is particularly noticeable in high-In-content InGaN/GaN quantum wells. For LEDs, especially those used in display applications, minimizing such color drift is essential.

Due to the presence of the QCSE effect, *c*-plane InGaN/GaN quantum well structures usually adopt a thin quantum well layer (2–3 nm) design to increase the overlap of the wave functions of the electron and hole. Additionally, they adopt a multi-quantum well layer design to increase the volume of the active region. However, the difference in the effective masses of the electron and hole leads to a difference in migration rates. This causes a nonuniform distribution of electrons and holes within the multi-quantum wells, ultimately affecting the luminous efficiency of the device. In addition, InGaN-based LED generally exhibits the highest quantum efficiency under low current injection. As the injection current increases, the quantum efficiency of the LED drops significantly, a phenomenon known as "efficiency droop". This droop not only limits the development of high-power LEDs but also causes a sharp increase in energy loss under high current operating conditions. When the emission wavelength moves toward the green/yellow spectrum, this issue of efficiency droop becomes more pronounced. Although the mechanism of efficiency droop is not fully understood, QCSE is generally considered to have a significant impact [2].

In *c*-plane InGaN/GaN quantum wells, the polarization field not only reduces the radiative recombination probability of electrons and holes but also lengthens the carrier lifetime. When the *c*-plane LED light source is used in a visible light communication system, QCSE severely limits the system's modulation bandwidth. In these systems, the direct modulation rate of the light source is primarily influenced by the carrier lifetime and the resistance–capacitance (RC) time constant. The carrier lifetime of blue light LEDs is typically in the 10 ns range, allowing for a direct modulation bandwidth of around 10 MHz. While increasing the injection current density can shorten the carrier lifetime, this approach to enhancing modulation bandwidth at the expense of LED quantum efficiency has limited effectiveness. Moreover, it can lead to increased energy consumption, reduced reliability, and heightened safety concerns under high current conditions.

6.1.2 Advantages of Nonpolar and Semipolar III-Nitrides

To mitigate the adverse effects of the QCSE effect on LEDs, researchers have adopted various methods, either to regulate the stress in the InGaN/GaN heterostructure [3, 4] or to shield the electric field at its interface [5, 6]. An effective solution involves offsetting the stacking direction of the InGaN/GaN heterostructure from the crystal *c*-axis direction, that is, to prepare III-nitride devices along nonpolar or semipolar directions. This approach can effectively avoid or reduce the impact of the polarization field on the InGaN/GaN heterostructure. The traditional (0001) plane (or *c*-plane) is usually referred to as the polar plane. In contrast, the (11-20) plane (or *a*-plane) and the (1-100) plane (or *m*-plane), which are perpendicular to the polar plane, are known as nonpolar planes. Other crystal planes that are inclined to the polar plane are classified as semipolar planes. Figure 6.1 illustrates the common nonpolar and semipolar planes in III-nitride materials with a wurtzite structure.

The control of the polarization field in the InGaN/GaN quantum well due to changes in growth direction is significant. As shown in Fig. 6.2, theoretical calculations indicate that as the angle ϑ between the growth direction and the crystal *c*-axis increases, the piezoelectric polarization intensity in the $In_xGa_{1-x}N$ layer grown on GaN decreases from its maximum value at the 0° polar direction. After reaching zero, the piezoelectric polarization intensity experiences a brief increase before gradually decreasing again, finally reaching zero at the 90° nonpolar direction [7]. In 2000, Waltereit and colleagues first reported the experimental study of nonpolar nitride heterojunctions [8]. They used a molecular beam epitaxy (MBE) system to prepare nonpolar *m*-plane GaN epitaxial films on gamma-phase lithium aluminate (γ-$LiAlO_2$) substrates. The GaN/aluminum gallium nitride (AlGaN) structure prepared based on this *m*-plane GaN showed luminescent characteristics that were unaffected by the piezoelectric polarization field. This work has drawn attention to the nonpolar and semipolar directions, leading to an increasing number of reports utilizing these orientations to control the polarization field in nitride heterostructures.

Due to the lack of in-plane symmetry, the surface luminescence of nonpolar and semipolar nitrides also shows certain polarization characteristics. In traditional *c-plane* InGaN/GaN quantum well structures, the in-plane biaxial strain is isotropic, leading to mixing between the $|X\rangle$ state and $|Y\rangle$ states in both the heavy hole band and the light hole band, which results in isotropic light emission along the growth direction (the crystal *c*-axis). However, in nonpolar or semipolar InGaN/GaN quantum

Fig. 6.1 Schematic diagram of polar, nonpolar, and semipolar planes in wurtzite structure

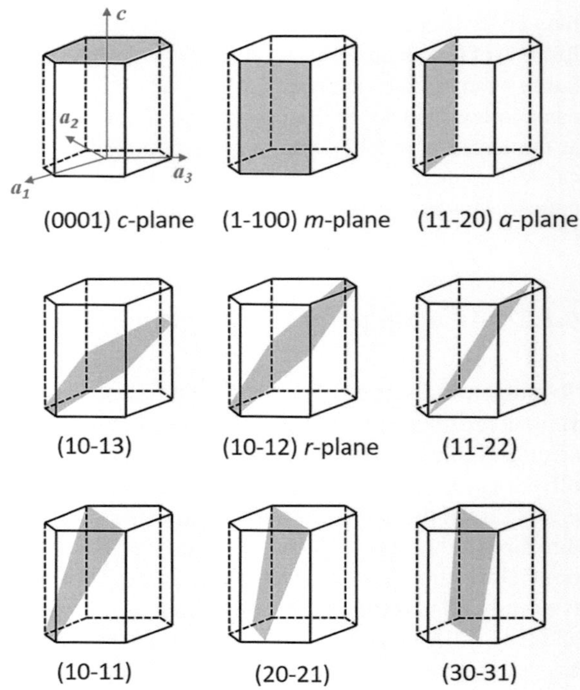

Fig. 6.2 The piezoelectric polarization intensity $P_{LZ'}^{PZ}$ in the $In_xGa_{1-x}N$ layer and the total polarization intensity $\Delta P_{Z'}$ vary with the angle ϑ between the growth direction and the *c*-axis of the crystal. The angles of some nonpolar and semipolar directions are also marked in the figure. Curves (1)–(4) correspond to In compositions of 0.05, 0.10, 0.15, and 0.20, respectively [8]

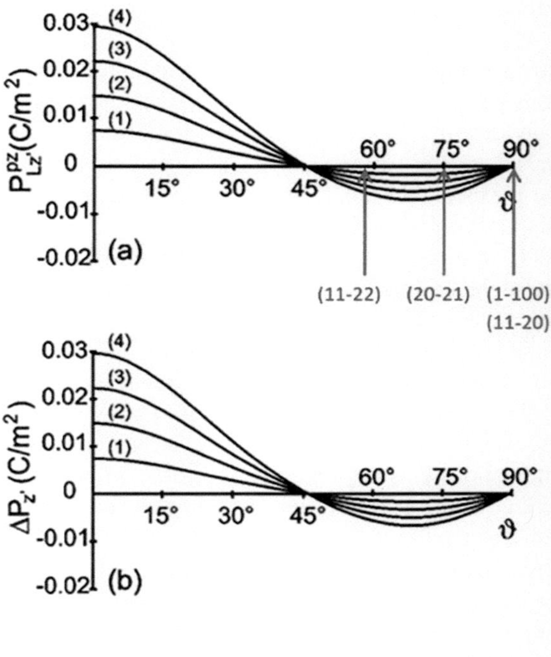

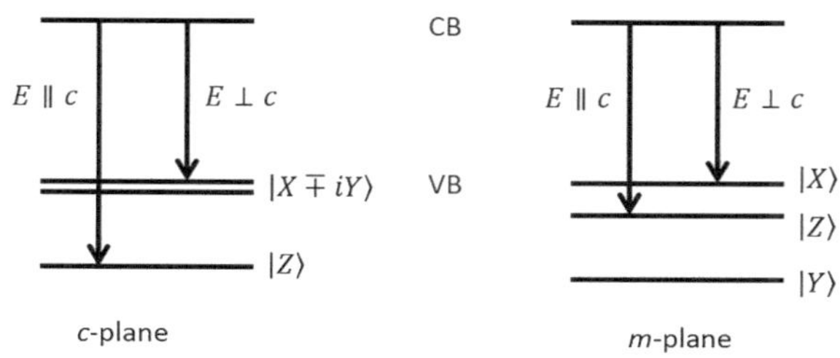

Fig. 6.3 Diagram of the energy bands of *c*-plane and *m-plane* InGaN quantum wells

well structures, the stress within the growth plane is anisotropic, causing displacements of the heavy hole band and light hole band in the valence band. As shown in Fig. 6.3, taking the *m*-plane InGaN/GaN quantum well as an example, when the composition in the quantum well is greater than 5%, the $|Y\rangle$ state energy level will be higher than the $|X\rangle$ state and $|Z\rangle$ state energy levels. In addition, the differing probabilities of hole occupancy between these states result in the emitted light in the growth direction exhibiting a certain degree of polarization. When the transition probability of the $|X\rangle$ state energy level exceeds that of the $|Z\rangle$ state, the electric field components of the emitted light in the normal direction are perpendicular to the *c*-axis ($E \perp c$). Currently, polarized light in the display field is mostly obtained through filtering, which leads to energy wastage. Using nonpolar and semipolar nitrides, along with other light sources that possess natural polarization properties, would be a promising alternative.

As the arrangement of Ga and N atoms on the surface changes, some nonpolar and semipolar samples have certain advantages in incorporating In atoms. A high growth temperature helps to improve the crystal quality. However, at a high growth temperature, the solubility of In atoms in GaN is very low. Thus, finding a balance between the low-temperature requirements for high In incorporation and the high-temperature requirements for crystal quality is key to achieving high In component InGaN/GaN quantum wells and overcoming the "green gap" issue. First-principles calculations have shown that the semipolar (11-22) plane GaN has a higher ability to incorporate In atoms, compared to the *c*-plane and nonpolar plane [9]. Based on this, Wernicke and colleagues conducted experimental research to study the In incorporation abilities of (0001), (10-10), (10-12), (11-22), (10-11), and (20-21) planes of GaN under the same growth conditions. They found that in the growth temperature range of 670–790 °C, the semipolar (10-11) and (11-22) planes showed the highest ability to incorporate In atoms, while the In component in the semipolar (20-21) plane was similar to that in the polar and nonpolar planes [10]. Zhao and others compared the luminescence of (10-10), (11-22), (20-21), and (30-3-1) plane InGaN/GaN quantum wells grown under the same conditions. They discovered that the (11-22) plane InGaN/GaN quantum well has a longer luminescence wavelength than other samples [11], indicating a higher In component. This suggests that, under the same growth conditions, the semipolar (11-22) plane has a stronger ability to incorporate In atoms, and the (11-22) plane holds great potential for obtaining long-wavelength InGaN-based quantum wells.

Regarding doping, some nonpolar planes and semipolar planes exhibit unique advantages. McLaurin and colleagues reported achieving a p-type gallium nitride (p-GaN) layer with a hole concentration of 7.2×10^8 cm^{-3} on the *m*-plane without annealing [12]. Similarly, Tsuchiya and Kaeding reported comparable results for the nonpolar (11-20) plane and semipolar (10-1-1) plane [13, 14]. These reports highlight the p-doping advantages of these nonpolar and semipolar planes. Theoretical research shows that as the angle between the growth direction and the crystal *c*-axis continues to increase, the effective mass of holes in InGaN/GaN quantum wells gradually decreases. This indicates that compared to the polar plane, holes in the semipolar and nonpolar planes have a higher migration speed.

6.1.3 Challenges in Nonpolar and Semipolar III Nitrides

Like *c*-plane GaN epitaxy, the large-scale epitaxy of nonpolar and semipolar GaN faces challenges due to a lack of suitable substrate materials. Although early efforts to grow nonpolar GaN were attempted on γ-$LiAlO_2$ substrates, the metal atoms in these oxide substrates exhibit strong diffusion under the high-temperature conditions required for the growth of III nitrides. Currently, the epitaxial substrates for nonpolar and semipolar GaN mainly include sapphire (Al_2O_3), silicon (Si), silicon carbide (SiC), and free-standing substrates. It is noteworthy that the highest performance blue and green nonpolar and semipolar GaN-based laser devices are all achieved on free-standing substrates. The homoepitaxy of free-standing substrates offers significant advantages in defect control. These free-standing substrates are typically cut from thick *c*-plane GaN film epitaxially by hydride vapor phase epitaxy (HVPE). The size of these substrates is generally around 10 mm, and their preparation is very costly. The high cost and limited size restrict their large-scale promotion and application.

Among heteroepitaxial substrates, the SiC substrate has a relatively small lattice mismatch with GaN (about 3.5%), but its acquisition cost is also high. The lattice constants of the spinel structure Al_2O_3 are $a = 4.758$ Å and $c = 12.991$ Å. For nonpolar *a*-plane GaN grown on *r*-plane sapphire substrates along the $[1\text{-}100]_{GaN} \parallel [11\text{-}20]_{sapphire}$ and the $[0001]_{GaN} \parallel [\text{-}1101]_{sapphire}$ directions, the lattice mismatches are 16.1% and 1.1%, respectively. For semipolar (11-22) plane GaN grown on *m*-plane sapphire substrate along the $[1\text{-}100]_{GaN} \parallel [11\text{-}20]_{sapphire}$ and the $[\text{-}1\text{-}123]_{GaN} \parallel [0001]_{sapphire}$ directions, the lattice mismatches are 16.1% and 6.3%, respectively. For the cubic structure of silicon, with a lattice constant of $a = 5.431$ Å, the lattice mismatch between GaN and the (111)-plane silicon is 16.9%. Compared with the first two substrates, the lattice mismatches between sapphire and silicon substrates with GaN are more significant. However, due to their mature preparation processes and ease of availability, sapphire and silicon substrates are widely used in the epitaxial research of nonpolar and semipolar III-nitrides. The discussion in this chapter will also focus on nonpolar and semipolar nitrides obtained on sapphire and silicon substrates.

Due to the lack of suitable substrate materials, heteroepitaxial nonpolar and semipolar GaN often contain a large number of line dislocations, with a line dislocation density of approximately 10^{10} cm^{-2}. Since line dislocations play a role in nonradiative recombination centers, their high density significantly reduces the device performance. Unlike the growth of *c*-plane GaN on sapphire substrates, where only the line dislocations are usually present, semipolar or nonpolar GaN crystals also contain a substantial number of basal-plane stacking faults (BSF), with a density of around 10^5 cm^{-1}. In *c*-plane GaN, BSFs are confined near the interface between GaN and sapphire. They are oriented perpendicular to the growth direction and cannot extend to the surface. Therefore, BSFs in *c*-plane GaN do not impact the performance of top devices. However, in nonpolar GaN, BSFs are parallel to the growth direction, and in semipolar GaN, they are inclined at an angle to the growth direction. As a result, BSFs in nonpolar and semipolar GaN can pass through all upper layers and top devices, finally extending to the sample surface. Both dislocations and stacking faults will affect the electrical and optical properties of nonpolar and semipolar GaN-based optoelectronic devices.

The surface morphology of nonpolar and semipolar nitrides may lead to a decrease in the quality of the epitaxial layer, limiting the device performance. Whether in homoepitaxy or heteroepitaxy, stripes along the *c*-axis direction can be observed on the surface of nonpolar nitrides, and stripes along the *c*-axis projection direction can be observed on the surface of semipolar nitrides [15–17]. Theoretical calculations show that the atomic diffusion on nonpolar and semipolar surfaces is anisotropic. During growth, atoms tend to move along the directions with lower diffusion kinetic barriers, leading to the undulating surface features observed in nonpolar and semipolar nitrides. Such surface morphology may cause uneven barrier interfaces in subsequent quantum well growth, and hence energy fluctuations. To overcome these negative effects from surface morphology, the sample surface can be chemically or mechanically polished before the growth of the device structure, which helps reduce the surface roughness [18–20].

6.2 Heteroepitaxy of Nonpolar and Semipolar Nitrides

6.2.1 Selective Area Epitaxy of Nonpolar and Semipolar Nitrides

Selective area epitaxy (SAE) is an epitaxial technique where material growth occurs in specific areas of the substrate by covering patterned mask materials (such as SiO_2, SiN_x, etc.) on the substrate surface. In 1994, Kato and colleagues first reported the growth of III nitrides using selective area epitaxy [21]. They prepared SiO_2 masks with patterns of stripes and dots on sapphire substrates and carried out GaN and AlGaN epitaxial growth. Zheleva and others used transmission electron microscopy (TEM) to observe the dislocation-blocking by the SiO_2 mask and found that the GaN

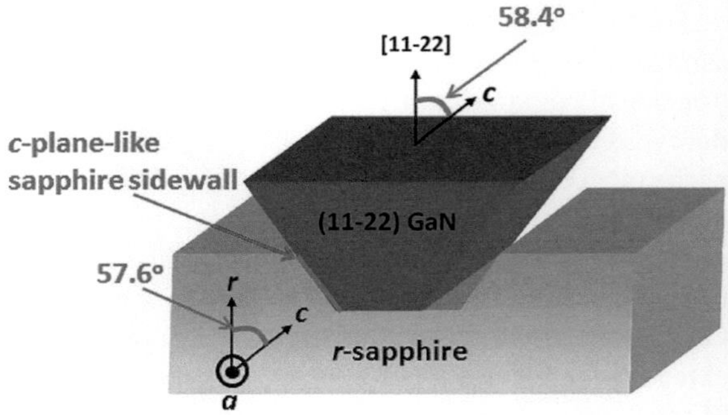

Fig. 6.4 Schematic diagram of epitaxial growth of semipolar (11-22) GaN on a patterned (11-23) *r*-plane sapphire substrate [26]

grown laterally on SiO_2 had a significantly reduced dislocation density, indicating that the SAE technique can effectively minimize dislocations in epitaxial materials [22]. Currently, in addition to effectively improving the crystal quality of GaN materials, the SAE technique also enables the controlled preparation of micro-/nano-structures and heteroepitaxy of GaN on specific semipolar orientation [23–25].

1. **Selective Area Epitaxy on Patterned Sapphire Substrate**

The growth of semipolar GaN on patterned sapphire substrates basically utilizes the epitaxial relationship between the (0001) plane sapphire and the (0001) plane GaN. The growth starts from the attempts to create micro-surfaces with (0001) or similar orientations on obliquely cut sapphire substrate, allowing GaN growth on these micro-surfaces. As shown in Fig. 6.4, Okada and colleagues proposed a method for growing semipolar (11-22) plane GaN on patterned (11-23) *r*-plane sapphire substrates. The angle between the *c*-axis and *r*-axis in *r*-plane sapphire is 57.6°, which is very close to the 58.4°angle between the *c*-axis and [11-22]-axis in GaN. Consequently, GaN film formed by coalescence of GaN islands growing on the *c*-plane areas on *r*-plane sapphire will have a semipolar (11-22) plane surface.

To prepare a micro-surface with a *c*-plane orientation on the (11-23) *r*-plane sapphire substrate, Okada and colleagues first used standard photolithography on an *r*-plane sapphire substrate with a 0.5° off-axis angle. A pattern was prepared with 3 μm-width stripes, aligned parallel to the *a*-axis of the sapphire crystal. Then, through reactive ion etching (RIE) and inductive coupled plasma (ICP) etching, grooves of about 1 μm depth were etched in the window area. One side wall of the groove is a pseudo *c*-plane. It should be noted that the surface of this side wall is not a true *c*-plane, but rather a pseudo *c*-plane with a slight tilt. Scanning electron microscopy (SEM) measurements showed an approximate 12.4° deviation angle between the pseudo *c*-plane sidewall and the sapphire substrate, which complicates subsequent epitaxial growth. Due to the chemical inertness of the sapphire substrate, it is difficult to control the groove patterns with specific tilt angles using standard dry etching techniques.

Chen et al. and Tendille et al. respectively reported the preparation of patterned (11-23) *r*-plane sapphire substrates using a chemical wet etching process [25, 27]. They first used standard photolithography to create SiO_2 mask patterns of specific sizes on the surface of the *r*-plane sapphire substrate. Chen et al. then performed selective wet etching using a 300 °C phosphoric acid (H_3PO_4) solution for 5 min of, followed by removal of the SiO_2 mask with a buffered oxide etch (NH_4F:HF = 6:1). Tendille et al. on the other hand, used a 1:3 mixture of 270 °C H_3PO_4 and sulfuric acid (H_2SO_4) for 30 min of selective wet etching, retaining the top SiO_2 layer to prevent the growth of (11-20) plane GaN on the *r*-plane sapphire substrate.

Using a patterned (11-23) *r*-plane sapphire substrate, semipolar (11-22) plane GaN is typically grown through a three-step process to achieve a smooth surface [26]. First, a thin low-temperature nucleation layer is deposited, followed by annealing at a higher temperature. This annealing step encourages the migration of the nucleation layer from other planes onto the pseudo *c*-plane sidewall, enhancing the growth of *c*-plane GaN on the pseudo *c*-plane sidewall. The second step uses high-temperature, high-pressure, and low V/III ratio conditions to promote the *c*-plane GaN growth, allowing it to dominate and effectively cover the defect area. Finally, growth conditions are adjusted to facilitate GaN coalescence between windows and to smooth the surface. The semipolar (11-22) plane prepared by this method has a full width at half maximum (FWHM) of the X-ray diffraction (XRD) rocking curve of 249 arcseconds along the [11-2-3] direction, and a FWHM of 326 arcseconds along the [1-100] direction [28].

The same approach can be applied to the epitaxial growth of semipolar (20-21) plane GaN, using a (22-43) plane sapphire substrate. In the (22-43) plane sapphire substrate, the angle between the *c*-axis and the [22-43] crystal axis is 74.64°, which is very close to the angle of 75.09° between the [0001] and (20-21) crystal axes in GaN. Leung et al. reported the study of epitaxial growth of semipolar (20-21) plane GaN on patterned (22-43) plane sapphire substrates [29]. As shown in Fig. 6.5, a groove pattern with a width of 3 μm, a depth of 0.5 μm, and a spacing of 3 μm was first prepared on the (22-43) sapphire substrate with a 0.45° off-axis angle, using standard photolithography and RIE etching processes. The substrate was then tilted at a specific angle, and a layer of SiO_2 was deposited on all surfaces except the *c*-plane sidewall. Finally, the patterned substrate was placed in a metal–organic chemical vapor deposition (MOCVD) growth chamber for a series of processes, including high-temperature nitridation treatment [19] and the deposition of a low-temperature aluminum nitride (AlN) buffer layer, for the epitaxial growth of semipolar (20-21) plane GaN. The epitaxially grown semipolar (20-21) plane GaN was examined with XRD $2\theta - \omega$ scanning over the range of 30°–90°, revealing only the (20-21) GaN peak and the (22-43) sapphire peak. As shown in Fig. 6.5c, SEM analysis indicated that the surface of the coalesced (20-21) plane GaN sample is uneven, comprising microfacets of (10-11), (10-1-1), and (10-10) orientations. For the semipolar (20-21) plane GaN prepared by this method, the FWHM values of the XRD rocking curves in the [10-14] direction and [11-20] direction are 192 arcseconds and 217 arcseconds, respectively [30].

2. **Selective Area Epitaxy on Patterned Silicon Substrate**

According to the epitaxial relationship, semipolar GaN with a large inclination angle (with respect to the *c*-plane) requires a high-index (11 h) silicon substrate, making

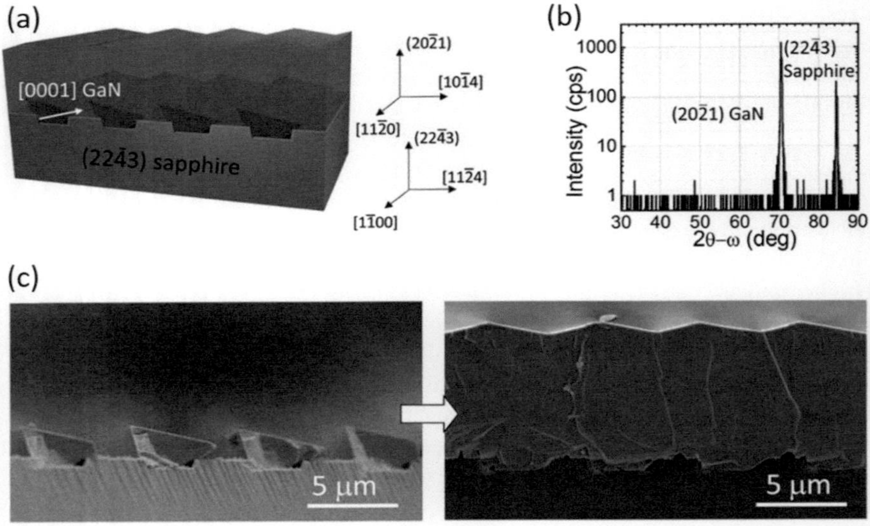

Fig. 6.5 **a** Schematic illustration of the crystal orientations of semipolar (20-21) plane GaN and (22-43) plane sapphire substrate. **b** XRD $2\theta - \omega$ scan curve of the epitaxial semipolar (20-21) plane GaN wafer. **c** and **d** SEM side views of the semipolar (20-21) plane GaN with epitaxial thicknesses of 300 nm and 8 μm, respectively [29]

it challenging to achieve epitaxy of semipolar GaN with these orientations on a flat silicon substrate. Instead, most epitaxy of semipolar GaN with a large inclination angle is usually performed on patterned silicon substrates. Compared with sapphire substrates, silicon substrates are more cost-effective and offer more advanced etching processes. However, at high temperatures, gallium atoms react with silicon, causing remelting erosion, which can lead to collapse during GaN growth. This remelting erosion issue has limited the development of silicon-based nitrides.

The angle between the (1-11) plane and the (113) plane in silicon is 58.5°, which is very close to the angle of 58.4° between the (11-22) plane and the (0001) plane in GaN. Tanikawa et al. reported the growth of semipolar (11-22) plane GaN on a patterned (113) plane silicon substrate [31]. Using standard photolithography, they first prepared [21-1] oriented SiO_2 strips on the (113) plane silicon substrate. A wet etching was then performed on the window areas between the SiO_2 strips, using a 40 °C potassium hydroxide (KOH) solution (25 wt%). Due to anisotropic etching, the final trench sidewalls are formed as (1-11) plane and (-11-1) plane, while the trench bottom consists of (113) plane or (011) plane (Fig. 6.6). In their study of (11-22) plane GaN on this patterned silicon substrate through MOCVD, it is found that a narrow window width and a deep trench depth helped inhibit GaN formation on the (-11-1) sidewall, allowing GaN to nucleate and grow only on the (1-11) sidewall. In addition, low V/III ratios and high-pressure growth conditions are more conducive to improving surface flatness and crystal quality.

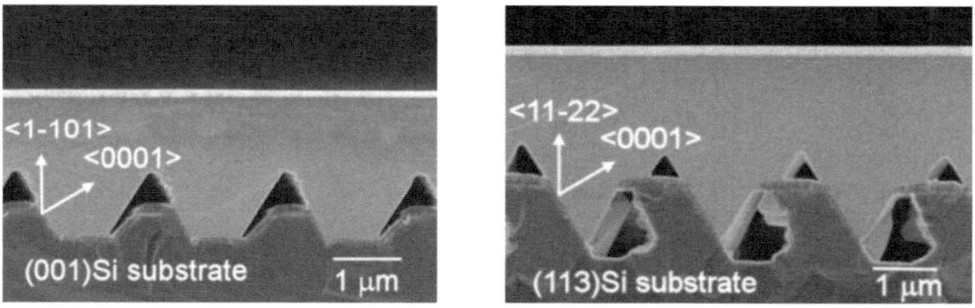

Fig. 6.6 SEM cross-sectional views of semipolar (1-101) GaN and (11-22) GaN epitaxially grown on patterned (001) and (113) silicon substrates, respectively [32]

At the same time, Tanikawa et al. also reported the growth of semipolar (1-101) plane GaN on a patterned (001) plane silicon substrate. The angle between the (111) plane and the (001) plane in silicon is 54.7°. By applying a 7.3° axial tilt to the (001) plane silicon substrate, the angle between the silicon (111) plane and the substrate surface becomes 62°, which matches the angle between the (1-101) plane and the (0001) plane in GaN. During the patterning step, a layer of SiO_2 was deposited on the (-1-11) plane sidewall to inhibit GaN growth there, allowing GaN growth only from the exposed (111) plane sidewall. Figure 6.6 shows the SEM cross-sectional views of semipolar (1-101) GaN overgrown on the patterned (001) silicon substrate, and semipolar (11-22) GaN on the patterned (113) silicon substrate [32].

In contrast to the trench-shaped patterned silicon substrate, Xiang et al. designed a pyramid-shaped patterned silicon substrate [33, 34]. As shown in Fig. 6.7, they used standard photolithography to create a SiO_2 pattern layer with a micropore array on the (113) plane of silicon. The patterned window areas were then wet etched for 18 min using a 30 °C KOH solution (25 wt%), forming a pyramid-shaped pit with (111), (1-11), (-1-11) and (-111) plane sidewalls. After removing the surface SiO_2 layer with a 10% HF solution, the substrate was tilted at a specific angle to deposit another 20 nm thick SiO_2 layer on all surfaces except the (1-11) sidewalls, using electron beam deposition. This re-covering process not only improved the crystal quality of the epitaxial GaN but also mitigated the remelting erosion issue during the GaN growth. The semipolar (11-22) GaN grown on this patterned substrate exhibited a FWHM value of 0.186° for the XRD rocking curve scanned along the c-projection, and a BSF density of $1 \times 10^4 cm^{-1}$.

6.2.2 *Epitaxial Lateral Overgrowth of Nonpolar and Semipolar GaN*

The epitaxial lateral overgrowth (ELOG) technology effectively improves the crystal quality of c-plane GaN in heteroepitaxy, ultimately facilitating the development

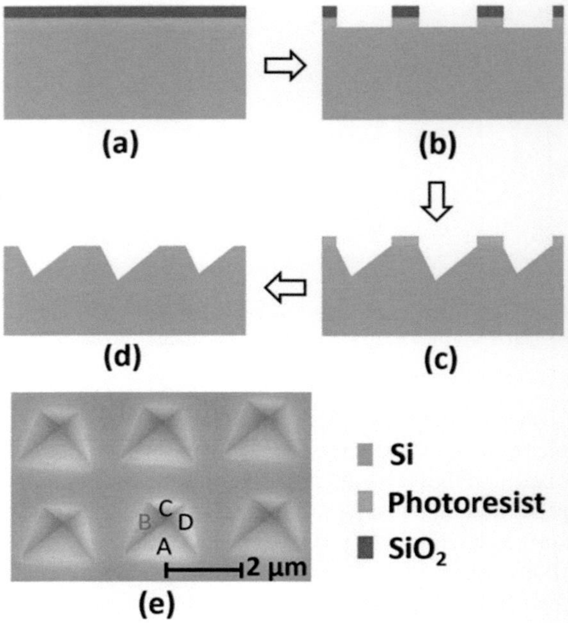

Fig. 6.7 a–d Schematic diagram of the preparation process of patterned (113) silicon substrate. **e** SEM top view of the patterned (113) silicon substrate, where the sidewalls marked A, B, C, and D correspond to (111), (1-11), (-1-11), and (-111) planes, respectively [33]

of InGaN-based blue lasers. This technology is also applied to the heteroepitaxial growth of nonpolar and semipolar GaN.

1. **ELOG on Strip Substrate Template**

Like the development of selective area epitaxy technology, the early lateral overgrowth of nonpolar and semipolar GaN was typically performed on strip-patterned substrate templates. As an example, Ni et al. utilized a substrate template prepared with the following process, for the lateral overgrowth of semipolar (11-22) plane GaN [35]. First, by using the low-temperature GaN buffer layer technology, a 1.5 µm thick (11-22) GaN template is directly grown on an *m*-plane sapphire substrate. A 140 nm thick SiO_2 layer is then deposited on its surface using a PECVD process. This template is subsequently processed into a strip pattern with a width of 10 µm and a spacing of 4 µm using standard photolithography and etching processes. For comparative studies, the authors prepared two types of strip orientations: one aligned along the [1-210] direction of the sapphire substrate and the other along the *c*-axis direction (also known as the GaN [1-100] direction and the GaN [11-2-3] direction). Figure 6.8 shows the SEM top views, SEM cross-sectional views, and TEM cross-sectional views of the semipolar (11-22) plane GaN grown on both types of substrate templates.

It was found that on the [1-210]-orientation stripe substrate template, the GaN epitaxially grown in the window area shows asymmetry (Fig. 6.8b). Its surface consists of the (0001) plane and the (11-20) plane, with the growth rate of GaN along the *c*-axis exceeding that along the *a*-axis. As a result, the GaN covering the

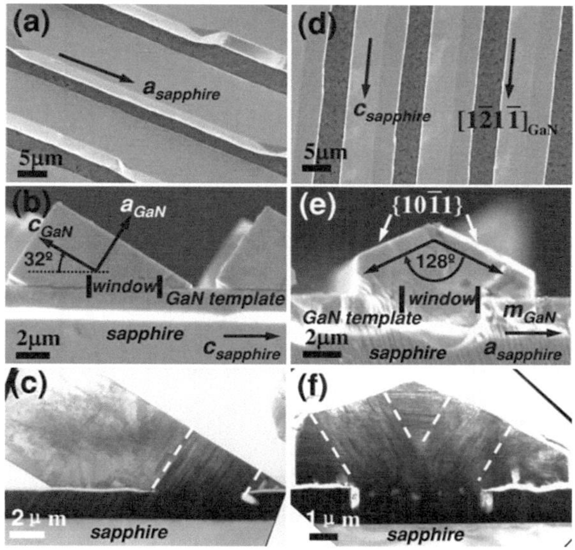

Fig. 6.8 Semipolar (11-22) GaN overgrown on stripe substrate templates with [1-210]-orientation stripes and *c*-orientation stripes. **a** and **d** SEM top views **b** and **e** SEM cross-sectional views **c** and **f** TEM cross-sectional views [35]

SiO_2 stripes mainly comes from the growth along the *c*-axis direction. TEM characterization revealed a significant reduction in both the dislocation density and the BSF density for the GaN covering the SiO_2 stripes. Furthermore, BSFs in the GaN grown in the window area can still propagate to the GaN surface. On the other hand, on the *c*-orientation stripe substrate template, the GaN epitaxially grown in the window area shows symmetry (Fig. 6.8e). Its surface is mainly composed of two {10–11} planes. Due to the slow growth rate of GaN along the *m*-direction, a relatively small area of the SiO_2 stripe is covered. This leads to a reduction in the efficiency of preventing defect propagation, which was further confirmed through TEM characterization. Bougrioua et al. reported the study of epitaxial nonpolar (11-20) plane and semipolar (11-22) plane GaN on stripe substrate templates with different orientations. It was found that for the nonpolar and semipolar GaN overgrown on the stripe substrate templates, the crystal quality of the GaN on the stripes oriented perpendicular to the *c*-axis is superior compared to the GaN on the stripes with other orientations [36]. By using a two-step growth method, Mierry et al. optimized the growth conditions to achieve a smooth semipolar (11-22) plane GaN on a stripe substrate template with a width of 7 µm, a gap of 3 µm, and oriented along the [1-100] direction [37]. The growth process consists of two steps: the first step uses a high V/III ratio (~3500) to obtain the GaN nucleation layer, while the second step uses a low V/III ratio (~500) and low-pressure growth conditions to promote *c*-plane growth until the layers merge to form a smooth surface. Using the same design for the stripe substrate template, Song et al. studied the effect of changing the stripe width (from 6 to 12 µm) on the crystal quality of the epitaxial (11-22) plane GaN [38]. It was found that, with the same three-step growth conditions, increasing the stripe width reduced the linewidth of the XRD rocking curve of the epitaxial GaN and increased the PL emission intensity of the sample. However, increasing the stripe width will also extend the time required to cover the SiO_2 stripes, which in turn increases the thickness of the GaN epitaxial layer. Additionally, this results in a reduction in the surface smoothness of the GaN epitaxial layer after merging.

2. **ELOG on Nanopillar Substrate Template**

Traditional strip substrate templates often suffer from the uniformity issue, with a significant difference in crystal quality between the window area and the nonwindow area. Moreover, a substrate with a large strip size requires a thicker epitaxial layer (10 ~ 20 μm) to achieve a fully merged surface. To address this, Xing et al. reported the ELOG of nonpolar and semipolar GaN on nanopillar substrate template [39, 40]. Figure 6.9 shows a fabrication process for the nanopillar substrate template for the ELOG of semipolar (11-20) plane GaN. A layer of 2.6 μm (11-20) plane GaN is first grown on an *r*-plane sapphire substrate using a two-step growth method, followed by the deposition of a 200 nm SiO_2 layer and a 10 nm nickel layer on its surface. Then, it is annealed at 850 °C for 1 min in a nitrogen atmosphere, forming a self-assembled nickel nanoparticle layer with a diameter of approximately 200 nm on the surface. Next, the nickel nanoparticles are used as a mask for etching the underlying SiO_2 layer using the RIE process. The resulting SiO_2 pattern layer is then used as a secondary mask to etch the GaN layer beneath it with the ICP process, creating nanopillars. It is worth noting that the SiO_2 layer on top of the nanopillars needs to be preserved during the process. Compared to nano-imprinting and other nano-pattern preparation techniques, this self-assembled nanoparticle method is highly cost-effective. Furthermore, by modifying the annealing conditions, the diameter of the nickel nanoparticles can be adjusted from 100 to 1000 nm.

Due to the small gap between the nanopillars, the GaN grown on the sidewalls can quickly meet and merge. Measurements indicate that the epitaxial GaN layer can coalesce at a thickness of about 1 μm, with a smooth surface forming at a thickness of 4 μm. This substrate significantly reduces the growth time compared to traditional strip-patterned substrates. The narrow gap also limits the growth at the bottom of the gap, promoting the overgrowth from the sidewalls, and forming air voids between the nanopillars after merging. These air gaps, along with the top SiO_2 layer, effectively prevent the propagation of defects in the GaN template. Figure 6.10a plots the variation in the FWHM of the XRD rocking curve with the azimuth angle for a nonpolar (11-20) plane GaN grown on a nanopillar substrate template, and that for a nonpolar (11-20) plane GaN directly grown on an *r*-plane sapphire substrate by the traditional two-step growth method. The results show a significant reduction in the FWHM for the GaN grown on the nanopillar substrate template, measured at 443 arcseconds and 345 arcseconds in the scan directions parallel and perpendicular to the *c*-axis, respectively.

The same method was also applied to study semipolar (11-22) GaN growth on an *m*-plane sapphire substrate. Figure 6.10b shows the research results from 2013. The FWHM of the rocking curve for semipolar (11-22) GaN grown on the nanopillar substrate template decreased significantly. In the direction parallel to the *c*-axis, the FWHM reduced from 715 to 460 arcseconds, while in the direction perpendicular to

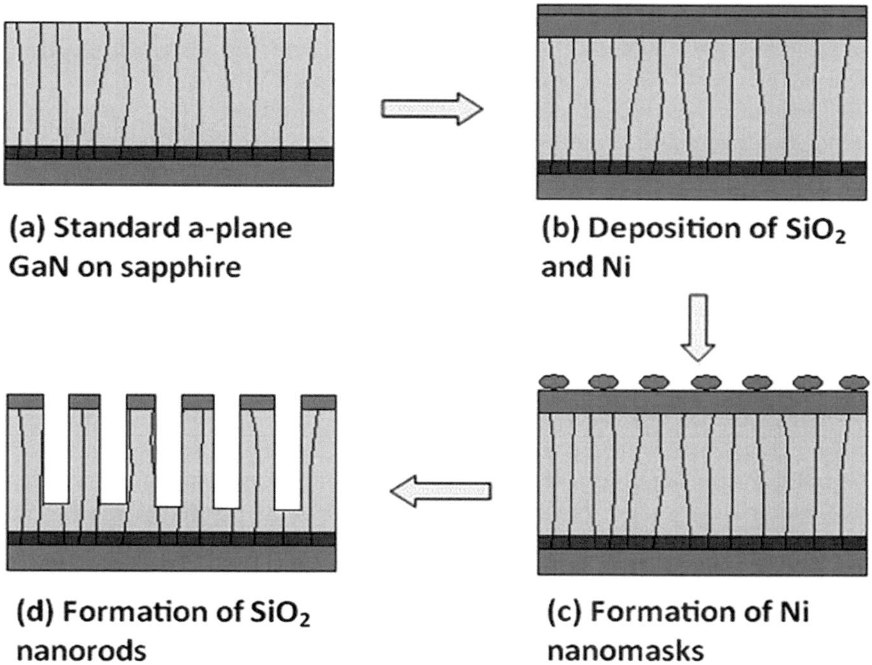

Fig. 6.9 Schematic of the nanopillar substrate template preparation process for semipolar (11–20) plane GaN growth using the self-assembled nickel nanoparticle mask method [39]

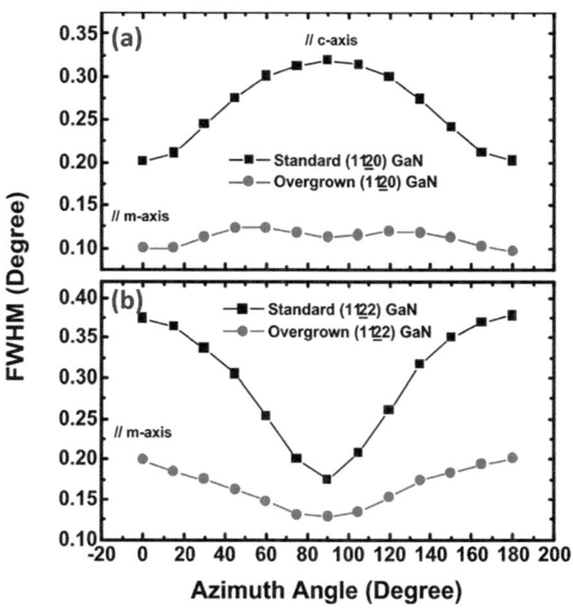

Fig. 6.10 **a** Comparison of the FWHMs of on-axis XRD rocking curves as a function of azimuth angle for the nonpolar (11-20) plane GaN grown by ELOG and directly grown on *r*-plane sapphire substrates. **b** Comparison of the FWHMs of on-axis XRD rocking curves as a function of azimuth angle for the semipolar (11-22) plane GaN grown by ELOG and directly grown on *m*-plane sapphire substrates [40]

the *c*-axis, it decreased from 1346 to 626 arcseconds. These results indicate that the epitaxial growth technology using the nanopillar substrate template can effectively improve the crystal quality of nonpolar and semipolar GaN.

Although the method of epitaxial growth on nanopillar substrate template is simple and effective, the formation of metal nanospheres involves a certain degree of randomness, making it challenging to precisely control the nanomask pattern. Additionally, the high-temperature annealing process limits the ability to obtain large-diameter (>1 μm) nickel nanospheres, which are essential for further improving the crystal quality of the epitaxial GaN.

3. **ELOG on Microrod Substrate Template**

To overcome the challenges of size limitations and precise control in ELOG on nanopillar substrate templates, the Sheffield team developed a new method for lateral overgrowth of nonpolar and semipolar nitrides on microrod substrate templates. This process begins with depositing approximately 200 nm of SiO_2 on the surface of the GaN template. Standard photolithography is then used to create a surface mask pattern, followed by the fabrication of microrod arrays using RIE and ICP dry etching processes, while preserving the top SiO_2 layer. This method enables precise control over the diameter, spacing, and array direction of the microrods. Figure 6.11a shows the microrod array substrate template for the overgrowth of semipolar (11-22) GaN on *m*-plane sapphire, with a microrod diameter of 3.5 μm and a spacing of 2 μm between adjacent microrods in the [-1100] and [-1-123] directions [41]. Figure 6.11b–d show the surface morphology of the semipolar (11-22) GaN after 1000, 2000, and 3000 s of growth on the microrod substrate template. Epitaxial growth was observed to start from the sidewalls of the microrods and mainly grow in the *c*-direction and the *a*-direction, with a small growth rate in the *m*-direction. By comparing the length, it was found that the growth in the *c*-direction was faster than that in the *a*-direction. When the adjacent growth fronts meet and merge, the growth begins to extend toward the surface, completing a second merging process above the microrods. This second merging covers the top SiO_2 layer on the microrods and continues until a flat surface is formed. Due to the significantly faster growth rate of GaN in the *c*-direction and its lower defect density, the first merging process effectively suppresses the propagation of defects in the epitaxial GaN. The semipolar (11-22) plane GaN grown by this method exhibits low XRD rocking curve FWHMs of 320 and 260 arcseconds along the [1–100] direction and the [-1-123] direction, respectively.

The arrangement direction of the microrod array has a significant impact on the crystal quality of the epitaxial GaN. As shown in Fig. 6.12c and d, two microrod arrays with the same size but different arrangement directions were prepared on the semipolar (11-22) plane GaN template. In Fig. 6.12c, the microrods are arranged in an ABAB… tight configuration along the *m*-direction, while in Fig. 6.12d, the microrods are arranged in the same position along the *m*-direction. These two arrays have a 45° rotation angle relative to each other. After the same growth time, the GaN films epitaxially grown on the two substrate templates present different surface morphologies. It was found that the semipolar (11-22) plane GaN (Sample A) grown

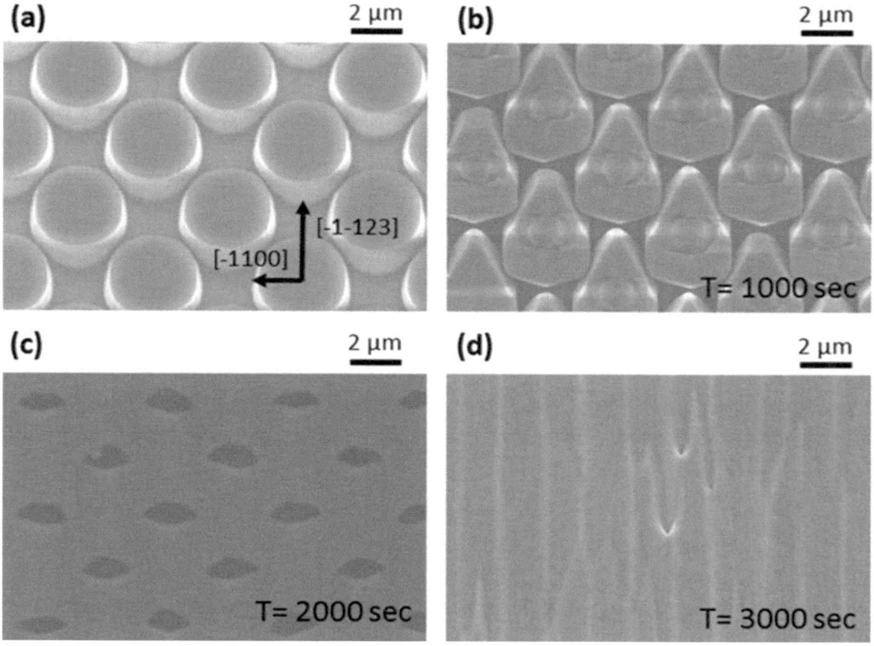

Fig. 6.11 **a** SEM top view of a typical microrod substrate template for the growth of semipolar (11-22) plane GaN **b–d**) Surface morphology of samples grown for 1000, 2000, and 3000 s, respectively, using the semipolar (11–22) plane GaN microrod substrate template [41]

on the template in Fig. 6.12c has a smoother surface. This is attributed to the fact that the growth rate of GaN along the *m*-direction is much slower than that along the *c*-direction or the *a*-direction. The tight arrangement along the *m*-direction helps to effectively compensate for this growth rate anisotropy, allowing the epitaxial GaN to merge quickly and form a smooth surface. The semipolar (11-22) plane GaN grown on the optimized microrod substrate template can form a flat surface within a thickness of 5 μm. The XRD rocking curve FWHM is below 360 arcseconds, approaching the crystal quality of standard *c*-plane GaN. This indicates that laterally epitaxial semipolar GaN can meet the requirements for high-quality devices in subsequent steps.

4. **Secondary Lateral Epitaxial Overgrowth**

To address the high defect density issue in the window area, a secondary lateral overgrowth technology has been proposed, which utilizes the ELOG technique twice. By redepositing and preparing mask patterns on the window area of the first ELOG GaN, the secondary ELOG process effectively blocks the propagation of defects in the window area, thereby further improving the crystal quality of the overgrown GaN. In 2018, Bai et al. reported a study on the secondary lateral epitaxial growth of nonpolar (11-20) plane GaN using a combination of stripe and microrod array

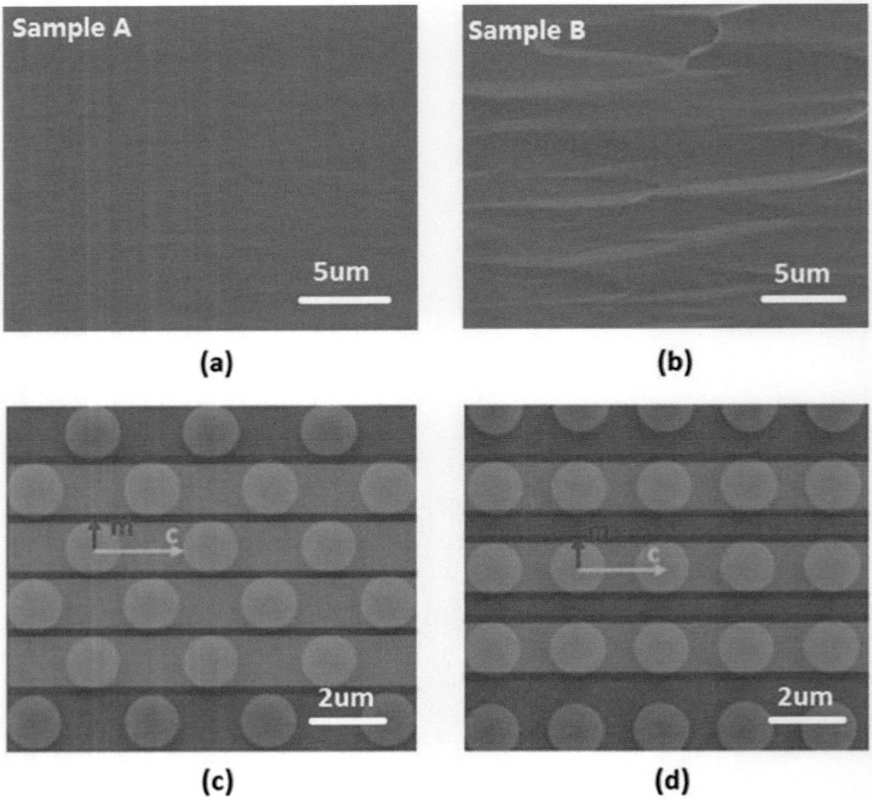

Fig. 6.12 **a** and **b** SEM images of the surface morphology of the semipolar (11-22) plane GaN ELOG on microrod substrate templates with different array arrangements. **c** and **d** Details of the array arrangements of the different microrod substrate templates

patterns [42]. As shown in Fig. 6.13c, the microrod array pattern was placed on top to leverage its ability to quickly coalesce and form a smooth surface. Figure 6.13d shows the FWHM values of on-axis XRD rocking curves against the azimuth angle of the nonpolar (11-20) plane GaN grown on three types of patterned substrate templates. The results indicated that secondary lateral overgrowth can effectively improve the crystal quality of nonpolar GaN. However, this technique involves the additional processes of pattern preparation and MOCVD growth, increasing both the procedure complexity and cost, which limits its widespread application.

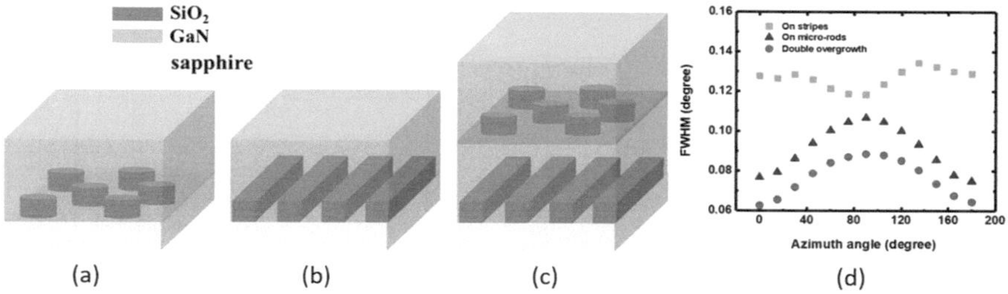

Fig. 6.13 **a–c** Schematic diagram of nonpolar (11-20) plane GaN ELOG on various substrate templates, including microrods, stripes, and a combination of stripes and microrods. **d** FWHM of XRD rocking curves as a function of azimuth angle of the nonpolar (11-20) plane GaN ELOG on those substrate templates [42]

6.2.3 Growth of Nonpolar and Semipolar InGaN/GaN Quantum Well

Nonpolar and semipolar InGaN/GaN quantum well structures can be directly grown on the high-quality GaN obtained from the above-mentioned selective epitaxy or lateral overgrowth. To further improve the quality of these InGaN/GaN quantum well structures, researchers have employed various methods to optimize the interface and the stress within the quantum wells.

Chemical mechanical polishing (CMP) is a process that combines chemical etching and mechanical force to treat the surface of semiconductor substrates, achieving a flat surface. In the growth of nonpolar and semipolar III-nitrides, CMP is commonly used to smooth out the surface undulations caused by the anisotropic growth rate. This helps reduce issues related to uneven thickness and component distribution in subsequent quantum well structure growth. Dinh et al. reported a study on the use of CMP technology to flatten the GaN epitaxial wafer, before the growth of semipolar (11-22) plane InGaN/GaN LED. As shown in Fig. 6.14, after CMP treatment, the surface roughness of the semipolar blue LED decreased from 30 to 4 nm ($50 \times 50\ \mu m^2$). Meanwhile, measurements indicated that the luminous uniformity and light output power were also improved for the semipolar LEDs with CMP treatment [18]. A research team from Yale reported on the application of CMP technology in the growth of semipolar (20-21) plane III-nitrides. CMP can reduce the influence of micro-facets on the surface of the semipolar (20-21) plane, achieving a surface roughness of less than 0.5 nm within a $15 \times 15\ \mu m^2$ area [19, 29].

Inserting an $In_xGa_{1\text{-}x}N$ single layer or an $In_xGa_{1\text{-}x}N$/GaN superlattice layer before growing nonpolar or semipolar InGaN/GaN quantum wells can reduce the stress within the quantum wells, thereby improving device performance. Zhao et al. studied the effect of the superlattice layer on the performance of semipolar (11-22) plane InGaN-based LEDs [43]. As shown in Fig. 6.15, they inserted a 15-layer $In_{0.03}Ga_{0.97}N$/GaN (3/6 nm) superlattice structure between the n-type gallium nitride

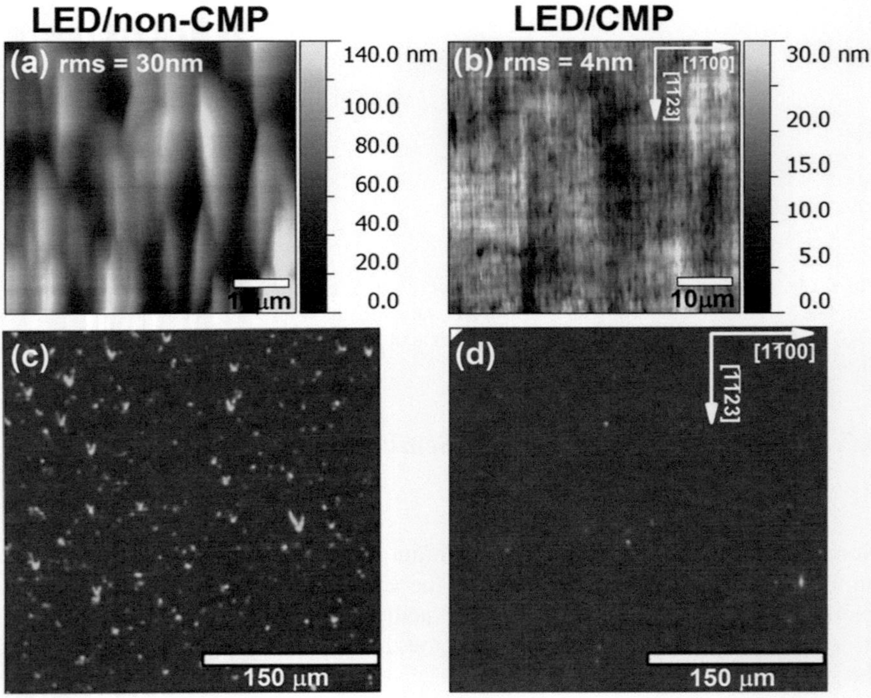

Fig. 6.14 **a** and **b** AFM images (50 × 50 μm^2) of the surface of 445 nm blue LEDs grown on (11-22) plane GaN substrates without and with CMP treatment; **c** and **d** Fluorescence images (320 × 320 μm^2) of those LEDs grown on (11-22) plane GaN substrates without and with CMP treatment [18]

(n-GaN) layer and the quantum well layer. Characterization revealed that this superlattice insertion layer increased the IQE of the green LED from 13 to 23%, resulting in an enhancement factor of about 1.8 times. This is slightly lower than the 3.5 times enhancement typically observed in traditional *c*-plane LED. The reduced enhancement is attributed to the additional mismatch dislocations caused by the insertion layer [44]. Khoury et al. observed additional mismatch dislocations in the semipolar (20-21) plane of an InGaN-based LED with a superlattice layer using TEM characterization. Most of the mismatch dislocations were found near the interface of the superlattice layer. The formation of these dislocations helps to release the stress in the emitting layer in advance, thereby reducing their impact on subsequent devices [45].

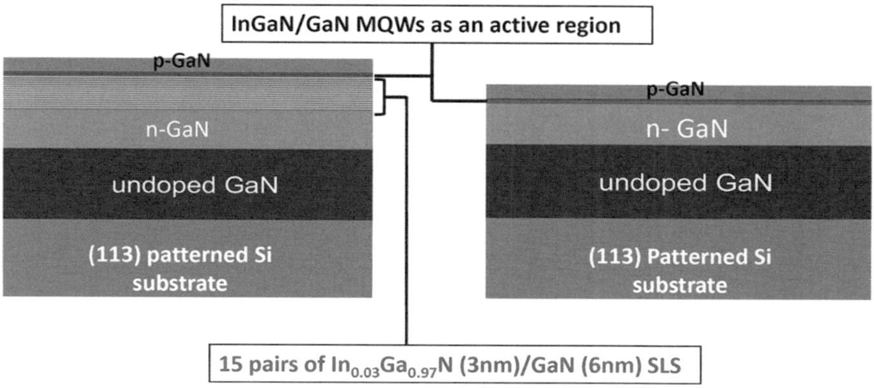

Fig. 6.15 Schematic diagram of the semipolar (11-22) plane LED structure grown on a patterned (113) plane silicon substrate, with and without the insertion of a superlattice layer [43]

6.3 Optical Properties of Nonpolar and Semipolar III-Nitrides

6.3.1 Emission Related to Stacking Fault

Photoluminescence (PL) spectroscopy involves irradiating a sample with light that has energy greater than the material's bandgap. The emitted light from the sample is then collected and analyzed to provide insights into the material's band structure and luminescent defect properties.

In addition to line dislocations, planar defects can also occur in nonpolar and semipolar III-nitrides, affecting the band structure and carrier transport. As shown in Fig. 6.16, when excited by a 325 nm laser at low temperatures, a weak peak appears near 3.40 eV in semipolar (11-22) plane GaN, in addition to the band-edge emission near 3.47 eV. This peak is usually attributed to the presence of BSF. The atomic layer arrangement in the BSF region consists of one or more layers of cubic phase GaN embedded within the hexagonal phase GaN. Since the bandgap of cubic phase GaN is slightly smaller than that of hexagonal phase GaN, the BSF region can be regarded as a quantum well structure. If there is a large number of BSFs in the sample, not only will a strong BSF-related peak be produced, but the intensity of the band-edge peak will also be suppressed. By comparing the intensity ratio between the band-edge peak and the BSF-related peak, the crystal quality of the GaN thin film can be estimated. Zhang et al. found that by controlling the size parameters in the microrod substrate template, the crystal quality of the semipolar (11-22) plane epitaxial GaN could be effectively improved [17].

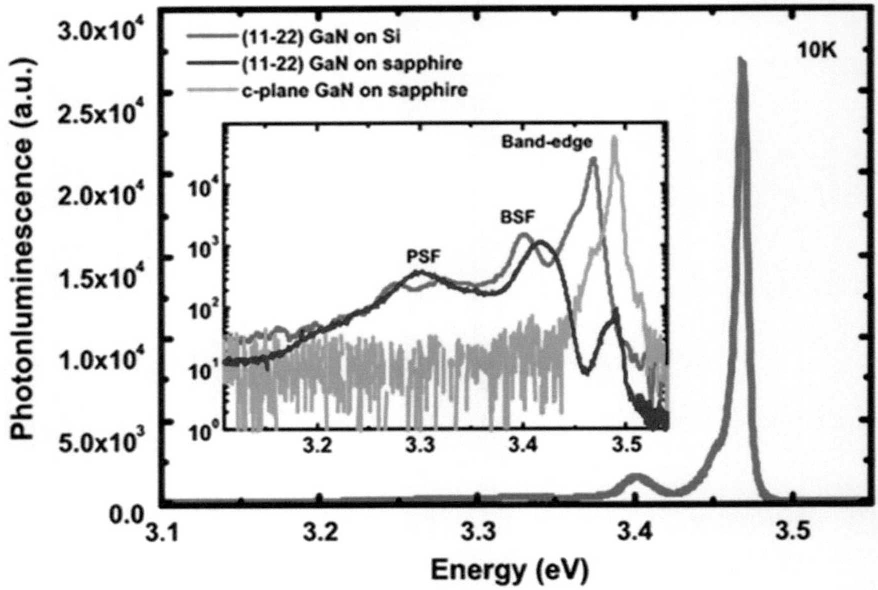

Fig. 6.16 Low-temperature PL spectra of semipolar (11-22) plane GaN grown on a patterned silicon substrate, semipolar (11-22) plane GaN on a sapphire substrate, and (0001) plane GaN on a sapphire substrate [34]

6.3.2 Reduced Impact from Piezoelectric Field

One of the most significant advantages of nonpolar and semipolar orientation, compared to the polar orientation, is the reduced internal electric field. This leads to distinct optical properties, unlike the *c*-plane, as verified among a number of characterization methods.

Electroluminescence (EL) spectroscopy involves injecting external current into the device and collecting the emitted light for testing and analysis. It is a common method applied to nitride LEDs. For the *c*-plane InGaN-based LED, during the EL measurement, it was observed that the EL peak blue-shifted as the injection current increased. This is due to the increase in the number of carriers in the active region with higher injection current, which partially shields the piezoelectric field in the active region, thereby alleviating the QCSE effect. Figure 6.17 depicts the changes in the EL spectra and peak wavelengths of semipolar (11-22) plane InGaN/GaN LED, with emission ranging from green to amber light, as the injection current is varied [46]. When the injection current increased from 10 to 100 mA, the peak positions of the green, yellow-green, and yellow (11-22) LEDs blue-shifted by 8 nm, 15 nm, and 19 nm, respectively. This blue-shift value is smaller than that observed in the corresponding *c*-plane device (where the *c*-plane green LED typically undergoes a blue-shift of about 13 nm under the same range of current variation). This suggests that the QCSE effect in the semipolar (11-22) plane LEDs is weaker than that in the *c*-plane LED device.

In addition to increasing the injection current to raise the number of carriers in the active layer, the carrier concentration can also be increased by enhancing the light irradiance, i.e., increasing the power of the excitation light. This measurement is more suitable for samples that have not formed good electrical contacts, such as InGaN/GaN quantum well structures. Figure 6.18 shows the power-dependent PL spectra of the nonpolar *a*-plane and *c*-plane InGaN/GaN MQW samples at low temperature. As the excitation power of the laser increased from 0.1 to 15 mW, the PL peak of the nonpolar *a*-plane sample remained relatively stable, while the PL peak position of the *c*-plane sample exhibited a significant blue shift.

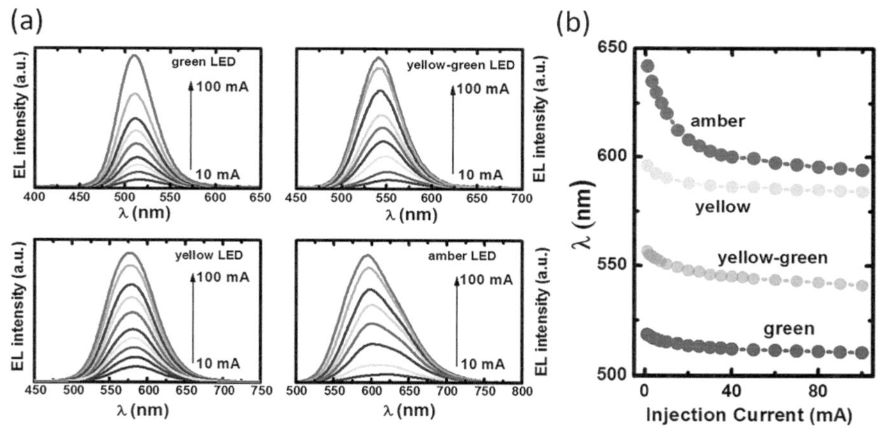

Fig. 6.17 **a** EL spectra and **b** EL peak wavelength of the semipolar (11-22) plane LEDs from green to amber light, under different injection currents [46]

Fig. 6.18 **a** and **b** PL spectra of nonpolar *a*-plane polar *c*-plane InGaN/GaN MQW samples at low temperature with varying excitation power densities

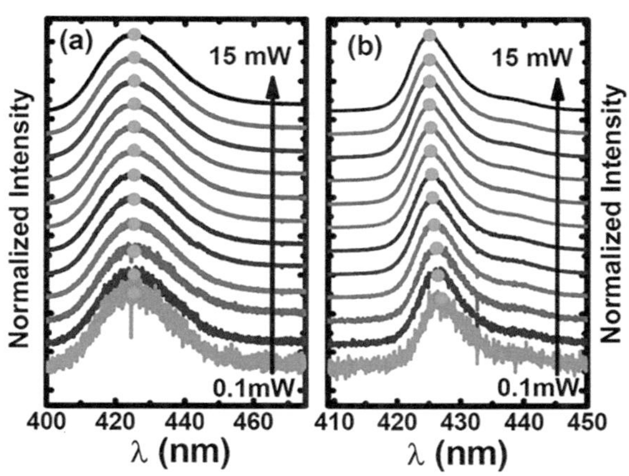

The piezoelectric field not only affects the luminescent properties of InGaN/GaN quantum wells but also alters their light absorption behavior. Photoluminescence excitation (PLE) spectroscopy can evaluate the material's absorption edge by observing the changes in the luminescence spectrum from the sample as the excitation light wavelength is continuously varied near the sample's band edge. Figure 6.19 depicts the PL and PLE spectra of a semipolar (11-22) plane and a *c*-plane InGaN/GaN MQW sample at low temperatures [47]. The PLE spectra were obtained by fixing the detection wavelength at the respective PL peak of the sample while varying the excitation light wavelength. It was found that for the two samples with similar emission wavelengths, although the FWHM of the PL spectrum for the *c*-plane sample is slightly smaller, the absorption edge related to the quantum well in the *c*-plane sample (ranging from 3.2 to 2.5 eV) is broader than that of the semipolar (11-22) plane sample. The broadening of the quantum well-related absorption edge in the polar *c*-plane sample can be attributed to the quantum-constrained Franz-Keldysh (QCFK) effect. Due to the influence of the internal polarization electric field, some transitions between forbidden energy levels in the luminescence process become allowed during absorption. The sharp quantum well-related absorption edge in the semipolar (11-22) plane sample suggests a weakened QCFK effect within the quantum well, indicating a reduced polarization electric field. For the absorption edge observed in PLE spectroscopy, a sigmoidal function can be used for quantitative analysis.

$$a = \frac{a_0}{\left[1 + \exp\left(\frac{E_B - E}{\Delta E}\right)\right]} \tag{6.1}$$

In the above formula, E_B is the effective absorption energy level, ΔE is the broadening parameter related to the Urbach tail energy, and a_0 is a constant. According to the formula, the quantum well-related absorption edges of the semipolar (11-22) plane and *c*-plane samples were fitted to be 2.634 eV and 2.813 eV, respectively. By combining these with their PL peaks, it can be concluded that the Stokes shifts

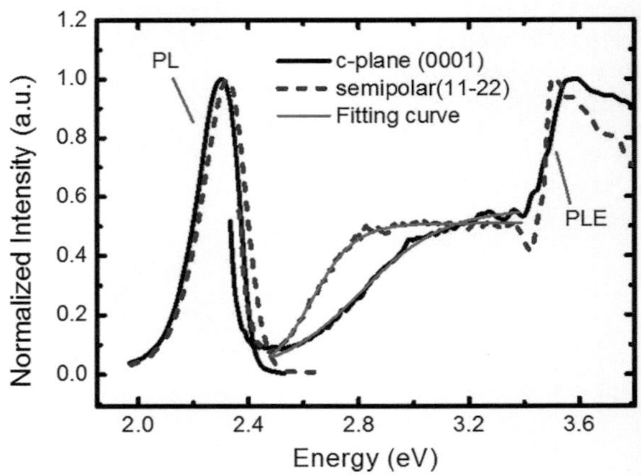

Fig. 6.19 PL and PLE spectra of a semipolar (11-22) plane and a *c*-plane InGaN/GaN MQW samples at low temperature [47]

for the semipolar (11-22) plane and the *c*-plane samples are 312 meV and 509 meV, respectively. This means that the weakening of the piezoelectric field also reduces the energy offset between the absorption edge and the emission peak of the sample.

The piezoelectric field not only alters the band structure in nitride materials but also affects the carrier distribution within them. Time-resolved photoluminescence (TRPL) is widely used to analyze the carrier lifetime and diffusion length in optoelectronic devices. The scatter plots in Fig. 6.20 present the TRPL decay trajectories of two groups of semipolar (11-22) plane and *c*-plane InGaN/GaN MQW samples with similar wavelengths at low temperature. The decay trajectories were recorded by fixing the detection wavelength at the PL peak of each sample. These decay trajectories can be fitted with a double exponential function:

$$I = A_1 e^{-\frac{t}{\tau_1}} + A_2 e^{-\frac{t}{\tau_2}} \tag{6.2}$$

where τ_1 and τ_2 represent the fast and slow components of the carrier lifetime, respectively, and A_1 and A_2 are fitting constants. Compared to the *c*-plane samples, semipolar (11-22) plane samples exhibit a smaller carrier lifetime that is reduced by an order of magnitude, indicating a weakening of the QCSE effect. As the *c*-plane sample moves from 490 to 538 nm, the carrier lifetime increases from 26.7 to 78.4 ns, a nearly threefold increase. In contrast, the carrier lifetime of the semipolar (11-22) plane sample only increases from 2.7 ns at 497 nm to 3.5 ns at 534 nm. This

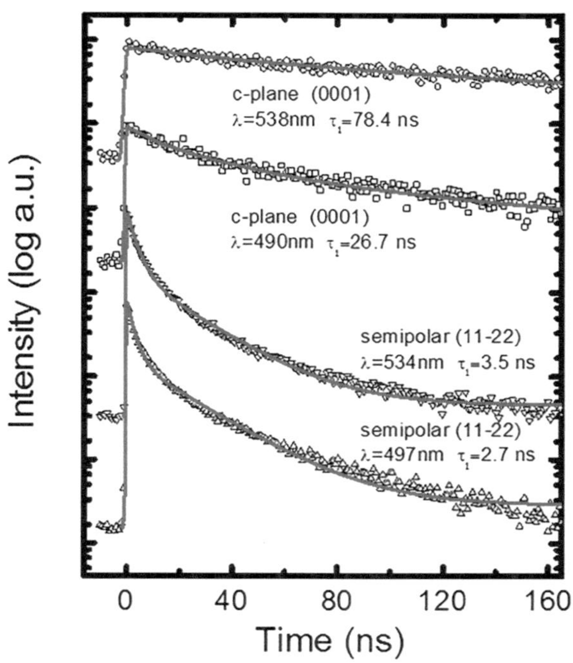

Fig. 6.20 Low-temperature TRPL decay trajectories of two groups of semipolar (11-22) plane and *c*-plane InGaN/GaN MQW samples with similar wavelengths. The red lines represent the fitting curves using a double exponential function [47]

indicates that the QCSE effect in the quantum wells of the semipolar (11-22) plane is much less pronounced with changing In composition compared to that in traditional *c*-plane samples. Owing to the extremely short carrier lifetime, semipolar (11-22) plane LEDs exhibit very high modulation speeds in the long-wavelength band, with modulation bandwidths of green, yellow, and amber LEDs reaching up to 540 MHz, 350 MHz, and 140 MHz, respectively [48].

6.3.3 Anisotropy

Unlike the in-plane isotropy of polar *c*-plane III-nitrides, nonpolar and semipolar nitrides exhibit anisotropy in the growth plane. A common test for anisotropy in nonpolar and semipolar nitrides is the azimuth scan of the on-axis XRD rocking curve, which investigates the change in FWHM as a function of the scanning azimuth. As shown in Fig. 6.21, for the on-axis XRD rocking curve, the FWHM value of the semipolar (11-22) plane GaN grown on a sapphire substrate changes periodically with the scanning azimuth. The FWHM value reaches a minimum in a specific direction (such as parallel to the *c*-axis projection). This anisotropy of the peak broadening is attributed to the tilt of the mosaic structure and the asymmetry of the domain sizes in nonpolar and semipolar nitrides [49, 50]. The former is related to the angular distribution of dislocations with a screw component, while the latter is influenced by lattice mismatch.

The carrier mobility in nonpolar and semipolar nitrides also exhibits a certain degree of anisotropy. As shown in Fig. 6.22, Hall measurements were performed on semipolar (11-22) plane GaN grown on various nanopillar substrate templates. The analysis revealed that the electron mobility along the [1-100] is greater than that along the [11-2-3] direction. Research shows that several factors, including in-plane electric fields, extended defects (such as BSF), and surface morphology, can affect

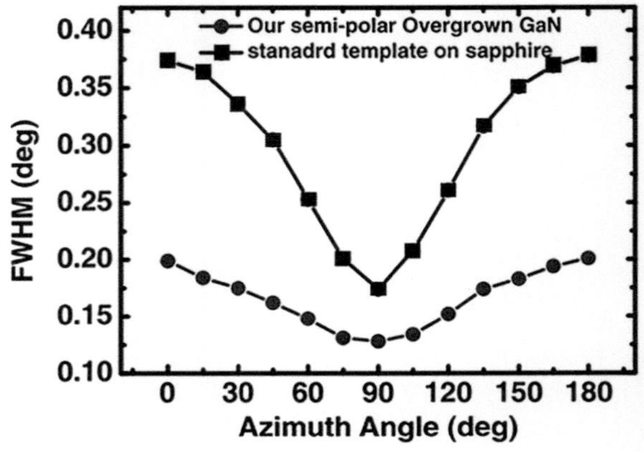

Fig. 6.21 FWHM values of on-axis XRD rocking curves as a function of azimuth angle for the semipolar (11-22) plane GaN films grown on nanopillar substrate template and standard sapphire substrate [51]

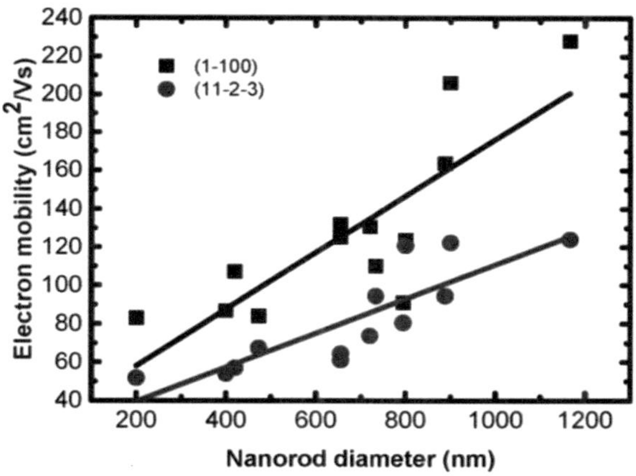

Fig. 6.22 Electron mobility along the [1-100] and [11-2-3] directions of semipolar (11-22) plane GaN grown on nanopillar substrate templates with different nanopillar diameters [52]

the carrier migration in nonpolar and semipolar nitrides, causing a smaller migration speed along the *c*-axis (or the *c*-projection).

The in-plane uneven biaxial strain in nonpolar and semipolar nitrides causes the splitting of the topmost valence band, leading to the anisotropy in the transition to the split subbands, and hence, polarization of the emitted light.

The degree of polarization (ρ) is usually defined as the ratio between the difference and the sum of the emission intensities in two orthogonal directions. For nonpolar *a*-plane nitrides, the directions parallel and perpendicular to the *c*-axis are commonly chosen.

$$\rho = \frac{I_{\perp c} - I_{\parallel c}}{I_{\perp c} + I_{\parallel c}} \tag{6.3}$$

Figure 6.23 shows the change in the integral intensity of two peaks in a nonpolar *a*-plane InGaN-based LED sample as the polarizer rotates. As the polarizer angle changes from 0° to 360°, the integral intensities of both peaks undergo two periodic changes [53]. As the In content in the sample increases, the strain-induced band-splitting becomes larger, leading to an increasing difference in transition probability and conversion energy. This results in a more pronounced polarization degree in the long-wavelength EL peak (448 nm) of the sample. It has also been reported that when the injection current of the device further increases, the polarization degree of the emitted light from the semipolar (11-22) LED sample decreases. This is speculated to be related to the band filling effect. Under high current injection conditions, the transition to the topmost subband will be saturated, reducing the difference in transition probabilities of different energy levels to some extent. As a result, the polarization degree of the emission peak decreases under high current injection conditions.

Fig. 6.23 Polarization degrees of two peaks (418 and 448 nm) in a nonpolar *a*-plane sample at various injection currents (5, 30, and 100 mA) [53]

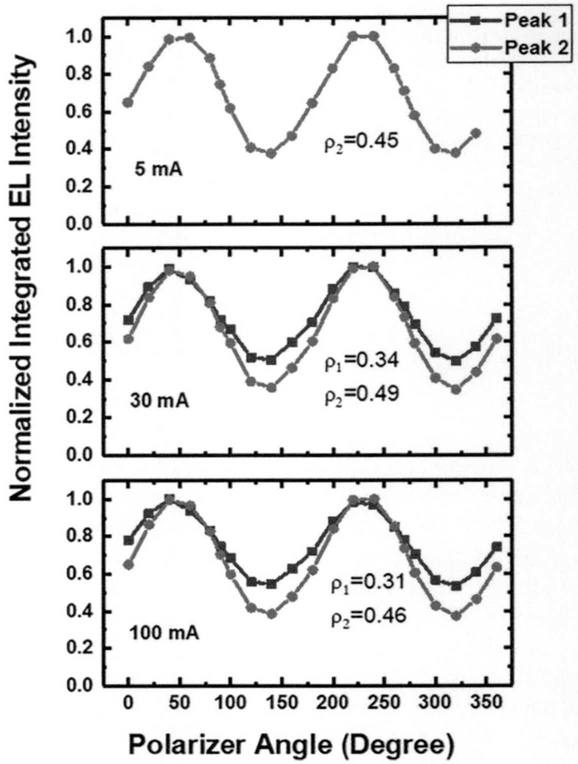

6.4 Summary

This chapter begins by discussing the unique advantages of nonpolar and semipolar nitrides, such as reducing the quantum-confined Stark effect, enabling the development of long-wavelength optoelectronic devices. It also addresses the challenges of obtaining high-quality nonpolar and semipolar GaN heteroepitaxial films. Although a significant amount of research has been conducted using sapphire and silicon substrates, this chapter primarily focuses on the selective area epitaxy and the epitaxial lateral overgrowth on patterned substrates. Additionally, techniques including chemical mechanical polishing and superlattice layers have also been introduced to improve the quality of InGaN/GaN quantum wells. The reduced piezoelectric field in nonpolar and semipolar InGaN/GaN quantum wells results in optical properties such as low color drift and short carrier lifetime, making nonpolar and semipolar nitrides highly promising for applications in fields of smart wearable devices, micro-nano displays, and visible light communication. At the same time, due to the tilt angle between the growth direction and the crystal *c*-axis, the isotropy in conventional *c*-plane III-nitrides is broken in the nonpolar and semipolar orientations, resulting in anisotropy in both structural and optical properties.

References

1. Yue H, Jinfeng Z, Jincheng Z (2013) Nitride wide bandgap semiconductor materials and electronic devices. Science Press
2. Auf der Maur M, Pecchia A, Penazzi G, Rodrigues W, Di Carlo A (2016) Phys Rev Lett 116:027401
3. Kim CS, Kim HG, Hong CH, Cho HK (2005) Appl Phys Lett 87:013502
4. Ryou JH, Lee W, Limb J, Yoo D, Liu JP, Dupuis RD, Wu ZH, Fischer AM, Ponce FA (2008) Appl Phys Lett 92:101113
5. Deguchi T, Shikanai A, Torii K, Sota T, Chichibu S, Nakamura S (1998) Appl Phys Lett 72:3329
6. Kivisaari P, Oksanen J, Tulkki J (2013) Appl Phys Lett 103:211118
7. Romanov AE, Baker TJ, Nakamura S, Speck JS (2006) J Appl Phys 100:023522
8. Waltereit P, Brandt O, Trampert A, Grahn H, Menniger J, Ramsteiner M, Reiche M, Ploog K (2000) Nature 406:865
9. Northrup JE (2009) Phys Rev B 79:041306
10. Wernicke T, Schade L, Netzel C, Rass J, Hoffmann V, Ploch S, Knauer A, Weyers M, Schwarz U, Kneissl M (2012) Semicond Sci Technol 27:024014
11. Zhao Y, Yan Q, Huang CY, Huang SC, Shan Hsu P, Tanaka S, Pan CC, Kawaguchi Y, Fujito K, Van de Walle CG, Speck JS, DenBaars SP, Nakamura S, Feezell D (2012) Appl Phys Lett 100:201108
12. McLaurin M, Mates TE, Speck JS (2005) Appl Phys Lett 86:262104
13. Tsuchiya Y, Okadome Y, Honshio A, Miyake Y, Kawashima T, Iwaya M, Kamiyama S, Amano H, Akasaki I (2005) Jpn J Appl Phys 44:L1516
14. Kaeding JF, Asamizu H, Sato H, Iza M, Mates TE, DenBaars SP, Speck JS, Nakamura S (2006) Appl Phys Lett 89:202104
15. Marcinkevičius S, Kelchner KM, Kuritzky LY, Nakamura S, DenBaars SP, Speck JS (2013) Appl Phys Lett 103:111107
16. Masui H, Asamizu H, Tyagi A, DeMille NF, Nakamura S, DenBaars SP (2009) Appl Phys Express 2:071002
17. Zhang Y, Bai J, Hou Y, Yu X, Gong Y, Smith RM, Wang T (2016) Appl Phys Lett 109:241906
18. Dinh DV, Akhter M, Presa S, Kozlowski G, O'Mahony D, Maaskant PP, Brunner F, Caliebe M, Weyers M, Scholz F, Corbett B, Parbrook PJ (2015) Phys Status Solidi 212:2196
19. Song J, Choi J, Xiong K, Xie Y, Cha JJ, Han J, Appl ACS (2017) Mater Interfaces 9:14088
20. Khoury M, Li H, Li P, Chow YC, Bonef B, Zhang H, Wong MS, Pinna S, Song J, Choi J, Speck JS, Nakamura S, DenBaars SP (2020) Nano Energy 67:104236
21. Kato Y, Kitamura S, Hiramatsu K, Sawaki N (1994) J Cryst Growth 144:133
22. Zheleva TS, Nam OH, Bremser MD, Davis RF (1997) Appl Phys Lett 71:2472
23. Wang T (2016) Semicond Sci Technol 31:093003
24. Xing K, Wang J, Wang L, Tao X, Chen S, Liang H (2019) Appl Phys Express 12:105501
25. Sekiguchi H, Kishino K, Kikuchi A (2008) Appl Phys Express 1:124002
26. Okada N, Kurisu A, Murakami K, Tadatomo K (2009) Appl Phys Express 2:091001
27. Chen HG, Ko TS, Ling SC, Lu TC, Kuo HC, Wang SC, Wu YH, Chang L (2007) Appl Phys Lett 91:021914
28. Tendille F, De Mierry P, Vennéguès P, Chenot S, Teisseire M (2014) J Cryst Growth 404:177
29. Leung B, Wang D, Kuo YS, Xiong K, Song J, Chen D, Park SH, Hong SY, Choi JW, Han J (2014) Appl Phys Lett 104:262105
30. Song J, Choi J, Zhang C, Deng Z, Xie Y, Han J, Appl ACS (2019) Mater Interfaces 11:33140
31. Tanikawa T, Hikosaka T, Honda Y, Yamaguchi M, Sawaki N (2008) Phys Status Solidi Curr Top Solid State Phys 5:2966
32. Hikosaka T, Tanikawa T, Honda Y, Yamaguchi M, Sawaki N (2008) Phys Status Solidi 5:2234
33. Yu X, Hou Y, Shen S, Bai J, Gong Y, Zhang Y, Wang T (2016) Phys Status Solidi 13:190
34. Bai J, Yu X, Gong Y, Hou YN, Zhang Y, Wang T (2015) Semicond Sci Technol 30:065012
35. Ni X, Özgür U, Baski AA, Morkoç H, Zhou L, Smith DJ, Tran CA (2007) Appl Phys Lett 90:182109

36. Bougrioua Z, Laügt M, Vennéguès P, Cestier I, Gühne T, Frayssinet E, Gibart P, Leroux M (2007) Phys Status Solidi 204:282
37. De Mierry P, Kriouche N, Nemoz M, Nataf G (2009) Appl Phys Lett 94:191903
38. Song KR, Lee JH, Han SH, Yi HR, Lee SN (2013) Mater Res Bull 48:5088
39. Xing K, Gong Y, Bai J, Wang T (2011) Appl Phys Lett 99:181907
40. Bai J, Gong Y, Xing K, Yu X, Wang T (2013) Appl Phys Lett 102:101906
41. Zhang Y, Bai J, Hou Y, Smith RM, Yu X, Gong Y, Wang T (2016) AIP Adv 6:025201
42. Bai J, Jiu L, Gong Y, Wang T (2018) Semicond Sci Technol 33:125023
43. Zhao X, Huang K, Bruckbauer J, Shen S, Zhu C, Fletcher P, Feng P, Cai Y, Bai J, Trager-Cowan C, Martin RW, Wang T (2020) Sci Rep 10:12650
44. Haller C, Carlin JF, Jacopin G, Martin D, Butté R, Grandjean N (2017) Appl Phys Lett 111:262101
45. Khoury M, Li H, Bonef B, Mates T, Wu F, Li P, Wong MS, Zhang H, Song J, Choi J, Speck JS, Nakamura S, DenBaars SP (2020) Opt Express 28:18150
46. Bai J, Xu B, Guzman FG, Xing K, Gong Y, Hou Y, Wang T (2015) Appl Phys Lett 107:261103
47. Zhang Y, Smith RM, Hou Y, Xu B, Gong Y, Bai J, Wang T (2016) Appl Phys Lett 108:031108
48. Haggar JI, Cai Y, Ghataora SS, Smith RM, Bai J, Wang T, Appl ACS (2020) Electron Mater 8:2363
49. Sun Q, Ko TS, Yerino CD, Zhang Y, Lee IH, Han J, Lu TC, Kuo H-C, Wang SC (2009) Jpn J Appl Phys 48:071002
50. Moram MA, Vickers ME (2009) Reports. Prog Phys 72:036502
51. Xing K, Gong Y, Yu X, Bai J, Wang T (2013) Jpn J Appl Phys 52:08JC03
52. Xu B, Yu X, Gong Y, Xing K, Bai J, Wang T (2015) Phys Status Solidi 252:1079
53. Bai J, Jiu L, Poyiatzis N, Fletcher P, Gong Y, Wang T (2019) Sci Rep 9:9770

Chapter 7
Construction and Characterization of ZnO Micro/Nanolasers

Junfeng Lu, Jun Dai, Wei Liu, Xiaoxuan Wang, Feifei Qin, Ru Wang, Zengliang Shi, Gangyi Zhu, and Chunxiang Xu

The first generation of semiconductors represented by silicon and the second generation represented by GaAs and InP have triggered the rapid development of microcomputers, integrated circuits, and modern optoelectronic integrated devices. In recent years, the third generation of wide bandgap semiconductors, represented by gallium nitride (GaN), has shown a wide range of application markets in the fields of lighting, shortwave lasers, ultraviolet (UV) optoelectronic detection, and power devices. Zinc oxide (ZnO) has a hexagonal wurtzite structure and an energy bandgap (3.37 eV) similar to GaN. Its exciton binding energy is as high as 60 meV, far greater than room-temperature thermal energy (26 meV); thus, it is expected to realize exciton-type shortwave light-emitting devices and low-threshold laser devices at room temperature. In 1997, Zikang Tang and others obtained the optically pumped stimulated emission of ZnO thin films at room temperature [1], and the *Science* magazine predicted the potential advantages of ZnO in the field of ultraviolet laser devices with the title "*Will UV Lasers Beat the Blues?*" [2]. Over the past 20 years, the research on ZnO has mainly focused on luminescence and lasing properties, design and construction of luminescent devices and optoelectronic detection devices, micro/nanostructures, and optoelectronic integration, which has achieved a series of important progress. A large amount of previous research has revealed the excitonic emission characteristics in ZnO photoluminescence (PL) and clarified the characteristic structure of excitonic spectra from multiple perspectives such as exciton concentration, exciton–exciton interaction, and exciton–phonon interaction, while a large number of literature reports have reported the impact of defect states on optical, electrical, and magnetic properties. It is worth noting that the ZnO micro/nanostructures with rich morphology have been obtained in material preparation, which provides an excellent material science basis for further research on micro/nanolasers [3–6]. In 2001, the F-P laser

J. Lu (✉) · J. Dai · W. Liu · X. Wang · F. Qin · R. Wang · Z. Shi · G. Zhu · C. Xu
Nanjing University of Aeronautics and Astronautics, Nanjing, China
e-mail: lujunfeng@nuaa.edu.cn

S. Xu (ed.), *Optical Characterization of Microstructures and Optoelectronic Devices Based on Wide Band Gap Semiconductors*, Wide Bandgap Semiconductors,
https://doi.org/10.1007/978-981-95-1928-6_7

in single ZnO nanowire was first discovered [7], and then Geralla and Zhu et al. reported the whispering gallery mode (WGM) microcavity laser, further promoting the research of ZnO lasers [8, 9]. Current research on ZnO micro/nanolasers has covered cavity construction, gain mechanism, mode control, etc., and achieving the lasing enhancement has been with the help of metal surface plasmons. Research on ZnO optoelectronic devices has expanded from early electroluminescent devices in ZnO thin film to the design, construction, and performance of micro/nano electrically pumped lasers, micro/nanostructured ultraviolet optoelectronic detectors, and other devices. This chapter will systematically review the optoelectronic properties and characterization methods of ZnO micro/nanostructures from the aspects of material preparation and characterization, fluorescence characteristics, stimulated emission, mode regulation, electroluminescence, etc.

7.1 Preparation and Characterization of ZnO Micro/Nanostructures

Based on the special wurtzite crystal structure and growth characteristics of ZnO, variously shaped ZnO micro/nanostructure single crystals can be obtained (such as microrods, nanorods, microbelts, microdisks, microspheres, etc.) by adopting different synthesis methods and controlling growth conditions, many of which are natural optical microcavities. The resonant modes observed in ZnO micro/nanostructures can be mainly summarized into the following three types: (1) random lasers, (2) Fabry–Pérot (F-P) lasers, and (3) whispering gallery mode (WGM) lasers. The WGM oscillation is a kind of resonant mode formed by confining the light wave in a ring or spherical resonator based on total internal reflection at the inner wall. It has superior characteristics such as high Q value, small mode volume, high energy density, and narrow linewidth, which are widely studied and applied in the research of optics, electronics, biochemical sensing detection, etc.

There are many methods for preparing ZnO micro/nanostructures, which can be divided into gas-phase, liquid-phase, and solid-phase methods according to different preparation environments. The gas-phase method refers to the process of converting the materials used for crystal growth into gas-phase atoms through sublimation, evaporation, decomposition, etc., and obtaining the required crystal micro/nanostructure through nucleation growth under appropriate conditions. The liquid-phase method uses a solution as a medium or carrier to cause a certain physical and chemical reaction of the reaction source and then crystallize and grow to form a micro/nanostructure material, mainly including hydrothermal method, sol–gel method, solvothermal method, microemulsion method, etc. The solid-phase method refers to the method of reacting solid powder to form micro/nanostructured materials under certain external energy, such as mechanical ball milling method, etc. At present, gas-phase and liquid-phase methods are mainly used in experimental research to prepare various ZnO micro/nanostructures.

7.1.1 Gas-Phase Method

The gas-phase method is usually carried out at a higher temperature, with the advantages of simple process, high product purity, good crystallinity, and good size controllability. It is widely used to prepare various high-crystalline-quality ZnO micro/nanostructures. According to different pathways of gas-phase conversion, the gas-phase preparation of ZnO micro/nanostructures mainly includes chemical vapor deposition (CVD), physical vapor deposition (PVD), metal–organic chemical vapor deposition (MOCVD), molecular beam epitaxy (MBE), thermal evaporation method, magnetron sputtering method (RF), thermal decomposition method, etc. This article mainly introduces the commonly used CVD method and its growth mechanism.

The CVD method refers to the process of transforming the intended crystal material into a gaseous state through sublimation, evaporation, and decomposition, which becomes saturated vapor under appropriate conditions, and then condenses and crystallizes to grow crystals. As a type of chemical vapor deposition technology, the MOCVD method carries metal–organic compounds and other reaction sources to the heated substrate above the reaction chamber through the carrier gas. As the temperature rises, a series of chemical and physical changes occur in the gas phase and at the gas–solid interface, and epitaxial growth finally occurs on the substrate surface. Since the process of crystal growth is spatially discontinuous and heterogeneous, the crystallization occurs only on the growth interface, and the advancement of the interface is controlled by the temperature gradient. In order to obtain high-quality crystals, a reasonable temperature-gradient distribution must be established in the melt growth system, which is related to the heat transfer mechanism in heat transport. The latent heat of crystallization released during crystal growth must be promptly transported away from the interface, and then the solidification process occurs. As shown in Fig. 7.1, zinc vapor is obtained through direct evaporation or chemical reaction where zinc or zinc compounds are served as raw materials, and then it reacts with oxygen to form various forms of nano-ZnO. The CVD method has many advantages: small particle size, uniformity, good dispersion, high chemical reaction activity, controllable preparation process, continuous process, and the ability to grow various forms of ZnO micro/nanostructures, including micro/nanowires, micro/nanorods, micro/nanodisks, micro/nanocombs, micro/nanobelts, etc. [10–18]. The advantage of the MOCVD method is that a variety of organometallic compounds can be selected as source materials, and therefore it can be used to grow a variety of compound semiconductors. It can not only prepare high-purity materials, but also precisely control the thickness, composition, and interface of the ultrathin-layer materials that are grown, and it can perform selective growth. The characteristics of this method are high supersaturation of the crystal, fast growth rate, good crystallinity of the sample, and easy control of the growth process, and it has been widely used in the preparation of ZnO thin films and ZnO nanomaterials [19].

According to various gas-phase reaction processes, the growth mechanism of ZnO can be divided into gas–liquid–solid (VLS) and gas–solid (VS) growth mechanisms. The VLS mechanism was first proposed by Wagner and Ellis in the 1960s and used

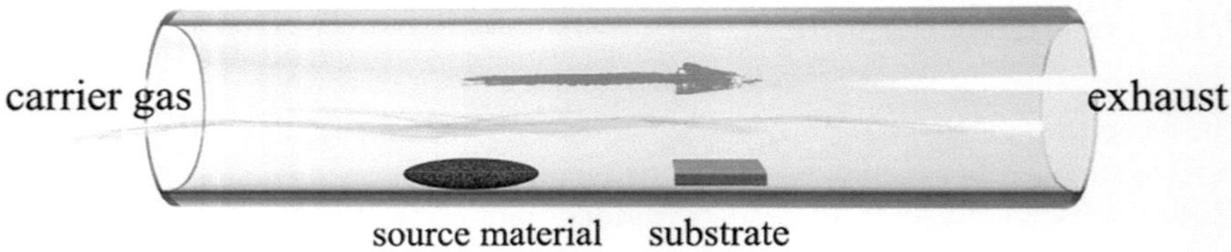

Fig. 7.1 Schematic diagram of CVD preparation

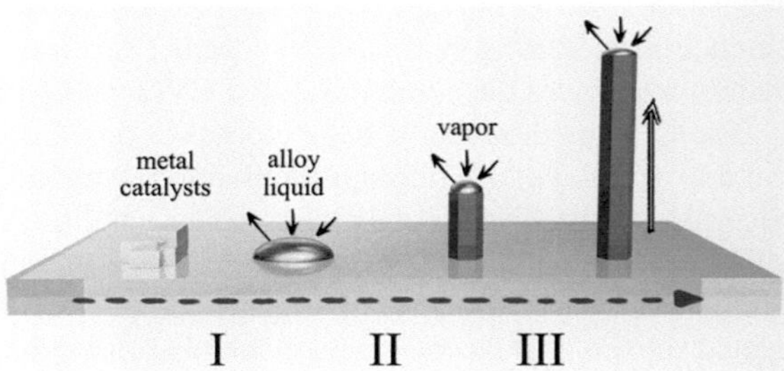

Fig. 7.2 VLS growth mechanism model [21]

for the growth of Si microwhiskers [20]. The preparation of whiskers by the VLS method needs to introduce the catalyst. The catalyst and the growth component form alloy liquid particles with a lower solidification point. At a certain temperature, the component vapor evaporated from the growth raw materials continuously dissolves in the alloy droplets to form liquid components. When the solute component in the liquid reaches supersaturation, nanomaterials precipitate along the preferred direction of the solid–liquid interface, thereby growing into a linear crystal, as shown in Fig. 7.2 [21]. In 2001, Yang Peidong's research group observed the VLS growth process of ZnO nanowires under gold catalysis using in situ transmission electron microscopy (TEM) [22].

The mixture of ZnO and carbon powder is a common formula for VLS growth. In the high-temperature zone, ZnO is reduced by carbon powder to form zinc vapor, which is transported to the low-temperature zone and mixed with liquid gold particles to form alloy liquid particles. When the zinc vapor is supersaturated, nanomaterials nucleate at the solid–liquid interface and grow epitaxially in the direction of low surface energy. In the VLS growth process, the morphology, density, position, etc., of the nanostructure can be controlled by selecting and changing the catalyst and growth conditions. For example, the change in morphology is mainly achieved by controlling factors such as supersaturation, temperature, and pressure. If the gas-phase supersaturation in the reactor is too low, it is difficult to form a nucleus and synthesize a crystal; if the gas-phase supersaturation is too high, the nucleation base unit will be adsorbed on the nucleus surface in the form of several molecular clusters, affecting the stable growth of the crystal and even forming polycrystalline agglomerates. Only when the gas-phase supersaturation is moderate will the gaseous base unit be adsorbed on the nucleus surface in the form of a single molecule, forming orderly adsorption, which is conducive to the directional and uniform growth of the crystal. A large number of theoretical and experimental studies on the growth kinetics of one-dimensional materials have been reported and concluded that the two-dimensional (2D) nucleation probability on the crystal surface can be expressed by the following formula [23]:

$$P_N = B\exp\left(-\frac{\pi\sigma^2}{k^2T^2\ln\alpha}\right) \tag{7.1}$$

where P_N is the nucleation probability, B is a constant, σ is the surface energy of the whisker, k is the Boltzmann constant, T is the absolute temperature, and $\alpha = \mathrm{p}/\mathrm{p}_0$ is the supersaturation of the growth atmosphere, where p is the actual vapor pressure and p_0 is the saturated vapor pressure at temperature T. The surface energy is related to the interface, where the surface energy of the low-index crystal face is lower. According to the above formula, the lower the surface energy, the greater the probability of 2D nucleation. On the other hand, the binding energy of atoms adsorbed on low-energy surfaces is low, and the probability of desorption is high. The competition and coordination of these two processes determine the shape of the crystal. Two parameters—temperature and supersaturation—controlled by process conditions are important factors affecting nucleation. Higher temperatures and greater supersaturation are more conducive to 2D nucleation, forming a flake structure. Conversely, lower temperatures and smaller supersaturation favor the growth of linear structures.

Unlike VLS, the VS mechanism does not require the introduction of a catalyst. First, the elemental gas is produced through thermal evaporation, chemical reduction, gas-phase reaction, etc., and then the gas is transported to the substrate area and nucleates on the substrate, leading to the growth of nanostructures. This way of growing whiskers is often explained as nucleation at microscopic defects (dislocations, twins, etc.) on the gas–solid interface, forming a one-dimensional structure. VS growth can be seen as Zn "self-catalyzed." In the absence of a catalyst, the preparation of one-dimensional nanostructures usually depends on the VS process, such as the preparation of ZnO nanobelts [24]. The growth model is shown in Fig. 7.3.

7.1.2 Liquid-Phase Method

The liquid-phase method can also obtain chemically uniform micro/nanostructures for materials with complex compositions. It has lower cost, larger output, convenient preparation, and does not require very harsh conditions, but the crystalline quality of the obtained materials is not high. Hydrothermal synthesis is the most used method for preparing ZnO micro/nanostructures by the liquid phase.

The hydrothermal method is a medium with water solution or steam, which has the advantages of low reaction temperature, low economic cost, and suitability for large-scale production [25, 26]. The hydrothermal growth process can generally be described by the "growth unit" theoretical model: In the initial transport stage of the

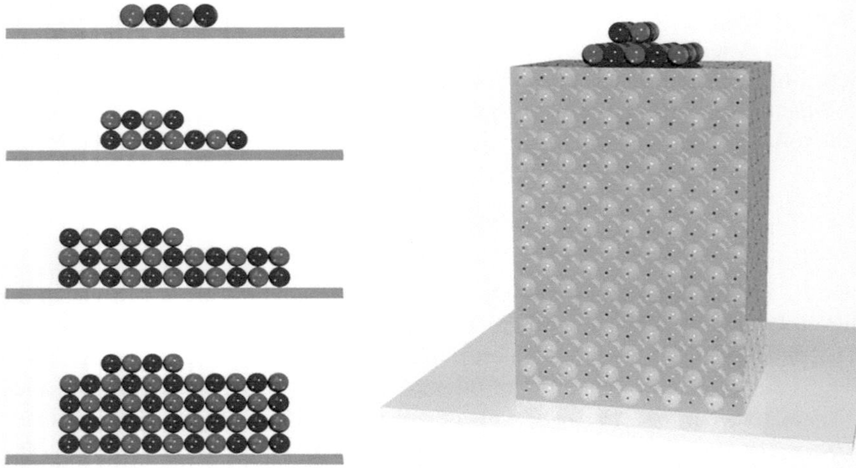

Fig. 7.3 VS growth mechanism model [24]

reaction, various substances can flow under the action of convection, and the dissolved ions and molecules can undergo a series of reactions, eventually forming nanomaterials with a certain morphology. For the preparation of ZnO micro/nanostructures, soluble zinc salts (such as zinc acetate or zinc nitrate, etc.) and weak alkalis (such as ammonia or hexamethylenetetramine, etc.) are generally used as reaction precursors. $Zn(OH)_2$ precipitate is formed in the water solution, and then ZnO is obtained through hydrolysis. The reaction process can be explained by the following chemical reaction equation:

$$\textit{Dissolution reaction}: (CH_2)_6N_4 + 6H_2O \leftrightarrow 6HCHO + 4NH_3 \tag{7.2}$$

$$\textit{Hydroxyl reaction}: NH_3 + H_2O \leftrightarrow NH_4^+ + OH^- \tag{7.3}$$

$$\textit{Oversaturation reaction}: Zn^{2+} + 4OH^- \rightarrow Zn(OH)_4^{2-} \tag{7.4}$$

$$\textit{Growth reaction}: Zn(OH)_4^{2-} \rightarrow ZnO + H_2O + 2OH^- \tag{7.5}$$

Boyle invented a two-step reaction method based on this, which involves obtaining a ZnO template layer on the substrate and conducting a chemical bath reaction. The ZnO nanowire array prepared by his method has good orientation [27]. Xu Chunxiang's research group synthesized ZnO nanorod array by using the hydrothermal method (as shown in Fig. 7.4a and b). The specific growth process is mainly divided into four steps. First, the magnetron sputtering method is used to sputter a layer of nanofilm on the substrate as the ZnO seed layer to reduce the thermodynamic potential barrier and lattice mismatch between the substrate and ZnO nanorods, promoting

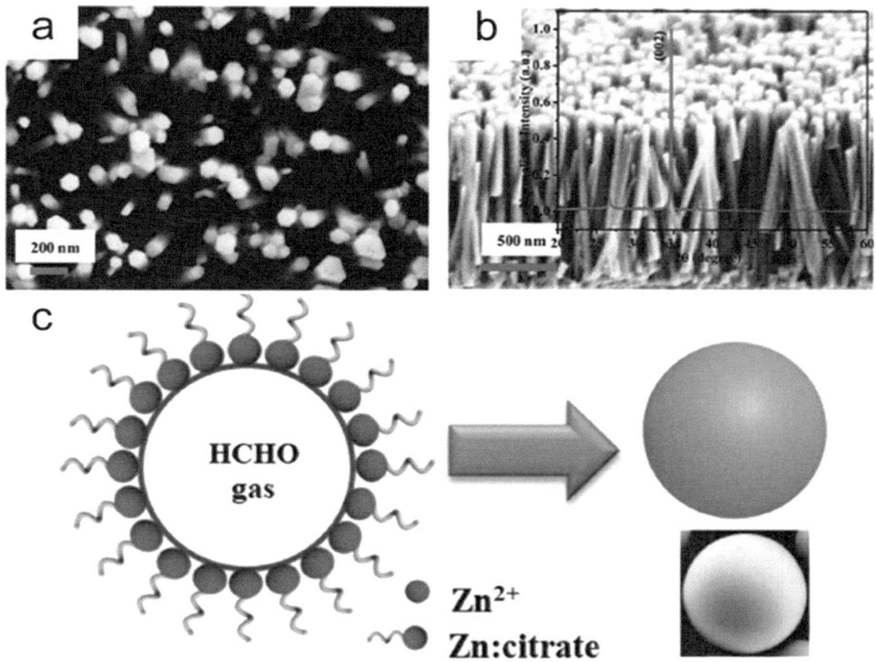

Fig. 7.4 Hydrothermal preparation of ZnO micro/nanostructures [28, 31]

lattice formation and inducing the growth of uniform ZnO nanorod arrays. Second, the precursor is evenly mixed in deionized water. Third, the treated substrate is placed in the precursor solution and used for the growth of ZnO nanorod arrays in a high-pressure reaction kettle. Finally, after a suitable temperature and time reaction, the sample is taken out, rinsed, and blown dry. The morphology of ZnO can be regulated by various methods, such as changing the concentration of reactants, solution pH, reaction time, and reaction temperature, etc. In addition, the research results indicate that trisodium citrate plays a key role in blocking, capping, and orienting during the growth of ZnO. The hydroxyl and carboxyl groups in the citrate ions can attract Zn^{2+} and control the growth direction of the ZnO structure, which can regulate the hydrothermal growth process to form a smooth ZnO microsphere structure (as shown in Fig. 7.4c) [28–30].

7.1.3 Characterization of ZnO Micro/Nanostructures

The development of modern materials science largely depends on the understanding of the relationship between material properties and their composition, structure, and microstructure. Therefore, the characterization technology at the microscopic level constitutes an important part of materials science. Since ZnO can easily form various micro/nanostructures, whose morphological structure has a direct impact on material properties, it is of great significance to carry out fine microstructure characterization. This article takes several typical ZnO micro/nanostructures as examples and introduces several main microstructure characterization techniques such as field emission scanning electron microscopy (FESEM), transmission electron microscopy (TEM), X-ray diffractometry (XRD), etc.

1. ZnO Micro/Nanorods

As a typical ZnO micro/nanostructure, there are many methods for preparing ZnO micro/nanorods. Figure 7.5 shows the scanning electron microscopy (SEM) and TEM images of ZnO nanowires prepared by the VLS method [22]. Figure 7.5a is the SEM image of the prepared ZnO nanowires when a gold thin layer is sputtered on the substrate as a catalyst. The diameter of the nanowires ranges from 40 to 70 nm, and the length is between 5 and 10 μm. By adjusting the thickness of the gold thin layer, the diameter of the prepared nanowires can also be controlled. Figure 7.5b is a TEM image of a single nanowire. It can be seen that a clear alloy particle exists at the top of the nanowire, which provides strong experimental evidence for the VLS growth mechanism. Figure 7.5c shows the high-resolution TEM (HRTEM) image of a single-crystal ZnO nanowire. It shows that the distance between the adjacent two crystal faces is about 0.26 nm, which corresponds to the spacing of (0002) crystal face, indicating that the nanowire grows along the <0001> direction.

Figure 7.6a and b shows the SEM images of ZnO submicron rods prepared by the CVD method, with a diameter of about 600 nm, a regular hexagonal cross section, and smooth and flat lateral surfaces [31, 32]. The corresponding HRTEM image is shown in Fig. 7.6c. It can be seen that it has a clear lattice structure with an interlayer distance of about 0.26 nm, which matches the interplanar spacing of the wurtzite ZnO (0002) plane. The selected area electron diffraction (SAED) pattern shown in the illustration indicates that the ZnO microrods prepared by this method have good single-crystal performance. In addition, the crystal structure of the prepared sample was analyzed by X-ray diffraction. As shown in Fig. 7.6d, the X-ray diffraction pattern of the sample clearly shows diffraction peaks at 31.8°, 34.3°, and 36.5°, corresponding to the (1000), (0002), and (1010) planes of the wurtzite structure ZnO, respectively. This is completely consistent with the standard diffraction pattern of ZnO (JCPDS, 36-1451). The highest intensity diffraction peak at 34.3° reflects

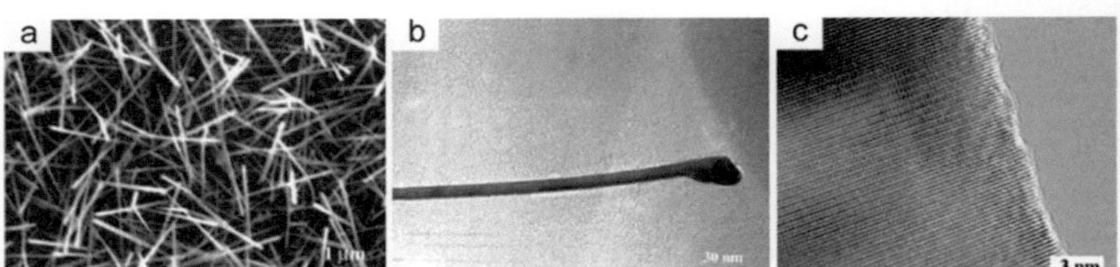

Fig. 7.5 Preparation and characterization of ZnO nanowires by VLS method [22]

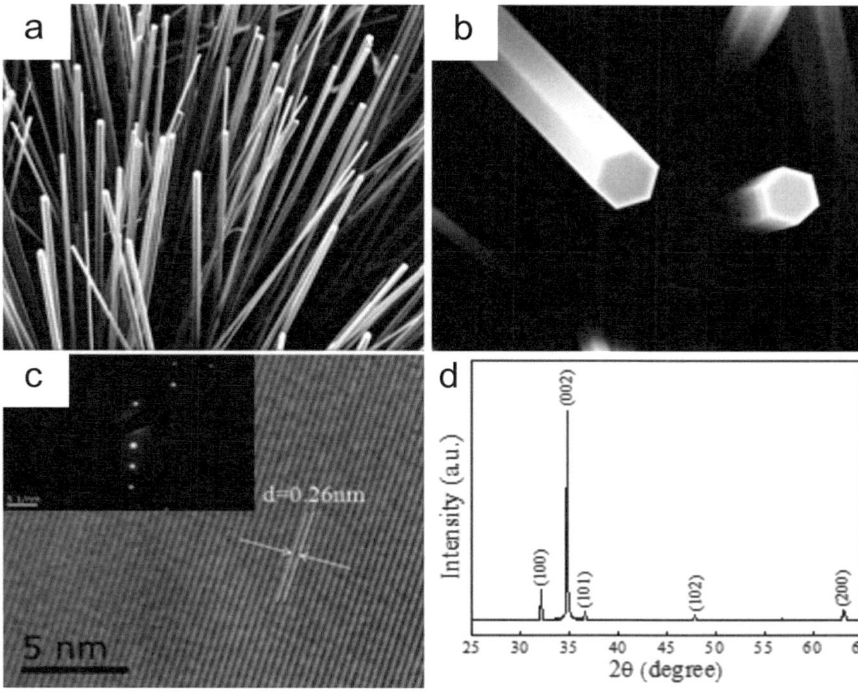

Fig. 7.6 Submicron-sized ZnO microrods prepared by CVD method [32, 33]

the growth of ZnO microrods mainly along the [0001] direction, which is consistent with the results observed by high-resolution transmission electron microscopy and selected area electron diffraction. Obviously, the extremely narrow diffraction peak width indicates that these ZnO microrods have high crystalline quality.

2. ZnO Nanobelts

The research group of Wang Zhonglin [34] first prepared polycrystalline ZnO nanobelts on a sapphire substrate using a high-temperature evaporation method, as shown in Fig. 7.7a. ZnO nanobelts are usually obtained by high-temperature evaporation of ZnO powder without a catalyst, with uniform width and rectangular cross section. As shown in Fig. 7.7c, the nanobelt grows along the $[2\bar{1}\bar{1}0]$ (*a*-axis), the side faces are $\pm(01\bar{1}0)$ faces, the top and bottom faces are $\pm(0001)$ faces, and the thickness is very thin (5–20 nm). If the surface polarized charge cannot be compensated during the growth of the nanobelt, the polar faces will connect to each other to minimize their area, eventually forming a ring, as shown in Fig. 7.7d. If the polar faces do not connect into a ring, the nanobelt will continue to bend and wind around, forming a nanospring, as shown in Fig. 7.7e. If the nanobelt is wound, the repulsion between the charged surfaces makes the nanospring elongate, while the elastic deformation force pulls the ring closer. The balance of the two forms an elastic nanospring, as shown in Fig. 7.7f.

3. ZnO Micro/Nanodisks

Due to their uniform symmetrical shape and large surface area, disk-shaped ZnO micro/nanostructures have potential application prospects in the fields of information storage, transducers, light emitters, high-sensitivity sensors, lasers, etc. The Xu Chunxiang's research group [14, 35, 36] obtained high-quality ZnO micro/nanodisks using the CVD method, as shown in Fig. 7.8a. From the HRTEM image of the

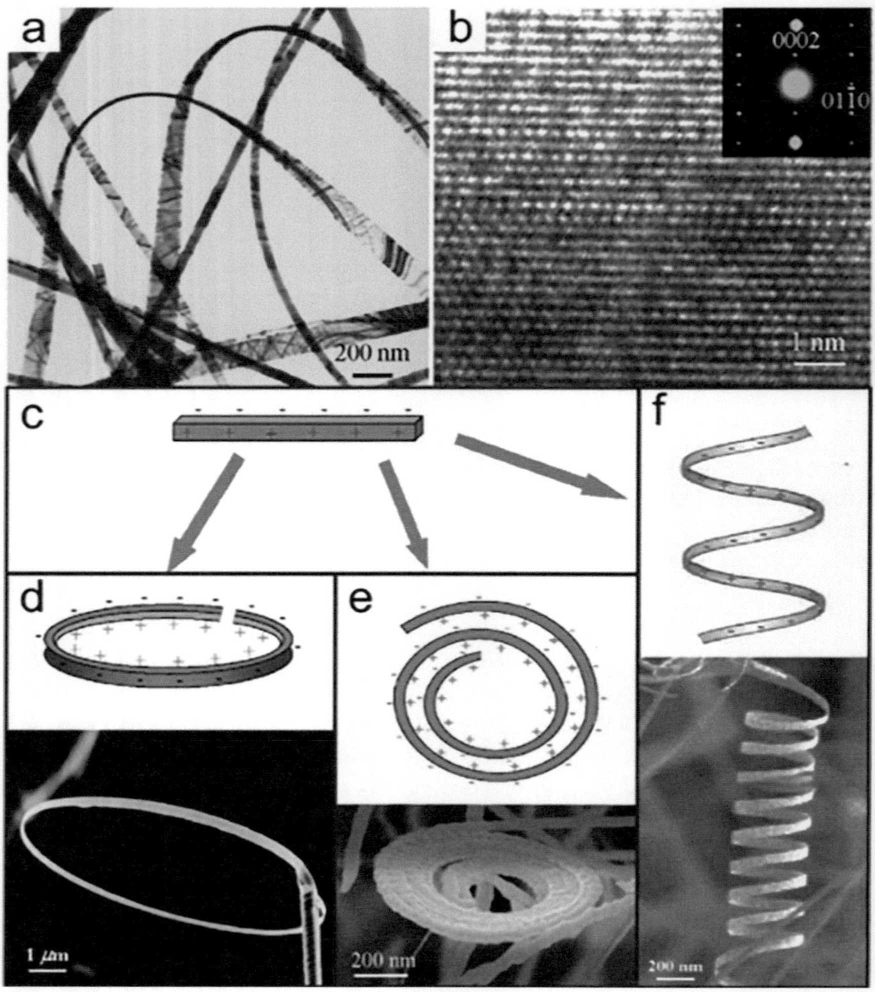

Fig. 7.7 Growth process of ZnO nanobelts and nanosprings [34]

nanodisk observed from the [0001] direction, it can be seen that the atoms are orderly arranged, forming a sixfold symmetrical projection structure, as shown in Fig. 7.8b. Along these six symmetrical directions, the lattice spacing corresponding to the $(10\bar{1}0)$ plane is 0.28 nm. This indicates that the ZnO nanodisk mainly grows along the $\pm[10\bar{1}0]$, $\pm[1\bar{1}00]$ and $\pm[1\bar{1}0]$ six symmetrical directions, and the growth along the [0001] direction is suppressed (Fig. 7.8c). As shown in Fig. 7.8d, the XRD results show that the symmetrical structure of the ZnO nanodisk has a typical wurtzite ZnO crystal structure, with lattice constants of a = 3.250 Å and c = 5.207 Å. Strong diffraction peaks appear at 31.8°, 34.3°, and 36.5°, corresponding to the $(10\ \bar{1}\ 0)$, (0002), and $(10\ \bar{1}\ 1)$ planes of wurtzite ZnO.

The growth of ZnO micrometer disks is not only related to a larger degree of supersaturation but also to a higher growth temperature. As the surface tension coefficient of the liquid decreases with increasing temperature, it is difficult for zinc vapor to condense into sufficiently large catalyst droplets at higher growth temperature. Therefore, it cannot nucleate through the VLS process but can only be controlled by the VS process, where gaseous oxygen atoms and zinc atoms directly form solid hexagonal columnar nuclei. Liquid zinc atoms are preferentially adsorbed on the (0001) plane of ZnO because of its lowest surface energy, forming a thin layer of liquid zinc that blocks the subsequent zinc and oxygen atoms from entering the (0001) plane, thereby inhibiting the growth of this plane. Although the growth rate of the six side

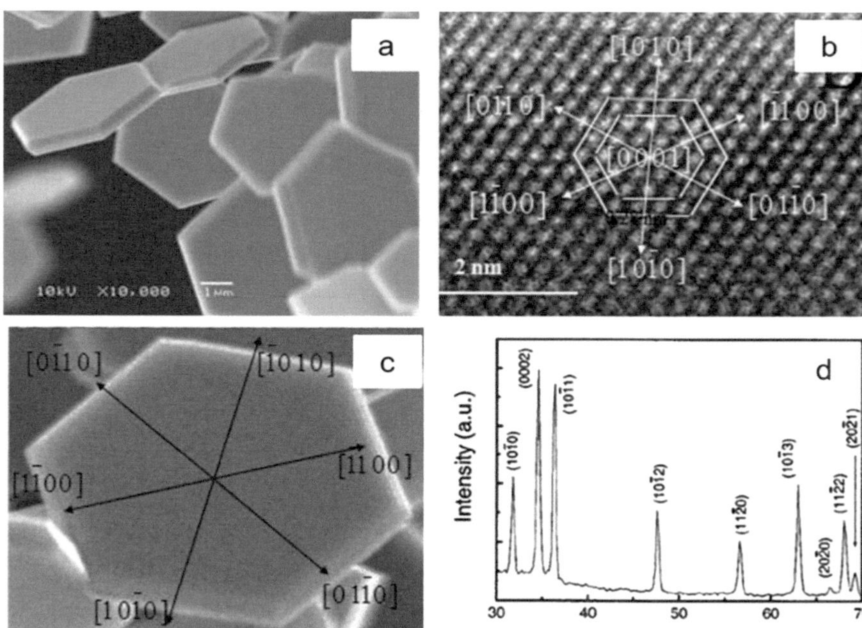

Fig. 7.8 ZnO nanodisks prepared by CVD method [14]

faces $\{10\overline{1}0\}$ may also be affected by the liquid zinc layer, this effect is much smaller than that of (0001) plane; therefore, the six side faces can grow rapidly. Since the growth rates of the six faces are the same, a sixfold symmetrical disc structure can be formed.

7.2 Luminescent Properties of ZnO

As is well known, the fluorescence spectrum of ZnO includes band-edge exciton emission in the ultraviolet region and deep-level defect emission in the visible light region. The intensity ratio of band-edge exciton emission and deep-level defect emission can be used to estimate the impact of defect energy levels on radiation. A deep understanding of its fluorescence characteristics is of guiding significance for further research on luminescent and laser devices. In this section, we summarize the fluorescence emission characteristics of ZnO from low temperature to room temperature, including bound exciton emission, free exciton emission, phonon sidebands, defect emission, electron–hole plasma (EHP) emission, and exciton–exciton scattering emission, etc. At the same time, we review its optical dielectric properties and carrier dynamics processes in combination with transient fluorescence and transient absorption spectroscopy techniques.

7.2.1 Low-Temperature Fluorescence

All direct bandgap semiconductor materials exhibit near-band-edge exciton recombination fluorescence radiation, whose mechanisms are closely related to defects and carrier concentration in the materials. Due to the presence of defects in semiconductor luminescent materials, donor and acceptor energy levels are formed in the bandgap. When excitons are bound by these donor and acceptor defects, corresponding bound exciton emissions are formed. To demonstrate the rich near-band-edge exciton recombination fluorescence radiation, the measurements should be performed at low-temperature conditions to reduce the effects of temperature on defects and excitons, thereby revealing a more obvious exciton fine spectral structure. A large number of literatures have studied the low-temperature and variable-temperature photoluminescence spectra on ZnO materials with different structures. The low-temperature photoluminescence spectrum of ZnO usually consists of multiple narrow band-edge emission peaks. Teke et al.'s research shows that in high-quality ZnO single crystals, neutral shallow donor-bound excitons usually dominate, manifesting as the corresponding sharp fluorescence characteristic peaks [37]. Due to crystal field and spin–orbit interaction, the valence band of wurtzite materials (such as ZnO) splits into three sub-bands. The holes on different valence sub-bands and electrons in the conduction band form corresponding A, B, and C excitons, as shown in Fig. 7.9a. Δ_{cf}

represents the splitting caused only by the crystal field, and Δ_{so} represents the splitting caused only by the spin–orbit interaction. The middle is a schematic diagram of the energy levels of the three types of excitons [38]. As shown in Fig. 7.9b, the low-temperature photoluminescence spectrum of ZnO single crystals presents a relatively wide exciton radiation peak, with exciton energy covering 3.348–3.374 eV. It displays many sharp fluorescence characteristic peaks, which come from different donor-bound excitons and acceptor-bound excitons [37]. In the figure, $D_n^0X_A$ represents the A exciton bound by the neutral donor energy level D_n; $D_n^0X_B$ represents the B exciton bound by the neutral donor energy level D_n; $A_n^0X_A$ represents the A exciton bound by the neutral acceptor energy level A_n. Due to the formation of defect energy levels in the bandgap by impurities and defect states, the corresponding bound exciton peaks are very rich. The exciton fluorescence of ZnO materials highly depends on the defect states introduced in the structure. Both single-crystal samples [37, 39–41] and nanostructures [42, 43] reflect their unique bound exciton fluorescence. Generally, the ZnO fluorescence spectrum only shows the A free exciton fluorescence peak near 3.378 eV. The B free exciton has a few meV higher energy than the A free exciton. A few studies have reported B exciton fluorescence peaks. For example, Hamby et al. reported the A free exciton peak at 3.378 eV and the B free exciton peak at 3.385 eV [39] in bulk ZnO single crystals at 20 K.

The bandgap of the semiconductor varies with temperature. In the low-temperature spectrum, the characteristic peaks of phonon replicas of excitons at different levels can be observed. When the energy of the exciton is dissipated in the form of phonons, there is a 72-meV phonon energy gap between the corresponding levels of phonon spectra. As shown in Fig. 7.10a, FX(A) is the free exciton of ZnO. There are two peaks on the left side of the spectrum, which are the first-level phonon replica of FX(A)-1LO and the second-level phonon replica of FX(A)-2LO. As the temperature rises, all emission peaks move toward the low-energy direction, showing an approximately asymmetric single-peak structure at room temperature. The change in exciton energy with temperature can be used to fit the relationship between the bandgap and temperature and estimate the bandgap at absolute zero. In addition, the bound excitons

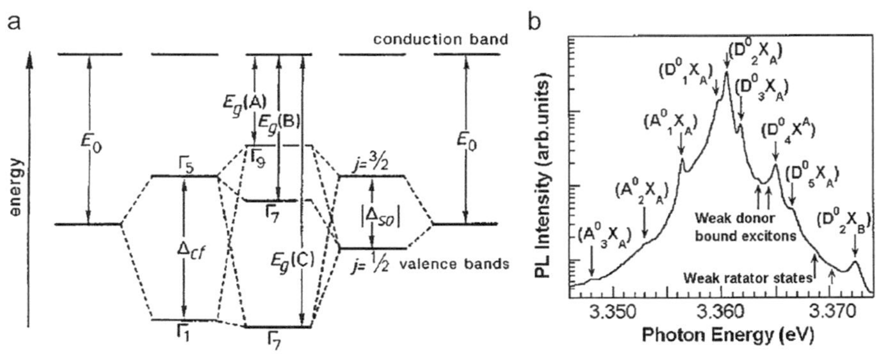

Fig. 7.9 ZnO bandgap splitting and low-temperature fluorescence characteristics [37]

existing at low temperatures are released from various defects due to thermal energy, and their corresponding bound exciton fluorescence peaks disappear. The intensity of the bound exciton peak quickly quenches, while the intensity of the free exciton peak and the phonon replicas relative to that of the bound exciton peak gradually increases; however, the overall fluorescence decreases as the temperature rises. The change in fluorescence intensity with temperature can be used to fit the exciton binding energy. Unlike bulk single crystals, the number of bound exciton peaks observed in ZnO nanostructures is usually less than in ZnO single crystals. For some nanostructures, only a wider radiation peak is observed even at lower temperatures, making it difficult to distinguish individual bound exciton peaks. Qiu et al. studied the near-band-edge excitonic emission of ZnO nanowires [43], as shown in Fig. 7.10b. In the spectrum at 6 K temperature, the emission peaks at 3.363, 3.367, and 3.357 eV are considered to be neutral donor-bound exciton luminescence; the shoulder peak at 3.353 eV is caused by radiation recombination related to acceptor-bound excitons; the peak at 3.342 eV may be due to the double-electron satellite peak, and the emission peak around 3.303 eV originates from the electron transition between the donor and the acceptor. The large surface-to-volume ratio of nanostructured ZnO also introduces some surface defects, which have a significant impact on luminescence. Grabowska and Wischmeier et al. found that there is very obvious surface defect state-bound exciton fluorescence in nano-ZnO at low temperatures [44, 45]. Lu et al. studied the photoluminescence spectrum of ZnO micrometer tower structures with temperature changes [46] and fitted the free exciton energy with the Varshni empirical formula, as shown in Fig. 7.11a.

$$E_{FX}(T) = E(0) - \frac{\alpha T^2}{T + \beta} \tag{7.6}$$

In the formula, $E(0) = 3.373$ eV is the bandgap at 0 K, $\alpha = 9.0 \times 10^{-4}$ eV K^{-1} reflects the change in the bandgap with temperature, indicating the linear change of $E(T)$ at high temperatures, and $\beta = 970$ K represents the quadratic change of $E(T)$ related to the Debye temperature at low temperatures. As shown in Fig. 7.11b, the change in the integrated intensity of near-band-edge photoluminescence with temperature can be fitted with the Arrhenius equation:

$$I(T) = \frac{I(0)}{1 + C\exp\left(-\frac{E_a}{kT}\right)} \tag{7.7}$$

In the formula, $I(0)$ is the luminescence intensity at 0 K, E_a is the exciton binding energy, and k is the Boltzmann constant. By fitting, the exciton binding energy of the ZnO microtower structure is approximately 57 meV.

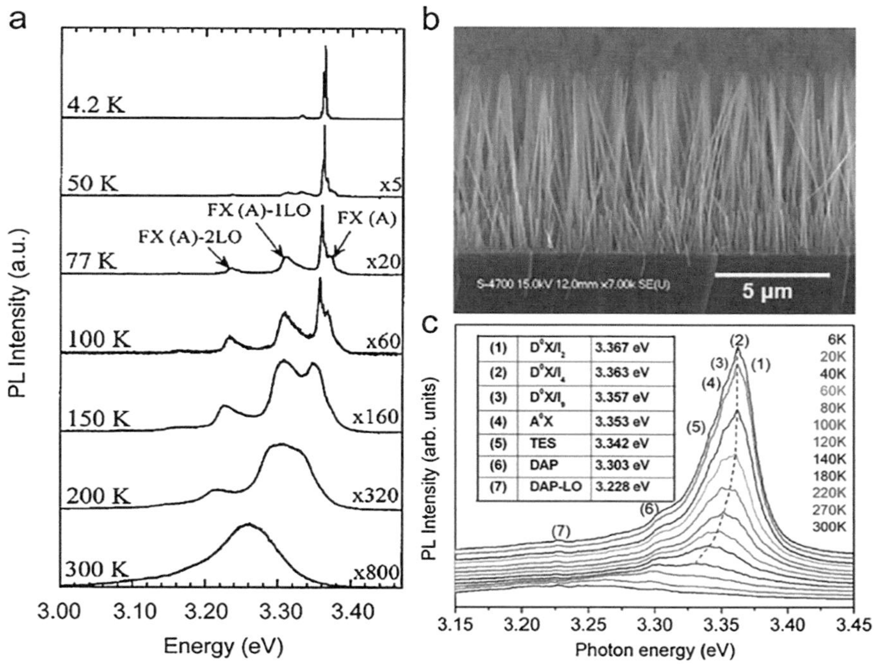

Fig. 7.10 ZnO temperature-dependent fluorescence spectrum [39, 43]

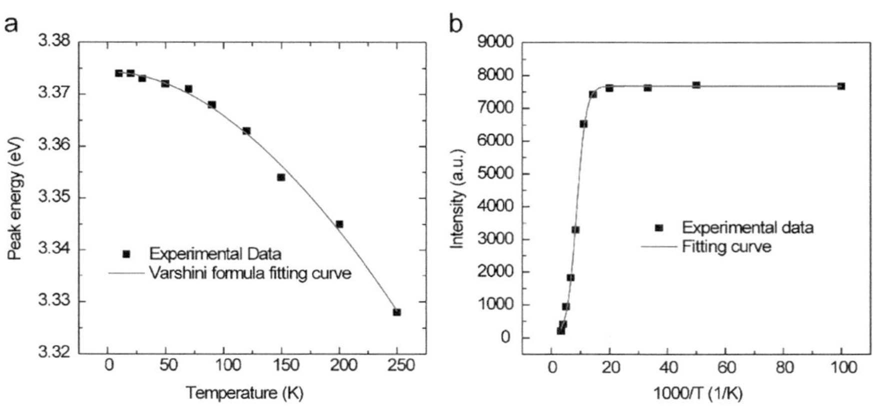

Fig. 7.11 Relationship between the energy of free excitons in ZnO and the intensity of near-band-edge radiation with temperature [46]

7.2.2 Room-Temperature Fluorescence

From the characteristics of the above ZnO low-temperature and variable-temperature fluorescence spectra, we can see that the fluorescence spectrum at room temperature is mainly composed of near-band-edge free exciton peaks and defect luminescence peaks at visible light region. The spectrum of ZnO from 3.1 to 1.653 eV is usually considered to be defect luminescence, and the wider luminescence band may come from the superposition of many different deep energy level defects. Common defects in the ZnO lattice mainly include zinc vacancies, oxygen vacancies, zinc interstitials, oxygen interstitials, and antisite intrinsic defects. These defects exist in different charge states and may form clusters with other defects, as shown in Fig. 7.12. For example, the V_oZn_i cluster formed by oxygen vacancies and zinc interstitials is one of the confirmed defect clusters, and its energy level is 2.16 eV below the conduction band [47]. These defects usually directly or indirectly affect the artificial doping effect, minority carrier lifetime, and luminescence efficiency in semiconductors [48]. These defects are also the reason why ZnO usually exhibits n-type conductivity. Generally, oxygen vacancies are common due to their lower formation energy, while zinc antisites, oxygen antisites, and oxygen interstitial defects with higher formation energy usually have lower concentrations. Vlasenko et al. produced zinc interstitial defects in ZnO under low-temperature conditions through electron radiation and found that zinc interstitial defects produce shallow donor energy levels by the means of optical methods and magnetic resonance [49]. First-principles calculations found that zinc vacancies can produce two acceptor energy levels: V_{Zn} and V_{Zn}^- are 0.1–0.2 and 0.9–1.2 eV higher than the valence band top, respectively [3, 48, 50]. There is still disagreement about the source of the common green luminescence band (2.4–2.5 eV) in undoped ZnO. Some people attribute it to the transition from the deep donor energy level of oxygen vacancies or zinc interstitial atoms to the valence band, while others believe it may be the transition from the shallow donor energy level to the zinc vacancy [51, 52]. Because the growth of ZnO nanostructures is usually carried out in a zinc-rich environment, oxygen vacancies and zinc interstitial atoms tend to dominate.

The yellow component in ZnO fluorescence is attributed to the luminescence of oxygen interstitial atom defects. The energy level of the general oxygen interstitial atom is approximately 2.28 eV below the conduction band, and the transition from the conduction band to this energy level produces yellow light radiation. The transition between the zinc interstitial atom energy level (0.22 eV below the conduction band) and the oxygen interstitial atom energy level results in orange light radiation [53]. In ZnO nanorods prepared by wet chemical methods, the yellow fluorescence part may also come from oxygen interstitial defects, but the desorption process of hydroxyl can transform the yellow defect luminescence into green and red by heat treatment of $Zn(OH)_2$ residues on the surface of ZnO nanostructures. The red–orange defect-state luminescence of ZnO may originate from the oxygen vacancies at the position of 1.65 eV below the conduction band [54, 55], or it may originate from energy level

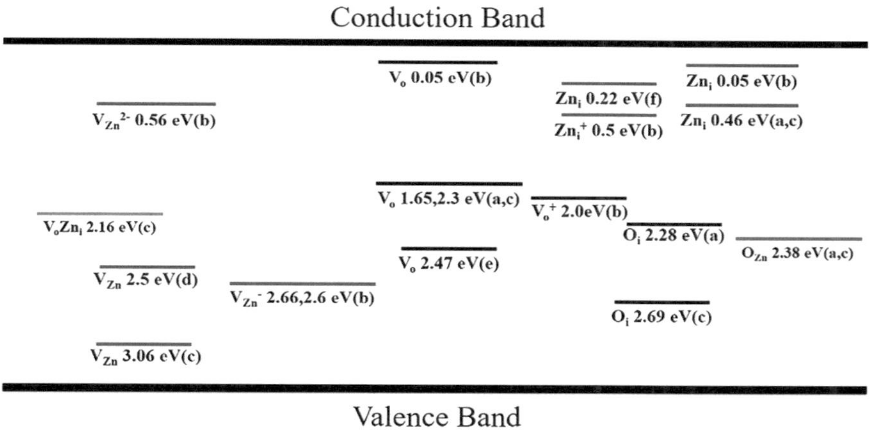

Fig. 7.12 Schematic diagram of defect energy levels in the ZnO bandgap [48, 53, 57–60]

of the oxygen interstitial atom at the positions above the valence band at 0.72 and 1.59 eV, theoretically leading to the transition emission from orange to red light [56].

7.2.3 Electron–Hole Plasma Recombination

In the photoluminescence process of ZnO, low excitation power leads to a lower carrier concentration, and excitonic radiation dominates. When the carrier concentration exceeds the critical Mott transition density, an electron–hole plasma (EHP) forms [61–63]. Therefore, time-resolved photoluminescence spectra, pump–probe transient absorption spectra, and other ultrafast spectra are utilized to reveal the carrier dynamics differences between exciton transitions and EHP recombination. Due to the rapid thermalization of EHP (<100 fs), the rise and decay time of EHP recombination (1–2 ps) is faster than that of exciton transition (>5 ps) [64–67]. EHP also leads to bandgap renormalization, which moves the conduction band edge down and the valence band edge up. Therefore, under the effect of bandgap renormalization, the bandgap of ZnO usually decreases, which can generally be distinguished by the change in fluorescence peak wavelength at different pumping powers, whether it belongs to exciton transition or EHP recombination [33, 68, 69]. Klingshirn et al. studied the source of stimulated radiation gain in bulk ZnO single crystals under different excitation intensities and temperatures. Their results show that at low temperatures (<200 K) and high excitation powers, EHP recombination dominates [70]. Versteegh et al. pointed out that the laser source of ZnO nanowires at room temperature is EHP recombination [71]. Nakamura et al.'s analysis of the random lasing characteristics of ZnO nanopowder within the temperature range of 20–300 K shows that the gain source of the lasing is the electron–hole plasma recombination

[72]. Figure 7.13 shows the photoluminescence spectra of ZnO nanopowders excited at different temperatures by low power ($\sim$5 J/m^2) and high power (357 J/m^2). With the increase of pumping energy flux, the shape of the emitted spectrum changes significantly. Under very weak pumping power, only a broad emission band is observed, which is the spontaneous emission caused by the electron–hole recombination near the band edge of ZnO nanopowders. At high pumping power, a narrow emission band composed of several peaks can be observed, and the intensity of the peaks changes with the pulse. Hendy et al. reported the formation process of EHP in ZnO single crystals [73]. A 266-nm pulse laser was used to excite ZnO single crystal and obtain a very high excitation density, thereby making the average exciton spacing r < 3 nm. In this case, the optical pump–terahertz probe transient spectrum shows that the photoconductivity is independent of frequency, as expected for a high-density electron–hole gas. At 0.6 THz, the excitation flux is about 1.5 J/cm^2. The transient changes in photoconductivity are shown in Fig. 7.14a. Throughout all pump–probe times, the imaginary photoconductivity is essentially zero, and the real part decays rapidly, with a half-life of 1.5 ps, which is 1–2 orders of magnitude faster than the formation of excitons. After the initial rapid decline, the decay of the photoconductivity noticeably slows down to 20 ps. Ignoring the charge diffusion between the pump pulse and the probe pulse and comparing the peak photoconductivity, it can be estimated that the plasma has reached a stable concentration of approximately 2 $\times$ 10^{24} m^{-3}. This stable concentration depends only on the initial excitation intensity, and Fig. 7.14b shows the decay when the excitation intensity is increased by a factor of 5. For a concentration less than 2–3 $\times$ 10^{24} m^{-3}, the Auger recombination rate decreases rapidly with decreasing density and is less than the exciton radiative recombination rate, which is equivalent to an exciton spacing of about 7 nm, the same order of magnitude as the Bohr radius. When the distance between the electron–hole pairs is comparable to the exciton Bohr radius, the plasma becomes stable. For ZnO, the high-density plasma decay in Fig. 7.14b is essentially independent of the lattice temperature. Moreover, in the low-concentration region, the decay caused by annihilation is faster than that of carrier cooling, indicating that annihilation occurs in the hot plasma before cooling.

7.2.4 Exciton–Exciton Scattering

Exciton–exciton inelastic scattering can generate excited states (P_n-band emission) or continuous states (P-band emission) [74]. Dai et al. studied the luminescent characteristics of ZnO tetrapod microstructures and found that they consist of two emission bands at 376.2 nm (3.296 eV) and 386.0 nm (3.212 eV), where the emission band at 376 nm is related to the recombination of free excitons, and the emission band at 386 nm is related to the P-band emission caused by inelastic exciton–exciton scattering [75]. Matsuzaki et al. reported on the phenomenon of ZnO luminescence mechanism from exciton–exciton scattering to exciton–electron scattering [76]. As

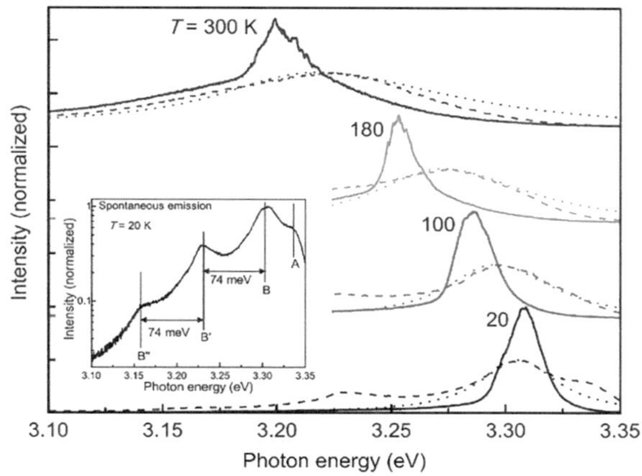

Fig. 7.13 Spontaneous (dashed line) and stimulated (solid line) emission of ZnO at different temperatures [72]

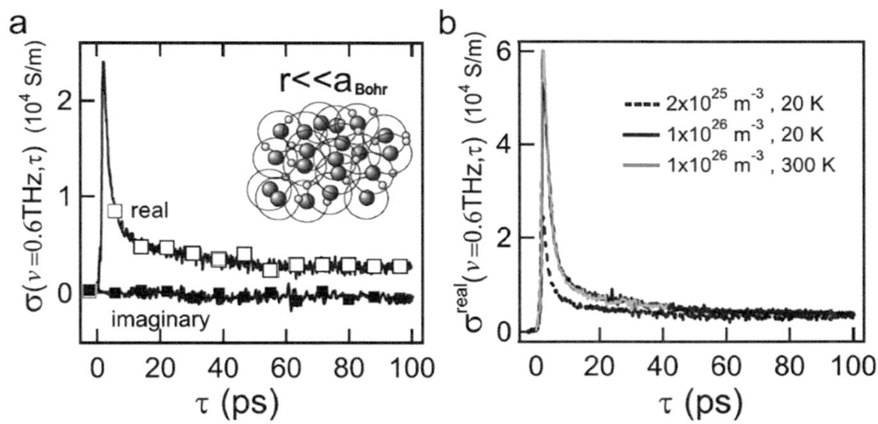

Fig. 7.14 Photoconductivity dynamics under high excitation density in ZnO single crystals [73]

shown in Fig. 7.15a, micrometer-scale ZnO thin films show a single redshifted photoluminescence peak in the temperature range of 3–300 K, with a full width at half maximum (FWHM) of 20–30 meV. Figure 7.15b gives the relationship between the photon energy of luminescent center and temperature, where the solid line is the free exciton transition energy of bulk ZnO reported by Wang et al. [77]. Figure 7.15c shows the relationship between the difference in free exciton energy and the photon energy of fluorescence center with temperature, where the red dashed line represents the relationship between the difference and temperature at $n = 2$ and $n = \infty$, which can be drawn from the following formula:

$$E_{ex}(T) - \hbar\omega_{\max}^{ex-ex}(T) = E_b\left(1 - \frac{1}{n^2}\right) + \frac{3}{2}k_B T \tag{7.8}$$

It can be clearly seen that the difference falls in the energy range of n = 2 and n = ∞ within the range of 3–150 K, proving that it originates from exciton–exciton scattering. When the temperature is higher than 150 K, the difference deviates from the exciton–exciton scattering region, showing a linear increase with an intercept of 0 and a slope of 5.21 × 10^{-4} eV/K. Exciton–electron scattering is the only observed zero-intercept, linear, temperature-related physical process. Therefore, the fluorescence radiation process changes from exciton–exciton scattering to exciton–electron scattering at a temperature of ~150 K.

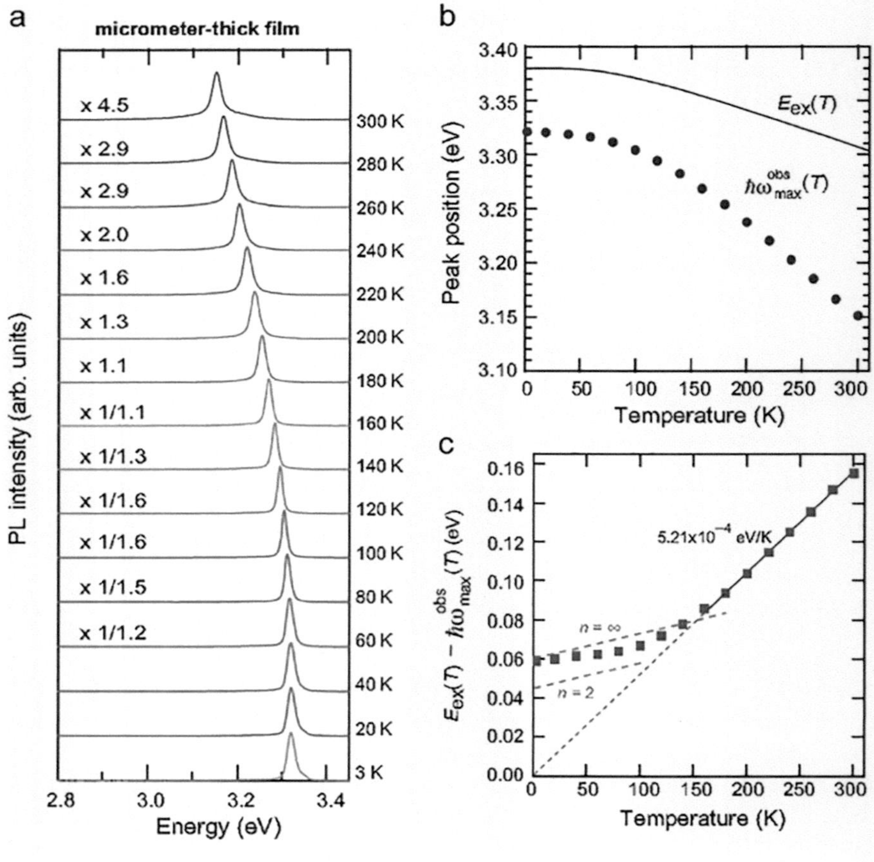

Fig. 7.15 Temperature-dependent laser spectra of micron-scale ZnO thin films [76]

7.2.5 *Time-Resolved Fluorescence*

Time-resolved fluorescence is widely used in the study of semiconductor luminescent materials, which can explain the fluorescence dynamics characteristics and their corresponding physical processes. People have conducted a large amount of research on the time-resolved spectra of ZnO materials, which serve to explain the influence of ZnO's structure, defects, and other factors on the luminescence process. Figure 7.16a shows the time-resolved photoluminescence spectra of ZnO nanoparticle defect luminescence with temperature changes, and the decay curve is fitted according to the Kohlrausch–Williams–Watts formula [78]:

$$n(t) = n_0 \exp\left[\left(-\frac{t}{\tau_k}\right)^{\beta_k}\right] \tag{7.9}$$

where n_0 represents the exciton density at $t = 0$, τ_k is the decay lifetime, $\beta_k = D/(D + 2)$ is the stretch index, and D is the spatial dimension of the exciton decay process. The average lifetime is $\langle\tau\rangle = \tau_k \beta_k^{-1} \Gamma\left(\beta_k^{-1}\right)$, where Γ is the gamma function. With the increase of temperature, the average decay time of defect emission from 3-nm nanoparticles decreases from 3194 ps at 10 k to 1888 ps at room temperature. Meanwhile, the dimension coefficient β_k increases from 0.58 to 0.69, which means that the photoluminescence relaxation becomes more dispersed as the diffusion relaxation rate increases. The thermal quenching process of the luminescence intensity of ZnO nanoparticle defect emission needs to be fitted with a two-step quenching Arrhenius model; the formula is as follows:

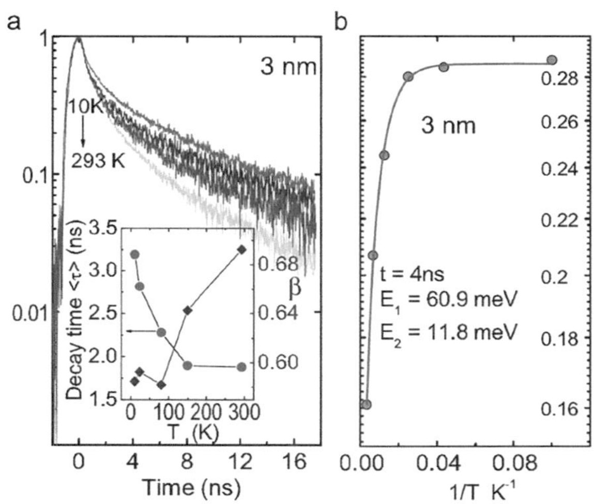

Fig. 7.16 Time-resolved photoluminescence spectra and thermal quenching curves of ZnO nanoparticles [78]

$$I(T) = \frac{I(0)}{1 + C_1 \exp\left(-\frac{E_1}{kT}\right) + C_2 \exp\left(-\frac{E_2}{kT}\right)} \tag{7.10}$$

As shown in Fig. 7.16b, the fitting gives $E_1 = 60.9 \pm 10.5$ meV, which is consistent with the exciton binding energy of bulk ZnO, and $E_2 = 11.8 \pm 1.2$ meV corresponds to the ionization energy of excitons bound by defects.

Compared with single crystals and high-quality thin films, ZnO nanostructures usually exhibit higher nonradiative defect density and shorter lifetimes. Zhao et al. studied the time-resolved fluorescence spectra of ZnO nanorods with different diameters. The results show that the decay time of excitons in nanorods largely depends on the diameter of the nanorods [79]. They derived a time constant related to the surface recombination rate through a double-exponential decay function, which decreases with the decrease of the nanorod diameter. They then suppressed the surface recombination channel through heat treatment, improving the luminescence performance of ZnO nanorods. The double-exponential decay curve of ZnO defect-state fluorescence can be explained by single-channel recombination and double-channel recombination models, allowing the defect-state concentration of ZnO nanowires to be obtained [80]. Richters et al. conducted a detailed study on the photoluminescence characteristics of ZnO nanowires and determined the characteristics of surface exciton fluorescence radiation [81]. Surface exciton recombination originates from a 20-nm-thick surface layer, and the recombination efficiency can be improved by coating an insulating layer on the nanowires. They proposed that surface exciton recombination fluorescence radiation can be distinguished by double-exponential recombination, fluorescence spectrum structure, and fluorescence dependence on nanowire diameter. Wischmeier et al. studied the photoluminescence dynamics of near-band-edge emission of surface excitons in ZnO nanowires under different luminescence wavelengths, temperatures, and excitation intensities [45]. The experimental results show that the asymmetric surface exciton emission band is caused by the wide energy distribution of the near-surface state. As shown in Fig. 7.17a, the initial double-exponential transient decay curve gradually becomes a single-exponential decay as the excitation intensity increases. Figure 7.17b shows that the fast decay lifetime τ_1 increases from 110 to 210 ps after the excitation intensity increases by 5 orders of magnitude, while the slow decay lifetime remains unchanged at 500 ps. The fluorescence decay of the donor-bound exciton is independent of the excitation intensity. This indicates that the surface exciton concentration increases as the excitation intensity increases, and the fast decay lifetime τ_1 shows a dependence on the surface exciton density. The analysis of temperature-dependent measurements shows that excitons are activated to the near-surface state at higher energies as the temperature rises, leading to an increase in the proportion of long decay-time components.

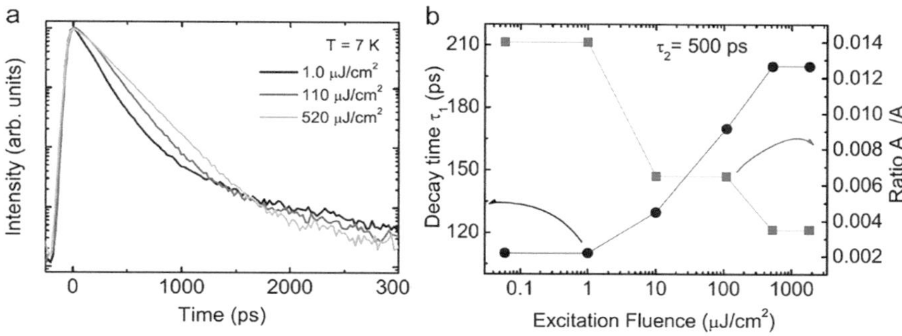

Fig. 7.17 Transient fluorescence characteristics of ZnO [45]

7.3 Stimulated Emission of ZnO

As a typical representative of wide bandgap semiconductor materials, ZnO materials with a direct bandgap of 3.37 eV and an exciton binding energy of 60 meV have great advantages in constructing optoelectronic interconnections and optoelectronic integrated devices. In recent years, a lot of research has been carried out on the stimulated emission of ZnO microstructures, and several important research results have been achieved. This section mainly introduces the stimulated emission characteristics in ZnO materials.

7.3.1 Random Lasers

Light randomly scatters in strong scattering media to form irregular optical resonance circuits, which is called a "random laser." At present, researchers have observed random lasers in ZnO gain media of different morphologies, such as disordered nanoparticles, polycrystalline thin films, powder thin films, nanocrystals, powders, etc. Cao et al. [82, 83] first reported the lasing behavior in ZnO powders. Their experimental results show that in the closely disordered ZnO crystals, when the pumping power exceeds the threshold, multiple discrete narrow linewidth (0.09 nm) radiation peaks can be observed, and the output light waves have coherence. A more classic example is the random laser reported by Yang et al. in ZnO nanoneedles in 2006 [84]. There are many ring cavities formed by closed light paths in ZnO nanoneedles. Along different paths, light has different scattering losses, eventually forming the random lasing output shown in Fig. 7.18. The result shows obvious randomness and chaos of oscillation modes. Random lasers are formed by different scattering paths, and the direction and spectral mode of lasing output are random. In other words, the mode structure of the random laser cannot be precisely controlled.

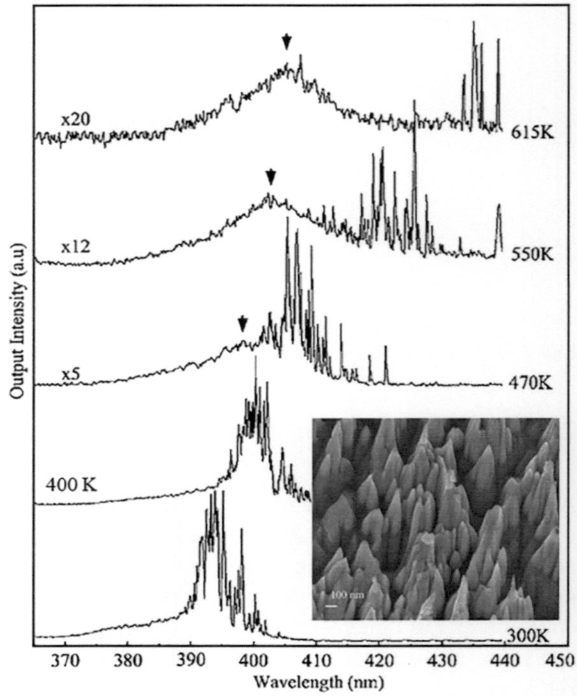

Fig. 7.18 Stimulated emission spectra of ZnO nanoneedles at different temperatures [84]

7.3.2 F-P Laser

In Fabry–Pérot (F-P) type optical microcavities, light oscillates back and forth between two parallel end faces, achieving gain amplification. Different wavelengths of lasing output can be achieved by adjusting the cavity size. In 1998, Tang and others from the Hong Kong University of Science and Technology [68] prepared ZnO microcrystalline thin films on sapphire substrates through MBE technology, achieving ultraviolet F-P lasers at room temperature. This ultraviolet stimulated emission is attributed to the radiative recombination process of exciton–exciton collisions. Another typical example is the laser in ZnO nanowires reported by Professor Yang Peidong [7] in 2001. Using a simple CVD method, ZnO nanorod arrays with adjustable diameters of 20–150 nm were prepared on Si substrates, as shown in Fig. 7.19b. The ZnO nanowires transitioned from spontaneous emission to stimulated emission at different pumping powers. The oscillation path is given in Fig. 7.19c.

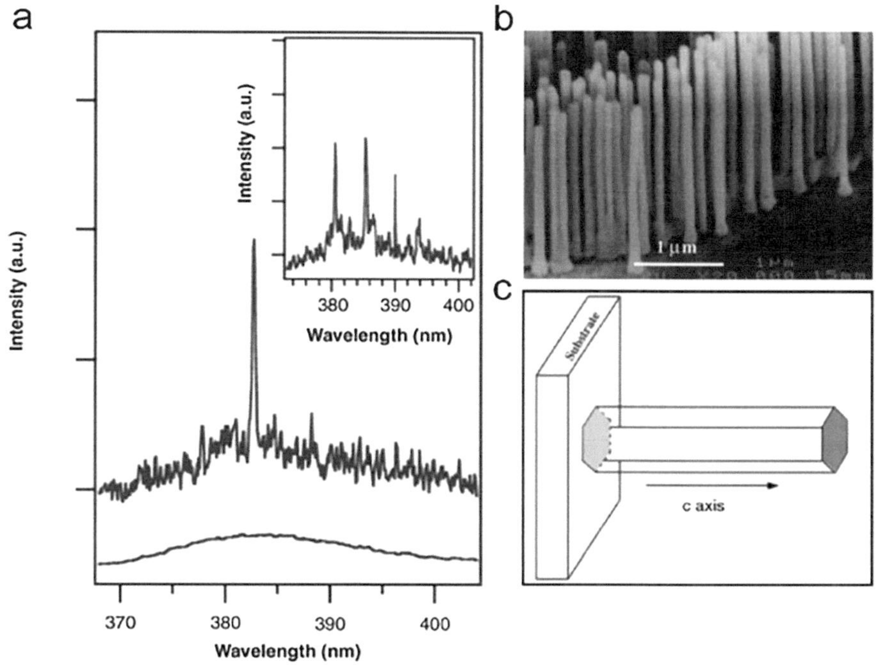

Fig. 7.19 Characterization of ZnO nanowire stimulated emission spectrum and morphology [7]

7.3.3 *WGM Laser*

Whispering Gallery mode (WGM) usually appears in circular or regular polygonal microcavities. Based on total internal reflection, light is confined to oscillate repeatedly in the microcavity, thereby achieving gain amplification and lasing output. ZnO micro/nanostructures are usually naturally regular hexagonal cross-section structures and have a high refractive index. Therefore, the light wave can continuously perform total internal reflection on the inner wall of the hexagonal cross-section microcavity when light enters the ZnO cavity, thereby obtaining ultraviolet WGM stimulated radiation. Wang and others from Auburn University in the United States [85] synthesized ZnO nanonails using the CVD method and achieved ultraviolet WGM lasers in ZnO nanonails with a diameter of 650 nm excited by 337-nm nanosecond laser pulse for the first time, with a threshold of about 17 MW/cm^2 and an FWHM of about 0.08 nm. In 2008, Czekalla and others from Germany [9] achieved WGM lasers at a low temperature of 10 K in a single ZnO microwire with a diameter of 6.4 μm, using a 266-nm nanosecond laser (10 ns, 20 Hz) as the excitation source. The mode structure is clear, the quality factor Q is as high as 3700, the FWHM is about 1 meV, and the lasing threshold is about 170 KW/cm^2. In 2009, Dai Jun [86] and others obtained WGM ultraviolet lasers at room temperature in ZnO microrods, clearly

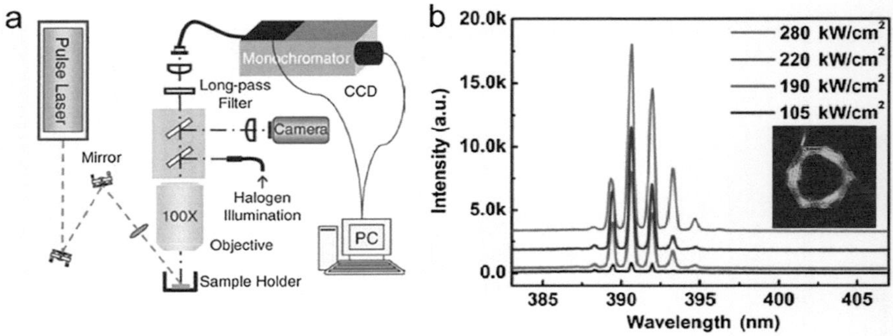

Fig. 7.20 Stimulated emission spectra of ZnO whispering gallery mode under different pump power densities [87]

showing the transition from spontaneous emission to stimulated emission. As shown in Fig. 7.20a, with the modification of the micro-area spectral system, the efficiency of excitation light and luminescence collection has greatly improved. Figure 7.20b shows the WGM laser measured in the hexagonal prism-shaped ZnO microrod by the research group of Xu Chunxiang in 2011 [87]. With the increase of excitation power, laser oscillation at around 392 nm can be observed. The inset in Fig. 7.20b also shows that the optical field energy is mainly concentrated on the inner surface of the hexagonal prism, realizing obvious spectral lasing by total internal reflection on the inner wall. In addition to single-photon processes, WGM lasers caused by multiphoton absorption have also been widely reported. For example, Chen et al. [88] reported a ZnO upconversion laser based on six-photon absorption in 2018.

7.3.4 *Surface Plasmon Coupling Enhanced WGM Laser*

In 1957, Ritchie et al. [89] discovered that when a high-energy electron beam passes through a metal medium, the quantized oscillation of metal free electrons in the positive ion background can be excited, which is a surface plasmon. Its essence is the collective oscillation of electrons generated under the action of the optical field. In recent years, the localized surface plasmons have been used to enhance semiconductor luminescence, and a large number of research results also have reported lasers enhanced by localized surface plasmons. For example, Okamoto et al. [90] reported the spontaneous emission of InGaN quantum well structures enhanced by metal surface plasmons, and Kwon et al. [91] used Ag surface plasmons to enhance the luminescence of InGaN/GaN light-emitting diodes (LEDs).

How to improve the laser quality of ZnO and achieve the luminescence enhancement of devices is a hot topic of research. Common methods include interface modification, enhancement of the interaction between light and matter based on microcavities, etc. [92, 93]. Previous studies have shown that ZnO microcavities modified by metal can achieve effective luminescence enhancement [94–98]. The schematic diagram of ZnO luminescence enhanced by metal surface plasmons is shown in Fig. 7.21. Since both the WGM microcavity and the metal surface plasmons can effectively confine the optical field energy on the material surface, the lasing emission enhanced by surface plasmons is realized through evanescent wave coupling. The enhancement of ZnO luminescence by localized metal surface plasmons first manifests as the enhancement of spontaneous emission luminescence [99]. Figure 7.22a shows the fluorescence of ZnO modified by Au reported by Lin Yi et al. [100]. In their work, the ultraviolet fluorescence is enhanced while the defect-related emission is almost suppressed to the noise level. In Fig. 7.22b, Lu Junfeng et al. achieved 170 times photoluminescence enhancement and 8–10 times lasing enhancement by modifying Al nanoparticles on the surface of ZnO microrods [101, 102]. Compared with Al nanoparticles, the surface plasmon resonance of Au is at 500–600 nm, corresponding to the defect-related emission of ZnO. Therefore, the typical feature of ZnO luminescence enhancement by Au modification is the enhancement of ultraviolet emission and the suppression of defect emission. The plasmon resonance peak of Al is close to the near-band-edge radiation of ZnO, so the enhancement effect of Al is more obvious. In addition to metals, graphene can also be used to improve the luminescence performance of semiconductors. For example, Sunget [103] reported the lasing enhancement in ZnO/graphene composite microcavities, attributing it to the surface plasmon enhancement effect of graphene. The research by Zhu Qiuxiang et al. [104] showed that the lasing enhancement in ZnO microcavities is more pronounced under the synergistic effect of Al nanoparticles and graphene surface plasmons.

Figure 7.22c–f summarizes the spectra of ZnO lasing enhancement obtained using the above materials. It can be seen that different metals all have the effect of enhancing ZnO luminescence. Similar to the fluorescence process, the lasing enhancement effect of Al and graphene is more pronounced than other metals. Due to the different optical responses of different metals [105, 106], the coupling mechanisms between metal surface plasmons and ZnO are different. Common coupling mechanisms include direct energy resonant coupling and surface plasmon-assisted electron transfer. For example, the coupling between metals with UV response such

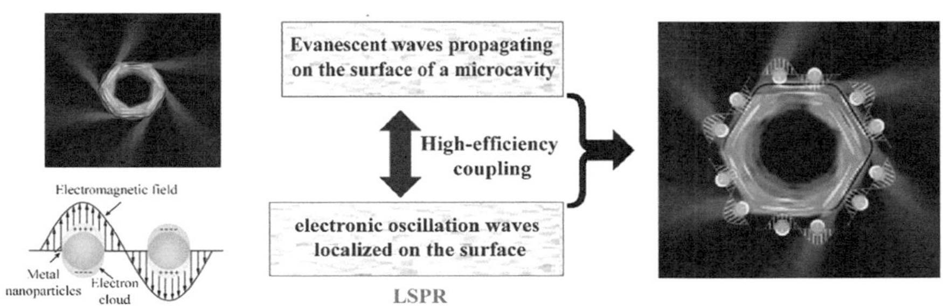

Fig. 7.21 Schematic diagram of surface plasmon enhanced ZnO luminescence

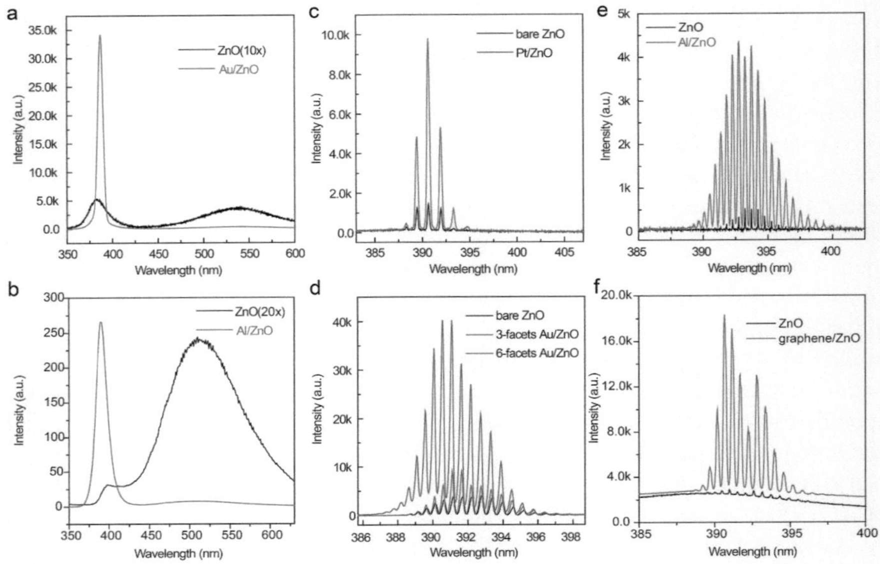

Fig. 7.22 Spectra of spontaneous and stimulated radiation of ZnO decorated with different plasmonic materials [87, 101, 102, 110, 111]

as Pt, Al, and ZnO is direct energy coupling. Due to their surface plasmon response in the visible light region, the extinction peak of Au, Ag, etc. is close to the defect light of ZnO, whose coupling with ZnO is essentially a localized surface plasmon-assisted electron transfer process. There is currently much controversy over surface plasmons in graphene, but an acceptable explanation is that there is energy coupling between graphene and ZnO. Because the π-electron resonance energy corresponding to surface plasmons in graphene is above 4.6 eV, which is much larger than the bandgap of ZnO, this energy coupling requires surface wrinkling to scatter the energy of surface plasmons [103, 107–109].

Experimental and theoretical studies have shown that the spatial localization and near-field enhancement characteristics of surface plasmons on metal surfaces are effective means to enhance semiconductor luminescence, but the mechanism of metal surface plasmon-enhanced luminescence is still controversial. On the one hand, the collective oscillation of free electrons after the excitation of surface plasmons allows high-energy peripheral free electrons to be directed to the semiconductor, achieving luminescence enhancement through electron transfer and recombination. On the other hand, the localized field enhancement characteristics caused by localized surface plasmons enhance the exciton recombination process around the metal, realizing the energy coupling process. The resonance of metal surface plasmons is affected by factors such as the type of metal, morphology, and excitation light source. The response range of different metals varies, and the differences in spectral enhancement effects caused by different coupling strengths are shown in Fig. 7.23, which

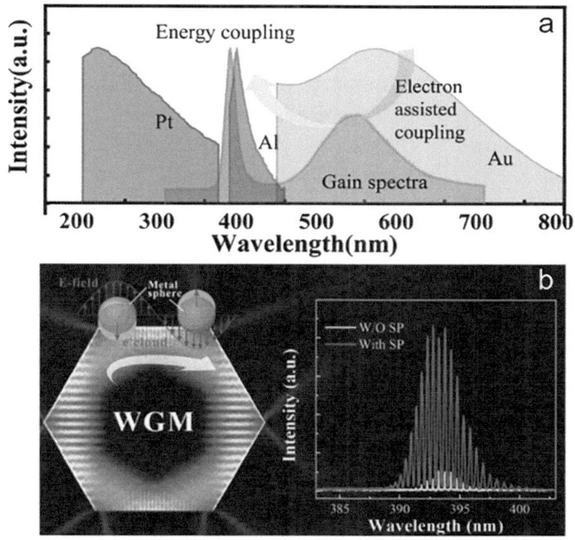

Fig. 7.23 Principle and experimental results of surface plasmon-enhanced ZnO luminescence [112]

is achieved by coupling the surface plasmons of different metals with the ZnO gain spectrum to enhance the intrinsic luminescence of ZnO and suppress defect-related emission. The lasing performance of the semiconductor microcavity enhanced by metal surface plasmons can also be regulated and optimized by changing the type of metal, size, etc., and introducing isolation layers to achieve the coupling of surface plasmons and excitons in the semiconductor microcavity.

7.3.5 Polariton Laser

A polariton is a kind of quasiparticle formed by the strong coupling of photons and certain polarized excitations (such as phonons, excitons, plasmas, etc.), including phonon polaritons, exciton polaritons, and surface plasmons.

1. Exciton Polariton Laser

In the 1950s, researchers predicted the existence of exciton polaritons, which are a new type of quasiparticle formed after strong coupling of photons and excitons in semiconductors and their microstructures. They exhibit a hybrid state of half-light and half-matter particles and have the characteristics of bosons [113]. Spectrally, they show anticrossing dispersion behavior, known as Rabi splitting. Their formation can lead to "Bose–Einstein" condensation, and the condensation of exciton polaritons in the microcavity represents a process that is easier to produce coherent light than particle number inversion [114]. Since polariton condensation does not require inversion, the pumping power of future polariton lasers may be much lower than that of photon lasers. However, strong coupling of photons and excitons is usually difficult to achieve in experiment, which must meet a condition: The probability of photons being reabsorbed by the gain medium must be very high, that is, the Rabi frequency must be greater than the damping rate (when the laser irradiates the material, the fluctuations in the particle population exists between energy levels, and its oscillation frequency is the Rabi frequency). Researchers have found that strong coupling is easy to achieve if the gain medium is constructed into a suitable resonant cavity. Strong coupling can be defined by the following three important parameters: the rate of energy exchange between the cavity mode and the exciton, the rate of light escaping from the empty cavity, and the rate of medium depolarization [115]. If the rate of energy exchange between the cavity mode and the exciton is greater than the sum of the other two rates, the excited state will be preserved in the system composed of the cavity mode and the exciton. At this time, the elementary excitations in the system are no longer separate cavity modes or excitons but exciton polaritons formed by the coupling of the two.

The earliest research on exciton polaritons focused mainly on low-dimensional structures such as semiconductor material quantum wells and quantum lines. Many interesting physical phenomena were discovered in these studies, but the coupling efficiency of photons and excitons was very low due to the lack of efficient coupling cavities. This greatly hindered the basic research of exciton polaritons. In 1992, Claude Weisbuch et al. [116] effectively restricted the light field in the quantum well through a high-reflectivity distributed Bragg reflector (DBR) structure, achieving strong coupling of excitons and the light field at low temperatures. The strong coupling region, marked by anticrossing dispersion behavior, was clearly observed and accurately measured in the experiment. This research result was undoubtedly surprising, which solved the coupling problem between the gain medium and the electromagnetic field and thus opened a new direction for microcavity exciton polaritons. In 1994, Houdm et al. [117] first used angle-resolved fluorescence spectroscopy to measure the in-plane dispersion of exciton polaritons, providing an effective research method for studying the dispersion of microcavity exciton polaritons. With the support of these two research methods, the study of microcavity exciton polaritons developed rapidly. The current research on exciton polariton radiation mainly involves: parametric oscillation, parametric amplification, Bose–Einstein condensation, and polariton lasers. The exciton polaritons observed in materials such as CdTe and GaAs are mainly achieved under low-temperature conditions. In order to achieve exciton polaritons at room temperature, materials with high exciton binding energy at room temperature have become the focus of research. Recently, researchers have observed stimulated emission of exciton polaritons in GaN microcavities at room temperature. ZnO material is also an ideal material for realizing exciton polaritons because of its higher exciton binding energy and regular natural microcavities, in which the relatively stable strong coupling between excitons and optical cavity modes can exist at room temperature.

Vugt et al. [118] utilized lasers or electron beams as spatially resolved excitation sources to study the optical cavity mode radiation in a single ZnO nanowire. As shown in Fig. 7.24a–d, they found that the F-P mode resonance did not form at the

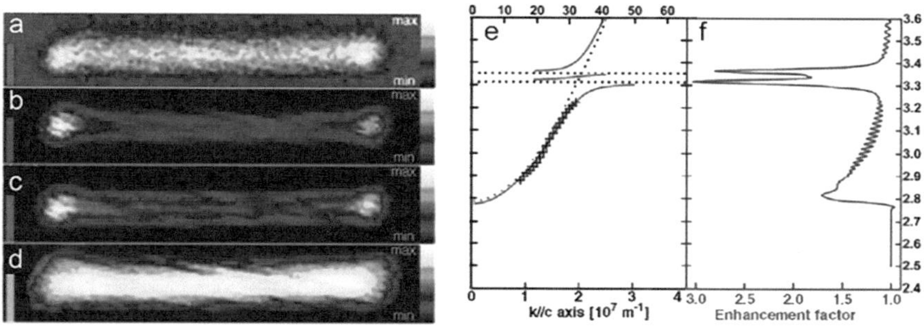

Fig. 7.24 Exciton polaritons in ZnO nanorods [118]

positions of the fluorescence peak (3.25–3.40 eV) and defect-related emission peak (less than 2.7 eV), but in the range less than the fluorescence peak (2.7–3.2 eV), the radiation intensity of ZnO at the two ends of the nanorod was the strongest, forming an effective F-P mode resonance in this cavity. As shown in Fig. 7.24b and c, the enhancement factor of the radiation intensity is defined as the ratio of the radiation intensity at the end to the radiation intensity at the rod body. Within the range of 2.7–3.2 eV, it can be seen that there are a series of resonance peaks, which are formed by the F-P cavity mode due to the reflection and oscillation of light at the two ends of the nanorod. The exciton polariton resonance mode simulated by the exciton polariton dispersion curve of ZnO and the F-P cavity mode formula in Fig. 7.24e matches very well with the results measured in Fig. 7.24f, indicating that the exciton polariton has been realized in ZnO nanorods.

The research group of Professor Chen Zhanghai from Fudan University [119] discovered the phenomenon of whispering gallery mode radiation in one leg of a tapered four-needle ZnO nanostructure (as shown in Fig. 7.25a). As shown in Fig. 7.25b, there are two types of radiation polarization states, transverse electric (TE) and transverse magnetic (TM), in ZnO, which mainly originate from A, B, and C excitons. The whispering gallery mode of TE polarization mainly comes from A and B excitons, while the whispering gallery mode of TM polarization mainly comes from B and C excitons. Scanning microfluorescence spectroscopy measurements found that as the cavity gradually narrows, the whispering gallery mode under the same mode gradually blueshifts. The relationship between the resonance energy of the whispering gallery mode E and the radius of the peripheral circle R can be used to explain this phenomenon:

$$R = \frac{hc}{3\sqrt{3}nE}\left(N + \frac{6}{\pi}\tan^{-1}\left(\beta\sqrt{3n^2 - 4}\right)\right) \tag{7.11}$$

Here, n is the refractive index, h is the Planck's constant, c is the speed of light in vacuum, and N is the series of cavity modes. For TE mode, $\beta = n$, and for TM mode, $\beta = 1/n$. From Formula (7.11), we can see that the resonance energy E is inversely

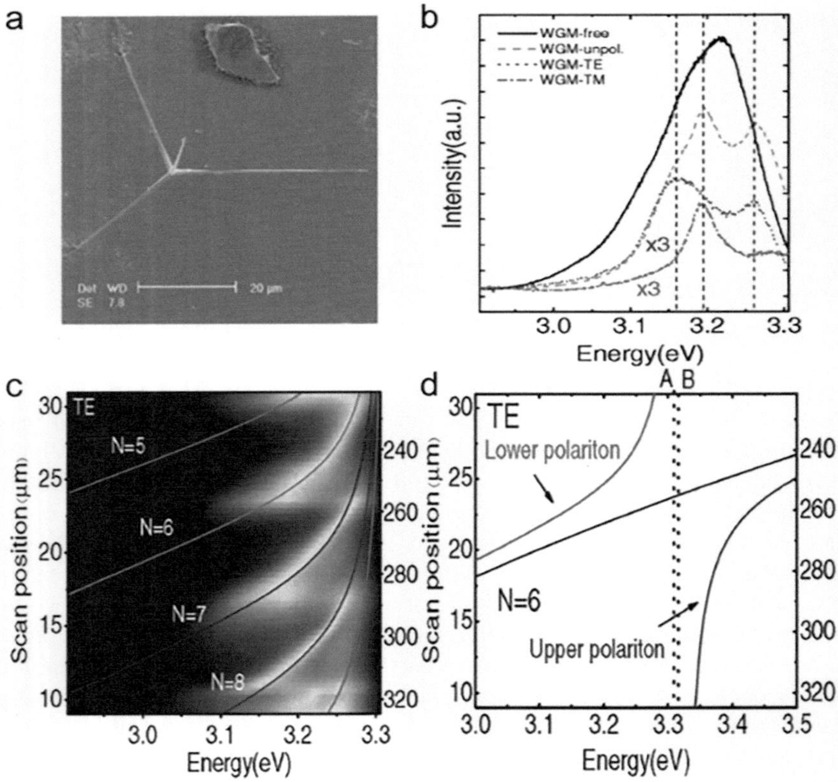

Fig. 7.25 Exciton polaritons of the cavity mode in ZnO microrods [121]

proportional to the radius *R* of the peripheral circle. As the diameter of the ZnO microrod decreases, the resonance energy *E* increases. Figure 7.25c shows the PL image of TE polarization, scanned in nanorods with diameters linearly decreasing from 320 to 220 nm. It was found that the resonance peak blueshifted, and the number of TE modes decreased from 8 to 5. As it approaches the exciton energy, the blueshift amplitude decreases, forming a nonlinear blueshift process. This is because the dispersion of *n* with wavelength is very small in the visible light spectrum region. In the band-edge luminescent region, strong coupling occurs between the cavity mode and the exciton near the exciton transition energy level to form exciton polaritons. This can be explained by the high refractive index produced by the exciton polariton dispersion near the exciton energy.

$$n^2 = \varepsilon(\omega, k) = \varepsilon_\infty \left(1 + \sum_{j=A,B,C} \Omega_j \frac{\omega_{j,L}^2 - \omega_{j,T}^2}{\omega_{j,T}^2 - \omega^2 - i\omega\gamma_j} \right) = \frac{c^2 k^2}{\omega^2} \tag{7.12}$$

where ε_∞ is the background dielectric constant, that is, the contribution of other transitions to the dielectric constant excluding the exciton transition. $\omega_{j,T}$ and $\omega_{j,L}$ are the transverse and longitudinal resonance frequencies when the wave vector is 0, and γ_j is the decay constant. Ω_j is the weight of the three types of excitons A, B, and C. Figure 7.25d are the blueshift results of the cavity mode in the linearly decreasing cavity, simulated using exciton polariton dispersion, which are completely consistent with the experimental results, indicating that effective cavity mode exciton polaritons have been generated in this cavity.

2. Plasmonic Nanolasers

The luminescence enhancement effect of metal surface plasmons on ZnO not only manifests in direct enhancement but also utilizes their high spatial locality to transform traditional photon oscillation into collective electron oscillation and confines it in a very small mode volume, thereby providing an effective strategy for realizing all-optical integration and developing smaller, faster, and more efficient nanophotonic devices. People have designed and constructed semiconductor gain media, insulating medium layers, and metal thin-film hybrid resonant cavities. By using semiconductor materials to provide sufficient gain to compensate for metal ohmic losses, combined with the high spatial locality of metal surface plasmons, they have broken through the optical diffraction limit and realized the output of plasmonic laser modes at the nanoscale [120–122].

Representatively, in 2014, Oulton et al. [123] designed and constructed an Ag/LiF/ZnO hybrid structure resonant cavity, obtaining a plasmonic hybrid ZnO nanowire laser output with a pulse time less than 800 fs, as shown in Fig. 7.26a. By comparing the stimulated radiation spectra of traditional photonic lasers and plasmonic lasers, it was found that the latter showed a significant blueshift compared to the former. They attributed this to the need for higher gain compensation and the reduction of nanowire size in the plasmonic hybrid resonant cavity, as shown in Fig. 7.26b. In addition, dual-pump technology can also be used to measure the transient response spectra of two different laser modes, as shown in Fig. 7.26c and d. The experimental results show that the physical processes such as accelerated spontaneous recombination, gain conversion, and recovery occur because the exciton energy of ZnO is close to the resonant frequency of the plasmonic laser of this composite structure, providing a suitable platform for exploring the interaction between light and matter in an accelerated state.

In 2017, Lu et al. [120] further compared and analyzed the input–output relationship, lasing mode, threshold, etc., of traditional photonic lasers and new plasmonic lasers. Combined with exciton coupling dynamics, the reason for the blueshift of the lasing mode has been successfully explained, which originated from their different gain mechanisms. As shown in Fig. 7.27, in the hybrid microcavity of Ag and ZnO, the stimulated spectrum shows a significant blueshift. The rapid coupling between Ag plasmons and ZnO excitons avoids the EHP effect caused by the high exciton concentration in the hybrid resonant cavity, which is the fundamental reason why the resonant peak of plasmonic nanolasers appears on the higher-energy side compared to that of traditional F-P lasers. Combined with time-resolved spectroscopy, the

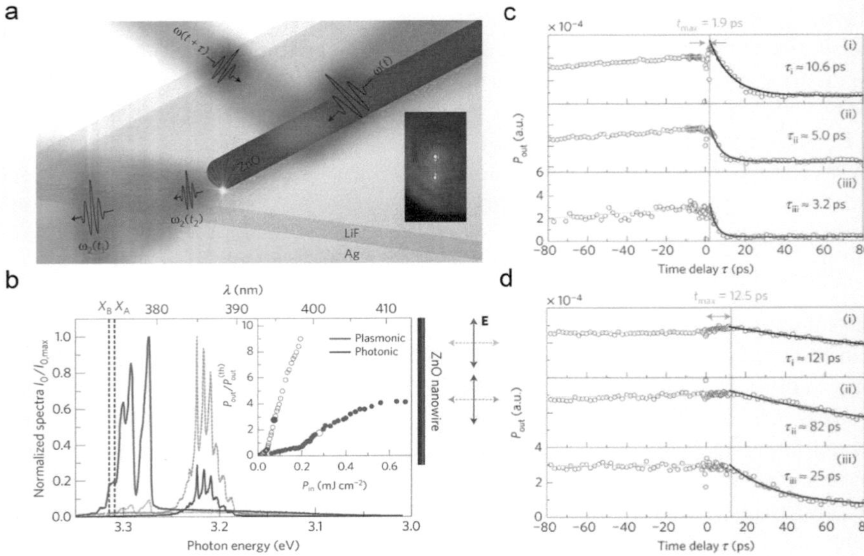

Fig. 7.26 Comparison of optical characteristics of traditional lasers and plasmonic lasers [123]

exciton recombination lifetime is 743.59 ps for traditional F-P nanolasers, whereas the coupling lifetime of plasmons and excitons is 17.60 ps. The Purcell factor can be estimated as high as 40, revealing the rapid energy conversion process from excitons to plasmons and also providing strong evidence for the essential difference in gain between the two types of nanolasers.

Ting-Chang Lu et al. [124] implemented a laser mode with clear ZnO plasmonic nanolaser by using a similar structure. By improving the flatness of the Ag thin film, the coupling efficiency between surface plasmon polaritons (SPPs) and the excitons in ZnO was increased, resulting in an effective increase in the laser quality factor and a significant increase in the exciton recombination rate, further reducing the threshold of the ZnO plasmon nanolaser (as shown in Fig. 7.28b). In addition, the research group conducted theoretical and experimental research on the ZnO nanolaser based

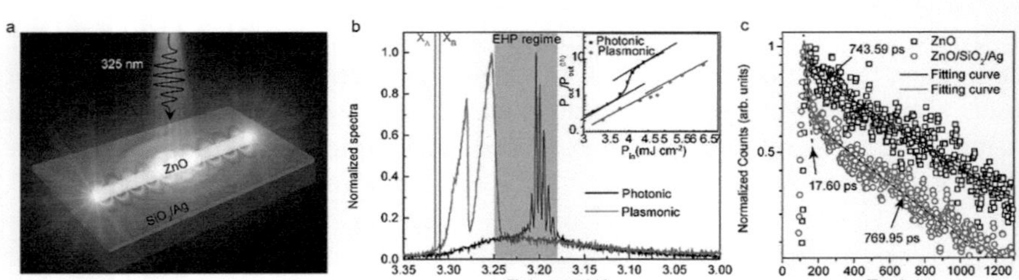

Fig. 7.27 Optical characteristics of traditional F-P lasers and plasmonic lasers [120]

on the Al metal. The matching of the coupling frequency of Al and ZnO in theory and the use of single-crystal Al thin film in experiment promoted the recombination of SPP and ZnO excitons in the nanolaser, achieving a lower lasing threshold, as shown in Fig. 7.28b [125, 126]. Also, the research group constructed a ZnO/graphene/insulator/metal layer (GIM) structure, which can change the dispersion characteristics of SPP and eliminate the constraints of the fixed plasmon properties of the metal. Due to the difference in Fermi energy levels, the addition of graphene will change the electron density on the metal surface. The electron density on the Al surface in the GIM structure increases, which leads to the plasma frequency of Al blueshift. Accordingly, the threshold of the Al-based ZnO nanowire laser is reduced by 50%, mainly due to the reduction in internal loss of the SPP mode. As shown in Fig. 7.28c, the Ag–GIM structure shows the opposite effect [127]. Compared to a flat SPP nanolaser, the research group proposed a ZnO nanolaser based on the pseudo-wedge SPP mode. As shown in Fig. 7.28d, it has an effective volume mode 100 times smaller than that of the flat SPP nanolaser. Its dispersion curve makes the group velocity of the pseudo-wedge SPP mode less than 1/165 of the speed of light, with a spontaneous radiation factor as high as 0.8 [128].

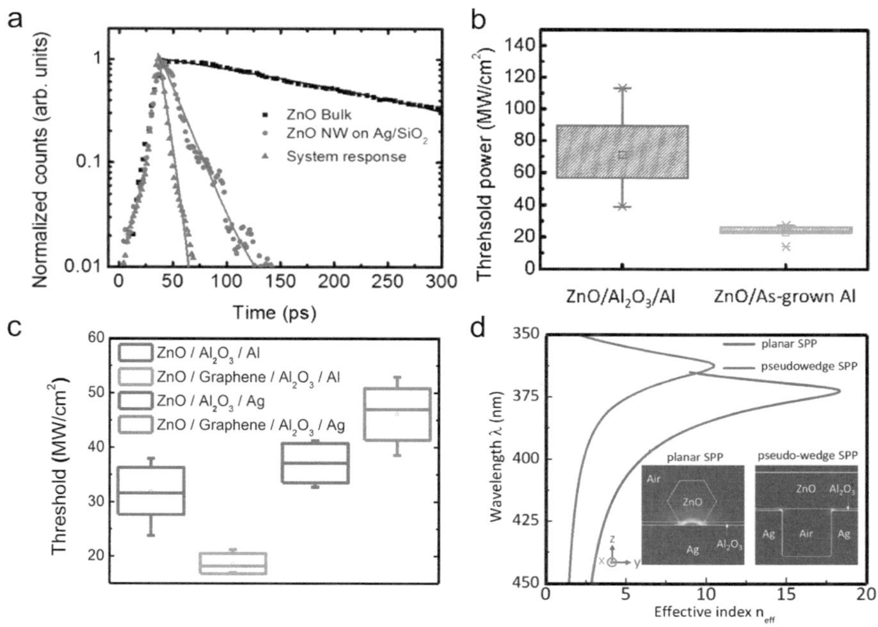

Fig. 7.28 Optical characteristics of plasmon lasers on various substrates at 77 K [128]

7.4 ZnO Laser Mode Control and Single-Mode Implementation

Mode control and single-mode manipulation have always been the focus and hotspots of research in the laser field. The third-generation wide bandgap semiconductor materials represented by ZnO have many applications in ultraviolet shortwave optoelectronic devices, especially laser devices [129, 130], and related laser mode control research work is also common [131, 132].

According to the basic conditions for laser generation, taking the classic Fabry–Pérot (F-P) cavity as an example, the light field travels a cycle in the resonant cavity, where the obtained gain is at least equal to the loss to maintain the field strength and phase unchanged [138]

$$\frac{2n_{eff}L2\pi}{\lambda} = N2\pi,\ N = 1, 2, 3, \ldots \tag{7.13}$$

where n_{eff} is the effective refractive index of the cavity, λ is the resonant wavelength, and N is the mode number, which is a positive integer. This defines the minimum cavity size that can satisfy the basic mode oscillation, which is $\lambda/2n_{eff}$. For the whispering gallery resonant cavity where the light wave forms an oscillation loop by multiple total reflections at the inner wall, its phase shift condition can be rewritten as:

$$\left[\exp\left(\frac{in2\pi L_p}{\lambda}\right)\right]r^m = 1, \tag{7.14}$$

where L_p is the propagation path length of a light wave to complete an oscillation loop, and m is the number of sides of the whispering gallery resonant cavity. Combining with the semiclassical plane wave model [133], the expression can be substituted into the simplified Fresnel formula:

$$r = \exp(-i2\theta), \tag{7.15}$$

$$\alpha = \tan\theta, \tag{7.16}$$

$$\alpha = \frac{\beta\sqrt{n^2\sin^2\theta_i - 1}}{\cos\theta_i}, \tag{7.17}$$

where θ_i is the angle of incidence, and n is the refractive index. For the TE mode, β is the refractive index n, while for the TM mode, β is the reciprocal of the refractive index.

Taking the ZnO microcavity with a wurtzite hexagonal structure as an example, its TE mode is much stronger than the TM mode. To simplify the discussion, only the TE mode should be considered here, that is, $\beta = 1/n$.

Therefore, the mode equation of the ZnO hexagonal resonant cavity can be obtained:

$$N = \frac{3\sqrt{3}nR}{\lambda} - \frac{6}{\pi}\tan^{-1}\left(n\sqrt{3n^2 - 4}\right), \tag{7.18}$$

where R is the radius of the resonant cavity. It can be seen that the wavelength control of the specific resonant cavity mode mainly depends on the refractive index and size of the cavity whether it is the classical F-P resonant cavity or the hexagonal whispering gallery resonant cavity. Of course, if there is no special requirement for the order of the cavity mode, different cavity mode selective outputs can also be achieved by controlling the optical gain region. Thus, various methods of controlling lasing modes and achieving single-mode output have been developed, such as using bandgap engineering [134, 135], active cavity self-absorption effect [136], Burstein-Moss (B-M) effect [137], etc., for mode control, and DBR/distributed feedback (DFB) [138], Vernier effect [87, 139], gain symmetry breaking [140, 141], control of resonant cavity size [32, 142], etc., for single-mode manipulation.

7.4.1 Cavity Size Design and Mode Regulation

For an active cavity, the fundamental way to achieve single-mode lasing is to increase the free spectral range, so that only a single mode appears within the gain region. According to the formula for deriving the mode spacing (that is, the free spectral range),

$$\Delta\lambda = \frac{\lambda^2}{L_P\left(n - \lambda\frac{dn}{d\lambda}\right)}, \tag{7.19}$$

where L_P is the propagation path length for a light wave to complete an oscillation loop, and $\left(n - \lambda\frac{dn}{d\lambda}\right)$ is the effective refractive index. For traditional F-P cavities, $L_P = 2L$, where L is the cavity length; for hexagonal whispering gallery cavities, $L_p = 3\sqrt{3}R$, where R is the radius of the circumscribed circle. Therefore, the mode spacing is inversely proportional to the cavity size; thus, the simplest and most direct way to achieve single-mode laser output is to reduce the size of the cavity. In this regard, Professor Yang Peidong from the University of California, Berkeley, and others [142] synthesized ZnO nanodisks with controllable diameters from 280 to 900 nm using a bottom-up approach and achieved single-mode laser emission in submicron-sized resonant cavities with diameters of 842 nm, 612 nm, and 491 nm, with lasing thresholds of 500 μJ/cm^2, 860 μJ/cm^2, and 1500 μJ/cm^2, respectively, as shown in Fig. 7.29.

In addition, Professor Xu Chunxiang's research group from Southeast University conducted systematic research on ZnO whispering gallery mode lasers regulated by cavity size [32]. As shown in Fig. 7.30, it demonstrates the variation of WGM laser

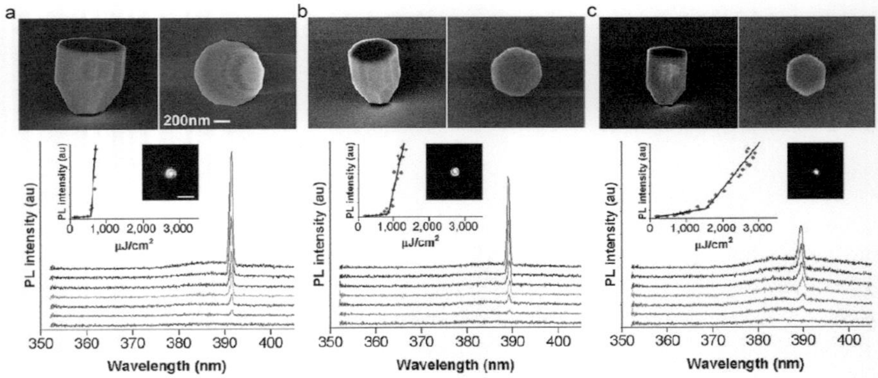

Fig. 7.29 Single-mode laser output manipulated by size [142]

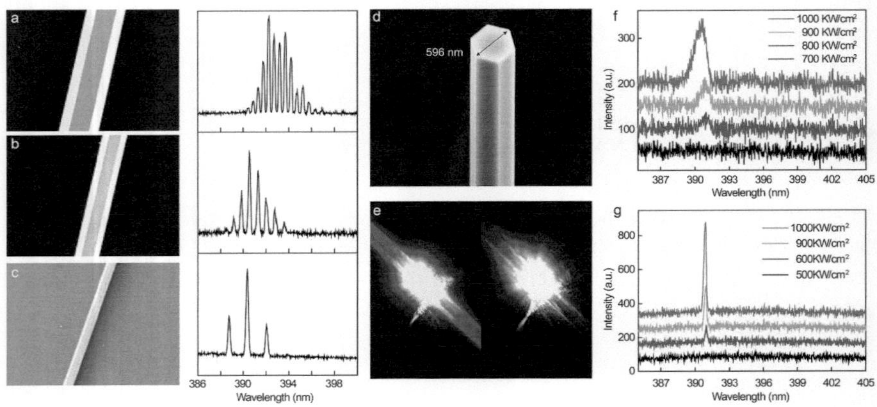

Fig. 7.30 Single-mode output based on size effect [32]

modes, mode spacing, and mode numbers with the size of the cavity. In the submicron-scale ZnO resonant cavity, single-mode laser output was achieved. Further, the optical loss of the microcavity was greatly reduced by utilizing the spatial confinement of the ZnO light field by graphene surface plasmons and the resonant coupling between graphene surface plasmons and whispering gallery modes, significantly improving the quality of the single-mode laser and reducing the lasing threshold.

7.4.2 *Microcavity Coupling and Single-Mode Operation*

It is well known that the reduction of the cavity size leads to a decrease in the ability to confine the light field, thus unable to obtain high-quality laser mode output. To solve this problem, Professor Xu Chunxiang and others achieved single-mode laser output by using the Vernier effect based on the coupling between two ZnO microcavities with slightly different diameters [87, 139]. When the two resonant cavities are effectively coupled, only the laser modes that exist in both independent resonant cavities are selected and enhanced due to the Vernier effect, while other modes gradually dissipate until they disappear. Eventually, the free spectral range extends beyond the spectral width of the gain medium, achieving single-mode laser output. The free spectral range of the coupled resonant cavity can be further derived from Eq. (7.19):

$$FSR_{12} = \frac{\lambda^2}{\frac{3\sqrt{3}}{2}\left(n - \lambda\frac{dn}{d\lambda}\right)|D_1 - D_2|}, \tag{7.20}$$

where D_1 and D_2 are the diameters of the two coupled cavities. If D_1 is close to D_2, then FSR_{12} is greatly expanded, far greater than the spectral range of a single microcavity, thus realizing the existence of only a single resonant mode within the gain spectrum range through the Vernier effect. As shown in Fig. 7.31, two ZnO resonators with diameters of 4.3 and 3.9 μm were selected, and two microrods were simultaneously excited at IV in Fig. 7.31a with a gap of 45 nm (meeting the premise of energy coupling). The optical fields in the two resonators strongly coupled, successfully achieving a single-mode laser output of 390.4 nm, with a side-mode suppression ratio of 19.7 dB. Compared with the method of directly reducing the cavity size to achieve single-mode lasing, this method avoids energy loss caused by the cavity size being too small, thus maintaining high-quality, low-threshold single-mode laser output. To further improve the performance of the single-mode laser, they constructed a Pt nanoparticle/ZnO micrometer comb composite structure to enhance the single-mode laser through localized surface plasmon coupling, while reducing the lasing threshold of single-mode output. The optimized composite resonator coupling cavity side-mode suppression ratio reached up to 20.5 dB; the lasing quality factor reached 2300; the lasing intensity increased by 7 times, and the threshold was reduced to 86 kW/cm^2.

7.4.3 Strain Dynamic Regulation

The above two regulation methods both rely on the predesign of the cavity structure and size. Once the cavity is formed, the position of the output wavelength is also fixed and cannot be further regulated. In 2018, Professor Wang Zhonglin from the Beijing Institute of Nanoenergy and Nanosystems, Chinese Academy of Sciences and Professor Xu Chunxiang from Southeast University [131] utilized the intrinsic property that the refractive index of the resonator is sensitive to external mechanical strain [143], thus realizing the dynamic and sustainable regulation of the output wavelength for the ZnO whispering gallery lasing by continuously applying external strain. Once the external force is removed, the cavity mode resonance wavelength

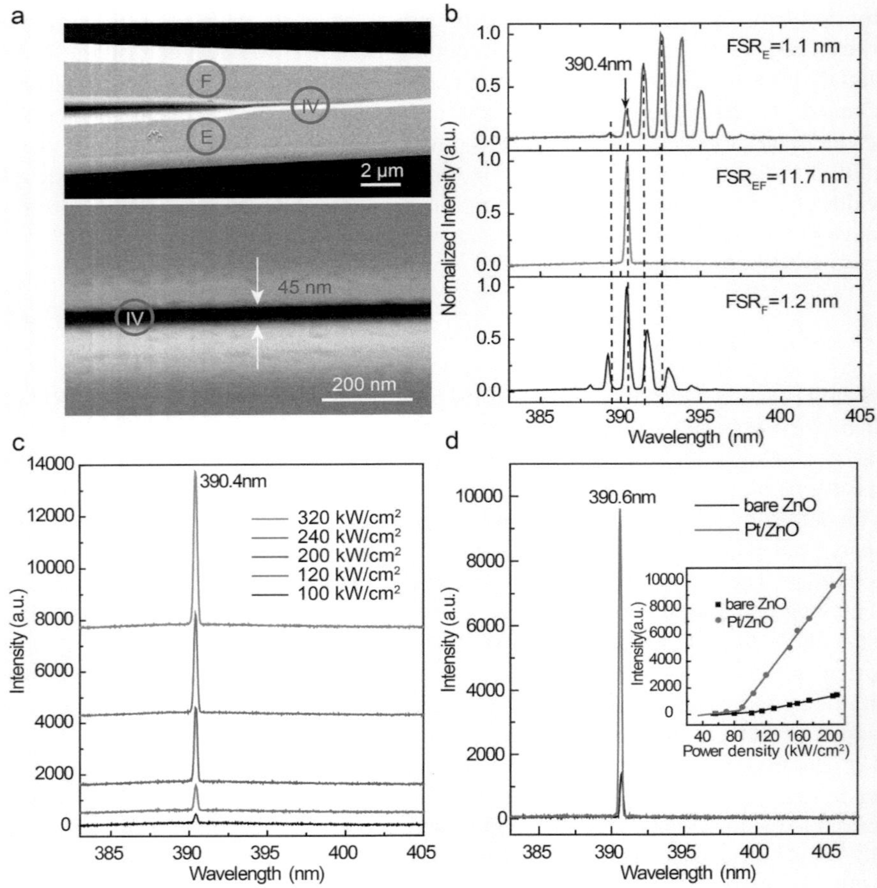

Fig. 7.31 Single-mode output based on the Vernier effect [87, 139]

also recovers. As shown in Fig. 7.32, a single ZnO microrod is fixed on the surface of a flexible substrate. A uniaxial tensile strain in the [0001] direction of the ZnO microrod can be obtained when the substrate is bent upward by a displacement stage, and the stimulated spectra under different tensile strains are synchronously measured. A significant redshift of the resonant wavelength for the cavity mode is observed.

Furthermore, in 2019 [132], they used the synergistic effect of piezoresistance and piezoelectric polarization to achieve a single-mode laser output and dynamic regulation in a single ZnO microrod with a diameter of 1.7 μm, as shown in Fig. 7.33. On the one hand, the piezoresistive effect can be used to regulate the gain spectrum line of the ZnO microcavity; on the other hand, the crystal polarization induced by external strain can give rise to a change in the cavity refractive index, thereby achieving dynamic regulation of the resonant cavity mode. Due to the difference in the regulation mechanisms, the change rates of the gain spectrum line and the resonant

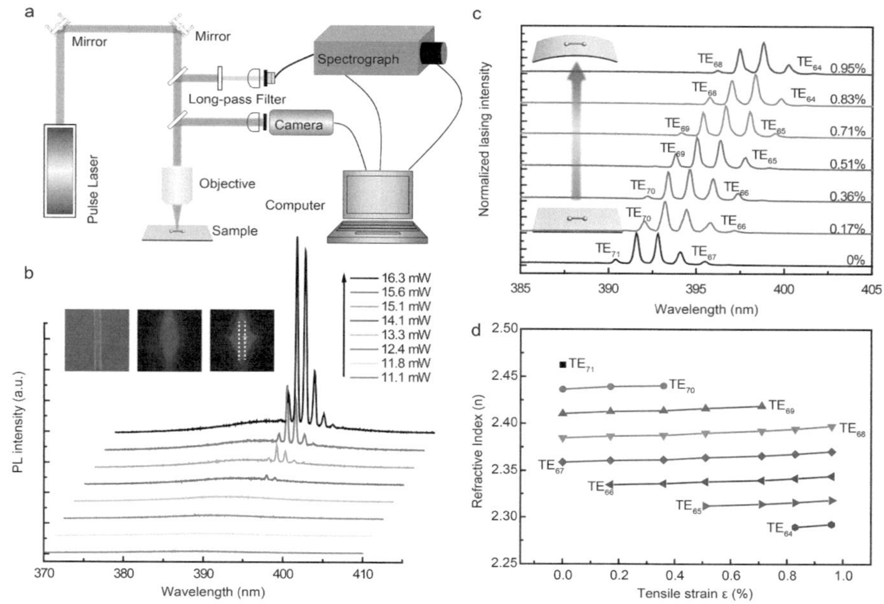

Fig. 7.32 Dynamic regulation of ZnO whispering gallery laser mode [131]

cavity mode under the applied strain are different, which provides the possibility for the dynamic selection and regulation of different cavity modes and ultimately achieves single-mode laser output. Moreover, under the action of external strain, with the growth and decay of different order modes, it is expected to obtain a wide range of adjustable single-mode laser outputs. From another perspective, this technology also provides a new strategy for strain sensing through color changes as a sensing medium.

7.5 ZnO-Based Electroluminescent Devices

High-efficiency semiconductor light-emitting diodes have become a prominent research topic in modern optoelectronics, among which GaN-based LEDs have shown their advantages in solid-state lighting, flat-panel displays, information storage, optical communication, microprocessing, and other fields [144–148]. ZnO, as a wide bandgap semiconductor material with a bandgap width of 3.37 eV and an exciton binding energy as high as 60 meV, is one of the ideal candidate materials for developing ultraviolet light-emitting diodes and laser diodes (LDs). ZnO electroluminescent devices have become a hot research topic in the past 20 years.

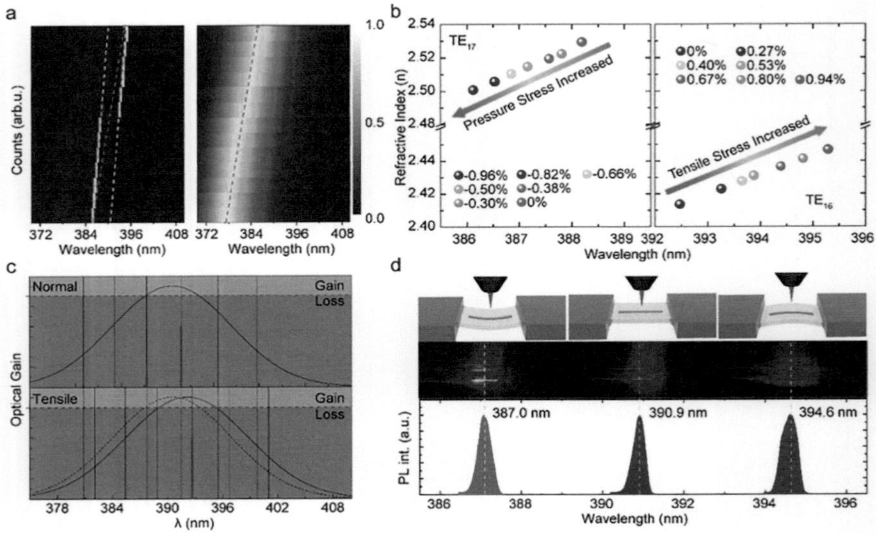

Fig. 7.33 Single-mode laser output and dynamic regulation based on piezoelectric and piezoresistive synergistic effects [132]

7.5.1 *ZnO-Based Homojunction Electroluminescent Devices*

For the design of electroluminescent devices, homojunction design is given priority, considering issues such as energy-level matching, carrier injection balance, and lattice matching. In 2005, Atsushi et al. prepared p-type ZnO by chemical vapor deposition and constructed a ZnO homojunction LED [149]. In 2006, Liu et al. reported that n-ZnO thin films and p-ZnO thin films were deposited on a ZnO substrate by vapor transport, thus realizing a ZnO homojunction LED structure [150]. Subsequently, Sun et al. reported the epitaxial growth of a homojunction structure on ZnO nanowires and obtained ultraviolet emission from ZnO [151]. Researchers from the Changchun Institute of Optics, Fine Mechanics and Physics of the Chinese Academy of Sciences studied the p-type MgZnO thin films prepared by the Li–N co-doping method and constructed a ZnO homojunction LED [152]. In 2011, the research group led by Professor Jianlin Liu at the University of California, Riverside, constructed a homojunction laser diode consisting of n-type ZnO thin films and p-type ZnO nanowires and achieved electrically pumped F-P lasing from ZnO nanowires, as shown in Fig. 7.34 [130]. In 2013, Huang et al. excited ZnO-based homojunctions with optical pumping and electrical pumping, respectively, and obtained random lasing behavior with thresholds of 300 kW/cm^2 and 40 mA [153]. Currently, a lot of exploration has been done on ZnO-based homojunction electroluminescent devices, but the stability of p-type ZnO is still a problem to be solved.

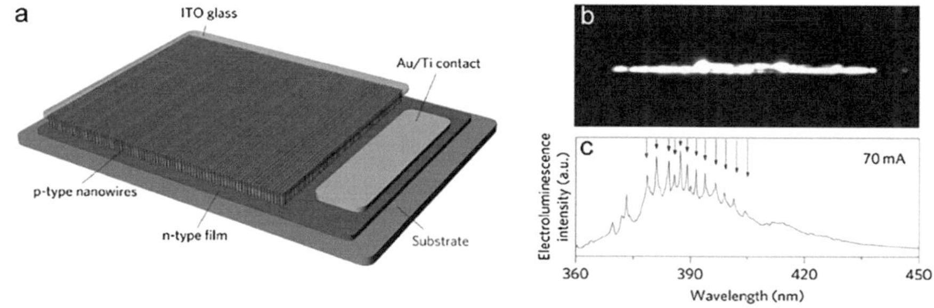

Fig. 7.34 ZnO homojunction laser diode [130]

7.5.2 *ZnO-Based Heterojunction Electroluminescent Devices*

Due to the bottleneck difficulty of p-type doping of ZnO materials, it has become an optional scheme to construct ZnO-based heterojunction structures, which has received widespread attention from researchers. P-type GaN material is an ideal hole injection material for constructing heterojunctions with ZnO because of its similar crystal structure to ZnO, close lattice constants, and low lattice mismatch. As shown in Fig. 7.35, the energy bands of ZnO and GaN are quite close, with electron affinities of 4.35 eV and 4.20 eV, respectively. According to the bandgap of ZnO of 3.37 eV and GaN of 3.4 eV, the conduction band mismatch of ZnO and GaN is calculated to be $\Delta E_c = 0.15$ eV, and the valence band mismatch is $\Delta E_v = 0.13$ eV. The mismatches of the valence band and the conduction band are close, and the mismatch values are very small [154]. In addition, the preparation process of GaN material is relatively mature and is often used as the substrate material for green, blue-violet, white light, and other light-emitting devices.

In 2009, the research group led by Professor Zhonglin Wang prepared n-type ZnO nanowires on a p-type GaN substrate to achieve high-brightness blue LEDs. The device structure and emission spectrum are shown in Fig. 7.36a, and the main

(a)

Materials	ZnO	GaN
Lattice structure	wurtzite	wurtzite
Lattice constant (nm)	a = 0.325 c = 0.521	a = 0.319 c = 0.519
Band gap (eV)	3.37	3.4
Lattice mismatch	1.9 %	

(b)

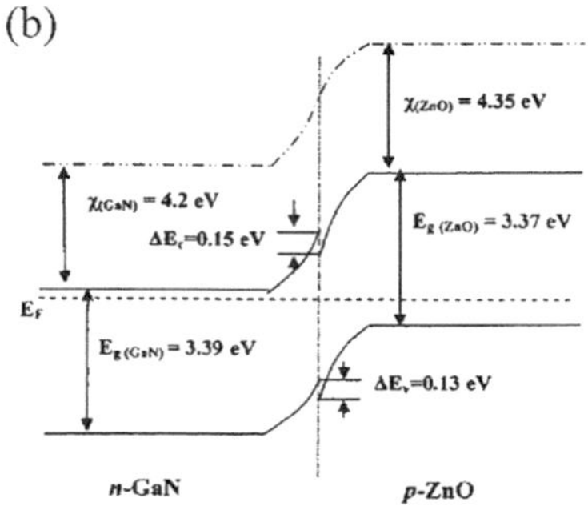

Fig. 7.35 Schematic diagram of lattice constant and band structure [154]

peak position shifts from 440 to 400 nm with the increase of current [155]. In 2014, the research group led by Professor Yue Zhang from the University of Science and Technology Beijing deposited high-quality ZnO thin films and GaN epitaxial thin films, respectively, through the radio frequency magnetron sputtering method, forming a heterojunction structure and realizing an n-ZnO/p-GaN heterojunction LED device that emits blue light under low voltage [156]. The specific device structure is shown in Fig. 7.36b. In 2015, Yang et al. [157] prepared ZnO nanorods with a certain tilt angle on a GaN substrate through oblique deposition technology. Its electroluminescence spectrum shows a strong violet emission at 390 nm and a weaker blue-green emission at 490 nm. The former originates from electron–hole transition recombination of ZnO near-band-edge, while the latter is attributed to the recombination at the ZnO/GaN interface.

In 2018, the research group of Xu Chunxiang at Southeast University [158] introduced an AlN electron-blocking layer into ZnO nanowire/GaN devices, producing a pure ultraviolet emission at 389 nm, as shown in Fig. 7.37. Furthermore, in 2020, they used n-type ZnO microwires and n-type GaN thin films to prepare a mutually compensating symmetrical double heterojunction, achieving an alternating current (AC) ultraviolet LED with an opening voltage of 12 V, with the same total light emission in both directions. The wavelength remains basically unchanged, effectively overcoming flicker, providing a new approach for the preparation of AC LEDs for future LED lighting, as shown in Fig. 7.38 [159].

Compared to LEDs, LDs require higher current density and injection efficiency. In 2009, Ma et al. achieved random lasing at about 380 nm under an injection voltage of 6 V based on the Au/SiO_2/ZnO structure, as shown in Fig. 7.39a [160]. In 2010, the research group of Shan Chongxin prepared the Au/i-ZnO/MgO/n-ZnO structure and achieved stable, low-threshold electrically pumped random lasing, with a threshold of only 6.5 mA, as shown in Fig. 7.39b [161]. In these studies, the mode of random lasing is uncertain, making it difficult to be widely used in the future.

In 2011, the research group of Xu Chunxiang at Southeast University fully exploited the structural advantage of ZnO microcavity, prepared an n-type ZnO

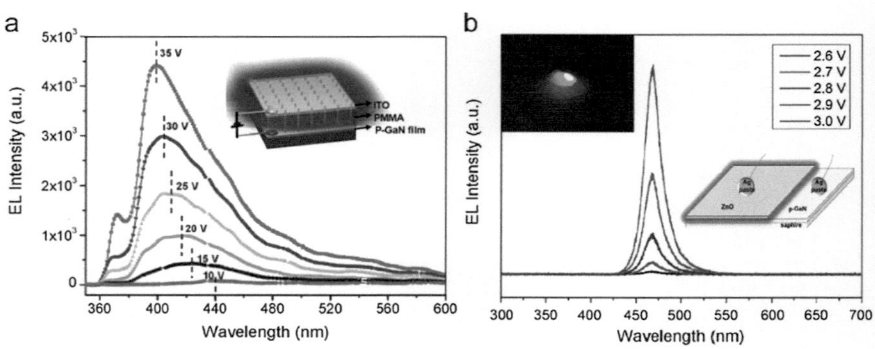

Fig. 7.36 n-ZnO/p-GaN heterojunction light-emitting diode [155, 156]

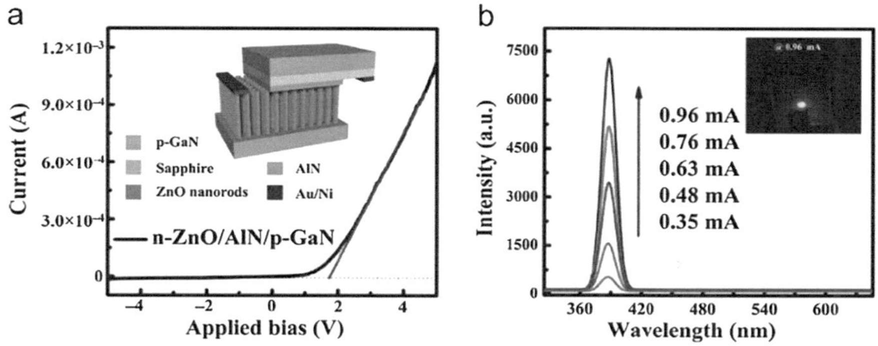

Fig. 7.37 n-ZnO/AlN/p-GaN heterojunction light-emitting diode [158]

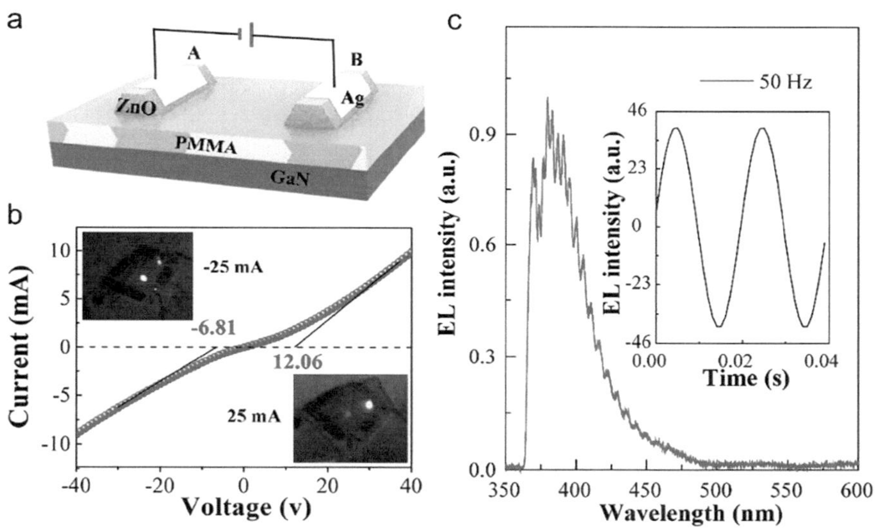

Fig. 7.38 n-ZnO/n-GaN/n-ZnO symmetrical double heterojunction AC light-emitting diode structure [159]

microrod and p-type GaN heterojunction, and achieved a mode-stable Fabry–Pérot cavity electrically pumped laser, with a laser threshold of 12 mA, as shown in Fig. 7.40 [129].

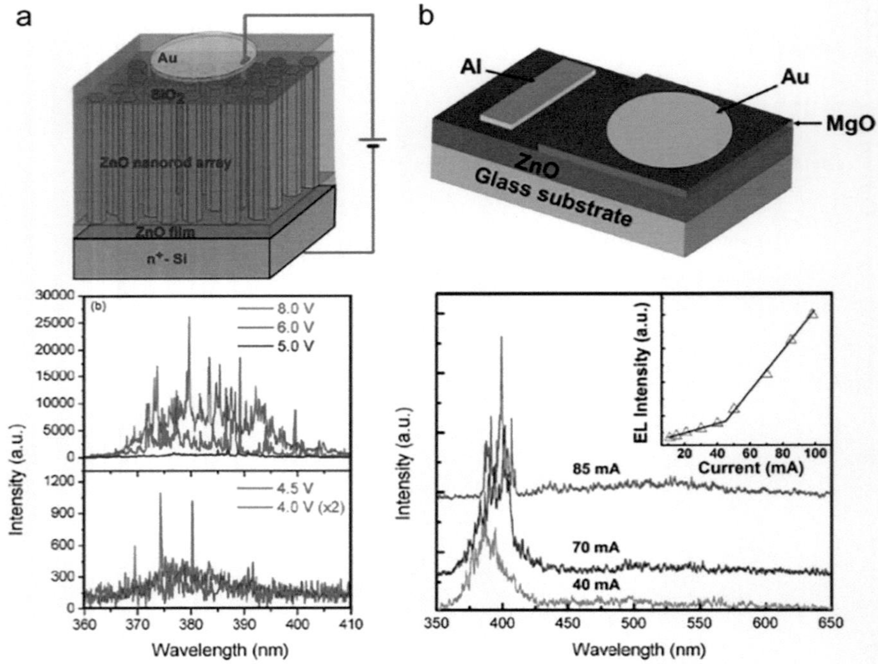

Fig. 7.39 **a** and **b** n-ZnO/p-GaN heterojunction laser diode [160, 161]

Fig. 7.40 n-ZnO/p-GaN heterojunction Fabry–Pérot microcavity laser diode [129]

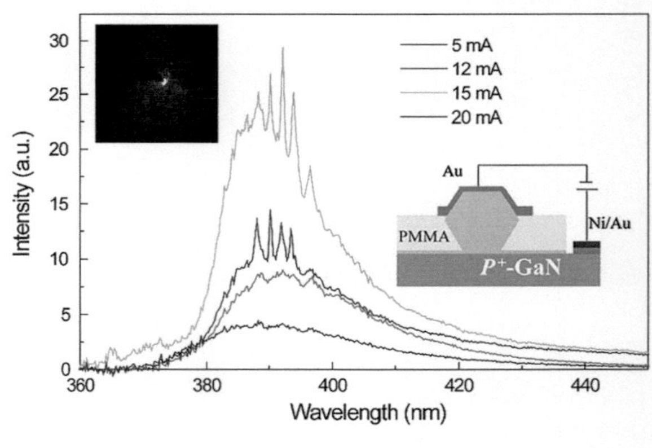

7.6 Summary

ZnO has achieved many important research results in shortwave optoelectronic devices, especially laser devices. From material synthesis, to performance regulation, to device construction, a preliminary exploration path has been formed. In terms of material synthesis, ZnO micro/nano single crystals with controllable morphology and size can be obtained by optimizing process conditions and realizing different modes of laser output. In terms of performance regulation, the laser output performance can be improved by introducing metal surface plasmons, and through cavity structure design and piezoelectric polarization effects, wavelength-dynamically tunable laser-mode output and single-mode operation can be achieved. In terms of device construction, a new path can be taken to replace p-type ZnO with p-type GaN to construct a heterostructure microlaser diode, achieving clear and controllable mode structure electric pump WGM laser output. However, to make breakthrough progress, the difficulty still lies in the preparation of p-type ZnO materials, which is the basis for subsequent functional applications. How can high-quality, stable p-type ZnO micro/nanostructure single crystals be achieved? It is still the focus and direction of effort in the field of science and technology. At present, many research groups are working together in response to this technical problem. From theoretical design to experimental demonstration, some progresses have been made through different means such as single-element doping, multielement mixed doping, and oxygen excess. In addition, ZnO also has good biocompatibility, which can be used in biosensor devices, and has made many representative research progress. Also, the transparent conductive film based on Al-doped ZnO (AZO) has been widely used in the electrodes of solar cells. Compared with indium tin oxide (ITO) conductive film, AZO has a shorter absorption band edge and advantages matching with amorphous silicon thin-film battery technology and has the value of large-area batch production, whose production line construction has entered a rapid expansion period. Therefore, we should not only focus on overcoming the bottleneck of p-type doping of ZnO materials but also pay attention to other physical properties of ZnO, such as piezoelectric effect, absorption of ultraviolet light, etc., to explore its applications in different fields, such as piezoelectric nanogenerators, stress sensors, sunscreen cosmetics, etc. Functional integration can be carried out to maximize its value and find new industrial application breakthroughs.

References

1. Zu P, Tang Z, Wong GK et al (1997) Ultraviolet spontaneous and stimulated emissions from ZnO microcrystallite thin films at room temperature. Solid State Commun 103(8):459–463
2. Service R E (1997) Materials science-Will UV lasers beat the blues? Science 276(5314):895–895
3. Kohan AF, Ceder G, Morgan D et al (2000) First-principles study of native point defects in ZnO. Phys Rev B 61(22):15019–15027
4. Park CH, Zhang SB, Wei SH (2002) Origin of p-type doping difficulty in ZnO: the impurity perspective. Phys Rev B 66(7):073202
5. Mccluskey MD, Jokela SJ (2009) Defects in ZnO. J Appl Phys 106(7):071101
6. Look DC, Farlow GC, Reunchan P et al (2005) Evidence for native-defect donors in n-type ZnO. Phys Rev Lett 95(22):225502
7. Huang MH, Mao S, Feick H et al (2001) Room-temperature ultraviolet nanowire nanolasers. Science 292(5523):1897–1899
8. Zhu G, Xu C, Zhu J et al (2009) Two-photon excited whispering-gallery mode ultraviolet laser from an individual ZnO microneedle. Appl Phys Lett 94(5):051106

9. Czekalla C, Sturm C, Schmidt-Grund R et al (2008) Whispering gallery mode lasing in zinc oxide microwires. Appl Phys Lett 92(24):241102
10. Yang PD, Yan HQ, Mao S et al (2002) Controlled growth of ZnO nanowires and their optical properties. Adv Func Mater 12(5):323–331
11. Pan ZW, Dai ZR, Wang ZL (2001) Nanobelts of semiconducting oxides. Science 291(5510):1947–1949
12. Li YB, Bando Y, Sato T et al (2002) ZnO nanobelts grown on Si substrate. Appl Phys Lett 81(1):144–146
13. Xu CX, Sun XW, Dong ZL et al (2004) Self-organized nanocomb of ZnO fabricated by Au-catalyzed vapor-phase transport. J Cryst Growth 270(3–4):498–504
14. Xu CX, Sun XW, Dong ZL et al (2004) Zinc oxide nanodisk. Appl Phys Lett 85(17):3878–3880
15. Xu CX, Sun XW, Dong ZL et al (2006) Zinc oxide hexagram whiskers. Appl Phys Lett 88(9):093101
16. Xu C, Dai J, Zhu G et al (2014) Whispering-gallery mode lasing in ZnO microcavities. Laser Photonics Rev 8(4):469–494
17. Wu JJ, Liu SC (2002) Low-temperature growth of well-aligned ZnO nanorods by chemical vapor deposition. Adv Mater 14(3):215–218
18. Yang JL, An SJ, Park WI et al (2004) Photocatalysis using ZnO thin films and nanoneedles grown by metal–organic chemical vapor deposition. Adv Mater 16(18):1661–1664
19. Wang XX, Xu CX, Qin FF et al (2018) Ultraviolet lasing in Zn-rich ZnO microspheres fabricated by laser ablation. Nanoscale 10(37):17852–17857
20. Wagner RS, Ellis WC (1964) Vapor-liquid-solid mechanism of single crystal growth. Appl Phys Lett 4(5):89–90
21. Wu YY, Yang PD (2001) Direct observation of vapor-liquid-solid nanowire growth. J Am Chem Soc 123(13):3165–3166
22. Huang MH, Wu YY, Feick H et al (2001) Catalytic growth of zinc oxide nanowires by vapor transport. Adv Mater 13(2):113–116
23. Blakely JM, Jackson KA (1962) Growth of crystal whiskers. J Chem Phys 37(2):428–430
24. Dai ZR, Pan ZW, Wang ZL (2003) Novel nanostructures of functional oxides synthesized by thermal evaporation. Adv Func Mater 13(1):9–24
25. Hu JT, Odom TW, Lieber CM (1999) Chemistry and physics in one dimension: synthesis and properties of nanowires and nanotubes. Acc Chem Res 32(5):435–445
26. Vayssieres L (2003) Growth of arrayed nanorods and nanowires of ZnO from aqueous solutions. Adv Mater 15(5):464–466
27. Boyle DS, Govender K, O'brien P (2002) Novel low temperature solution deposition of perpendicularly orientated rods of ZnO: substrate effects and evidence of the importance of counter-ions in the control of crystallite growth. Chem Commun (1):80–81
28. Liu Y, Xu C, Zhu Z et al (2018) Controllable fabrication of ZnO microspheres for whispering gallery mode microcavity. Cryst Growth Des 18(9):5279–5286
29. Kuo CL, Kuo TJ, Huang MH (2005) Hydrothermal synthesis of ZnO microspheres and hexagonal microrods with sheetlike and platelike nanostructures. J Phys Chem B 109(43):20115–20121
30. Tian ZRR, Voigt JA, Liu J et al (2003) Complex and oriented ZnO nanostructures. Nat Mater 2(12):821–826
31. Dong X, Xu C, Yang C et al (2019) Photoelectrochemical response to glutathione in Au-decorated ZnO nanorod array. J Mater Chem C 7(19):5624–5629
32. Li J, Lin Y, Lu J et al (2015) Single mode ZnO whispering-gallery submicron cavity and graphene improved lasing performance. ACS Nano 9(7):6794–6800
33. Sidiropoulos TP, Röder R, Geburt S et al (2014) Ultrafast plasmonic nanowire lasers near the surface plasmon frequency. Nat Phys 10(11):870–876
34. Wang X, Vlasenko L, Pearton S et al (2009) Oxygen and zinc vacancies in as-grown ZnO single crystals. J Phys D Appl Phys 42(17):175411
35. Dai J, Xu CX, Xu XY et al (2013) Controllable fabrication and optical properties of Sn-doped ZnO hexagonal microdisk for whispering gallery mode microlaser. APL Mater 1(3):032105

36. Zhu Q, Lu J, Wang Y et al (2016) Burstein-moss effect behind Au surface plasmon enhanced intrinsic emission of ZnO microdisks. Sci Rep 6:36194
37. Teke A, Özgür Ü, Doğan S et al (2004) Excitonic fine structure and recombination dynamics in single-crystalline ZnO. Phys Rev B 70(19):195207
38. Voon LLY, Willatzen M, Cardona M et al (1996) Terms linear in k in the band structure of wurtzite-type semiconductors. Phys Rev B 53(16):10703
39. Hamby D, Lucca D, Klopfstein M et al (2003) Temperature dependent exciton photoluminescence of bulk ZnO. J Appl Phys 93(6):3214–3217
40. Reynolds D, Look DC, Jogai B et al (1999) Valence-band ordering in ZnO. Phys Rev B 60(4):2340
41. Boemare C, Monteiro T, Soares M et al (2001) Photoluminescence studies in ZnO samples. Physica B 308:985–988
42. Meyer B, Alves H, Hofmann D et al (2004) Bound exciton and donor-acceptor pair recombinations in ZnO. Phys Status Solidi (B) 241(2):231–260
43. Qiu J, Li X, He W et al (2009) The growth mechanism and optical properties of ultralong ZnO nanorod arrays with a high aspect ratio by a preheating hydrothermal method. Nanotechnology 20(15):155603
44. Grabowska J, Meaney A, Nanda KK et al (2005) Surface excitonic emission and quenching effects in ZnO nanowire/nanowall systems: limiting effects on device potential. Phys Rev B 71(11):115439
45. Wischmeier L, Voss T, Rückmann I et al (2006) Dynamics of surface-excitonic emission in ZnO nanowires. Phys Rev B 74(19):195333
46. Lu J, Zhu Q, Zhu Z et al (2016) Plasmon-mediated exciton-phonon coupling in a ZnO microtower cavity. J Mater Chem C 4(33):7718–7723
47. Djurišić AB, Leung YH (2006) Optical properties of ZnO nanostructures. Small 2(8–9):944–961
48. Janotti A, Van De Walle CG (2007) Native point defects in ZnO. Phys Rev B 76(16):165202
49. Vlasenko L, Watkins G (2005) Optical detection of electron paramagnetic resonance for intrinsic defects produced in ZnO by 2.5-MeV electron irradiation in situ at 4.2 K. Phys Rev B 72(3):035203
50. Janotti A, Van De Walle CG (2005) Oxygen vacancies in ZnO. Appl Phys Lett 87(12):122102
51. Willander M, Nur O, Sadaf JR et al (2010) Luminescence from zinc oxide nanostructures and polymers and their hybrid devices. Materials 3(4):2643–2667
52. Vanheusden K, Seager C, Warren WT et al (1996) Correlation between photoluminescence and oxygen vacancies in ZnO phosphors. Appl Phys Lett 68(3):403–405
53. Djurišić A, Leung Y, Tam K et al (2006) Green, yellow, and orange defect emission from ZnO nanostructures: influence of excitation wavelength. Appl Phys Lett 88(10):103107
54. Li D, Leung Y, Djurišić A et al (2004) Different origins of visible luminescence in ZnO nanostructures fabricated by the chemical and evaporation methods. Appl Phys Lett 85(9):1601–1603
55. Alvi N, Ul Hasan K, Nur O et al (2011) The origin of the red emission in n-ZnO nanotubes/p-GaN white light emitting diodes. Nanoscale Res Lett 6(1):130
56. Greene LE, Law M, Goldberger J et al (2003) Low-temperature wafer-scale production of ZnO nanowire arrays. Angew Chem 115(26):3139–3142
57. Lima S, Sigoli F, Jafelicci M Jr et al (2001) Luminescent properties and lattice defects correlation on zinc oxide. Int J Inorg Mater 3(7):749–754
58. Anantachaisilp S, Smith SM, Ton-That C et al (2014) Tailoring deep level surface defects in ZnO nanorods for high sensitivity ammonia gas sensing. J Phys Chem C 118(46):27150–27156
59. Wang D, Zhang T (2009) Study on the defects of ZnO nanowire. Solid State Commun 149(43–44):1947–1949
60. Kayaci F, Vempati S, Donmez I et al (2014) Role of zinc interstitials and oxygen vacancies of ZnO in photocatalysis: a bottom-up approach to control defect density. Nanoscale 6(17):10224–10234

61. Wolff P (1962) Theory of the band structure of very degenerate semiconductors. Phys Rev 126(2):405
62. Göbel G (1974) Recombination without k-selection rules in dense electron-hole plasmas in high-purity GaAs lasers. Appl Phys Lett 24(10):492–494
63. Reynolds D, Look DC, Jogai B (2000) Combined effects of screening and band gap renormalization on the energy of optical transitions in ZnO and GaN. J Appl Phys 88(10):5760–5763
64. Takeda J, Jinnouchi H, Kurita S et al (2002) Dynamics of photoexcited high density carriers in ZnO epitaxial thin films. Phys Status Solidi (B) 229(2):877–880
65. Johnson JC, Knutsen KP, Yan H et al (2004) Ultrafast carrier dynamics in single ZnO nanowire and nanoribbon lasers. Nano Lett 4(2):197–204
66. Yamamoto A, Kido T, Goto T et al (1999) Dynamics of photoexcited carriers in ZnO epitaxial thin films. Appl Phys Lett 75(4):469–471
67. Leung Y, Kwok W, Djurišić A et al (2005) Time-resolved study of stimulated emission in ZnO tetrapod nanowires. Nanotechnology 16(4):579
68. Tang Z, Wong GK, Yu P et al (1998) Room-temperature ultraviolet laser emission from self-assembled ZnO microcrystallite thin films. Appl Phys Lett 72(25):3270–3272
69. Klingshirn C (1992) Properties of the electron-hole plasma in II–VI semiconductors. J Cryst Growth 117(1–4):753–757
70. Klingshirn C, Hauschild R, Fallert J et al (2007) Room-temperature stimulated emission of ZnO: alternatives to excitonic lasing. Phys Rev B 75(11):115203
71. Versteegh MA, Vanmaekelbergh D, Dijkhuis JI (2012) Room-temperature laser emission of ZnO nanowires explained by many-body theory. Phys Rev Lett 108(15):157402
72. Nakamura T, Firdaus K, Adachi S (2012) Electron-hole plasma lasing in a ZnO random laser. Phys Rev B 86(20):205103
73. Hendry E, Koeberg M, Bonn M (2007) Exciton and electron-hole plasma formation dynamics in ZnO. Phys Rev B 76(4):045214
74. Sun H, Makino T, Tuan N et al (2000) Stimulated emission induced by exciton-exciton scattering in ZnO/ZnMgO multiquantum wells up to room temperature. Appl Phys Lett 77(26):4250–4252
75. Chen R, Ling B, Sun XW et al (2011) Room temperature excitonic whispering gallery mode lasing from high-quality hexagonal ZnO microdisks. Adv Mater 23(19):2199–2204
76. Matsuzaki R, Soma H, Fukuoka K et al (2017) Purely excitonic lasing in ZnO microcrystals: temperature-induced transition between exciton-exciton and exciton-electron scattering. Phys Rev B 96(12):125306
77. Wang L, Giles N (2003) Temperature dependence of the free-exciton transition energy in zinc oxide by photoluminescence excitation spectroscopy. J Appl Phys 94(2):973–978
78. Musa I, Massuyeau F, Cario L et al (2011) Temperature and size dependence of time-resolved exciton recombination in ZnO quantum dots. Appl Phys Lett 99(24):243107
79. Zhao Q, Yang L-L, Willander M et al (2008) Surface recombination in ZnO nanorods grown by chemical bath deposition. J Appl Phys 104(7):073526
80. Lettieri S, Amato LS, Maddalena P et al (2009) Recombination dynamics of deep defect states in zinc oxide nanowires. Nanotechnology 20(17):175706
81. Richters JP, Wischmeier L, Rückmann I et al (2009) Surface excitonic recombination dynamics in ZnO nanowires. Phys Status Solidi C 6(2):560–563
82. Cao H, Xu J, Zhang D et al (2000) Spatial confinement of laser light in active random media. Phys Rev Lett 84(24):5584
83. Cao H, Zhao Y, Ho S et al (1999) Random laser action in semiconductor powder. Phys Rev Lett 82(11):2278
84. Yang HY, Lau SP, Yu SF et al (2006) High-temperature random lasing in ZnO nanoneedles. Appl Phys Lett 89(1):011103
85. Wang D, Seo H, Tin C-C et al (2006) Lasing in whispering gallery mode in ZnO nanonails. J Appl Phys 99(9):093112

86. Dai J, Xu CX, Zheng K et al (2009) Whispering gallery-mode lasing in ZnO microrods at room temperature. Appl Phys Lett 95(24):241110
87. Wang Y, Qin F, Lu J et al (2017) Plasmon enhancement for Vernier coupled single-mode lasing from ZnO/Pt hybrid microcavities. Nano Res 10(10):3447–3456
88. Chen A, Zhu H, Wu Y et al (2018) Low-threshold whispering-gallery mode upconversion lasing via simultaneous six-photon absorption. Advanced Optical Materials 6(17):1800407
89. Ritchie R (1957) Plasma losses by fast electrons in thin films. Phys Rev 106(5):874
90. Okamoto K, Niki I, Shvartser A et al (2004) Surface-plasmon-enhanced light emitters based on InGaN quantum wells. Nat Mater 3(9):601–605
91. Kwon MK, Kim JY, Kim BH et al (2008) Surface-plasmon-enhanced light-emitting diodes. Adv Mater 20(7):1253–1257
92. Lai Y-Y, Lan Y-P, Lu T-C (2013) Strong light-matter interaction in ZnO microcavities. Light-Science & Applications 2:e76
93. Franke H, Sturm C, Schmidt-Grund R et al (2012) Ballistic propagation of exciton-polariton condensates in a ZnO-based microcavity. New J Phys 14(1):013037
94. Brewster MM, Zhou X, Lim SK et al (2011) Role of Au in the growth and nanoscale optical properties of ZnO nanowires. J Phys Chem Lett 2(6):586–591
95. Lin JM, Lin HY, Cheng CL et al (2006) Giant enhancement of bandgap emission of ZnO nanorods by platinum nanoparticles. Nanotechnology 17(17):4391–4394
96. You JB, Zhang XW, Fan YM et al (2007) Surface plasmon enhanced ultraviolet emission from ZnO films deposited on Ag/Si(001) by magnetron sputtering. Appl Phys Lett 91(23):231907
97. Lee M-K, Kim TG, Kim W et al (2008) Surface plasmon resonance (SPR) electron and energy transfer in noble metal-zinc oxide composite nanocrystals. J Phys Chem C 112(27):10079–10082
98. Im J, Singh J, Soares JW et al (2011) Synthesis and optical properties of dithiol-linked ZnO/gold nanoparticle composites. J Phys Chem C 115(21):10518–10523
99. Lawrie BJ, Haglund RF Jr, Mu R (2009) Enhancement of ZnO photoluminescence by localized and propagating surface plasmons. Opt Express 17(4):2565–2572
100. Lin HY, Cheng CL, Chou YY et al (2006) Enhancement of band gap emission stimulated by defect loss. Opt Express 14(6):2372–2379
101. Lu J, Xu C, Dai J et al (2015) Plasmon-enhanced whispering gallery mode lasing from hexagonal Al/ZnO microcavity. ACS Photonics 2(1):73–77
102. Lu J, Li J, Xu C et al (2014) Direct resonant coupling of Al surface plasmon for ultraviolet photoluminescence enhancement of ZnO microrods. ACS Appl Mater Interfaces 6(20):18301–18305
103. Hwang SW, Shin DH, Kim CO et al (2010) Plasmon-enhanced ultraviolet photoluminescence from hybrid structures of graphene/ZnO films. Phys Rev Lett 105(12):127403
104. Zhu Q, Qin F, Lu J et al (2017) Synergistic graphene/aluminum surface plasmon coupling for zinc oxide lasing improvement. Nano Res 10(6):1996–2004
105. Langhammer C, Yuan Z, Zorić I et al (2006) Plasmonic properties of supported Pt and Pd nanostructures. Nano Lett 6(4):833–838
106. Torma P, Barnes WL (2015) Strong coupling between surface plasmon polaritons and emitters: a review. Rep Prog Phys 78(1):013901
107. Liu R, Fu XW, Meng J et al (2013) Graphene plasmon enhanced photoluminescence in ZnO microwires. Nanoscale 5(12):5294–5298
108. Liu Y, Willis RF, Emtsev KV et al (2008) Plasmon dispersion and damping in electrically isolated two-dimensional charge sheets. Phys Rev B 78(20):201403
109. Ding W, Hsu L-Y, Heaps CW et al (2018) Plasmon-coupled resonance energy transfer II: exploring the peaks and dips in the electromagnetic coupling factor. J Phys Chem C 122(39):22650–22659
110. Wang YY, Xu CX, Li JT et al (2015) Improved whispering-gallery mode lasing of ZnO microtubes assisted by the localized surface plasmon resonance of Au nanoparticles. Sci Adv Mater 7(6):1156–1162

111. Lin Y, Xu CX, Li JT et al (2013) Localized surface plasmon resonance-enhanced two-photon excited ultraviolet emission of Au-decorated ZnO nanorod arrays. Adv Opt Mater 1(12):940–945
112. Xu C, Qin F, Zhu Q et al (2018) Plasmon-enhanced ZnO whispering-gallery mode lasing. Nano Res 11(6):3050–3064
113. Liu X, Galfsky T, Sun Z et al (2015) Strong light-matter coupling in two-dimensional atomic crystals. Nat Photonics 9(1):30–34
114. Kavokin A, Malpuech G, Laussy FP (2003) Polariton laser and polariton superfluidity in microcavities. Phys Lett A 306(4):187–199
115. Cao E, Lin W, Sun M et al (2018) Exciton-plasmon coupling interactions: from principle to applications. Nanophotonics 7(1):145–167
116. Weisbuch C, Nishioka M, Ishikawa A et al (1992) Observation of the coupled exciton-photon mode splitting in a semiconductor quantum microcavity. Phys Rev Lett 69(23):3314–3317
117. Houdre R, Weisbuch C, Stanley RP et al (1994) Measurement of cavity-polariton dispersion curve from angle-resolved photoluminescence experiments. Phys Rev Lett 73(15):2043–2046
118. Van Vugt LK, Ruhle S, Ravindran P et al (2006) Exciton polaritons confined in a ZnO nanowire cavity. Phys Rev Lett 97(14):147401
119. Sun L, Chen Z, Ren Q et al (2008) Direct observation of whispering gallery mode polaritons and their dispersion in a ZnO tapered microcavity. Phys Rev Lett 100(15):156403
120. Lu J, Jiang M, Wei M et al (2017) Plasmon-induced accelerated exciton recombination dynamics in ZnO/Ag hybrid nanolasers. ACS Photonics 4(10):2419–2424
121. Oulton RF, Sorger VJ, Zentgraf T et al (2009) Plasmon lasers at deep subwavelength scale. Nature 461(7264):629–632
122. Ma R-M, Oulton RF, Sorger VJ et al (2011) Room-temperature sub-diffraction-limited plasmon laser by total internal reflection. Nat Mater 10(2):110–113
123. Oulton RF, Sorger VJ, Genov DA et al (2008) A hybrid plasmonic waveguide for subwavelength confinement and long-range propagation. Nat Photonics 2(8):496–500
124. Chou Y-H, Chou B-T, Chiang C-K et al (2015) Ultrastrong mode confinement in ZnO surface plasmon nanolasers. ACS Nano 9(4):3978–3983
125. Chou Y-H, Wu Y-M, Hong K-B et al (2016) High-operation-temperature plasmonic nanolasers on single-crystalline aluminum. Nano Lett 16(5):3179–3186
126. Chou Y-H, Hong K-B, Chung Y-C et al (2017) Metal for plasmonic ultraviolet laser: Al or Ag? IEEE J Sel Top Quantum Electron 23(6):1–7
127. Li H, Li J-H, Hong K-B et al (2019) Plasmonic nanolasers enhanced by hybrid graphene-insulator-metal structures. Nano Lett 19(8):5017–5024
128. Chou Y-H, Hong K-B, Chang C-T et al (2018) Ultracompact pseudowedge plasmonic lasers and laser arrays. Nano Lett 18(2):747–753
129. Dai J, Xu CX, Sun XW (2011) ZnO-microrod/p-GaN heterostructured whispering-gallery-mode microlaser diodes. Adv Mater 23(35):4115–4119
130. Chu S, Wang G, Zhou W et al (2011) Electrically pumped waveguide lasing from ZnO nanowires. Nat Nanotechnol 6(8):506–510
131. Lu J, Xu C, Li F et al (2018) Piezoelectric effect tuning on ZnO microwire whispering-gallery mode lasing. ACS Nano 12(12):11899–11906
132. Lu J, Yang Z, Li F et al (2019) Dynamic regulating of single-mode lasing in ZnO microcavity by piezoelectric effect. Mater Today 24:33–40
133. Wiersig J (2003) Hexagonal dielectric resonators and microcrystal lasers. Phys Rev A 67(2):023807
134. Qian F, Li Y, Gradečak S et al (2008) Multi-quantum-well nanowire heterostructures for wavelength-controlled lasers. Nat Mater 7(9):701–706
135. Yang Z, Wang D, Meng C et al (2014) Broadly defining lasing wavelengths in single bandgap-graded semiconductor nanowires. Nano Lett 14(6):3153–3159
136. Liu X, Zhang Q, Xiong Q et al (2013) Tailoring the lasing modes in semiconductor nanowire cavities using intrinsic self-absorption. Nano Lett 13(3):1080–1085

137. Liu X, Zhang Q, Yip JN et al (2013) Wavelength tunable single nanowire lasers based on surface plasmon polariton enhanced Burstein-Moss effect. Nano Lett 13(11):5336–5343
138. Ghafouri-Shiraz H (2003) Distributed feedback laser diodes and optical tunable filters. Wiley & Sons
139. Wang Y, Xu C, Jiang M et al (2016) Lasing mode regulation and single-mode realization in ZnO whispering gallery microcavities by the Vernier effect. Nanoscale 8(37):16631–16639
140. Hodaei H, Miri M-A, Heinrich M et al (2014) Parity-time-symmetric microring lasers. Science 346(6212):975–978
141. Feng L, Wong ZJ, Ma R-M et al (2014) Single-mode laser by parity-time symmetry breaking. Science 346(6212):972–975
142. Gargas DJ, Moore MC, Ni A et al (2010) Whispering gallery mode lasing from zinc oxide hexagonal nanodisks. ACS Nano 4(6):3270–3276
143. Vedam K, Davis T (1969) Pressure dependence of the refractive indices of the hexagonal crystals beryl, α-CdS, α-ZnS, and ZnO. Phys Rev 181(3):1196
144. Kim MH, Schubert MF, Dai Q et al (2007) Origin of efficiency droop in GaN-based light-emitting diodes. Appl Phys Lett 91(18):183507
145. Yeh HJJ, Smith JS (1994) Fluidic self-assembly for the integration of GaAs light-emitting-diodes on Si substrates. IEEE Photonics Technol Lett 6(6):706–708
146. Ikeda M, Hayakawa T, Yamagiwa S et al (1979) Fabrication of 6h-sic light-emitting-diodes by a rotation dipping technique-electroluminescence mechanisms. J Appl Phys 50(12):8215–8225
147. Kennes K, Coutino-Gonzalez E, Martin C et al (2017) Silver zeolite composites-based LEDs: a novel solid-state lighting approach. Adv Func Mater 27(14):1606411
148. Chang CW, Tan WC, Lu ML et al (2014) Electrically and optically readable light emitting memories. Sci Rep 4(1):1–6
149. Tsukazaki A, Ohtomo A, Onuma T et al (2005) Repeated temperature modulation epitaxy for p-type doping and light-emitting diode based on ZnO. Nat Mater 4(1):42–46
150. Liu W, Gu SL, Ye JD et al (2006) Blue-yellow ZnO homostructural light-emitting diode realized by metalorganic chemical vapor deposition technique. Appl Phys Lett 88(9):092101
151. Sun XW, Ling B, Zhao JL et al (2009) Ultraviolet emission from a ZnO rod homojunction light-emitting diode. Appl Phys Lett 95(13):133124
152. Liu JS, Shan CX, Shen H et al (2012) ZnO light-emitting devices with a lifetime of 6.8 hours. Appl Phys Lett 101(1):011106
153. Huang J, Chu S, Kong JY et al (2013) ZnO p-n homojunction random laser diode based on nitrogen-doped p-type nanowires. Adv Opt Mater 1(2):179–185
154. Hwang DK, Kang SH, Lim JH et al (2005) P-ZnO/n-GaN heterostructure ZnO light-emitting diodes. Appl Phys Lett 86(22):222101
155. Zhang XM, Lu MY, Zhang Y et al (2009) Fabrication of a high-brightness blue-light-emitting diode using a ZnO-nanowire array grown on p-GaN thin film. Adv Mater 21(27):2767–2770
156. Shen YW, Chen X, Yan XQ et al (2014) Low-voltage blue light emission from n-ZnO/p-GaN heterojunction formed by RF magnetron sputtering method. Curr Appl Phys 14(3):345–348
157. Yang ZP, Xie ZH, Lin CC et al (2015) Slanted n-ZnO nanorod arrays/p-GaN light-emitting diodes with strong ultraviolet emissions. Opt Mater Express 5(2):399–407
158. You D, Xu C, Qin F et al (2018) Interface control for pure ultraviolet electroluminescence from nano-ZnO-based heterojunction devices. Sci Bull 63(1):38–45
159. Liu W, Li ZX, Shi ZL et al (2020) Symmetrical Bi-heterojunction alternating current ultraviolet light-emitting diode. IEEE Electron Device Lett 41(2):252–255
160. Ma X, Pan J, Chen P et al (2009) Room temperature electrically pumped ultraviolet random lasing from ZnO nanorod arrays on Si. Opt Express 17(16):14426–14433
161. Zhu H, Shan CX, Zhang JY et al (2010) Low-threshold electrically pumped random lasers. Adv Mater 22(16):1877–1881

Chapter 8
Gallium Oxide Solar-Blind Deep-Ultraviolet Detectors

Weihua Tang and Zeng Liu

8.1 Overview

As early as the mid-nineteenth century, scientists began to notice the electromagnetic waves in the shortwave direction of visible light in the whole electromagnetic spectrum and called the range between 10 and 400 nm the ultraviolet (UV) region. This region is divided into four parts: UVA (380–315 nm), UVB (315–280 nm), UVC (280–200 nm), and vacuum ultraviolet (VUV, 200–10 nm). Due to the Earth's ozone layer's absorption of ultraviolet light below 280 nm, the atmosphere contains almost no ultraviolet light below 280 nm; thus, this band of ultraviolet light is also called solar-blind light; this makes the communication accuracy in this band extremely high. Solar-blind ultraviolet detection technology plays an irreplaceable role in high-security ultraviolet communication, low false-alarm-rate missile warning, and low-detection-limit ultraviolet single-photon counting system applications. It has a wide range of applications in solar-blind ultraviolet communication, missile warning tracking, rocket plume detection, quantum information, space exploration, high-energy physics, corona detection, environmental monitoring, and biomedical and other military and civilian fields.

Solar-blind deep-ultraviolet detectors are optoelectronic devices that could convert the energy of incident deep-ultraviolet light into electrical signals. The solar-blind ultraviolet detectors currently under development mainly use the quantum photoelectric effect of semiconductors to achieve the detection of ultraviolet light. The main materials are $Al_xGa_{1-x}N$, $Zn_xMg_{1-x}O$, diamond, and gallium oxide (Ga_2O_3), which are wide bandgap semiconductors. Among these materials, it is difficult to find

W. Tang (✉)
Nanjing University of Posts and Telecommunications, Nanjing, China
e-mail: whtang@njupt.edu.cn

Z. Liu
Inner Mongolia University, Hohhot, China
e-mail: zengliu@imu.edu.cn

S. Xu (ed.), *Optical Characterization of Microstructures and Optoelectronic Devices Based on Wide Band Gap Semiconductors*, Wide Bandgap Semiconductors,
https://doi.org/10.1007/978-981-95-1928-6_8

cheap intrinsic substrates for $Al_xGa_{1-x}N$, and heteroepitaxy can easily lead to lattice mismatch and interlayer stress in heteroepitaxial layers; the growth temperature is high, and epitaxy is difficult. The main disadvantage of $Zn_xMg_{1-x}O$ is that when the bandgap of the material is greater than 4.5 electron volts (eV), the material is prone to phase separation, making it difficult to obtain a single wurtzite structure of $Zn_xMg_{1-x}O$. Diamond is also not the best choice for solar-blind ultraviolet detectors due to its high preparation cost and single bandgap (response band below 225 nm). The bandgap width of Ga_2O_3 is about 4.9 eV, and its corresponding absorption edge lies at 240–280 nm, which is exactly in the solar-blind zone; its bandgap and optical absorption edge can be further adjusted by doping. Coupled with its excellent chemical stability and thermal stability and the continuously maturing single-crystal preparation technology, Ga_2O_3 is gradually becoming an ideal material for the construction of solar-blind ultraviolet photodetectors; it has now attracted the attention of many researchers around the world. Our research group is one of the earliest to engage in the research of Ga_2O_3 solar-blind detectors and is considered by the "Gallium Oxide Technology Assessment Report" [1] of the US Air Force Research Laboratory to be the most active research group in the field of solar-blind detector research. Based on our group's research, this chapter first introduces the basic physical properties of Ga_2O_3 materials, common growth methods, and the basic principles and main parameters of the detector. We focus on our research group's achievements in the field of solar-blind detectors on Ga_2O_3 thin-film metal–semiconductor–metal (MSM), heterojunction, and Schottky photodiode detectors, mainly involving the photoelectric characterization and analysis of related devices.

8.2 Basic Principles of Gallium Oxide and Its Detectors

8.2.1 Physical Properties and Preparation Techniques of Gallium Oxide Materials

Ga_2O_3, as a typical wide bandgap semiconductor, has a bandgap width of ~4.9 eV, and it has five polymorphs (polycrystalline or variants), namely α-Ga_2O_3, ε-Ga_2O_3, β-Ga_2O_3, γ-Ga_2O_3, and δ-Ga_2O_3. The conversion conditions between them are shown in Fig. 8.1a, and the other four phases of Ga_2O_3 all convert to β-Ga_2O_3 under the corresponding conditions. β-Ga_2O_3 has the best thermodynamic and chemical stability, and this chapter focuses on β-Ga_2O_3 solar-blind ultraviolet detectors. As shown in Fig. 8.1b, its optical absorption edge is about 4.8 eV.

Gallium oxide has a high Baliga's figure of merit and Johnson's figure of merit, which indicates that gallium oxide semiconductors can theoretically achieve very low direct-current (DC) losses when used in low-frequency DC devices. Moreover, it is suitable for use in high-frequency, high-power devices. Most importantly, the bandgap of Ga_2O_3 is around 4.9 eV, which coincides with strong absorption in the solar-blind ultraviolet region of the electromagnetic spectrum, making it extremely

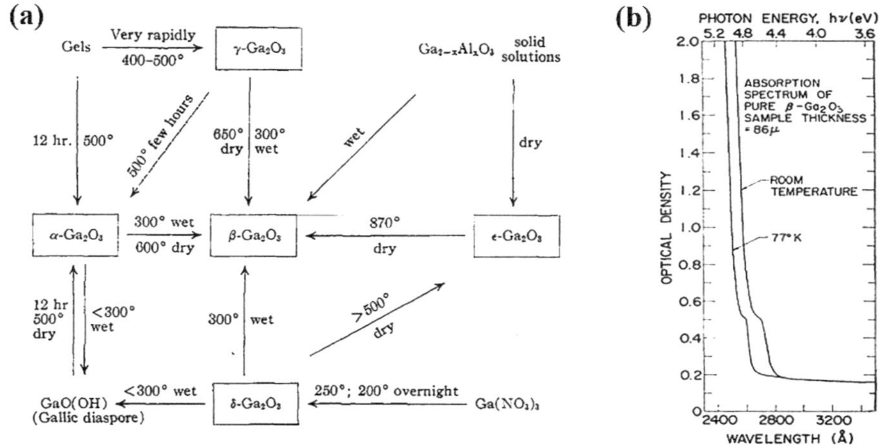

Fig. 8.1 **a** Conversion diagram and conversion conditions between the five polymorphs of Ga_2O_3 [2] and **b** absorption edge of pure β-Ga_2O_3 [3]

suitable for the preparation of solar-blind ultraviolet detectors [4–9]. Ga_2O_3 bulk single-crystal substrates can be grown by methods similar to those used to prepare sapphire crystals, such as the pulling method, the Bridgman method, and the floating zone method. At present, 2-inch β-Ga_2O_3 single-crystal substrates grown by the Bridgman method can be mass produced, and 4-inch and 6-inch β-Ga_2O_3 single-crystal substrates have also been reported [10]. Ga_2O_3 thin films can also be prepared by techniques such as magnetron sputtering, metal–organic chemical vapor deposition (MOCVD), pulsed laser deposition (PLD), molecular beam epitaxy (MBE), and halide vapor-phase epitaxy (HVPE).

8.2.2 Main Performance Parameters of Detectors

The working principle of a solar-blind ultraviolet photodetector based on Ga_2O_3 wide bandgap semiconductor is to convert the incident light signal into an electrical signal (usually current or voltage) by using its bandgap. When gallium oxide is irradiated by ultraviolet light with a wavelength higher than the optical absorption edge of it, some electrons in the valence band absorb energy and transition to the conduction band to generate nonequilibrium carriers. The nonequilibrium carriers form a photocurrent under the action of an external bias and are read by an external circuit. After long-term exploration and research on detectors, researchers have developed a mature detector evaluation index system. The quality factors used to measure its detection performance mainly include: quantum efficiency and spectral response, light–dark ratio, response speed, linearity, suppression ratio, signal-to-noise ratio (SNR), noise-equivalent power (NEP) and detectivity, etc. The following sections introduce these quality factors.

1. **Responsivity and Quantum Efficiency**

Responsivity and external quantum efficiency are important parameters characterizing the performance of the detector, and they can be converted to each other. Among them, the quantum efficiency is the first parameter to be considered for a detector. The external quantum efficiency refers to the ratio of the number of electron–hole pairs that form a photocurrent and are collected to the number of incident photons, *represented by the symbol*η_{ext}. Responsivity is an important basis for reflecting the photoelectric conversion ability of a photodetector to a light signal of a certain wavelength. It is defined as the magnitude of the photocurrent generated in the external circuit of the detector after a unit of incident light power acts on the detector and is represented by the letter R. The higher the responsivity, the stronger the photoelectric conversion ability of the device. The same photodetector has different responsivity to light of different wavelengths. The spectral response of the material refers to the relationship between the responsivity of the material within a certain wavelength range and the wavelength of the incident light. Its value is the ratio of the photoelectric current generated by the device to the power of the incident light. At the same time, the responsivity of the detector is also related to the bias applied to the device. As the bias increases, the responsivity of the device will also increase accordingly. The calculation formulas for external quantum efficiency and responsivity are as follows [11]:

$$\eta_{ext} = \frac{\frac{I_{ph}}{q}}{\frac{P_{in}}{h\nu}} = \frac{I_{ph}}{q} \cdot \frac{h\nu}{P_{in}} \tag{8.1}$$

$$R = I_{ph}/P_{in} = \frac{q\eta_{ext}}{h\upsilon} = \frac{\eta_{ext}\lambda}{1.24}(A/W), \quad R = \frac{\left(I_{ph} - I_d\right)}{P_{in} \cdot S} \tag{8.2}$$

where I_{ph} is the photocurrent, P_{in} is the incident light intensity, S is the illumination area, I_d is the dark current, η_{ext} is the external quantum efficiency, and λ is the wavelength of the incident light (μm).

2. **Photocurrent, Dark Current, and Light-to-Dark Ratio**

The dark current refers to the current output of the photodetector under a certain bias in a dark environment. The magnitude of the dark current mainly depends on the conductivity of the thin-film material and the bias. A larger dark current brings more serious noise, so we hope that the dark current of the photodetector is as small as possible. The photocurrent refers to the current caused by the directional movement of photogenerated carriers excited by light radiation of a certain wavelength under a certain bias applied to the photodetector. The magnitude of the photocurrent mainly depends on the light power, the photosensitive characteristics of the thin-film material, and the magnitude of the bias applied during testing. It is an important basis for evaluating the performance of photodetectors. The light-to-dark current ratio is the ratio of the photocurrent to the dark current. Within an acceptable range, we hope that the higher the light-to-dark current ratio, the better, because a higher light-to-dark current ratio means that the detector has less noise when working.

3. **Response Time**

Response time and recovery time are parameters that represent the transient characteristics of photodetectors. It reflects the response speed of the detector to rapidly changing incident light signals; that is, at the moment of applying and removing light, the change in the output current of the detector is often not synchronized with the light signal, and there is a delay between them, which is the response time, represented by the letter τ. At the moment of removing the light, the output current of the detector satisfies:

$$i_{ph}(t) = i_{ph0}\exp\left(-\frac{t}{\tau}\right) \tag{8.3}$$

In the formula, i_{ph} (t) and i_{ph0} represent the instantaneous photocurrent value of the photodetector at time t and the photocurrent value of the photodetector at the moment when the light disappears, respectively. According to this formula, the decay time is the time it takes for i_{ph} (t) to drop from the peak to 1/e of i_{ph0}, represented by τ_d. Similarly, the rise time is the time it takes for i_{ph} (t) to rise from the lowest point to (e-1)/e of i_{ph0}, represented by τ_r. For square-wave pulse signals, the rise time (τ_r) and decay time (τ_d) correspond to the time it takes for the output signal to go from 10 to 90% of the peak value. The bandwidth reflects the high-frequency response characteristics of the detector. In the application of photodetectors, we usually hope that the response time of the photodetector is as short as possible.

4. **Signal-to-Noise Ratio, Noise-Equivalent Power, and Detectivity**

The minimum radiation power that a detector can detect is mainly limited by some form of noise, which may come from the detector itself or from external radiation disturbances or subsequent electronic systems. Generally speaking, the noise of the detector can be divided into two major categories: radiation noise and internal noise of the detector. Radiation noise includes signal disturbance noise and background disturbance noise. The signal-to-noise ratio (SNR) is used to characterize the relative strength of the required signal to the background noise, which is defined as the ratio of the effective signal power to the noise power, or the ratio of the square of its signal amplitude.

$$SDR(dB) = 10\log_{10}\frac{P_{signal}}{P_{noise}} \tag{8.4}$$

Noise-equivalent power (NEP) and detectivity are currently two important related parameters used to measure the noise performance of detector systems. NEP is the measure of the smallest light signal that the detector can detect. Its value can be calculated by the incident light power required by the device under the unit signal-to-noise ratio. The smaller the NEP, the weaker the signal the detector can detect. Detectivity is the reciprocal of the noise-equivalent power, represented by the symbol D. Since Jones found that both the noise-equivalent power and detectivity are functions of the square root of the detector's effective area A and the amplifier bandwidth Δf, in order to eliminate the above effects, the normalized detectivity D^* was proposed. Their formulas are shown below [12]:

$$NEP = \frac{\sqrt{\overline{I}_n^2}}{R} \text{ (Watt)} \tag{8.5}$$

$$D = \frac{1}{NEP}; \; D^* = D \cdot (A \cdot \Delta f)^{\frac{1}{2}} \tag{8.6}$$

In the formula, $\sqrt{\overline{I}_n^2}$ represents the root-mean-square of the noise current, R is the responsivity of the device, A is the effective area of the detector, and Δf is the amplifier bandwidth. In general, the basic requirements for detector performance are high sensitivity, low noise, large bandwidth, high reliability, and low cost. Depending on the specific application, a balance must be struck between these performance indicators, especially gain and bandwidth, which are important parameters that need to be weighed to determine which type of device is suitable.

In order to evaluate the performance of the detector and understand the range of applications of the device, we need to test and calculate the main parameters of the detector. The main equipment required for photoelectric performance testing of the detector includes digital source meter, light source, and timing-shutter component. In our research group's testing platform, we mainly use the Keithley 2450 m and Keithley 4200 semiconductor tester in conjunction with a standard mercury-lamp light source with a wavelength of 254 nm and a timing-shutter component to control the ultraviolet-light source. The digital source meter provides the bias required for the sample device and collects the output electrical signal of the solar-blind ultraviolet detector. This testing platform can test the I–V characteristics and response time of the sample device under the conditions of 254-nm ultraviolet-light illumination and without ultraviolet-light illumination and further calculate performance parameters such as responsivity and light–dark current ratio to evaluate the performance of the solar-blind ultraviolet photodetector. In addition, different wavelengths of light are separated out through an ultraviolet–visible spectrometer to test the spectral response of the device.

8.3 Gallium Oxide Solar-Blind Detectors

Currently, many research units at home and abroad have carried out related research on gallium oxide solar-blind ultraviolet detectors. Generally speaking, semiconductor photodetectors are mainly divided into the following three structures: (1) metal–semiconductor–metal (MSM) structure photoconductive detectors,

(2) Schottky barrier detectors, and (3) heterojunction photodetectors. Different structures have their own advantages and disadvantages, and we introduce the preparation and photoelectric testing of these three types of gallium oxide-based solar-blind detectors, respectively.

8.3.1 Photoconductive Detectors

The conductivity in semiconductor materials increases due to the generation of photogenerated carriers excited by external light sources. The photodetector prepared based on this photoconductive effect is a photoconductive detector [13]. As shown in Fig. 8.2a, it is a β-Ga_2O_3 detector with a three-finger fork electrode contact; the finger length is 2800 microns, and the finger spacing is 200 microns [14]. Figure 8.2b is the ultraviolet–visible light absorption spectrum used to determine its wavelength absorption region. It can be clearly seen that the absorption intensity of this device in the wavelength range less than 250 nm is much greater than the absorption intensity of the wavelength greater than 250 nm, which is exactly at the edge of the solar-blind ultraviolet region; with the help of the optical absorption equation [15]:

$$(\alpha hv)^2 = A\left(hv - E_g\right) \tag{8.7}$$

where α is the absorption coefficient, hv is the energy of the incident photon, and E_g is the bandgap width. In the above equation, the $(\alpha hv)^2$ and hv are linearly extrapolated in the linear region of the curve, and the intersection with the horizontal axis is the bandgap width of β-Ga_2O_3, which is also the minimum energy of incident photons sufficient to excite photogenerated carriers inside β-Ga_2O_3.

As shown in Fig. 8.2c, the detector has the lowest current under no-light conditions (dark), and the current output by 365-nm light irradiation is also very small. However, the output current under 254-nm light irradiation is significantly increased compared to no-light conditions, because the absorption intensity of gallium oxide for this wavelength of light is much greater than 365 nm. The test was conducted at room temperature. Figure 8.2d shows the output current of the gallium oxide detector over time. The time required to rise from the initial output value to the stable value is 0.86 s, and the time to fall from the stable value to the initial output value is 1.21 s, that is, the rise and fall times of the detector are 0.86 s and 1.21 s, respectively.

The carrier transport mechanism in the β-Ga_2O_3 detector is shown in Fig. 8.3. The solid line represents the generation of photogenerated carriers, and the dashed line represents the recombination of carriers. Under the excitation of incident ultraviolet light, carriers transition from the valence band (process 1) and defect conduction band (process 2) to the conduction band, generating photogenerated electron–hole pairs. Under the irradiation of 254-nm ultraviolet light, this photogenerated carrier behavior mainly occurs in the form of process 1, while under the irradiation of 365-nm ultraviolet light, this behavior only occurs in the form of process 2. For the photoexcitation of semiconductors, process 1 is the main pathway. Therefore, the

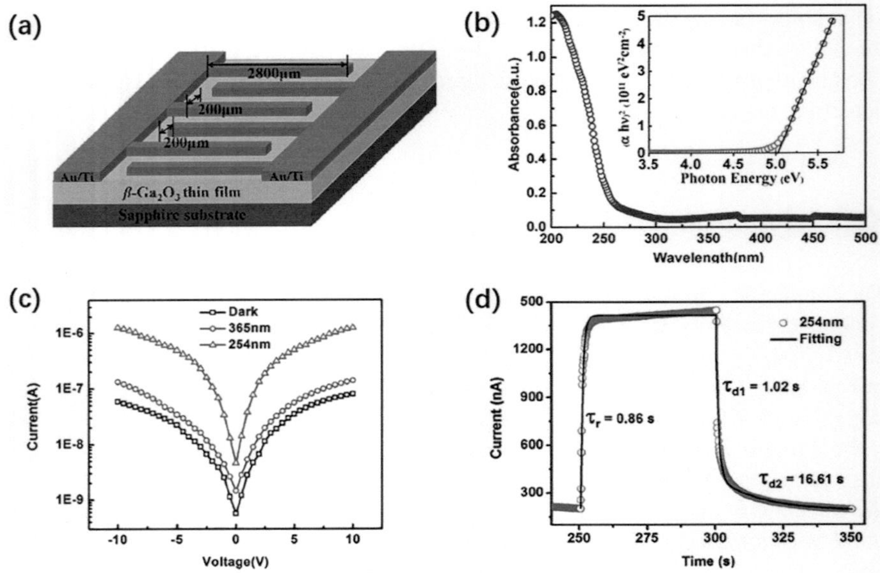

Fig. 8.2 Schematic diagram of the gallium oxide semi-gold fork electrode solar-blind detector, **b** ultraviolet–visible light absorption spectrum, **c** current–voltage (I–V) characteristic curve under dark conditions, 254 and 365-nm ultraviolet-light irradiation, and **d** response time

output current under 254-nm irradiation is greater than the output current under 365-nm irradiation. In addition, some photogenerated carriers may also be captured by the trapped states in the β-Ga_2O_3 film. When the incident light source is turned off, the excited electrons and holes in the conduction band in β-Ga_2O_3 recombine either in the recombination center (process 3) or in the band annihilation process (process 4). This process happens quickly, so the current drops quickly. At the same time, the captured carriers are also released.

For intrinsic oxide semiconductors, the study of oxygen vacancies as electron donors has a long history, and it is inevitable that oxide semiconductors are still difficult to be used as highly insulating materials due to the conductivity caused by oxygen vacancies. Moreover, the presence of oxygen vacancies often leads to a persistent photoconductivity phenomenon, which is not conducive to application in fast actual photoelectric response devices. The doping of Mn elements can be used as a convenient and reliable method to prepare high-insulation films for oxide semiconductor devices [16]. Undoped and Mn-doped β-Ga_2O_3 were grown on α-Al_2O_3 (0001) substrates, and $(GaMn)_2O_3$ epitaxial films were obtained by L-MBE technology through alternating deposition of Ga_2O_3 and Mn films. As shown in Fig. 8.4a, for Ga_2O_3 and $(GaMn)_2O_3$ films, besides the substrate peak, there is only the Ga_2O_3 monoclinic (−201) peak, which indicates that the film growth is carried out in a stable way with a preferred $(\overline{2}01)$ crystal plane orientation. For $(GaMn)_2O_3$ films, the shift of these peaks to smaller 2θ values compared to undoped Ga_2O_3

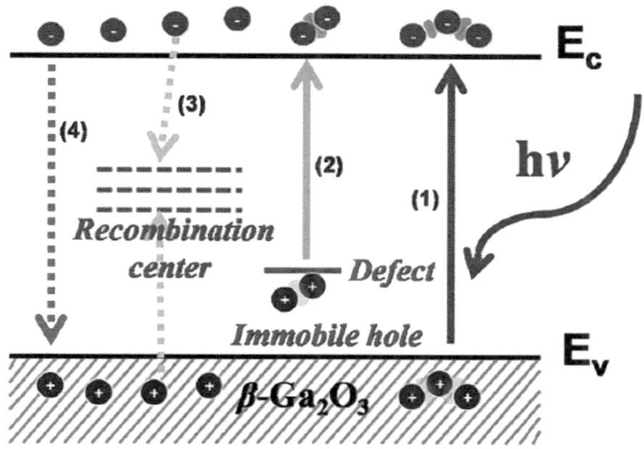

Fig. 8.3 Carrier transport mechanism in β-Ga_2O_3 detector

films indicates an increase in lattice constants, which can be attributed to the ionic radius difference between Mn and Ga (Mn^{2+}, Mn^{3+} and Ga^{3+} ions); their ionic radii are 0.83, 0.64, and 0.62 Å, respectively, and the Mn doping concentration is 28.4 at%. As shown in Fig. 8.4b, the reflection high-energy electron diffraction (RHEED) pattern with bright, clear linear stripes indicates a very smooth, single-phase film surface.

As shown in Fig. 8.5a and b, the photodetector is under the illumination of 254-nm light; the slope of the I–V curve of the Ga_2O_3 and $(GaMn)_2O_3$ thin-film detectors sharply increases in both cases. As shown in Fig. 8.5c and d, the photoresponse process is good. τ_r and τ_d are the time constants of the rising edge and the decaying edge, respectively. We noticed that both the current rise and decay processes are composed of two components: One is a fast-response component, and the other is a slow-response component. Usually, the fast-response part can be attributed to the rapid change in carrier concentration, while the slow-response part is caused by the

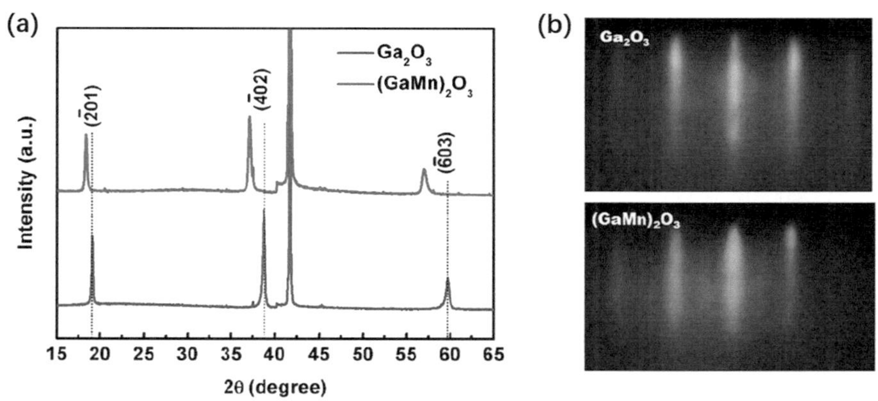

Fig. 8.4 β-Ga_2O_3 and $(GaMn)_2O_3$ films' **a** XRD and **b** RHEED images

capture and release of carriers due to defects such as oxygen vacancies in the β-Ga_2O_3 thin film. For example, when the input light power is 150 μW/cm ^{2} and the voltage is 10 V, the decay process is fast, and the τ_d of the $(GaMn)_2O_3$ thin-film detector is 0.28 s, while the decay process of the Ga_2O_3 photodetector is slow, composed of two component parts, namely $\tau_{d1} = 0.47$ s and $\tau_{d2} = 6.87$ s. Under the same voltage and the same light intensity, the response speed of the $(GaMn)_2O_3$ thin-film detector is much faster than that of the Ga_2O_3 photodetector.

Surface plasmon polaritons (SPPs) are a type of electromagnetic wave, formed by the oscillation of electrons near the surface of a metal, and are used to improve the photoelectric performance of many optoelectronic devices [17, 18]. Figure 8.6a, b, and c shows the I–V characteristic curves of the bare β-Ga_2O_3 thin-film photodetector and the 10-s and 20-s (gold nanoparticle) AuNPs/β-Ga_2O_3 composite thin-film photodetectors, respectively. The corresponding detector structure schematic prototype is shown in the lower-right inset. Using light with wavelengths of 254 and 532 nm as the light source, we can see that as the size of the gold nanoparticles increases, the current also increases. At the same time, under the same illumination of 254-nm and 532-nm light, the current increases with the increase in bias. Under the illumination of 532-nm wavelength light, the current does not significantly increase, indicating that β-Ga_2O_3 films are insensitive to light with a wavelength of 532 nm.

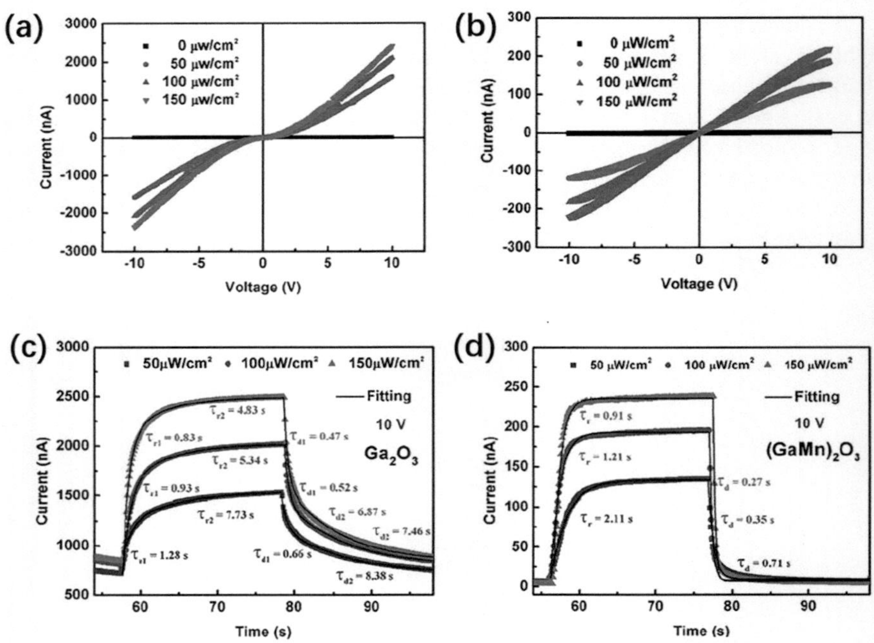

Fig. 8.5 β-Ga_2O_3 thin film's **a** I–V output curve and **c** current output over time; $(GaMn)_2O_3$ thin film's **b** I–V output curve and **d** current output over time

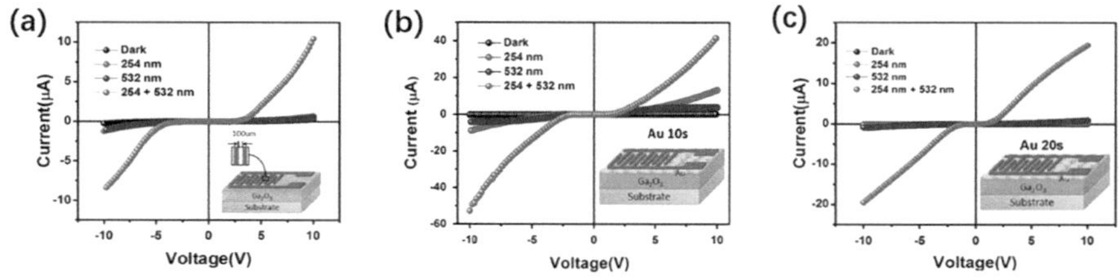

Fig. 8.6 **a** β-Ga_2O_3 film, **b** Au(10 s)/β-Ga_2O_3 film, and **c** Au(20 s)/β-Ga_2O_3 film I–V output curves

Under 1 mW/cm^2 light intensity, 254-nm wavelength light irradiation, and 5-V bias, the light response of β-Ga_2O_3 films of different thicknesses over time is shown in Fig. 8.7a [19, 20]. For β-Ga_2O_3 with thicknesses of 90, 180, 270, 303, 360, 450, and 540 nm, the photocurrent first increases and then decreases, with the maximum photocurrent output from the β-Ga_2O_3 film of 303 nm thickness. This is mainly because when the film thickness is less than 303 nm, the incident light cannot be completely absorbed by the film; when the film thickness is greater than 303 nm, the electric field distribution between the positive and negative electrodes becomes sparse. For the preparation of responsive detectors, the preparation of thin-film materials of appropriate thickness is a prerequisite for achieving excellent performance devices. Figure 8.7b shows the $(I_{photo} - I_{dark})/I_{dark}$ of ultraviolet photodetectors of different thickness β-Ga_2O_3 films. The $(I_{photo} - I_{dark})/I_{dark}$ of the 303-nm-thick β-Ga_2O_3 film photodetector is as high as 16000, which has a good application prospect in ultraviolet band photodetectors.

Ohmic and Schottky contacts, as the two basic forms of electronic device electrode contacts, play an important decisive role in the carrier transport process in the device. Figure 8.8a shows the detector of Ohmic and Schottky contacts prepared on the same β-Ga_2O_3 film by the same microfabrication technology [21], with the electrode being a square electrode with a side length of 400 microns and an electrode spacing of 200

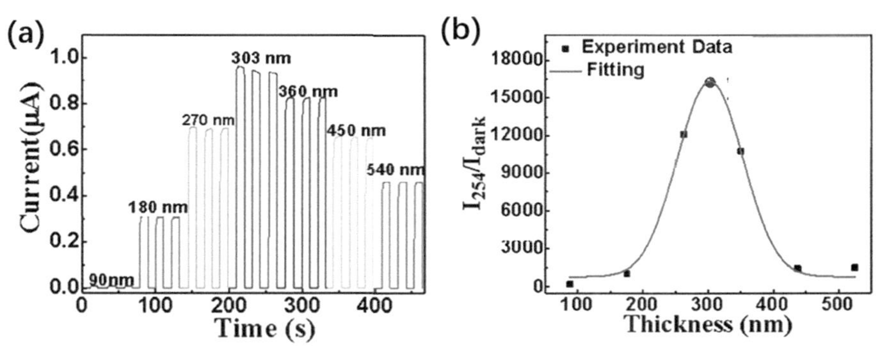

Fig. 8.7 **a** Current output over time for films of different thicknesses and **b** light–dark current ratio

microns. Figure 8.8b is the X-ray diffraction pattern (XRD) of the β-Ga_2O_3 film, which is a $(\bar{2}01)$ oriented single-crystal thin film.

As shown in Fig. 8.9a, under dark conditions, the output current of the Schottky contact device is lower than that of the Ohmic contact device. In Fig. 8.9b, for both Ohmic and Schottky contact devices, the output current excited by 365-nm incident light is slightly increased compared to the output current under dark conditions, but under 254-nm wavelength irradiation, the output current increases by four orders of magnitude, as discussed earlier in this chapter. It is worth noting that the photo–dark current ratio of the Schottky contact device is larger, which is largely due to its low dark current.

For a photodetector, the continuous periodic switching process of the photocurrent over time is a means to characterize the stable operation of the device, and the gallium oxide detector is no exception. As shown in Fig. 8.10a and b, they are the current outputs of Ohmic and Schottky contact β-Ga_2O_3 detectors over time, showing stable and similar changes. Figure 8.10c shows the relationship between the photocurrent

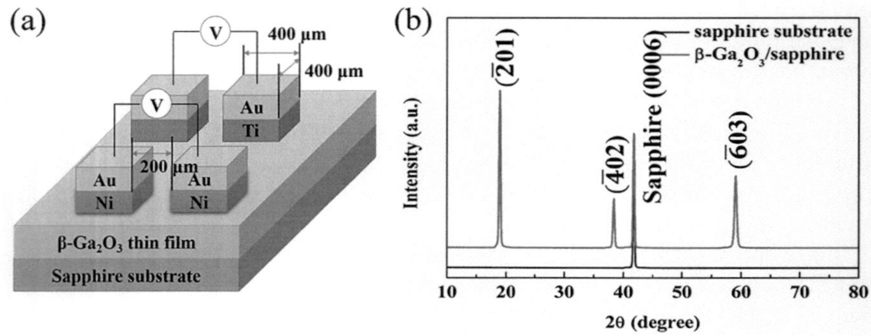

Fig. 8.8 **a** β-Ga_2O_3 schematic diagram of Ohmic and Schottky contact detectors and **b** β-Ga_2O_3 X-ray diffraction diagram

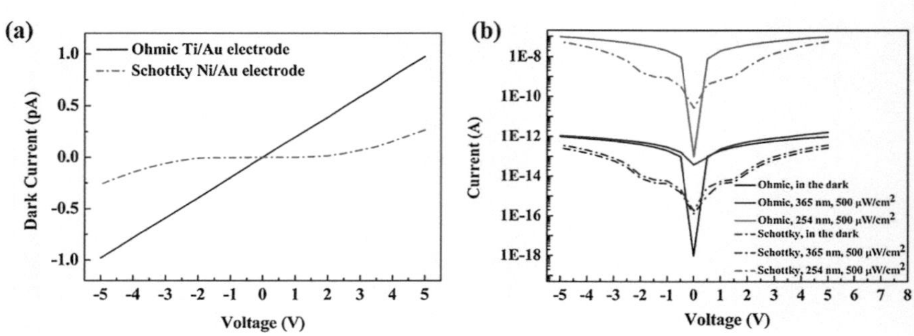

Fig. 8.9 **a** β-Ga_2O_3 linear current–voltage curve of Ohmic and Schottky contact detectors, **b** under dark conditions, 365-nm and 254-nm wavelength light illumination, the logarithmic current–voltage curve of β-Ga_2O_3 Ohmic and Schottky contact detectors

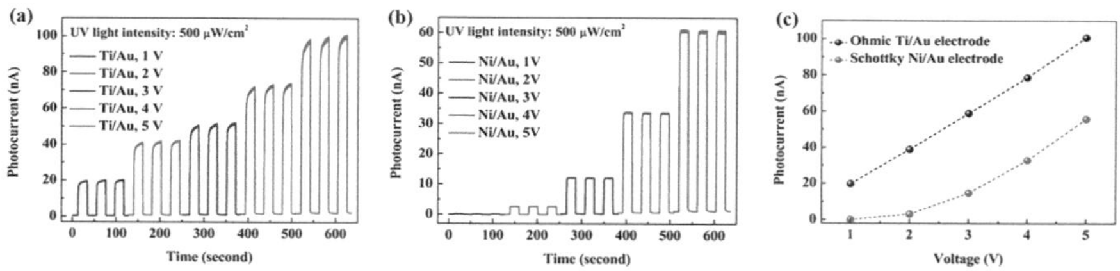

Fig. 8.10 **a** Continuous light response over time of Ohmic and **b** Schottky contact devices under incident light wavelength of 254 nm and light intensity of 500 μW/cm^2; **c** relationship between the photocurrent and voltage of Ohmic and Schottky contact devices

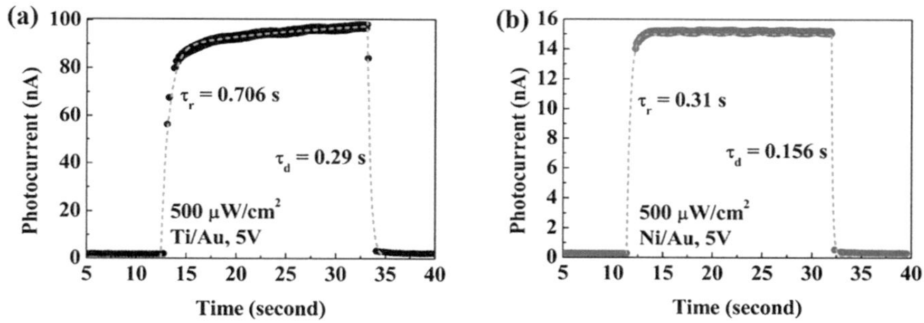

Fig. 8.11 **a** Response time of Ohmic and **b** Schottky contact devices, including rise time and fall time; incident wavelength is 254 nm, light intensity is 500 μW/cm^2, and bias voltage is 5 V

of the two types of devices and the applied voltage; as the applied voltage increases, the Ohmic and Schottky contact devices grow linearly and rectifyingly. In addition, as shown in Fig. 8.11a and b, due to the effective constraint of the Schottky junction potential barrier on the transport of carriers, the response speed of the Schottky contact device is greater than that of the Ohmic contact device, that is, the response time of the Schottky contact device is shorter.

8.3.2 Schottky Barrier Diode Detectors

In the previous section, we discussed the β-Ga_2O_3 detector with symmetric barrier electrode contacts. This section discusses the performance advantages and specific optical and electrical characterization of β-Ga_2O_3 detectors with asymmetric barrier contacts. By using metals with different work functions as the two electrodes of the β-Ga_2O_3 detector, for a fixed β-Ga_2O_3 electron affinity, different metal–semiconductor interface barriers are formed at the two ends of β-Ga_2O_3; that is, an asymmetric barrier β-Ga_2O_3 Schottky barrier diode type detector is formed. The prominent feature of this type of detector is low dark current and fast light response speed [22]. The following

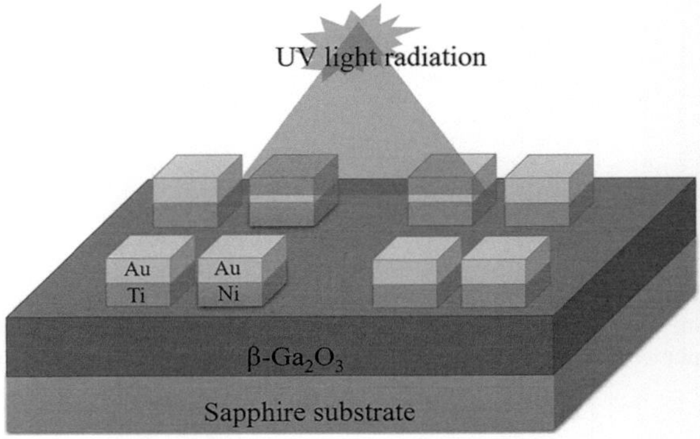

Fig. 8.12 Schematic diagram of the structure of the β-Ga_2O_3 Schottky barrier diode type detector with asymmetric barriers

takes the nickel (Ni)-β-Ga_2O_3-titanium (Ti) structure Schottky barrier diode type detector [23] as an example for explanation. The basic device structure is shown in Fig. 8.12; the metal work functions of Ni and Ti are about 5.15 electron volts (eV) and 4.33 eV [24]; the electron affinity of unintentionally doped β-Ga_2O_3 is about 4.00 eV [25]. Therefore, according to the Schottky–Mott principle [26, 27], when Ni and Ti are deposited as contact electrodes on the surface of β-Ga_2O_3, they form asymmetric barriers of about 1.15 eV and 0.33 eV with β-Ga_2O_3, respectively.

Figure 8.13 shows the spectral response of the β-Ga_2O_3 Schottky barrier diode type detector with asymmetric barriers, with a test wavelength range of 200–700 nm (from ultraviolet to visible light range). The main purpose of this test is to determine the wavelength selectivity of the device, which is the main performance indicator for achieving the precise detection range of the detector. Due to the bandgap width determined by β-Ga_2O_3, it has obvious solar-blind discrimination ability; that is, its light response in the solar-blind region is much higher than the response to other wavelengths. The detector shown in the figure has the highest light response to solar-blind ultraviolet light with a wavelength of 250 nm, up to 144.46 A/W. Compared to the light response to visible light, this device shows a 4.8×10^3 solar-blind/visible light selective incidence ratio, that is, the wavelength-selective ratio of the prepared asymmetric barrier β-Ga_2O_3 Schottky barrier diode type detector for solar-blind and visible light exceeds three orders of magnitude.

As mentioned above, the two ends of the asymmetric barrier β-Ga_2O_3 Schottky barrier diode type detector have different barrier sizes. When such a device is connected to a DC circuit, its electrical output curve should also be asymmetric; that is, it has a rectifying effect. This effect was first discovered in sulfides by German scientist Braun in 1874. As shown in Fig. 8.14a, whether in dark or light conditions, the output current of the device under forward and reverse bias is different, which is asymmetric. This indicates that the asymmetric barrier β-Ga_2O_3 Schottky barrier diode type detector mentioned here has a certain rectifying effect and also verifies the existence of an asymmetric barrier in the device. They have different constraints on the carriers under forward and reverse bias. As can be seen from Fig. 8.14b, the photocurrent of this detector is five orders of magnitude larger than the dark current.

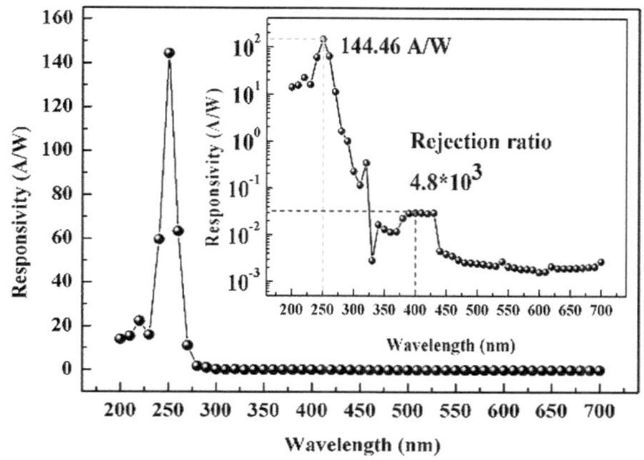

Fig. 8.13 Spectral response of the asymmetric barrier β-Ga_2O_3 Schottky barrier diode type detector

In addition to the strong absorption of solar-blind ultraviolet light by β-Ga_2O_3 material, the effective constraint of the asymmetric interface barrier on the carrier results in a low dark current under dark conditions, which is also one of the key purposes of preparing such devices. As shown in Fig. 8.14c, whether under forward or reverse bias, larger voltage values produce higher photoresponsivity, larger detectivity, and higher external quantum efficiency. Under the irradiation of 254-nm solar-blind ultraviolet light with a bias of 10 V and 220 μW/cm^2, the photoresponsivity, detectivity, and external quantum efficiency of the device are 132.3 A/W, 7.29×10^{14} Jones, and 64711%, respectively. In addition, in Fig. 8.14d, we can see that the device can still operate normally under zero-bias conditions, and the output current under illumination has a linear relationship with the light intensity. This type of detector is called a self-powered detector, mainly due to the effective separation of electron–hole pairs caused by the built-in electric field. The specific mechanism will be explained in the next section along with the heterojunction type detector.

Typically, the light response of a test detector is measured using an irradiation lamp (such as an light-emitting diode [LED]) that can emit light of a certain wavelength. However, it is generally not guaranteed that the light source has a pure wavelength, and there is always a certain delay when the lamp is switched on and off. This can lead to an underestimation of the detector's response speed and response time. As shown in Fig. 8.15a and b, a high-frequency laser is used as the light source here, specifically a KrF laser with a wavelength of 248 nm as the radiation source, with a laser flux of 1 J/cm^2 and a frequency of 5 Hz. Such a high-frequency, fast laser causes the output signal to switch quickly, requiring a high-frequency oscilloscope to collect the output signal. Here, an oscilloscope with a sampling frequency of 2.5 GHz is used. Through calculation, it can be found that the rise and fall times of the device are 9.7 ns and 77 microseconds, respectively.

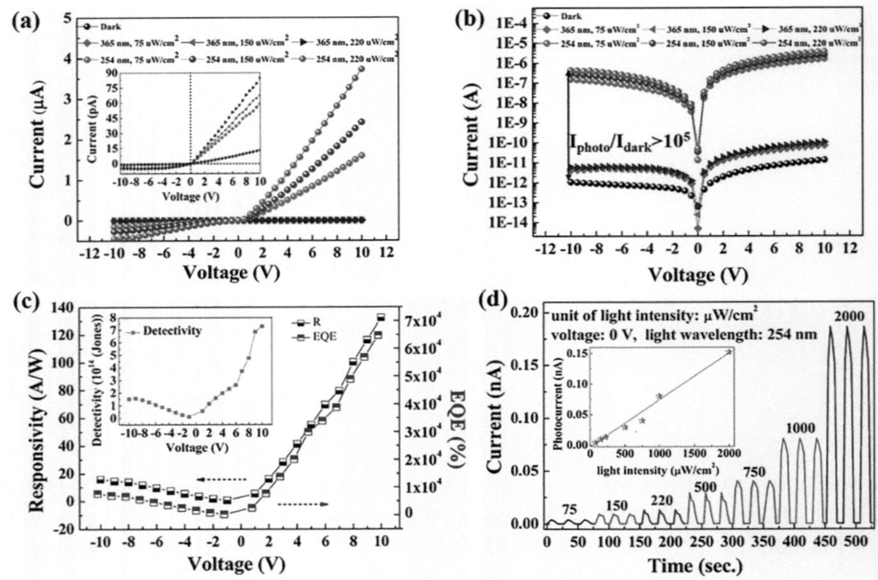

Fig. 8.14 **a** Linear current–voltage output curve, **b** logarithmic current–voltage output curve, **c** photoresponsivity, detectivity, and external quantum efficiency as a function of voltage, and **d** current output over time under no external power drive (zero bias) conditions of the β-Ga_2O_3 Schottky barrier diode type detector

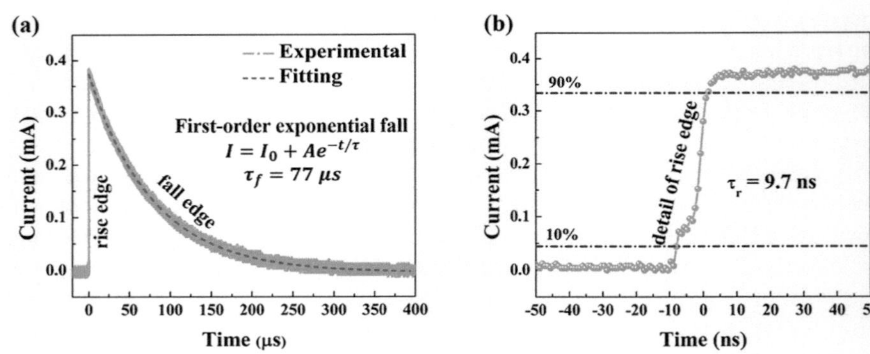

Fig. 8.15 The **a** fall time and **b** rise time of the β-Ga_2O_3 Schottky barrier diode type detector, with a light source of 248-nm KrF laser, and the output signal collected by an oscilloscope with a sampling frequency of 2.5 GHz

8.3.3 Heterojunction Diode Detectors

The heterostructure detector based on Ga_2O_3 can effectively control the transport of carriers due to the construction of the interface barrier, showing many excellent performances, such as low dark current, strong sensitivity, high photoresponse, self-power supply, etc. Fig. 8.16a and c shows the single-layer graphene/Ga_2O_3 [28] and p-Si/Ga_2O_3 [29] heterojunction solar-blind ultraviolet detectors, respectively. As can be seen from Fig. 8.16b and d, the photoresponse and detectivity of their devices are far superior to various detector results of Ga_2O_3. The lattice matching is the theoretical basis for the preparation of heterojunction devices. Table 8.1 summarizes the matching of gallium oxide with other semiconductors. The following introduces the many performance improvements of Ga_2O_3 heterojunction detectors from the perspectives of testing methods, data characterization, and result analysis.

Figure 8.17a is the XRD scan of the β-Ga_2O_3/Ga:ZnO heterojunction; apart from the (00 l) direction diffraction peak of the Ga:ZnO substrate, all the diffraction peaks are from the monoclinic β-Ga_2O_3 orientation of ($\overline{2}$ 01). As shown in the illustration of Fig. 8.17a, the distinct stripe-like reflection high-energy electron diffraction (RHEED) pattern reveals the good epitaxial properties of the β-Ga_2O_3 thin film grown on Ga:ZnO. High-quality β-Ga_2O_3 and Ga:ZnO thin films are the basis for preparing detectors with good performance. As a typical heterostructure device, Fig. 8.17b shows the rectification characteristics and structural schematic of the β-Ga_2O_3/Ga:ZnO heterojunction [30], with a rectification ratio exceeding 10^6, indicating its excellent rectification performance. To test the deep-ultraviolet light response of the formed heterojunction detector, incident light with wavelengths of 254 nm and 365 nm was used, with light intensities of 50 μW/cm^2 and 57 μW/cm^2,

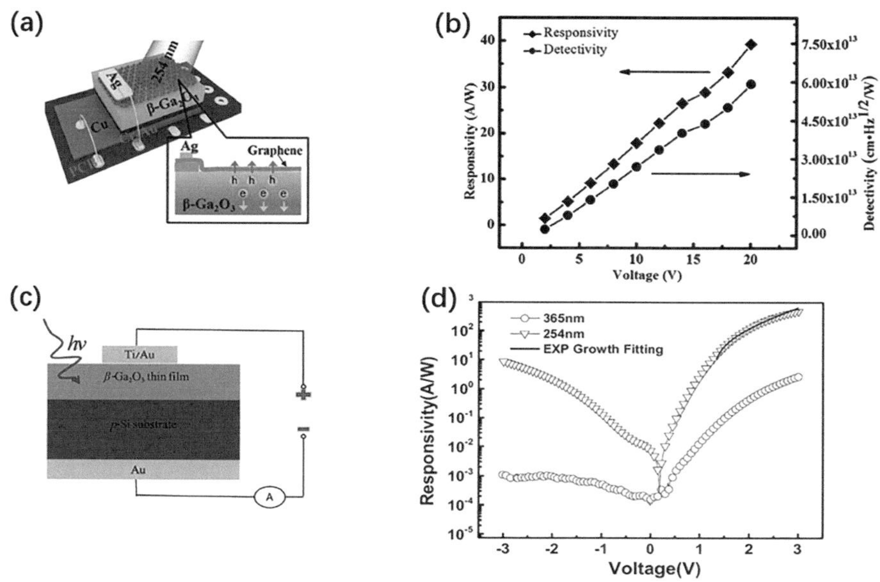

Fig. 8.16 **a** Single-layer graphene/Ga_2O_3 heterojunction solar-blind ultraviolet detector schematic, **b** response and detectivity under different biases; **c** p-Si/Ga_2O_3 heterojunction solar-blind ultraviolet detector schematic, **d** response under different biases

Table 8.1 Mismatch of gallium oxide with other semiconductors

Chemical formula	Ga_2O_3	GaN	ZnO	MgO	MgO	SiC	AlN	MoS_2	ZnS	Si	Graphene	Al_2O_3	GaAs	GaAs
Structure	Monoclinic (β)	Wurtzite (Hexagonal crystal system)	Wurtzite (Hexagonal crystal system)	Triclinic (*P*6$_3$/*mmc*)	Cubic (*Fm*3*m*)	Hexagonal (4*H*/6*H*)	Wurtzite	Hexagonal	Cubic (Zinc blende)	Diamond	Monolayer	Corundum (*α*)	Zinc blende	Zinc blende
Crystal plane	(−2 0 1)	(0 0 1)	(0 0 1)	(0 0 1)	(0 1 1)	(0 0 1)	(0 0 1)	(0 0 1)	(0 1 1)	(0 1 1)	(0 0 1)	(0 0 1)	(0 1 1)	(1 1 1)
Lattice parameters	$a = 14.97$ $b = 3.09$	$a = 5.508$ $b = 3.18$	$a = 5.628$ $b = 3.25$	$a = 5.198$ $b = 3.0015$	$a = 4.31$ $b = 3.047$	$a = 5.33$ $b = 3.078$	$a = 5.418$ $b = 3.128$	$a = 5.484$ $b = 3.166$	$a = 5.409$ $b = 3.825$	$a = 5.43$ $b = 3.84$	$a = 2.46$ $b = 4.26$	$a = 4.759$ $b = 8.243$	$a = 5.735$ $b = 8.111$	$a = 14.048$ $b = 8.111$
Supercell		$3a \times 1b$	$3a \times 1b$	$3a \times 1b$	$4a \times 1b$	$3a \times 1b$	$3a \times 1b$	$3a \times 1b$	$3a \times 3b$	$3a \times 3b$	$6a \times 3b$	$3a \times 2b$	$3a \times 2b$	$1a \times 2b$
Supercell mismatch degree	$1a \times 1b$	9.396% 2.83%	11.34% 4.92%	3.99% 3%	13.17% 1.41%	6.38% 0.39%	7.89% 1.22%	9.01% 2.4%						
	$1a \times 4b$								7.72% 7.712%	8.1% 7.29%	1.42% 3.29%			
	$1a \times 5b$											4.85% 6.28%	12.99% 4.76%	6.56% 4.76%

respectively, and the ultraviolet light was vertically irradiated on the device surface. As shown in Fig. 8.17c, when the detector is irradiated with 254-nm ultraviolet light, the photocurrent density significantly increases over a wide voltage range. At a reverse bias of −10 V, the current density increases from 0.93 μA/cm^2 to 230 μA/cm^2, and the on/off ratio reaches 255. In contrast, the photocurrent density did not show significant changes under 365-nm ultraviolet-light radiation, indicating that the detector has good wavelength selectivity for 254 and 365 nm. It is also worth noting that at zero bias, the dark current density reached 0.15 nA/cm^2 and the photocurrent density was 38.3 nA/cm^2; therefore, this β-Ga_2O_3/Ga:ZnO heterojunction can be used to make self-powered solar-blind deep-ultraviolet photodetectors. To better help readers understand the generation and transmission mechanism of photocurrent in the self-powered β-Ga_2O_3/Ga:ZnO heterojunction, Fig. 8.17d is the energy band diagram of this heterojunction detector. The bandgaps of β-Ga_2O_3 and Ga:ZnO are 4.9 eV and 3.3 eV, respectively. When the heterojunction is formed, due to the drift of electrons, a space charge region will form near the β-Ga_2O_3/Ga:ZnO interface, thereby forming a built-in electric field to balance the electron drift. Under the irradiation of 254-nm solar-blind deep-ultraviolet light, the electric field in the depletion layer can act as an effective internal driving force to separate the photogenerated electron–hole pairs, leading to an increase in photocurrent.

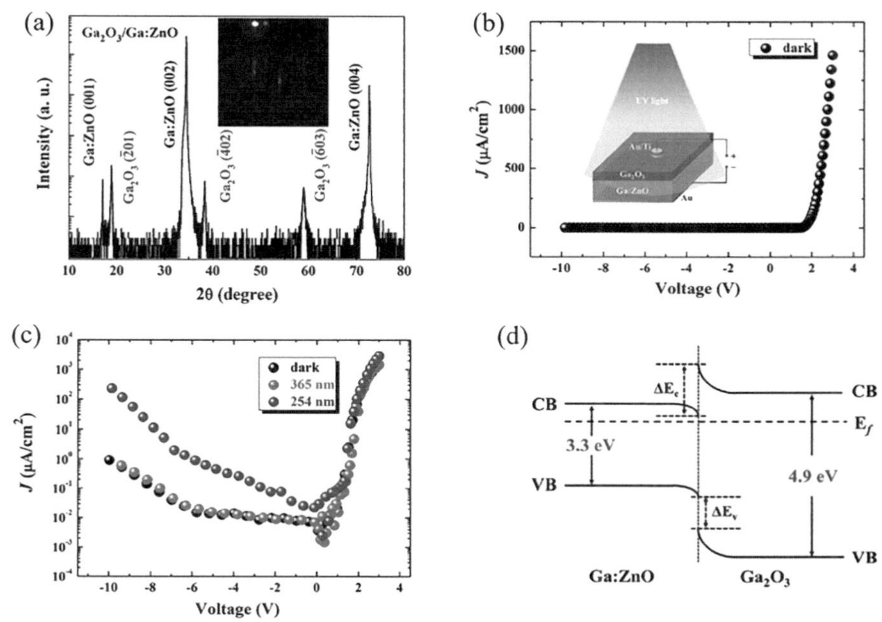

Fig. 8.17 Ga_2O_3/Ga:ZnO heterojunction's **a** X-ray diffraction diagram, **b** current–voltage linear output curve under dark conditions, **c** current–voltage logarithmic output curve under dark conditions, 254-nm and 365-nm wavelength ultraviolet-light irradiation, and **d** energy band diagram

To test the wavelength selectivity of the prepared device, a β-Ga_2O_3/Ga:ZnO heterojunction detector was tested under zero bias using a xenon lamp monochromator. Its spectral response rate (R_λ) varies between 220 and 500 nm as shown in Fig. 8.18a; when the wavelength starts to be less than 310 nm, the light response significantly increases. At 260 nm, R_λ peaks at about 0.76 mA/W. The R_λ value in the deep-ultraviolet region is two orders of magnitude higher than the incident ratio in the visible region, indicating that the device has a good response and wavelength selection ratio to solar-blind ultraviolet light. To evaluate the response speed of the β-Ga_2O_3/Ga:ZnO heterojunction photodetector, its light response over time was studied under zero bias by periodically turning on and off the ultraviolet lamp (light intensity is 50 μW/cm^2). The increase and decrease process of the current density involves fitting the double-exponential relaxation equation of the light response curve, which is as follows: $J = J_0 + Ae^{-t/\tau_1} + Be^{-t/\tau_2}$, where J_0 is the steady-state current density, A and B are constants, t is time, and τ_1 and τ_2 are two relaxation time constants. The rising and falling edges are usually composed of two components: One is a fast-response component, and the other is a slow-response component. Generally speaking, the fast-response component can be attributed to the rapid change in carrier density, while the slow-response component is due to the capture and release of carriers caused by defects. Under zero bias, the heterojunction shows almost the same and repeatable response after multiple illumination cycles. As shown in Fig. 8.18b, J254 instantly goes from 0.15 nA/cm^2 to a steady value of 38.3 nA/cm^2, with a rise time of 0.179 s and a fall time of 0.272 s.

In addition, the heterojunction detector also has a certain improvement and enhancement effect on the light response compared to the single-material detector [31]. Figure 8.19 shows the transient response of the Ga_2O_3/CuSCN heterojunction detector, Ga_2O_3, and CuSCN metal–semiconductor–metal (MSM) devices, the three prototype devices tested under a 1-V bias and 254-nm solar-blind ultraviolet illumination intensity of 1000 μW/cm^2. Obviously, compared with the 12 pA of CuSCN and 8 pA of Ga_2O_3, the heterojunction device shows a superstrong photocurrent of about 2 nA. In addition, an overshoot behavior of the transient photocurrent was

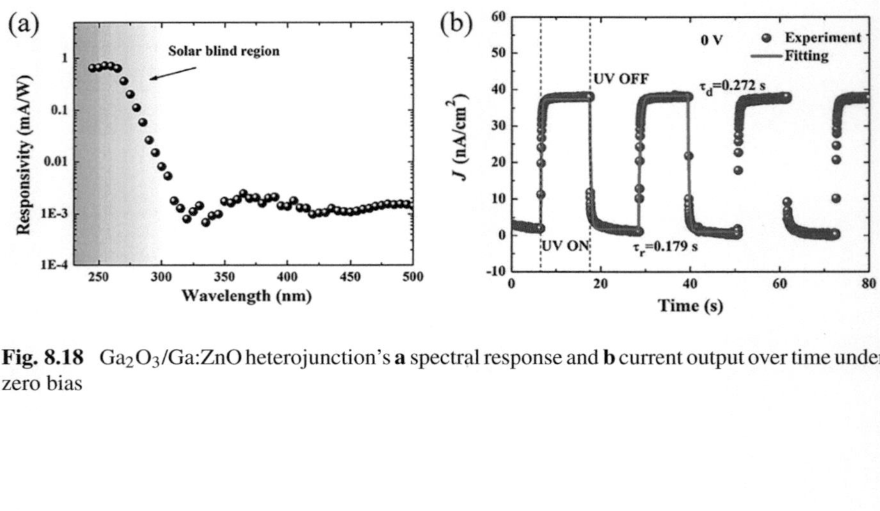

Fig. 8.18 Ga_2O_3/Ga:ZnO heterojunction's **a** spectral response and **b** current output over time under zero bias

observed in the Ga_2O_3 single device, which is the same as the report of the Ni/Au electrode Ga_2O_3-based MSM photodetector [32]. The photocurrent overshoot indicates that a large number of instantaneous accumulated nonequilibrium carriers are generated after illumination; under the joint action of nonradiative transition and radiative transition, some nonequilibrium carriers recombine before being driven to separate to both sides of the electrode by the electric field; the photocurrent of the sample begins to decrease, until a stable photocurrent is formed. For a more detailed comparison of the light response, the quantitative study of the rise and fall process of the current can be represented by fitting the time-varying photocurrent curve using a second-order exponential equation, as shown below: $I = I_0 + Ae^{-t/\tau_1} + Be^{-t/\tau_2}$, where I_0 is the steady-state photocurrent, t is time, A and B are constants, and τ_1 and τ_2 are two relaxation time constants. In addition, the 10–90% rise (or decay) law was used for a better rise time fit of the overshoot photocurrent in Ga_2O_3 devices. The time constants for the rising edge and falling edge are denoted as τ_r and τ_d, respectively. The fitting results of the rise and decay times are shown in the figure. The rise time (τ_{r1}) and decay time (τ_{d1}) of the Ga_2O_3/CuSCN heterojunction photodetector are 0.19 s and 0.16 s, respectively, faster than the rise and fall times of CuSCN and Ga_2O_3 devices, indicating that the construction of n-Ga_2O_3/p-CuSCN heterojunction can effectively improve the response speed.

In order to further understand the key role of core–shell heterojunction in improving photoelectric performance, the band diagram and photoelectric mechanism schematic of In-Ga_2O_3/CuSCN-In structure are shown in Fig. 8.20. The

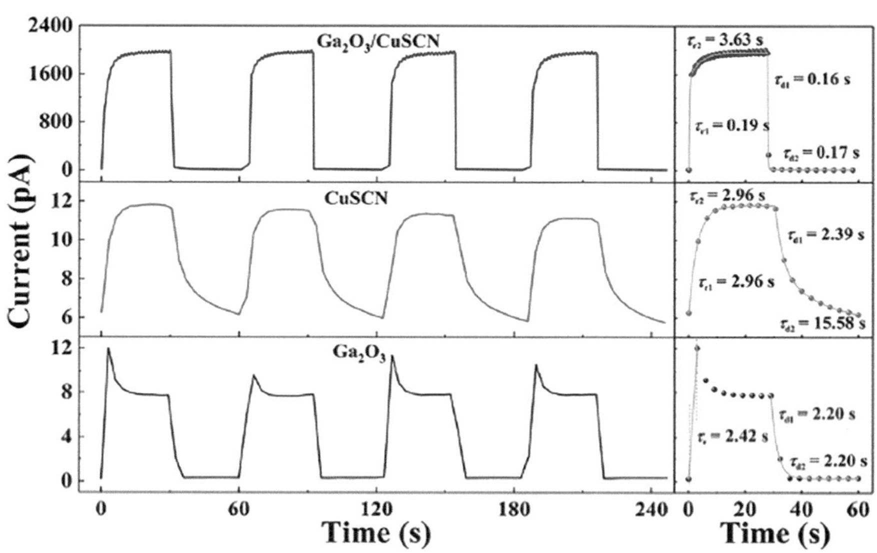

Fig. 8.19 Three device structures: In-Ga_2O_3/CuSCN-In, In-CuSCN-In, and In-Ga_2O_3-In under 254-nm wavelength ultraviolet-light irradiation, 1000 $\mu W/cm^2$ light intensity, and 1-V applied voltage

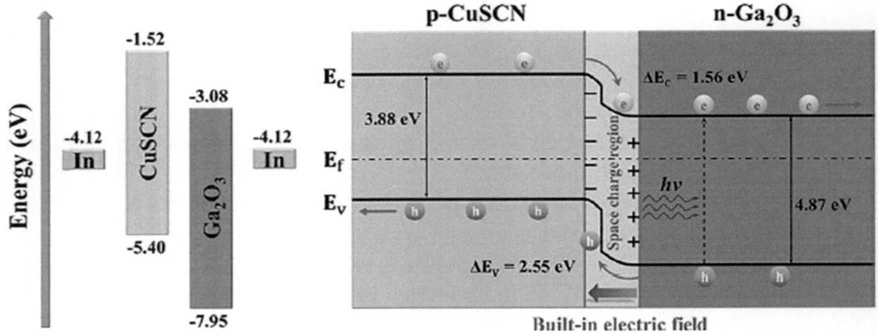

Fig. 8.20 n-Ga_2O_3/p-CuSCN heterojunction detector band structure and working mechanism

band energy level position of In-Ga_2O_3/CuSCN-In heterojunction device is a typical II-type n-Ga_2O_3/p-CuSCN heterojunction. The basic working mechanism of this photodetector is as follows:

a. As shown in Fig. 8.20, when Ga_2O_3 and CuSCN form a p–n junction, due to the diffusion movement of electrons, a space charge region is formed at the interface, and an internal electric field is generated, maintaining a balance between electron drift movement and diffusion movement.
b. Under dark conditions or nonresponsive wavelength (for example, 365 nm) light irradiation, the severe shortage of carriers in the heterojunction structure and the charge loss caused by the diffusion of carriers at the p–n interface have obtained a higher barrier, resulting in ultra-low current and nearly linear current.
c. Under solar-blind ultraviolet radiation, a large number of electron–hole pairs are generated in the active layer of Ga_2O_3, forming a suitable bandgap. The gradually increasing light-induced carrier concentration near the interface reduces the depletion region, thereby reducing the barrier height. In the space charge region, electrons on the conduction band move from p-type CuSCN to n-type Ga_2O_3, while the light-induced holes move in the opposite direction, which allows the carriers to separate quickly under the drive of the built-in electric field.
d. At the same time, the forward bias enhances the migration of charge carriers by suppressing the barrier and depletion layer, while the reverse bias inhibits the migration of charge carriers by promoting the width of the barrier and depletion layer. This is beneficial for the n-Ga_2O_3/p-CuSCN heterojunction to enhance the rectifying characteristics of the photocurrent under forward bias.
e. In addition, the core–shell structure has excellent light capture ability. This radial p–n junction structure provides a way to reduce the migration distance of photogenerated minority carriers to the collective electrode, thereby reducing charge recombination and improving carrier aggregation, which further enhances the photocurrent.
f. Finally, electrons and holes are collected by the indium electrode, forming a loop current.

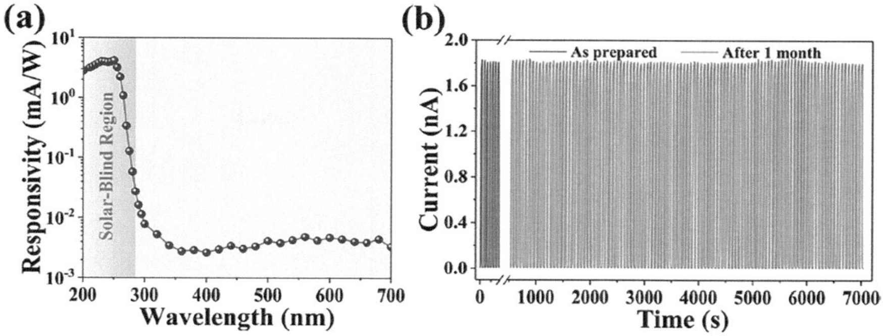

Fig. 8.21 n-Ga_2O_3/p-CuSCN heterojunction detector's **a** spectral response and **b** stability test

As shown in Fig. 8.21a, the spectral response of the heterojunction photodetector in the 200–700 nm range was directly measured with a monochromator at a bias of 5 V. Its peak response is located at 250 nm, which corresponds exactly to the bandgap (4.87 eV) of Ga_2O_3. The bandgap (3.88 eV) and absorption peak (303 nm) of CuSCN do not have a noticeable response peak, indicating that the light response of this heterojunction is mainly dominated by Ga_2O_3. The spectral response has a clear cutoff wavelength at 270–280 nm, proving that this heterojunction device can indeed be used as a photodetector with a high wavelength selectivity ratio. Figure 8.21b depicts the 100-cycle switching performance of the heterojunction photodetector after being stored in a laboratory at room temperature in air for 1 month. The device has significant stability and repeatability in switching between on and off states, even lasting for one month, and the photocurrent has almost no decay compared to the prepared device. Such excellent stability and reproducibility are due to the excellent chemical durability of the Ga_2O_3 single crystal and CuSCN thin film.

8.3.4 Array Detectors

Figure 8.22a shows the schematic diagram of the fabricated photodetector [33]. The electrode fingers are 10 μm wide and 500 μm long, with a 10 μm gap. The horizontal electrode linewidth is 50 μm, the length is 5000 μm, and the distance between the two lines is 1000 μm. SiO_2 is deposited by heat treatment. The film serves as an insulating layer, and then the insulating block is graphically processed by methods such as photolithography and reactive ion etching (RIE). As shown in Fig. 8.22b, it is the photodetector device of the prepared microstructure. Figure 8.22c shows the device encapsulated in a dual-row straight-insert ceramic package, so the photodetector array and circuit are not affected by the surrounding environment.

In order to measure the photodetector array, a 300-W xenon lamp was used as the excitation light source. The light scattered by the monochromator was calibrated with a calibrator to adjust the ultraviolet-enhanced silicon diode and optical power

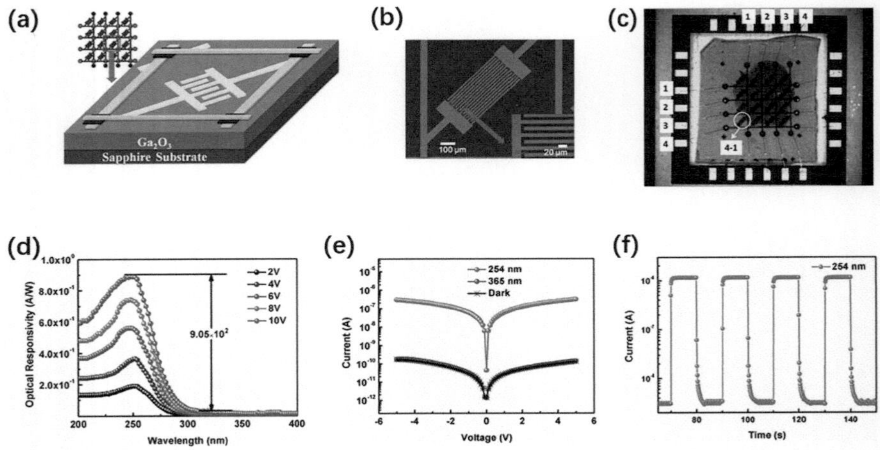

Fig. 8.22 Photodetector array: **a** schematic diagram, **b** microstructure diagram, **c** ceramic packaging, **d** spectral response diagram, **e** light–dark current I–V output curve, and **f** output current changing with time

meter, and then it was collimated to the front of the photodetector. The photocurrent was then activated and recorded by the amplifier. As shown in Fig. 8.22d, the prepared photodetector responds flatly in the short wavelength range, and a sharp area appears at 250 nm, indicating that the prepared β-Ga_2O_3 photodetector array has a strong response to the solar-blind light. Figure 8.22e shows the measured I–V, which clearly shows that the photoexcited current significantly increases by four orders of magnitude compared to the dark current, revealing that the device has a high signal-to-noise ratio. The I–V curve measured under 365-nm illumination did not show a significant increase, indicating that the photodetector is insensitive to 365-nm light. In order to study its stability and light switch characteristics, as shown in Fig. 8.22f, when the light intensity is 200 $\mu W/cm^2$ and the bias is 5 V, the photocurrent of the photodetector appears to open and close periodically with time. According to the measurement results, the dark current is about 1×10^{-11} A; the photocurrent instantly increases to a stable value of about 1.15×10^{-6} A, and the light–dark current ratio is about 10^5. The repeatability of the device indicates its good stability.

The uniformity of the photodetector unit can explain its feasibility in the application of the solar-blind detector array [34]. The growth conditions of the thin film in this experiment are the same as the deposition conditions of the gallium oxide thin film in Chap. 4. The deposition time of the thin film is about 2 h, and the thickness of the thin film is about 920 nm tested by the ellipsometer. After the thin film was deposited, we first characterized the prepared thin film, and the test results are shown in Fig. 8.23. From the XRD test results in Fig. 8.23a, we can see that the gallium oxide thin film grown on the sapphire substrate by MOCVD has sharp diffraction peaks at 18.92°, 38.32°, and 58.96°. Comparing with the Powder Diffraction File

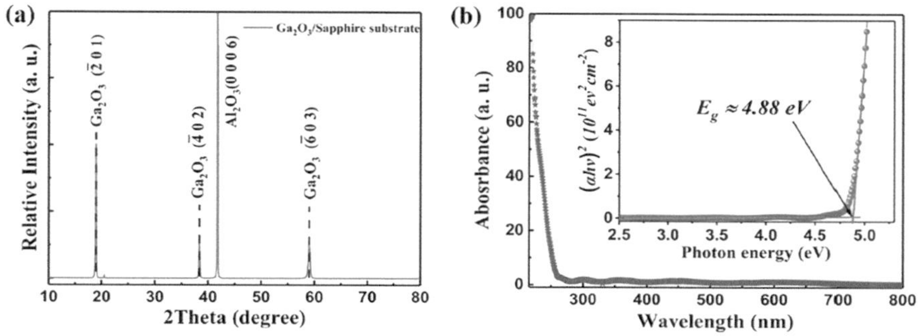

Fig. 8.23 XRD pattern of pure gallium oxide thin film grown on sapphire substrate (**a**) and UV–visible absorption spectrum (**b**), with the inset showing the $(\alpha h\nu)^2$–hν relationship diagram of the thin film

(PDF) card (No. 43-1012), we know that these diffraction peaks correspond to the three crystal planes (−2 0 1), (−4 0 2), and (−6 0 3) of β-Ga_2O_3. The sharp and slender diffraction peaks of high intensity indicate good crystalline quality of the thin film. Similarly, the UV–visible absorption spectrum of the thin film is shown in Fig. 8.23b. As can be seen from the figure, this thin film has very low absorbance in the visible light region and in the UVA and UVB regions, and significant optical absorption in the UVC band. By processing the graph of the $(\alpha h\nu)^2$–$h\nu$ relationship, we can further obtain the optical bandgap value of this thin film as 4.88 eV.

After depositing the gallium oxide thin film, we further fabricated gallium oxide detector arrays [35] using microfabrication. The optical micrographs of the device at different magnifications are shown in Fig. 8.24. From left to right in the figure, the magnification increases. As can be seen from the figure, the 16 × 4 array detector consists of long row wires and symmetrical transverse wires on both sides of each row. The detector units for detecting UV irradiation signals are connected in the middle region of the transverse wires. The arms of the detector units are vertically connected to the transverse wires. The linewidths of the transverse wires and row wires are 10 μm and 20 μm, respectively. Each detector unit has fork fingers with a length of 190 μm and a width of 10 μm. The gap between adjacent fork fingers is 20 μm, and the gap between the edge of the fork fingers and the arm is 10 μm. By calculation, the effective irradiation area of a single detector element is 2.06×10^{-4} cm^2. To increase the pixel density of the detector, the detector pixels of adjacent rows are staggered, with the pixels of odd rows arranged in one column and the pixels of even rows arranged in another column. The top-right and bottom-right images are 100 × and 200 × magnified images of the device, respectively. As can be seen from the figure, the gaps between the fork fingers of the device are intact, with no obvious defects or adhesion. Moreover, since no insulating layer is needed in the array and no multiple etching is required, the microfabrication difficulty and process complexity are relatively low. Therefore, it has significant advantages in terms of device fabrication cost and yield.

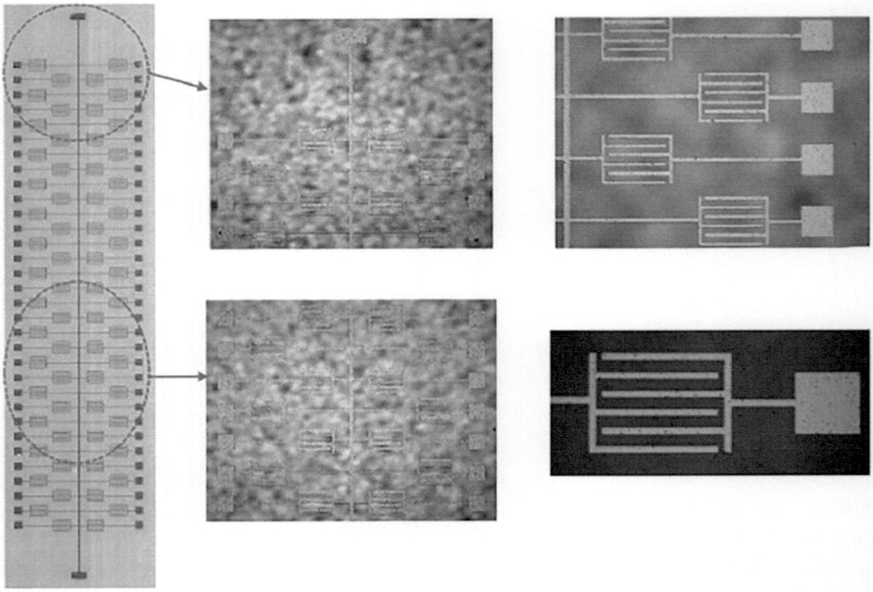

Fig. 8.24 Industrial low-magnification microscope image of the fabricated device (left) and optical microscope images of the device at different magnifications (middle and right)

To test the photoelectric response performance of the device under solar-blind UV irradiation, we first used a 4200 semiconductor test system to test the *I–V* characteristics of a single detector unit in the device under dark conditions and at different light intensities The test results are shown in Fig. 8.25a. The inset in Fig. 8.25a is a magnified image of the *I–V* curve of the device under dark conditions. As can be seen from the inset, in the absence of external light, the current of the sample increases linearly with the increase in voltage, indicating that there is good ohmic contact between the deposited metal electrode and the gallium oxide thin film. When the applied bias of the device was 10 V, the dark current of the device was 1.88 pA. This indicates that the device has a high resistance under dark conditions, and the device's own dark current and noise are low. Subsequently, when the device was exposed to a light intensity of 200 $\mu W/cm^2$, after being irradiated by a 254-nm ultraviolet light, the photocurrent of the device significantly increased. When the device bias was 10 V, the current in the device was 10.04 μA, and its light–dark ratio was approximately 5.34×10^6. Subsequently, as the light intensity increased, the photoresponse current of the device almost linearly increased. When the light intensity reached 2000 $\mu W/cm^2$, the photocurrent of the device under a 10-V bias increased to 57.5 μA, and the corresponding light–dark ratio was 3.06×10^7. According to Formula (8.2), we can further calculate the responsivity of the device at different voltages, as shown in Fig. 8.25b. From the figure, it can be seen that under a light intensity of 200 $\mu W/cm^2$, the responsivity of the device gradually increased from 2.36 A/W (0.1 V) to

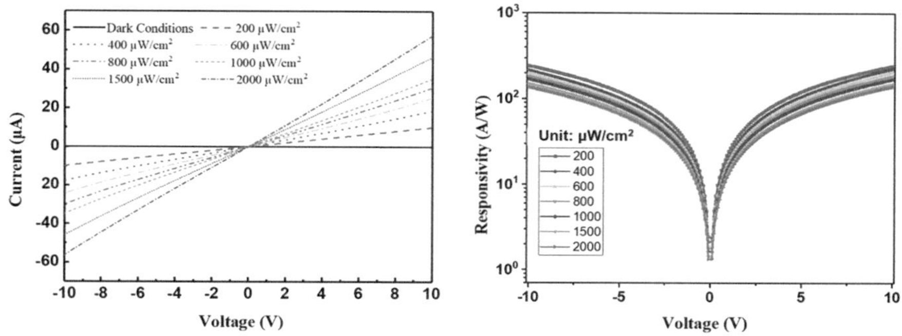

Fig. 8.25 **a** Linear I–V curves of the device under dark conditions and different 254-nm light intensities, where the light intensity gradually increases from 200 μW/cm^2 to 2000 μW/cm^2, with the inset showing the magnified I–V curve of the device under dark conditions; **b** responsivity–voltage curves of the detector under different light intensities

243.66 A/W (10 V) with the increase of voltage. As the light intensity increased, the responsivity of the device gradually decreased to 139.56 A/W. This may also be due to the self-heating effect in the gallium oxide thin film at a higher nonequilibrium carrier concentration [18].

In order to further test the photoelectric response characteristics of the sample, we continued to test the dynamic response current–time (*I–t*) curve of the device under the conditions of constant external bias and periodic switching of 254-nm irradiation; the results are shown in Fig. 8.26. Figure 8.26a shows the transient response characteristic curves of the sample under different positive and negative biases under the condition of periodic 254-nm irradiation with a light intensity of 1000 μW/cm^2. From the figure, it can be seen that under a 2-V bias, when the 254-nm ultraviolet light irradiates the sample, the current of the device quickly rises to 7.14 μA and remains stable in subsequent light exposure. When the ultraviolet light is turned off, the photocurrent of the device immediately decreases, and the whole process stably cycles with the switching of the ultraviolet light. When we change the external bias of the sample, the current under light exposure of the device correspondingly shows a proportional change, and the device can maintain stable operation under all external biases, indicating that the device has good photoelectric transient response performance under all external biases. Figure 8.26b shows the current–time curves of the sample under different light intensities at a constant external bias. From the figure, we can also see that under different light intensities, the device has a fast response speed and stable photoresponse characteristics. As the light intensity increases, its photoresponse current increases almost linearly, indicating that the detector has good linearity under different light intensities. In order to obtain the response speed of this detector element, we extracted a cycle from the *I–t* curve obtained under the test conditions of 10-V bias and 1000 μW/cm^2 light intensity of this detector element and used the double-exponential relaxation equation (Formula (8.4)) to perform nonlinear fitting on the rising and falling edges of this curve [20]; the fitting results are shown in Fig. 8.26c. As can be seen from the figure, by fitting the rising edge of the curve, we can obtain the response time of the detector element to the ultraviolet radiation irradiated on the thin-film surface as $\tau_{r1} = 0.23$ s and $\tau_{r2} = 0.54$ s. By fitting the falling edge of the curve, we can obtain the response time of the detector element after the ultraviolet irradiation is turned off as $\tau_{d1} = 0.05$ s and $\tau_{d2} = 0.15$ s. The fitted curve (red dashed line) and the experimental data (dark blue dotted line) overlap very well, indicating that the response time obtained from the fitting is the response time of the device. Therefore, the response time of this detector element to the external ultraviolet irradiation on/off is $\tau_r = 0.23$ s and $\tau_d = 0.05$ s. Finally, we tested the stability of the device's photoelectric response to external ultraviolet irradiation. As can be seen from Fig. 8.26d, after multiple on/off cycles, the device still has high photoresponse characteristics, and the photoresponse current of the device remains stable during the test, indicating that the device has good stability.

After testing the photoelectric response performance of this detector unit to 254-nm illumination, we sequentially tested the photoelectric performance of all detector elements on this 16 × 4 line array; the results are shown in Fig. 8.27. The figure

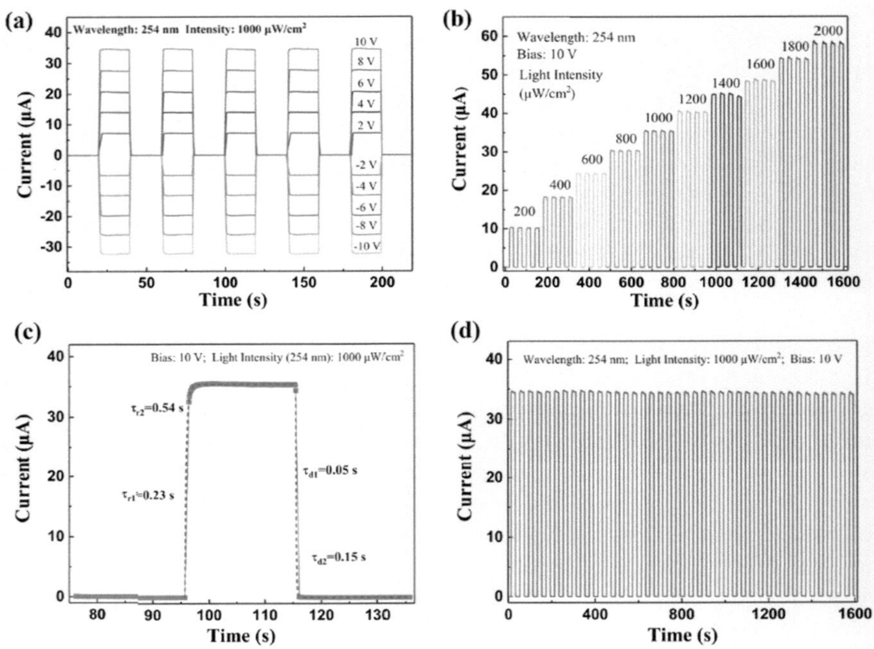

Fig. 8.26 Photoelectric response characteristics (I–t curve) of a single detector element in the line array: **a** current–time photoelectric response characteristics of the device under different biases at a light intensity of 1000 μW/cm^2; **b** I–t curve of the sample under different light intensities at a constant bias of 10 V; **c** response time fitting diagram of the device under a bias of 10 V and light intensity of 1000 μW/cm^2 during a single ultraviolet light on/off cycle; and **d** multiple cycle diagram of the device under a bias of 10 V and illumination conditions of 1000 μW/cm^2

Fig. 8.27 Under the conditions of 10-V external bias and 1000 μW/cm^2 of 254-nm illumination, the response current and responsivity of each pixel in the 16 × 4 linear array detector

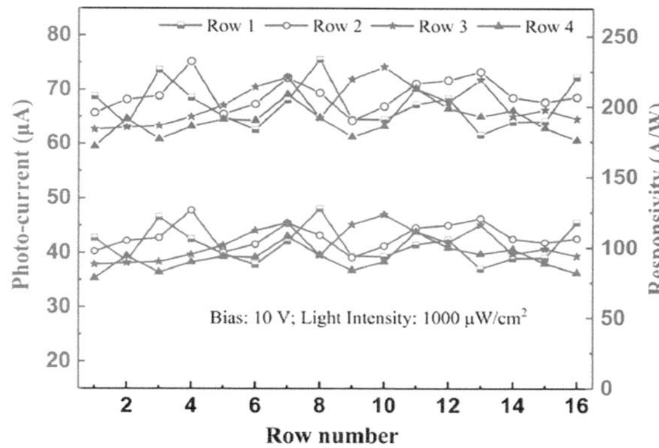

shows the photoresponse current (red dotted line) and corresponding responsivity (blue dotted line) of each column of detector elements under the test conditions of 10-V bias and 1000 μW/cm^2 light intensity. From this, we can see that the photoresponse current of each detector pixel under these test conditions fluctuates between 35 and 45 μA, and the responsivity is between 170 and 230 A/W. This indicates that the photoelectric response performance of each detector element of the gallium oxide line array detector prepared in this experiment has good uniformity.

8.4 Summary

This chapter first introduces the research significance of solar-blind detectors and compares and explains the outstanding advantages of gallium oxide-based solar-blind detectors in materials and device preparation under the current research background. Then, a brief but comprehensive introduction is given to the basic material properties of gallium oxide, as well as the growth technology of gallium oxide single-crystal substrates and thin films. Furthermore, a brief description of the main performance parameters of the detector is given. Finally, with the device structure classification of photoconductive type, Schottky barrier diode type, and heterojunction diode type, the preparation of the detector, material characterization, electrical testing, optical testing, and the means of analyzing test data are specifically introduced. This part includes symmetric barrier ohmic or Schottky contact gallium oxide detectors, asymmetric barrier metal–semiconductor Schottky junction gallium oxide detectors, and semiconductor–semiconductor heterojunction gallium oxide detectors. Among these, the latter is further divided into inorganic p–n junction, organic–inorganic hybrid junction, and wide band junction type detectors. Gallium oxide, as a typical material with great potential, has unique advantages in the field of solar-blind detection. Internationally, the research work on gallium oxide materials and related devices has been highly valued. At present, after more than 10 years of efforts by scientific researchers from major research institutes in China, fruitful results and progress have been made. There is confidence and a long way to go for gallium oxide to step onto the stage of industrial application.

References

1. Bayraktaroglu B (2017) Assessment of gallium oxide technology
2. Roy R, Hill VG, Osborn EF (1952) J Am Chem Soc 74(3):719–722
3. Tippins HH (1965) Phys Rev 140(1A):A316–A319
4. Daoyou G, Peigang L, Chen Zhengwei W, Zhenping TW (2019) Acta Physica Sinica 68(7):078501
5. Guo D, Guo Q, Chen Z, Wu Z, Li P, Tang W (2019) Mater Today Phys 11:100157
6. Liu Z, Li P, Zhi Y, Wang X, Chu X, Tang W (2019) Chin Phys B 28(1):017105
7. Guo D (2016) Research on gallium oxide photoelectric detection and information storage devices. Ph.D. thesis, Beijing University of Posts and Telecommunications
8. Cui W (2018) Research on the growth and electric field regulated photoelectric properties of Ga_2O_3 epitaxial thin films. Ph.D. thesis, Beijing University of Posts and Telecommunications
9. Zhao X (2018) Study on Ga_2O_3 doped thin film crystal phase, carrier regulation and day blindness detector. Ph.D. thesis, Beijing University of Posts and Telecommunications
10. Mastro MA, Kuramata A, Calkins J, Kim J, Ren F, Pearton SJ (2017) ECS. J Solid State Sci Technol 6(5):P356–P359
11. Razeghi M, Rogalski A (1996) J Appl Phys 79(10):7433–7473
12. Fang Y, Armin A, Meredith P, Huang J (2019) Nat Photonics 13:1–4
13. Garrido JA, Monroy E, Izpura I, Munoz E (1998) Semicond Sci Technol 13:563–568
14. Guo DY et al (2014) Opt Mater Express 4(5):1067–1076
15. Kumar SS, Rubio EJ, Noor-A-Alam M, Martinez G, Manandhar S, Shutthanandan V, Thevuthasan S, Ramana CV (2013) J Phys Chem C 117(8):4194–4200
16. Guo D, Li P, Wu Z, Cui W, Zhao X, Lei M, Li L, Tang W (2016) Sci Rep 6:24190
17. An Y, Chu X, Huang Y, Zhi Y, Guo D, Li P, Wu Z, Tang W (2016) Prog Nat Sci 26:65–68
18. An Y (2017) Study on Ga_2O_3 heterojunction and Au nanoparticle composite enhanced day blindness UV detector. Ph.D. thesis, Beijing University of Posts and Telecommunications
19. Wang X, Chen Z, Guo D, Zhang X, Wu Z, Li P, Tang W (2018) Opt Mater Express 8(9):2918–2926
20. Wang X (2020) Study on the growth and device application of self-supporting gallium oxide thin film. Ph.D. thesis, Beijing University of Posts and Telecommunications
21. Liu Z, Zhi Y, Li S, Liu Y, Tang X, Yan Z, Li P, Li X, Guo D, Wu Z, Tang W (2020) J Phys D Appl Phys 53:085105
22. Liang S, Sheng H, Liu Y, Huo Z, Lu Y, Shen H (2001) J Cryst Growth 225:110–113
23. Liu Z, Wang X, Liu Y, Guo D, Li S, Yan Z, Tan C-K, Li W, Li P, Tang W (2019) J Mater Chem C 7:13920–13929
24. Michaelson HB (1977) J Appl Phys 48(11):4729–4733
25. Mohamed M, Irmscher K, Janowitz C, Galazka Z, Manzke R, Fornari R (2012) Appl Phys Lett 101:132106
26. Schottky W (1939) Z Phys 113:367
27. Mott NF (1939) Proc R Soc A 171:27
28. Kong W-Y, Wu G-A, Wang K-Y, Zhang T-F, Zou Y-F, Wang D-D, Luo L-B (2016) Adv Mater 28:10725–10731
29. Guo XC, Hao NH, Guo DY, Wu ZP, An YH, Chu XL, Li LH, Li PG, Lei M, Tang WH (2016) J Alloys Compd 660:136–140
30. Wu Z, Jiao L, Wang X, Guo D, Li W, Li L, Huang F, Tang W (2017) J Mater Chem C 5(34):8688–8693

31. Li S, Guo D, Li P, Wang X, Wang Y, Yan Z, Liu Z, Zhi Y, Huang Y, Wu Z, Tang W (2019) ACS Appl Mater Interfaces 11:35105–35114
32. Oh S, Kim C-K, Kim J (2017) ACS Photonics 5:1123–1128
33. Peng Y, Zhang Y, Chen Z, Guo D, Zhang X, Li P, Wu Z, Tang W (2018) IEEE Photonics Technol Lett 30(11):993–996
34. Chen Y-C, Lu Y-J, Liu Q, Lin C-N, Guo J, Zang J-H, Tian Y-Z, Shan C-X (2019) J Mater Chem C 7:2557–2562
35. Zhi Y-S, Liu Z, Zhang S-H, Li S, Yan Z-Y, Li P-G, Tang W-H (2021) IEEE Trans Electron Devices 68:3435–3438

Chapter 9
Electrical and Optical Properties of Gallium-Doped Zinc Oxides

Rongxin Wang and Shijie Xu

9.1 Introduction

In 1907, Bädeker first reported transparent conductive cadmium oxide (CdO) materials, initiating the research and application of transparent conductive oxides (TCO). Currently, these TCO materials play a significant role in the field of optoelectronics, including solar cells, liquid crystal displays, organic and inorganic light-emitting diodes, lasers, detectors, and sensors, especially in the preparation of electrodes for these functional devices. Therefore, the electrical and optical properties of these TCO materials have received widespread attention and have become one of the hot topics currently being investigated.

Since Rupprecht first studied In_2O_3 semiconductor materials in 1954, tin-doped indium oxide ($Sn:In_2O_3$ or simply denoted as ITO) has become the most valuable and widely used electrode material with high transmittance and low resistivity within the visible light range. For example, the touchscreen ITO market has doubled from 3.4 billion in 2008 to 6.4 billion in 2013, almost doubling in just five years. The price of indium (In), the main metal element contained in ITO materials, has also continuously increased, that is, from just a few tens of US dollars per kg to 140–170 US dollars per kg in June 2003 and to 1010–1070 dollars per kg in March 2005. In fact, the content of In in the earth's crust is quite low, and the proven reserves worldwide are only about 130,000 tons. The continuous mining and consuming are causing the reserves of In on the earth to decrease day by day. In addition, the

R. Wang (✉)
Suzhou Institute of Nano-Tech and Nano-Bionics, Chinese Academy of Sciences, Suzhou, China
e-mail: rxwang2008@sinano.ac.cn

S. Xu
Department of Physics, The University of Hong Kong, Hong Kong, China

Department of Optical Science and Engineering, College of Future Information Technology, Fudan University, Shanghai, China

S. Xu (ed.), *Optical Characterization of Microstructures and Optoelectronic Devices Based on Wide Band Gap Semiconductors*, Wide Bandgap Semiconductors,
https://doi.org/10.1007/978-981-95-1928-6_9

difficulty of purifying high-purity In raw materials has made In a scarce material, strongly affecting the stable supply of ITO. To solve the problems of In shortage and high purification difficulty, seeking for alternative materials has naturally become an urgent and hot topic.

In the transparent conductive thin film electrode materials of the semiconductor devices currently being developed, they are generally composed of oxides of metal elements such as zinc (Zn), cadmium (Cd), indium (In), and tin (Sn). Although In-doped CdO (In:CdO), Cd_2SnO_4, $CdSnO_3$, and $CdIn_2O_4$ and other Cd-containing transparent oxides have low resistivity and can be used as transparent conductive thin film electrode materials, the toxicity of Cd has made it lose its practical value [1]. ZnO material with an exciton binding energy of 60 meV is a direct bandgap semiconductor with a bandgap of 3.37 eV at room temperature. In general, the as-grown ZnO shows n-type conductivity, and sometimes ZnO becomes a degenerate semiconductor with free electron concentration up to 10^{20} cm^{-3}, whose conductivity is closer to that of metals. Therefore, the low-cost, environmentally friendly ZnO has become one of the new hotspots in the research of transparent conductive materials.

Carrier concentration of unintentionally doped ZnO is around $10^{17} \sim 10^{20}$ cm^{-3}, but its thin film resistivity is relatively high, generally in the range of $10 \sim 10^{-2}$ $\Omega \cdot cm$ [2]. The resistivity can be further reduced by doping impurities of group III elements, such as boron (B), aluminum (Al), and gallium (Ga), making it a practical material with good optical transmittance and electrical conductivity [3–10]. For example, the resistivity of aluminum-doped zinc oxide (AZO) is as low as 10^{-5} $\Omega \cdot cm$ with a light transmittance of 88%. The resistivity of Ga:ZnO (GZO) can be as low as 1.4×10^{-4} $\Omega \cdot cm$ [4], and the light transmittance in the visible light band exceeds 90% [5]. The resistivity of B:ZnO and In:ZnO can also be as low as 10^{-4} $\Omega \cdot cm$ with a light transmittance > 85% [3], comparable to those of ITO materials. Among the abovementioned doping elements, the radii of Ga ions and covalent bonds are 0.62 Å and 1.26 Å, respectively, which are very close to the radii of Zn ions and covalent bonds (0.74 and 1.25 Å), much smaller than the radii of Al ions and covalent bonds (0.54 and 1.18 Å) and In ions and covalent bonds (0.81 and 1.44 Å). At the same time, the length (1.92 Å) of the Ga-O covalent bond is also very close to that of the Zn-O covalent bond (1.97 Å) [6]. Therefore, introducing Ga into the crystal structure of ZnO causes the smallest lattice distortion and can also avoid the risk of the electrical performance deteriorating of AZO thin films due to the high chemical activity of Al during the film formation process [2].

Furthermore, GZO also has other unique characteristics, including ease of pattern etching, high thermal stability, and excellent adhesion with substrate materials such as glass, silicon, sapphire, gallium nitride, and even organic materials (polyimide [PI], polyethylene terephthalate [PET], and polycarbonate [PC]). However, it is still crucial to prepare GZO thin films that meet the optoelectronic performance requirements of the device and ensure their long-term stability and reliability for applying transparent electrodes. At present, there are various methods to prepare GZO thin films, including radio frequency (RF) and direct current (DC) sputtering, pulsed laser deposition (PLD), plasma-enhanced molecular-beam epitaxy (PEMBE), metal–organic chemical vapor deposition (MOCVD), low-pressure chemical vapor deposition (LPCVD), atomic layer deposition (ALD), organic solution deposition, and liquid spin coating. Each preparation method requires different control parameters and conditions. For example, when depositing GZO thin films using physical preparation equipment such as sputtering, PLD, and molecular-beam epitaxy (MBE), it is necessary to control the deposition rate, substrate temperature, oxygen partial pressure, argon/oxygen flow ratio, target-sample distance, and source (Ga, Zn) temperature. When preparing GZO thin films using chemical vapor deposition and chemical solution methods, it is necessary to pay attention to the ratio of solution components, the flow rate of the carrier gas source, reaction temperature, reaction time, and spin coating speed. Regardless of preparation technology and deposition conditions used, they need to be combined with applications. By optimizing the preparation conditions and methods, the microstructure and chemical composition ratio of the deposited GZO thin films can be improved, with the aim of obtaining GZO thin films with ideal high transparency and low resistance. In addition to the preparation methods mentioned above, there are multiple choices for the source materials needed to obtain GZO thin films, including various types of liquid source materials used for chemical vapor deposition and chemical solution methods, solid metal sources (Zn, Ga) or oxide materials (zinc oxide, gallium oxide, mixtures of zinc oxide + gallium oxide, etc.) used for deposition in various vacuum and non-vacuum equipments. Furthermore, whether the GZO thin film is prepared by chemical or physical methods, annealing can be used to further improve the film performance. The annealing temperature, type of atmosphere, and duration time can adjust the original states of oxygen, zinc, and gallium elements in the GZO thin films. These annealing conditions may also change the internal microstructures of the films, for instance, reduce vacancy density and increase Ga substitution quantity via prompting element interdiffusion etc., accompanied by the merging of polycrystalline grains, growth of particle size, and reduction of crystal defects, etc. All these changes undoubtedly greatly affect the performance of the GZO thin film.

This chapter is organized as follows: In Sect. 9.2, a general description of the electrical and optical properties of GZO films is given. The effect of the work function of GZO is also discussed. Section 9.3 concentrates on reviewing GZO films grown using sputtering and solution phase deposition. The reason for concentrating on these two growth techniques generally is that sputtering technique is a method commonly used in the industry and solution deposition technique is low cost. In this section, a detailed discussion is given concerning the effect of preparation parameters (sputtering pressure, sputtering power, sputtering mode, and substrate temperature), post-annealing conditions on chemical composition, work function, microstructure, optoelectronic properties, and their relationship. In Sect. 9.4, an overview of GZO applications is given.

9.2 Electrical and Optical Properties: General Aspects

9.2.1 Electrical Properties of GZO

GZO thin film material is an n-type degenerate semiconductor. Basically, its electrical resistivity ρ depends on the concentration of carriers and mobility μ of the relevant free carrier, following the principle below:

$$\rho = 1/qn\mu \tag{9.1}$$

where q is the electron charge. From the above formula, it can be known that the resistivity ρ is inversely proportional to the carrier concentration n. That is, increasing the number of carriers in the material itself can effectively reduce the resistivity and enhance the conductivity of the material. GZO is composed of the main material ZnO doped by donor impurities Ga atoms. So it is necessary to first understand the electrical properties of the main material ZnO. Generally, unintentionally doped ZnO exhibits n-type conductivity, mainly due to oxygen vacancies V_O, interstitial Zn_i atoms, and other defects such as noncontrollable impurity hydrogen H. Özgür et al. [2] proposed through theoretical calculations that although the formation energy of oxygen vacancies is even lower than that of Zn interstitials Zn_i, the interstitial Zn_i atoms with ionization energy of 30–50 meV play a more dominant effect as shallow donor impurities [11]. The role of oxygen vacancies V_O in conductivity is very complex and has been identified as a negative-U defect. With the increasing Fermi level, a transition from +2 to the neutral occurs. Having an ionization energy of 100 meV, it can be at neutral state at low temperatures, or as a negative divalent oxygen which can act as double ionized donor, contributing two free electrons to the conduction band, or form a capture center via adsorption. The energy barrier of this capture center increases, hindering the further movement of electrons, hence resulting in the scattering mobility reduction of ionized impurities. Unintentionally incorporated hydrogen H into ZnO material has multiple effects [12]: (1) formation of defects located at interstitial positions H_i, (2) formation of shallow trap donor V_O-H via mutually binding with oxygen vacancy V_O, (3) passivating dangling bonds and zinc vacancies V_{Zn} at grain boundaries between polycrystalline grains and sample surface as acceptors, and (4) introduction of the plasma effect which effectively increases carrier concentration from 1×10^{17} cm^{-3} to 1×10^{22} cm^{-3} and decreases the resistivity from $5 \times 10^3\ \Omega \cdot$ cm to $1 \times 10^{-4}\ \Omega \cdot$ cm. The types and densities of the defects inside GZO thin film are dependent on preparation methods and growth conditions. It can also be adjusted through a lots of treatment methods such as high temperature, irradiation, and bombardment to reconstruct the crystal grain structure of the material, to reduce the internal defect density, in turn to enhance the reliability, and to improve the optoelectronic performance of the material.

Effect of Ga doping is discussed now. It is known that the doping of donor impurity gallium is mainly to improve n-type conductivity of ZnO. The carrier concentration of GZO thin film can be adjusted by changing concentration of Ga impurity [4]. If Ga atoms effectively substitute Zn at lattice positions, the effective incorporation of Ga atoms has achieved a rapid increase in electron concentration, which can be increased from 1.33×10^{18} cm^{-3} to 1.13×10^{20} cm^{-3}. The role of Ga in GZO is shown as follows [3, 4, 8]:

(1) A shallow donor contributing free electrons to the conduction band: With the gradual increase in the concentration of Ga doping, the continuous increase in the number of Ga occupying lattice positions for replacing Zn makes the Fermi level continuously move toward the conduction band and even makes GZO be degenerate semiconductor. That provides more free electrons, resulting in higher conductivity. However, when the concentration of Ga doping increases continuously beyond the limitation of the solid solubility, some Ga atoms will locate at interstitial positions. The theoretically expected maximum doping concentration is about 1.1×10^{21} cm^{-3} [13]. There are many experimental values reported in the literature. There is a certain difference from the theoretical expectation, that is, as high as 1.46×10^{22} cm^{-3} value [8].
(2) Effective reduction in lattice distortion introduced by V_{Zn} defects since the Ga-O bond length is less than the Zn-O bond length: Usually, there exists high density of V_{Zn} in GZO material. Such high density of defects causes lattice distortion, making structural stability worse, which hinders the improvement of conductivity.
(3) Interstitial Ga atoms acting as point defects: They can be trapping centers, in turn tend to reduce the mobility of free electrons and lead to a reduced conductivity.
(4) Formation of Ga_{Zn}-V_{Zn} and Ga_{Zn}-O_i composite defects, showing acceptor behavior to compensate carrier concentration.

According to Eq. (9.1), resistivity ρ is related to the two factors, namely the carrier concentration and carrier mobility μ. Honda et al. [14] and Banerjee et al. [15] proposed that increasing the carrier mobility can also improve the conductivity of the material. However, the carrier mobility and concentration are interrelated by a relationship $\mu \propto n^{-\frac{2}{3}}$. It is thus necessary to optimize and trade off carrier concentration and mobility simultaneously for obtaining a higher conductivity. During transport of carriers, the mobility of carriers may be changed by various scattering mechanisms. These scatterings mainly include lattice vibration scattering μ_{ph}, ionized impurity scattering μ_i, and polycrystalline grain boundary scattering μ_g. The total mobility μ is expressed as follows:

$$\frac{1}{\mu} = \frac{1}{\mu_{ph}} + \frac{1}{\mu_i} + \frac{1}{\mu_g} \tag{9.2}$$

For the wide bandgap semiconductors, lattice vibration scattering μ_{ph} predominately comes from optical phonons and acoustic phonons at working temperature, which is almost independent on impurity defects. In ZnO material, the optical phonon scattering due to the polar field interaction μ_{po} and the acoustic phonon scatterings through the piezoelectric potential barrier μ_{pe} and lattice deformation potential barrier μ_{ac} have been identified. These three types of phonon scatterings follow Matthiessen's rule [16, 17]:

$$\frac{1}{\mu_{ph}} = \frac{1}{\mu_{po}} + \frac{1}{\mu_{ac}} + \frac{1}{\mu_{pe}} \tag{9.3}$$

In polar semiconductors like ZnO, due to their unique ionic bonds, under the action of lattice vibration-mediated polarized electric fields, charge movement causes polarized LO phonon scattering. The polarized optical phonon scattering μ_{po} is characterized by Howarth–Sondheimer [17]:

$$\mu_{po} = \frac{2^{3/2}\pi\hbar^2\left(e^{T_{PO}/T} - 1\right)\chi(T_{PO}/T)}{q(kT)^{1/2}m^{*3/2}\left(\varepsilon_1^{-1} - \varepsilon_0^{-1}\right)} \tag{9.4}$$

where ε_1 is high-frequency dielectric constant, m^* is effective mass, and T_{po} is the polar optical temperature. For ZnO material, T_{po} is 837 K and $T_{po}/T \geq 2.8$ in the temperature range of $T = 0-300\text{K}$. The numerical function $\chi\left(T_{po}/T\right)$ can be written as follows [17]:

$$\chi\left(T_{po}/T\right) = \frac{1}{1 + \exp\left(-0.6T_{po}/T\right)} \tag{9.5}$$

The acoustic phonon scattering stems from the deformation potential induced by strain associated with acoustic phonons. The scattering rate increases with the wave vectors of the phonons. The acoustic phonon term μ_{ac} may be estimated with following equation [17]:

$$\mu_{ac} = \frac{\pi\hbar^4 c_l}{(2)^{1/2}qkTm^{*5/2}E_1^2}E_f(n)^{-1/2} \tag{9.6}$$

where E_1 is the acoustic potential barrier and c_l is the longitudinal elastic constant. The Fermi energy satisfies the following well-known formula [17]:

$$E_f(n) = \frac{\hbar^2}{2m^*}\left(3\pi^2 n\right)^{2/3} \tag{9.7}$$

In the absence of inversion symmetry, strain associated with phonons generates an electric field to influence electrostatic potential barrier scattering μ_{pe} [17]:

$$\mu_{pe}(n, T) = \frac{2^{3/2}\pi\hbar^2\varepsilon_0}{qkTm^{*3/2}P_{pe}^2}E_f(n)^{1/2} \tag{9.8}$$

Ionized impurity scattering originates from deflection of free carriers by the long-range Coulomb potential of the charged centers and depends on the crystal quality of the sample. Various defects, such as impurity dopants, impurity clusters, edge dislocations, screw dislocations, stacking faults, line defects, and higher charge states of ionized donors (self-doping by oxygen vacancies, or extrinsic dopants on interstitial sites), cause local perturbation of the band edge to disturb the motion of electrons.

When the intrinsic structural defect density in GZO materials is very high, these defects can capture electrons in the conduction band of n-type materials and become neutral or negatively charged trap centers or lines, eventually creating a space-charge region. The scatterings by these defects hamper electrons' transport or form another current path, resulting in mobility reduction [2]. For high concentration of ionized Ga donors in GZO, ionized impurity scattering μ_i can be referred to Brooks–Herring formula [16, 17]:

$$\mu_i = \mu_{io}(n)\frac{n}{Z^2 N_i} \tag{9.9}$$

where Z is the charge number of ionized donors N_D and acceptors N_A, $N_i = N_D + N_A = 2N_A + n$. $\mu_{io}(n)$ can be calculated with the following expression:

$$\mu_{i0}(n) = \frac{24\pi^3\varepsilon_0^2\hbar^3}{q^3 m^{*2}} \frac{1}{\ln\left(1 + \frac{3^{1/3}4\pi^{8/3}\varepsilon_0\hbar^2 n^{1/3}}{q^2 m^*}\right) - \frac{\frac{3^{1/3}4\pi^{8/3}\varepsilon_0\hbar^2 n^{1/3}}{q^2 m^*}}{1+\frac{3^{1/3}4\pi^{8/3}\varepsilon_0\hbar^2 n^{1/3}}{q^2 m^*}}} \tag{9.10}$$

where ε_0 represents the static dielectric constant, h is the Planck constant, and m^* is the reduced effective mass of electrons and holes:

$$\frac{1}{m^*} = \frac{1}{m_e^*} + \frac{1}{m_h^*} \tag{9.11}$$

where m_e^* and m_h^* are effective masses of the conduction band electrons and valence band holes, respectively. Their values in ZnO are 0.28 m_0 and 0.59 m_0 [18]. It has been reported that value of m^* varies with the concentration of carriers. For instance, when the Hall carrier concentration is between 1.02×10^{21} cm^{-3} and 1.23×10^{21} cm^{-3}, m^* is between 0.40 and 0.46 m_0 [19]. These values are much larger than those obtained using Formula (9.11), and further exploration is still required.

Scattering of polycrystalline grain boundary defects μ_g is related to crystal structure. Transparent conductive GZO is divided into single crystal and polycrystalline structures. Most of transparent electrodes are prepared using polycrystalline GZO materials. Polycrystals are composed of many small single-crystal grains, each with its own surface. There are grain boundaries between adjacent crystal grains, presenting surface/interface effects as barriers. The carrier transport behavior is thus influenced by both scatterings within grains and at grain boundaries. The higher the density of these crystal grains, the more the interference. The migration of carriers is affected by the size of these crystal grains L, the quality of the single crystal inside the grains, and the average free path $l_e\left(l_e = \frac{h\mu}{2q}\sqrt[3]{\frac{3n}{\pi}}\right)$ of free electrons. The available reports show that the values of l_e show some discrepancies. For example, Li et al.'s study reveals that l_e fluctuates within the range of $1 \sim 1.9$ nm with different film thicknesses [5], while it also swings within the range of $3.8 \sim 5.2$ nm due to the addition of oxygen [13]. When L is greater than l_e, due to the coexistence of phonons, point defects, dislocation defects, etc., inside small crystal particles, all scatterings from phonons and defects jointly limit the value of l_e The ingrain scattering of crystal particles determines the collision frequency ω_C, and the scattering is expressed as $\mu_g = \frac{q}{m_e^* \omega_c}$. That means that the grain boundaries hardly disturb the transport process of the carriers, and hence, the grain boundary scattering hardly plays a leading role [13]. When L is close to l_e, the carrier transport in the thin film will be cross the interface of crystal grains a few times. The carriers shuttle between many crystal grains. Therefore, the density of grain boundaries of crystal grains changes the speed of carriers, restraining their transport behavior. Under such circumstances, the mobility is determined by the grain boundary scattering [19]. μ_g controlled by the grain boundary defect scattering is expressed as follows [17]:

$$\mu_g = \frac{q}{m^*}\frac{l_e}{v_{Fm}(n)} = \frac{q}{\hbar}\frac{l_e}{\left(3\pi^2 n\right)^{1/3}} \tag{9.12}$$

Increasing the size of the crystal grains, reducing the density of the grain boundaries, and adjusting the axial direction of the small crystal grains all can enhance μ_g and hence improve the conductivity. The number of trapped electrons at grain boundaries varies with electron concentration, thereby raising the height of the energy barrier and increasing blocking effect of electrons in the grain boundary region [19]. Seto et al. [20] developed a model in which carrier mobility in polycrystalline semiconductors is limited by polycrystalline grain boundary and the density of defect traps.

The type of scattering mechanism can be determined with the help of variable temperature Hall measurements [4, 5]. If resistivity of the material decreases monotonically with increasing temperature, it has the characteristics of a semiconductor with a negative resistance effect. Nevertheless, resistivity of the material changes non-monotonically with temperature, showing a decrease first and then an increase. At lower temperatures, the conducting characteristics of heavily doped degenerate semiconductors manifest mainly weakly localized. As the temperature rises, phonon scattering gradually becomes the main mechanism, and the material then has metallic conducting characteristics. In thin film crystals, especially in polycrystalline materials, there exist many defects, such as dislocations, vacancies, and/or grain boundaries; electrons may transport in diffusion way rather than ballistic motion, leading to a great increase in the randomness of electron scattering. The conductivity can be expressed as follows [4]:

$$\sigma = \sigma_B \left[1 - \frac{C}{(k_f l_e)^2} \left(1 - \frac{l_e}{l_i} \right) \right] \tag{9.13}$$

where σ_B and k_f represent the Boltzmann conductance and Fermi wave vector, respectively. The inelastic diffusion length l_i ($l_i = \sqrt{D_{diff} \tau_i}$) and the diffusion coefficient D_{diff} both are related to temperature [4]. It decreases with increasing temperature. The term in the second bracket of Formula (9.13) represents the weak localization of carriers caused by constructive interference of scattered electrons, which is affected by the density and type of defects. It is also related to the concentration of Ga doped into the material. The two behaviors of material resistivity changing with temperature due to different scattering mechanisms can be observed in GZO thin films with different thicknesses and different Ga doping concentrations [4, 19]. For example, when the film thickness is relatively thin, due to the lattice mismatch between GZO film and the substrate, larger stress causes lattice distortion and generates defects such as dislocations, leading to a decrease in the crystal quality of the material itself and eventually reducing the conductivity. As the film thickness increases, the effect of substrate mismatch gradually weakens. As a result, the stress is gradually released and the defects are reduced. Meanwhile, the crystal grains grow larger. The crystal quality and the conductivity both are improved [5, 19]. Properly heavy doping of GZO makes the electrons in a weakly localized state. After degenerate energy band is formed, the metal-like transport behavior can be observed [4, 5]. As the Ga concentration further increases, the number of Ga at substitutional and interstitial positions increases, intensifying the deformation of lattice and accordingly changing the transport behavior of carriers. Therefore, the quality of thin film material and the concentration of carriers will determine the scattering mechanisms and hence conductivity.

In order to improve the conductivity of GZO materials, people have explored various methods, including optimizing film thickness, choosing different preparation methods and deposition conditions, changing source ratios, adjusting substrate materials, and post-high-temperature annealing. Some researchers pointed out that when the film thickness is less than 110 nm, the concentration of carriers basically remains unchanged at 2.8×10^{20} cm^{-3}. When the film thickness is greater than 110 nm, the concentration of carriers slowly decreases with increasing thickness [16]. For GZO films with different Ga ratios, the lowest resistivity of 1.4×10^{-4} $\Omega \cdot$ cm is achieved for the 5% Ga ratio [4]. If oxygen is added during the film deposition process, excessive oxygen not only enhances the compensation for point defects V_{Zn} but also reaches the crystal boundary and defect sites through diffusion where excessive oxygen can be adsorbed to form O_2^- capture centers. These centers hinder further electron transport, reduce the carrier mobility, and eventually cause the conductivity to deteriorate. For example, when GZO film was prepared by magnetron sputtering, increase in oxygen partial pressure exhibited such a phenomenon, decreasing the carrier concentration from 7×10^{20} cm^{-3} to 4×10^{20} cm^{-3}. The carrier concentration of GZO films deposited with plasma assistance also shows a similar changing trend with oxygen flow. On the contrary, the introduction of H_2 shows a complex effect, including reducing oxygen content and significantly reducing the resistivity from $1.94 \times 10^{-2}\ \Omega \cdot cm$ to $5.69 \times 10^{-4}\ \Omega \cdot cm$ [12]. Yamada et al. [21] pointed out that after annealing, GZO thin film materials not only become more uniform but also exhibit changes in following aspects: (1) a reduction in defects along the *c* axis, (2) defects along the *c* axis direction cannot be completely eliminated, and (3) improvement in the mobility by suppressing various defects, such as point defects, interstitial atoms, vacancies, dislocations associated with Zn and O, crystal grain size, and grain boundaries. Therefore, the microstructure of material can be further changed by optimizing annealing conditions, including annealing temperature, atmosphere, and duration, to obtain good crystal quality and excellent conductivity.

In addition, co-doping with multiple elements is also one of the ways to regulate the electrical properties of GZO materials. There are two purposes of co-doping with multiple elements [2, 22]:

(1) To improve n-type conductivity: By using H, F, Al, In, and other elements to co-dope with Ga, the resistivity of GZO materials can reach the order of $10^{-4}\ \Omega \cdot cm$, while the electron mobility is within the range of $14\ cm^2/V \cdot s$ ~$21\ cm^2/V \cdot s$ and the carrier concentration is in the range of $10^{20}\ cm^{-3}$ ~$10^{21}\ cm^{-3}$.
(2) To obtain p-type materials: Theoretically, the prerequisite for achieving p-type conductivity is to replace oxygen with Group V elements or replace zinc with Group I elements. In practice, Group I elements tend to become donors in interstitial positions rather than acceptors in substitution forms. Nitrogen in Group V elements does not easily form N_{Zn} defects, while P and As are more likely to form complex donor-like P_{Zn}, As_{Zn} defects [2]. Yamamoto and Yoshida [22] proposed to obtain p-type materials by co-doping with N and Ga. The strong association between N and Ga not only lowers the acceptor energy level but also increases the solubility by nearly 400 times [2]. Experiments have confirmed that N can act as an acceptor impurity in ZnO materials with the presence of Ga. For example, the p-type carrier concentration can reach $5 \times 10^{19}\ cm^{-3}$ [23]. Ryu et al. [24] obtained p-type materials by depositing ZnO on a GaAs substrate because As and Ga can enter ZnO from the substrate and jointly act as a doping source. When P and Ga are co-doped into ZnO, the formation probability of oxygen vacancy reduces, and p-type conductivity can also be obtained if P occupies the position of oxygen and becomes an acceptor.

Ion implantation technology is another traditional method to change resistivity. By increasing the dose of N^+ ion implantation, the resistivity of GZO thin films has been demonstrated to linearly increase from $1.2 \times 10^{-3}\ \Omega \cdot cm$ to $3.7 \times 10^4\ \Omega \cdot cm$. However, it is found that even using a concentration as high as $4.5 \times 10^{21}\ cm^{-3}$ of N^+ ion implantation, one cannot obtain p-type materials because implanted N^+ ions just play a compensation role.

9.2.2 Optical Properties of GZO

Being as a transparent conductive material, optical transparency of GZO material is another key consideration in addition to the aforementioned electrical characteristics. Transmittance of material can be described by $T = \frac{I}{I_0} = e^{-\alpha t}$ expression, which is related to thickness t of thin film material and depends on absorption coefficient α of material. Usually, α is a function of wavelength. For GZO semiconducting materials with a direct bandgap, the transmission window spans a very wide wavelength range, from ultraviolet (UV) light, visible light, to near-infrared (IR) light. Transparency is closely related to carrier concentration and crystal structure (single crystal, polycrystal, and amorphous structure). Transparency of GZO materials drops significantly in near-infrared region due to a plasmonic resonance effect of electron gas, while a strong light absorption occurs near ultraviolet region due to the large bandgap edge of GZO. The absorption coefficient of GZO satisfies the following formula:

$$\alpha \propto \left(h\nu - E_g\right)^{1/2} \tag{9.14}$$

where E_g is optical bandgap of material and $h\nu$ is photon energy. By fitting experimental transmission curve with $\alpha^2 \propto h\nu$, the optical bandgap of thin film material can be obtained. Some scholars believe that for heavily doped semiconductors, this simple method is quite inappropriate and needs to be modified with a broadening parameter Γ arising from electron–electron interactions and ionized impurity scattering [9]. Some scholars disagree with the correction, leading to a variety of opinions. In addition, carrier concentration of material affects optical bandgap value in the following ways:

(i) An increase in carrier concentration causes broadening of the optical bandgap [18, 25]. For GZO semiconducting materials, by probing transmission spectrum of GZO thin films, one can observe a variation trend of absorption edge with the doping concentration of Ga atoms or annealing process at high temperatures. When the carrier concentration increases and the Fermi level shifts to higher energy, the absorption edge of the GZO thin film blueshifts (i.e., shifts toward the short-wave direction) [4, 25]. Such blueshift is known as the famous Burstein-Moss (BM) effect in heavily doped n-type semiconductors, which can be written as follows [26, 27]:

$$\Delta E_g^{BM} = E_g - E_{g0} = \frac{\hbar^2 \left(3\pi^2 n\right)^{2/3}}{2m^*} \tag{9.15}$$

When all the energy levels below the conduction band are occupied, the offset of the BM effect increases. For GZO thin film materials, there have been reports that the bandgap values E_g calculated from the absorption spectra range from 3.25 to 4.05 eV, which is directly related to the Ga doping ratio, film thickness, and preparation methods. Some researchers have also proposed that understanding the bandgap variation from the perspective of band filling is not very accurate because there exist other effects.

(ii) When the carrier concentration is very high, many-body effects including electron exchange interactions, minority–carrier correlations, and carrier–ion correlations cause the bandgap narrowing [18, 25, 28]. Exchange and correlation of an electron with surrounding hole effectively lowers the energy of the electron, causing the conduction band downward shift. The higher the doping concentration, the greater the density of free carriers, and the offset expansion intensifies. Similarly, the more electrons surrounding the holes captured by the donor electrons, the lower their own energy. The repulsion with the positively charged donor ions further strengthens, causing the valence band to shift upward. The bandgap further narrows. In the case with very small effective electron mass and large shielding length, there are many-body effects such as energy exchange between multiple majority carriers (electron–electron), electron–hole, electron–impurity, and impurity–impurity interactions in heavy doping n-type semiconductor materials. All these many-body interactions result in the bandgap renormalization (BGR) and narrowing [18], which manifests as follows [28]:

$$\Delta E_g^{BGR} = K_1 n_e^{1/3} + K_2 n_e^{1/4}\left(\frac{m_e^*}{m_0}\right) + K_3 n_e^{1/2}\left(\frac{m_0}{m_e^*}\right)^{1/2}\left(1 + \frac{m_h^*}{m_e^*}\right) \quad (9.16)$$

where the parameters K_1, K_2, and K_3 are coefficients, with values of 2.66×10^8 eV · cm, 8.53×10^7 eV · cm$^{3/4}$, and 5.96×10^{12} eV · cm$^{3/2}$ [18]. Combining the above BM effect and BGR effect, the change in bandgap is as follows:

$$\Delta E_g = \Delta E_g^{BM} + \Delta E_g^{BGR} \quad (9.17)$$

For heavily doped GZO films, the system becomes very complex. When a high density of substitutional defects Ga_{Zn} exists, lattice vibration states will be adjusted. On the other hand, quasi-localized anti-gallium metal and oxygen atom binding state hybridize with the conduction band of ZnO. A band tail forms within the bandgap. In the theoretical calculations of BM shift, the non-parabolicity of the conduction band should be considered, which results in a concentration-dependent effective mass. The electron-concentration-dependent effective mass can be expressed as $m_{non}^* = m_0^*\sqrt{\left[1 + \frac{2P\hbar^2(3\pi^2 n)^{2/3}}{\pi^2 m^*}\right]}$, where P is a non-parabolic parameter [25]. Therefore, the non-parabolicity effect of the energy band will also affect the lattice dynamics. The offset energy will inevitably be further strengthened. In addition, for polycrystalline GZO films, the doped Ga atoms are not only simple metal phase or substitutional semiconductor phase but also will combine with oxygen to form gallium oxide phases. The normally stable gallium oxide single crystal has a wider bandgap (4.7 eV). Therefore, the coexistence of zinc oxide and gallium oxide alloy states in polycrystalline materials also intricately changes the band structure and the properties of the material.

9.2.3 *Work Function of GZO*

When contact electrodes are prepared for semiconductor devices, the work function of adopted material is of great concern. The value of work function is directly related to height of contact potential barrier between electrode and semiconductor, affecting the function and characteristics of device. Such contact is either a Schottky contact or an ohmic contact. Therefore, when GZO films are used as contact electrodes, it is necessary to know the value of the work function of GZO film material as well as the various factors influencing work function.

Owing to variations in Ga doping concentration, preparation method, post-growth treatment condition, and production batch, there usually exist some differences in surface and internal microstructure of the prepared GZO films, which make the value of the GZO work function variable in the range of 3.3 ~ 5.3 eV. With the help of different treatment methods, such as removing surface contamination, improving crystal quality of film, changing charge state of Ga dopants, and adjusting chemical ratio between surface polar terminal atoms (oxygen, Ga/Zn), the work function of GZO films can be modified. For example, Ratcliff et al. obtained 3.6 eV work function for the GZO film deposited on glass substrate by sputtering [29]. Then, they used argon ion bombardment, potassium hydroxide etching, hydrogen iodide etching, and oxygen ion reaction methods to treat the surfaces of GZO films for achieving 3.3 eV, 3.8 eV, 3.9 eV, and 4.5 eV work function of the treated films, respectively. It is reported by other researchers that by changing annealing temperature, the work function of the GZO film was adjusted from 4.34 eV for the unannealed sample to 4.37 eV after annealing at 500 °C, 4.65 eV after annealing at 600 °C, and 4.77 eV after annealing at 700 °C.

9.3 Effect of Preparation Methods and Conditions on Optical and Electrical Properties of GZO Film Materials

As mentioned earlier, GZO thin films can be prepared by both vapor deposition and liquid methods. By controlling preparation conditions and annealing parameters, the microstructure of GZO thin film can be influenced, and hence, the optical and electrical properties can be adjusted. Detailed descriptions will be given below.

9.3.1 Deposition Method

I. Vapor-Phase Deposition Method

So far different types of vapor-phase deposition methods, including RF and DC sputtering, PLD, MBE, PEMBE, and other physical vapor-phase depositions, as well as MOCVD, LPCVD, ALD, and other chemical vapor-phase deposition, have been developed. Herein, we focus on the preparation of GZO thin films by magnetron sputtering deposition method. Detailed description on the influence of preparation conditions on the microstructure, chemical composition ratio, and many other properties of GZO thin films, thereby changing the optical transmittance and electrical characteristics of the thin films, will be given.

a. Substrate Effect

GZO thin films can be deposited on a variety of transparent, opaque, soft, and hard substrates, including glass, silicon, sapphire, GaN, ZrO_2 (Y_2O_3), $SrTiO_3$, $ScAlMgO_4$, ceramics, flexible organic PET, and so on. The most used substrates were glasses on which the GZO thin films were deposited, exhibiting the carrier concentration of 6.38×10^{20} cm^{-3}, mobility of 21.69 cm^2/V · s, resistivity on the order of 10^{-4} Ω · cm, and light transmittance exceeding 87%, even up to 95%. The performance of GZO thin films on flexible substrates is slightly inferior. For instance, the resistivity of GZO on PET can also reach 1.3×10^{-3} Ω · cm, with a light transmittance of 93%.

When GZO is deposited on a heterogeneous material substrate, the difference in both lattice constants and thermal expansion coefficients causes film strain and introduces defects, showing the influence of substrate on the properties of GZO thin films. For example, Gabás et al. [30] used RF sputtering to deposit GZO on p-Si at 300 °C and observed the formation of a 3-nm amorphous transition layer at the interface between GZO and Si. Moreover, they found that Zn and Ga not only diffused into the substrate but also formed a Zn-Si compound. For the growth of GZO epitaxial thin films with MOCVD, the quality of the epitaxial materials on (111) ZrO_2 (Y_2O_3) and (111) $SrTiO_3$ substrates is higher than that on the sapphire substrate. The Ga content in GZO films grown on (111) ZrO_2 (Y_2O_3) substrates can exceed 7.5%, achieving a higher solubility. When grown on sapphire substrates, the Ga content is less than 3% [10]. As GZO film thickness increases, the effects of lattice mismatch and stress weaken, which in turn not only decreases the defect density and lattice distortions, but also increases the crystallinity of the film [19]. For example, the (002) main diffraction peak intensity became intense and full width at half maximum (FWHM), as well as the proportions of other diffraction peaks including (102), (200), (004), and (202), showed variations [5, 19]. The mobility of the thin film was improved by an order of magnitude, and the conductivity reached 5.44×10^{-4} Ω · cm [31].

Figure 9.1 shows the $\theta/2\theta$ X-ray diffraction (XRD) pattern of GZO thin films deposited on silicon and sapphire substrates. It is observed that the (002) diffraction peak intensity of the GZO film deposited on sapphire is higher than that of the GZO

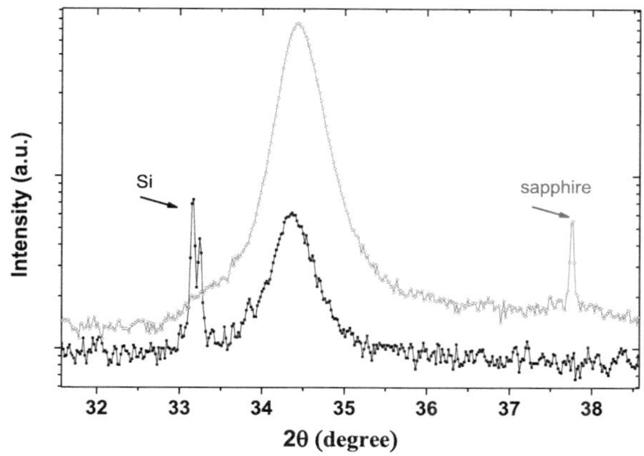

Fig. 9.1 XRD θ/2θ pattern of the GZO thin films deposited on silicon and sapphire substrates in the same batch

film deposited on (100) silicon. It can also be found that the fitted (002) peak position of the GZO film on sapphire is located at 34.43°, and its FWHM $\Delta_{(2\theta)}= 0.368°$. The (002) peak position of the GZO film on (100) silicon is at 34.36°, and the FWHM $\Delta_{(2\theta)}= 0.418°$, which is close to the (002) peak position (34.36°) of the GZO film on glass, while the FWHM $\Delta_{(2\theta)}= 0.342°$. Considering the structures of sapphire, silicon, and glass substrate materials, sapphire has a hexagonal structure, while silicon has a diamond structure, and glass is an amorphous material. Comparatively speaking, GZO has a closer lattice structure with sapphire, and hence, the lattice mismatch between GZO and sapphire is smaller, resulting in growth of a higher quality thin film. When the GZO thin film is deposited on amorphous glass, the stress is released more fully, and the quality of the obtained GZO thin film is better than that of the GZO thin film deposited on silicon.

b. DC and RF Sputtering Effects

We used two different principles of magnetron sputtering modes, namely RF and DC sputtering, to deposit GZO thin films on quartz substrates under pure Ar atmosphere, under the identical power and pressure. The chemical valence states of Zn and Ga, the Zn/Ga ratio, oxygen and zinc vacancies, and metal oxygen-deficient states in the GZO films deposited under RF and DC modes are revealed to be different. The films prepared under RF and DC possess resistivities of $9.4 \times 10^{-4}\ \Omega \cdot \text{cm}$ and $1.4 \times 10^{-3}\ \Omega \cdot \text{cm}$, carrier concentrations of $7.5 \times 10^{20}\ \text{cm}^{-3}$ and $6.8 \times 10^{20}\ \text{cm}^{-3}$, and mobility of $8.7\ \text{cm}^2/\text{V} \cdot \text{s}\ \text{cm}^{-3}$ and $7.9\ \text{cm}^2/\text{V} \cdot \text{s}$, respectively. Figure 9.2 shows the transmittance spectra of GZO thin films prepared under the two deposition modes. The transmission window ranges of the two samples are significantly different. The transmittance and transmission wavelength range of the GZO thin film obtained by the RF mode are much larger than those of the GZO thin film deposited by the DC mode. In the ultraviolet region, the transmittance of the GZO thin film deposited under the RF mode is significantly higher, with a transmittance of over 51% at 300 nm. And it shows two absorption characteristics below and above 300 nm, while the GZO thin film deposited under the DC mode has a transmittance of only 6%

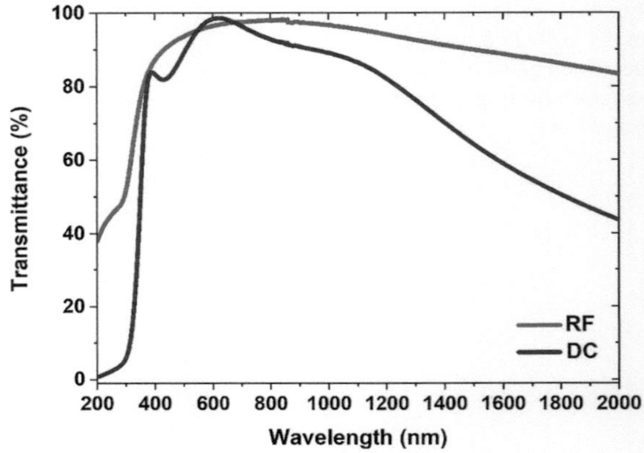

Fig. 9.2 Transmittance curves of GZO thin films deposited by RF and DC magnetron sputtering

at 300 nm, with a clear absorption edge only appearing at 300 nm. In the infrared region, the transmittance of the GZO thin film deposited under the DC mode is significantly lower than that of the GZO thin film deposited under the RF mode, reflecting the difference in the plasmon resonance effect from the free electron gas, which is consistent with the resistivity data.

c. Sputtering Power Effect

Figure 9.3 shows measured transmittance curves of the GZO thin films prepared under different RF sputtering powers. In the ultraviolet wave range, two different absorption characteristics from 200 to 300 nm and from 300 to 370 nm are clearly observed. According to Eq. (9.14), the optical bandgap values can be determined from measured absorption spectra of prepared GZO materials. The transmittance at 300 nm first decreases and then increases with the increase in sputtering power. This is mainly because of Ga doping effect, which is very consistent with the plasma effect shown by the transmission characteristics in the infrared region. From the XRD pattern of GZO thin films deposited under different sputtering powers, the trend of changes in the intensity, peak position, and FWHM of the main peak (002) can also be clearly observed. With the increase in power, the intensity of (002) increased by nearly fourfold, and the FWHM $\Delta_{(2\theta)}$ also decreased from 0.68° at 50 W RF sputtering power to 0.45° at 70 W RF sputtering power. When the RF sputtering power was further increased to 80 W, the FWHM $\Delta_{(2\theta)}$ slightly increased to 0.51°. The fitted peak position also showed the same trend of change, that is, from 34.2° (50W) to 34.08° (60W), 33.96° (70W), and finally to 33.99° (80W). The change in the main diffraction peak position reflects the variation in both stress and lattice parameters of the GZO thin films. During the sputtering of GZO thin films, substitutions of Zn^{2+} by Ga^{3+} can introduce lattice distortion due to the difference in the atomic radii of Zn and Ga. The enhancement of the diffraction peak intensity shows the improvement of crystal quality, and the carrier mobility of the thin films increases from 4.8 $cm^2/V \cdot s$ to 8.69 $cm^2/V \cdot s$ due to the increase in RF sputtering power. At the same time, the resistivity is greatly reduced from 6.38 $\Omega \cdot cm$ to 2.47×10^{-3} $\Omega \cdot cm$. The increase in

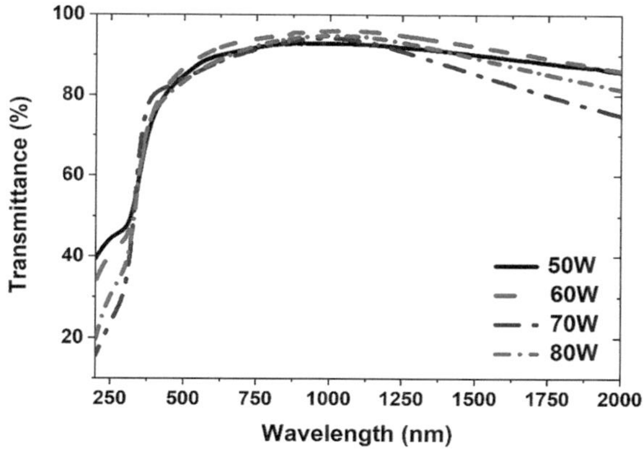

Fig. 9.3 Transmittance curves of the GZO films deposited under different RF sputtering powers

sputtering power also increases grain size, leading to an improvement in the crystal quality of the thin film, and even some adjusting in the crystal orientation. These results are consistent with those reported by other researchers. Under the premise of applying appropriate sputtering power, Zn:Ga ratio in the GZO film can be controlled by utilizing differential sputtering efficiency of these elements from the Ga_2O_3:ZnO mixing target material and saturated vapor pressure of Ga and Zn themselves in the sputtering chamber. The optical and electric characteristics of GZO thin films were finally controlled [7]. It is found that under different RF sputtering powers, the Ga:Zn ratio can be changed from 1.81 to 2.28. Shinde et al.'s study showed that only about 2% of Ga^{3+} replaces Zn^{2+} in ZnO, and the remaining Ga will occupy the interstitial position, becoming a neutral defect and resulting in deterioration of both electrical and optical properties of the GZO film [32].

d. Gas Pressure Effect

In terms of important role of atmosphere in sputtering, Kon et al. [33] compared the characteristics of GZO films prepared with three different inert process gases, namely Ar, Ne, and Kr. They found that as the process gas pressure increases, the kinetic energy of the three sputtering particles of Ga, Zn, and O all decreases, with Kr having the lowest energy. The GZO film prepared in the Ne atmosphere has the worst quality, while the highest sputtering rate can be obtained in the Ar gas. Table 9.1 lists the optoelectronic characteristics and microstructure parameters of the GZO films prepared under different Ar pressures. As the working pressure inside the chamber gradually rises, the carrier concentration continuously increases from 6.26×10^{19} cm^{-3} to 4.63×10^{20} cm^{-3} and the resistivity decreases by two orders of magnitude, that is, from 3.1×10^{-2} $\Omega \cdot$ cm to 8.2×10^{-4} $\Omega \cdot$ cm. However, the change in optical transmittance exhibits the opposite behavior. That is, the transmittance decreases as the Ar pressure increases, and the transmission window width gradually narrows. The transmittance reduction is more evident in both ultraviolet and infrared regions, that is, the transmittance at 250 nm down from 61 to 38% and the transmittance at 2000 nm from 98 to 83%. In Cuong et al.'s investigation

[34], they found that not only the (002) diffraction peak intensity in XRD and the polycrystalline grain size both increase as the working pressure decreases but also the proportion of Ga in the film fluctuates within the range of 3.5 ~ 4.4%. Assunção et al. [35] reported that as the Ar flow rate increases, the deposition rate decreases from 28 to 18 nm/min. Excellent conductivity of $2.6 \times 10^{-4}\ \Omega \cdot cm$ can be obtained by optimizing the Ar pressure, and the near-infrared transmittance also increases accordingly.

When preparing GZO films by sputtering, other types of reactive gases can also be introduced into chambers in addition to inert gases. Oxygen is one of the important reactive gases [36]. By adjusting Ar:O ratio, microstructure, defect type, and surface roughness of GZO film can be changed. For example, zinc vacancies (V_{Zn}) are dominant defects in GZO film prepared in the O-rich conditions, while zinc interstitials (Zn_i) are more abundant in the Zn-rich conditions. In an oxygen-rich condition, the formation order of intrinsic defects is unraveled to be $V_{Zn} > O_i > V_O$. While under an oxygen-deficient environment, the formation order of intrinsic defects is changed as $V_O > V_{Zn} > O_i$. The introduction of oxygen can reduce the number of oxygen vacancies, increase the amount of oxygen adsorbed at the grain boundaries of polycrystalline grains, and promote generating various defects such as O_i, V_{Zn}, and Ga_{Zn}. The change in environment will also result in the formation of more complex acceptor-type defects such as $Ga_{Zn}{-}V_{Zn}$ and $Ga_{Zn}{-}O_i$, which partially compensate for intrinsic defects. At the same time, it will enhance the oxidation reaction of Ga and Zn metal atoms. Because Zn ions are more likely to form chemical bonds with oxygen than Ga ions, the ratio of Zn:O:Ga will inevitably change (i.e., $Zn^{2+}{-}(1 + 3y)O^{2-}{-}2yGa^{3+}$). Available experimental data show that the ratio of Zn:O in GZO film deposited under low oxygen pressure is 1:0.9, and the film shows a narrower light transmission window [36]. As the oxygen partial pressure increases, oxygen vacancies and some defects are moderately suppressed, which in turn affects the carrier concentration and mobility as well as the electrical conductivity and optical transparency of the film.

Hydrogen is another major sputtering reaction gas [12]. Introducing hydrogen into GZO thin film during thin film preparation process can produce donor impurities or passivate dangling bonds and acceptors with negative charges V_{Zn}. Hydrogen may

Table 9.1 Optoelectronic characteristics of GZO films under different magnetron sputtering gas pressures

Pressure (mTorr)	Resistivity ($\times 10^{-3}\ \Omega \cdot cm$)	Mobility ($cm^2/V \cdot s$)	Carrier concentration ($\times 10^{20}\ cm^{-3}$)	Transmission (%)		XRD (002) FWHM $\Delta_{(2\theta)}$
				700 nm	500 nm	
1.8	56.2	3.7	0.4	95.2	93	1.01
2.2	4.1	6.4	2.3	95.9	90	0.65
2.6	2.7	6.3	2.0	93.3	87	0.63
3.2	1.8	4.8	4.4	92.3	85	1.01
3.9	3.5	2.7	3.4	90	82	0.84

also react with oxygen as $H_2 + O_2 \rightarrow H_2O$, which reduces the oxygen adsorbed at the surface and grain boundaries, lowers the trap defects, and enhances the migration ability of the carriers. At the same time, it will cause Zn and Ga to lose oxygen for creating vacancies and reintroducing defects and resulting in complex effects. For example, compared with GZO thin films deposited in pure argon atmosphere using DC sputtering, the introduction of hydrogen gas causes more shift of the (002) peak position with increasing deposition temperature. By optimizing the H_2:Ar ratio, GZO thin film with carrier concentration of 1.177×10^{21} cm^{-3}, mobility of 8.87 cm^2/V · s, resistivity as low as 3.40×10^{-4} Ω · cm, and transmittance of 90.03% was obtained.

e. The Role of Substrate Temperature in GZO Thin Film Deposition

Controlling substrate temperature during GZO thin film deposition is also one of the important parameters affecting the optical transmittance and electrical conductivity of thin film. Most of the experimental results show that the carrier mobility and the crystallization quality of films deposited at high temperatures is significantly higher than those at low temperatures. By observing the θ/2θ XRD pattern in Fig. 9.4, it can be found that the (002) peak changes with the deposition temperature from room temperature to 400 °C. The peak intensity first increases and then decreases. The XRD (002) intensity of GZO films deposited at 100 °C is the strongest, which is consistent with the published research results. For example, Wu et al. [37] observed that as the deposition temperature increased from room temperature to 300 °C, the (002) diffraction peak intensity increased and the FWHM decreased. Lee et al. [6] found that Ga^{3+} substitution occurred when the substrate temperature exceeded 300 °C, and the peak intensity ratio between (002) and (103) directions began to decrease from 100 °C. At 500 °C, the crystal main diffraction peak changed from (002) to (103). The oxygen deficiency in X-ray photoelectron spectroscopy (XPS) spectra slightly decreased with the increase in substrate temperature. When the substrate temperature exceeded 400 °C, the crystal quality decreased due to the formation of more oxygen vacancies, zinc vacancies, and other deep-level defects. Under the vacuum conditions, the vapor pressure of Zn is greater than that of Ga at the same temperature. As the substrate temperature increases, the probability of Zn volatilizing from the sample surface continues to increase, far greater than the probability of Ga leaving the sample surface. As a result, the content of Ga in the thin film increases. For example, the Ga content in the film deposited at 300 °C is twice as high as that deposited at low temperatures. Moderately increasing the substrate temperature and the Ga content in the GZO film is beneficial for more Ga atoms to gain energy to replace the Zn atoms at lattice positions, which increases the carrier concentration and improves crystal quality, film uniformity, and carrier mobility. In addition, element binding energy, chemical composition ratio, stress state, and crystal orientation are sensitive to deposition substrate temperature.

In addition to the above-discussed deposition conditions, the elemental ratio of the target material also plays an important role in the characteristics of the GZO film prepared by the vapor deposition method. The change in the Ga content in the GZO film prepared by the different Ga:Zn ratios in the target material not only causes

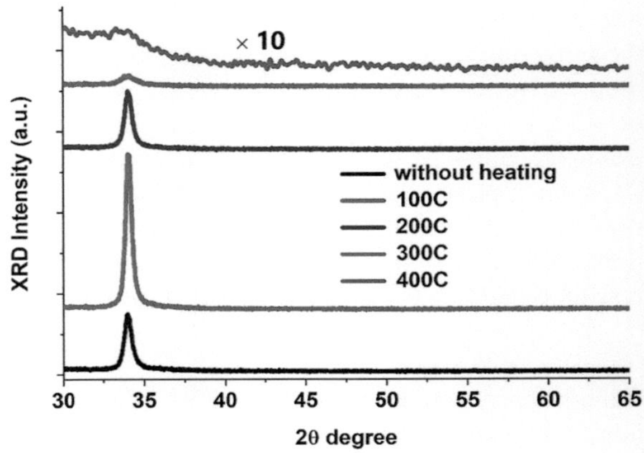

Fig. 9.4 XRD pattern of the GZO films deposited at different substrate temperatures

fluctuation of the optical bandgap value in the range of 3.58 ~ 3.74 eV but also affects the type of defects, carrier concentration, resistivity, transmittance, etc., of the film.

II. Solution Preparation Method

The solution method for preparing GZO thin film is a low-temperature economical method commonly used in poly(methyl methacrylate) (PMMA), polycarbonate (hard transparent plastic), PC, PI, PET, etc., flexible substrates at temperatures <130 °C. Generally, crystal particle size of film prepared at low temperatures is mostly on nanometer scale. Because thermal annealing at temperatures >400 °C is not suitable for these organic flexible substrates, resistivity and transparency of GZO films obtained at low temperatures are slightly inferior to those of GZO films obtained by vacuum methods and chemical vapor deposition. There have been many reports on how to prepare GZO thin films using solution methods. Here, we simply summarize the basic processes and steps of preparing GZO by solution method:

a. Solution Preparation

The solution for preparing GZO film should contain two types of raw materials:

(1) Zn and Ga source materials, including zinc acetate, zinc nitrate, zinc nitrate hydrate, acetylacetone gallium, gallium nitrate, gallium nitrate hydrate, and GaN_3O_9
(2) Auxiliary solution reagents, including isopropanol, ethanolamine, ethanol, glycerin, triethanolamine, urotropine, ammonia water, and sodium citrate.

The processes of preparing the solution are as follows: First, the solid substances containing Zn and Ga are added to the solvent and stirred evenly. It is dissolved under heating or at room temperature so that a solution containing only Zn or Ga substance is obtained. Then, auxiliary reagents are added to adjust the viscosity and functionality of the solution. Next, according to the Ga:Zn ratio requirements, the contents of Zn and Ga in the solution should be controlled to prepare a solution containing both Zn and Ga substances. In addition to stirring evenly and heating during the solution preparation process, the PH value of the solution can also be controlled. By adjusting the concentration, time, and temperature of the prepared solution, the size of the crystal particles and the appropriate Ga:Zn ratio can be selected and controlled during the subsequent GZO film formation, and the preparation of the solution can be completed.

b. Coating

The solution method can be used to obtain GZO thin films on the substrate in the following ways: (1) immersing substrate material in a prepared solution, waiting for GZO thin film to deposit and form; (2) rotating substrate, dropping prepared solution and spin-coating it on the substrate surface; (3) letting substrate stationary, and applying the prepared solution dropwise to cover the substrate surface; and (4) using tools and equipment to spray onto substrate surface. Among these preparation methods, spin-coating is a better way to obtain films with uniform performance and consistent thickness than immersion deposition, drop coating, spray coating, etc. The overall uniformity of the film on horizontally placed substrate is better than that of substrate tilted at an angle. The formation mechanism of GZO thin film varies with different parameters such as fluidity of liquid, coating (deposition) amount, coating speed, coating time, coating method, dwell time, and environmental temperature [38]. These preparation parameters will affect the consistency, uniformity, repeatability, and quality of GZO thin film, thereby modifying the optical and electrical properties of the thin film. For example, the thickness, density, flatness, crystal orientation, and peak intensity of GZO thin films obtained by top-to-bottom and bottom-to-top plasma spraying changed a lot. The thickness of the film deposited by inclined placement and downstream deposition is larger, and the uniformity becomes worse. The situation even becomes more serious as the inclination angle increases, which causes the film characteristics to fluctuate within a certain range, for instance, carrier concentration variation in the range of 4.15 cm^{-3} ~ 8.71 × 10^{20} cm^{-3}, mobility in the range of 3.47 $cm^2/V \cdot s$ ~ 8.32 $cm^2/V \cdot s$, resistivity in a large range of 8.7 $\Omega \cdot cm$ ~ 45.8 × 10^{-4} $\Omega \cdot cm$, and transmittance in the range of 85 ~ 88.7%.

c. Annealing Treatment

In order to improve the performance of GZO thin film prepared by solution methods, GZO film needs to undergo certain temperature treatment. Different treatment temperatures need to be selected for different substrate materials. In addition, annealing atmospheres, annealing times, heating and cooling rates, and other process parameters also should be carefully considered to optimize the characteristics of GZO thin film.

9.3.2 Annealing Treatment

As-grown GZO thin films are generally polycrystalline materials, and the size of their crystal grains is directly related to the preparation conditions. High-temperature annealing is an effective method to provide additional energy to atoms/particles and to improve film quality. On the one hand, Ga atoms in interstitial positions can migrate to Zn^{2+} lattice positions to complete the substitution of atoms and doping. On the other hand, it can help improve the compactness and quality of film via merging fine grains, increasing grain size, and eliminating various vacancies, interstitial atoms, and grain gaps in the film. The annealing temperature, atmosphere, and time duration determine the chemical valence state, defect concentration, microstructure, crystal quality, scattering effect, and eventually optoelectronic properties of the film. The following will discuss and make explanations in detail.

I. The Influence of Annealing Temperature

Figure 9.5 shows atomic force microscope (AFM) morphologies of the sputtered GZO thin films before and after annealing in an oxygen environment. From the figures, one can observe the changes in surface roughness and crystal grain size before and after annealing. The surface flatness Rq of the film before annealing is 1.0 nm, with the very small crystal grain size. After annealing at 400 °C, the surface flatness Rq decreases to 0.58 nm, indicating that the roughness is improved and the surface becomes smoother. As the annealing temperature further increases, the surface roughness of the film gradually increases, with a surface flatness Rq of 0.66 nm at 550 °C and 0.72 nm at 700 °C. During high-temperature annealing, the atoms in the film gain energy and then diffuse and migrate, resulting in change in the degree of crystallization and quality of the film. The higher the annealing temperature the sample experienced, the stronger the merging of many small grains occurred in the film. Meanwhile, the grain size enlarged and the number of grain boundaries greatly reduced. However, the high-temperature annealing also makes the surface rougher due to the enlargement of grains. At the same time, the defect density within the film material decreases to some extent.

Figure 9.6 shows the room temperature photoluminescence (PL) spectra of the GZO film as-grown and annealed at two different temperatures. From strongly enhanced intensity ratio between the band-edge peak at 406 nm (~3.05 eV) and defect-related peaks at longer wavelengths, it can be judged that the thermal annealing can significantly improve the quality of the film. The as-grown (unannealed) sample shows a very broad emission band from 500 to 850 nm, which is even stronger than the band-edge emission peak at 406 nm and the 437 nm sub-peak. After annealing at 550 °C, the relative intensities of the emission peaks have changed significantly. I_{406nm} intensity is now higher than the intensity of the sample defect emission peak; that is, the ratio of $I_{650\text{-}700\ nm}/I_{406nm}$ has decreased. Further increasing the annealing temperature to 700 °C, the 650 ~ 700 nm broad band has almost disappeared. Besides the band-edge peak at 406 nm, the emission peak at 510 nm has become more evident, which may reflect the thermal activation of Ga dopants. Thermal annealing at 700 °C makes more Ga atoms migrate to the Zn lattice positions, acting as effective dopants. With the increase of annealing temperature in an oxygen atmosphere, the emission from oxygen vacancy is significantly suppressed, indicating improvement in the crystal quality of the film.

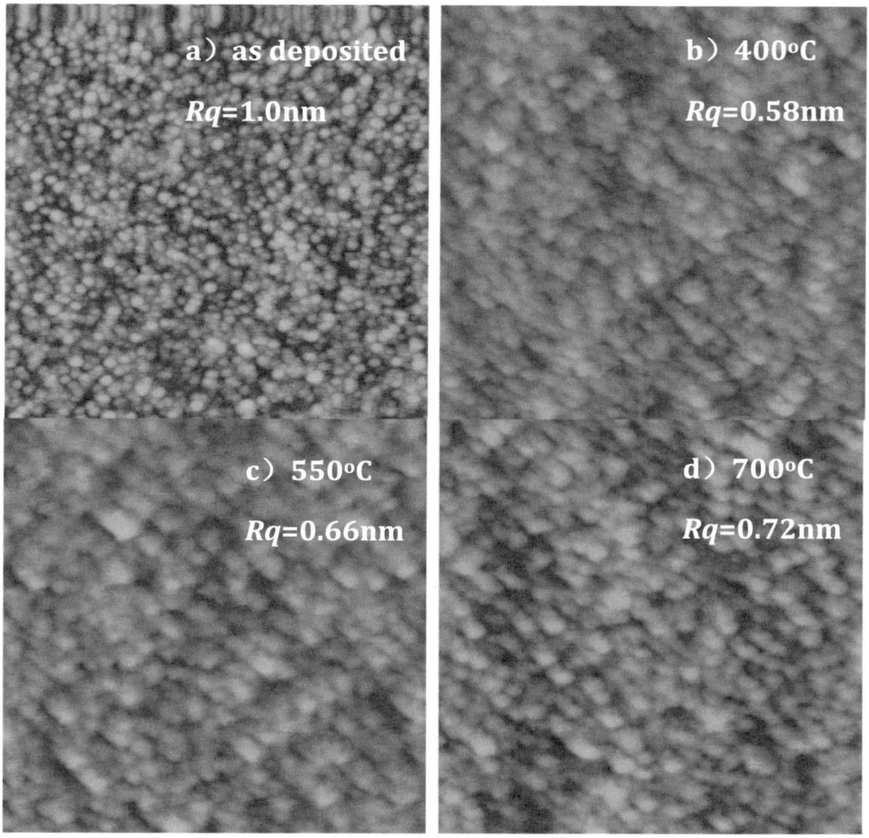

Fig. 9.5 AFM images of GZO thin films annealed at different temperatures

Fig. 9.6 Room-temperature PL spectra of GZO thin films on silicon substrates before and after annealing

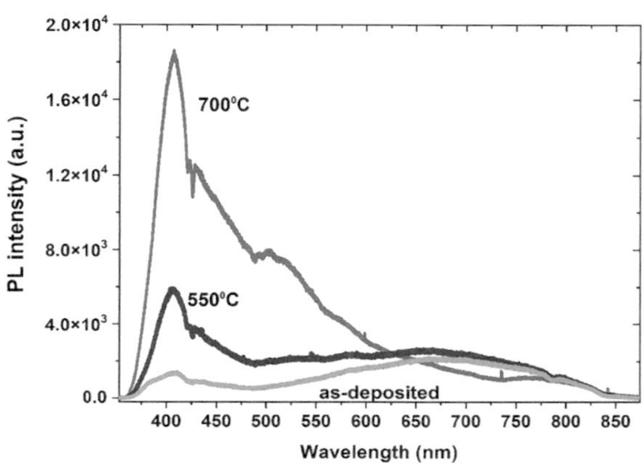

Figure 9.7 shows the XPS O *1 s* (filled squares) and 1021.6 eV Zn *2p* (solid line) spectral lines of the as-grown and annealed sample. O *1 s* spectral lines of the sample before and after annealing can be analyzed via deconvoluted into three respective peaks associated with the core line of 1021.6 eV Zn $2p_{3/2}$ in the state of Zn^{2+} within the ZnO wurtzite matrix structure. The peak $P_{O\text{-}I}$ at ~530.0 eV with a FWHM of 1.2–1.4 eV is attributed to the O^{2-} ion. The 530.8 eV $P_{O\text{-}II}$ peak is identified as the oxygen-deficient phase due to the incorporation of the impurity Ga donating two electrons, while the 531.9 eV $P_{O\text{-}III}$ peak is of surface-adsorbed O/OH or C-O/C-OH substances. The changes in the ratios of $P_{O\text{-}III}$, $P_{O\text{-}II}$, and $P_{O\text{-}III}$ in GZO samples can be used to understand and analyze the chemical state and even microstructure changes in the material. As the annealing temperature increases, the ratio $P_{O\text{-}III}/(P_{O\text{-}I} + P_{O\text{-}II} + P_{O\text{-}III})$ decreases, indicating a gradual decrease in surface adsorption. The ratio $P_{O\text{-}I}/P_{O\text{-}II}$ becomes the highest ratio after the GZO film is annealed at 700 °C, which means that the higher the annealing temperature is undertaken, the more likely the oxidation reaction occurs. Due to the occurrence of oxidization chemical reactions in the film during the thermal annealing process, the oxygen vacancies are significantly reduced. The AFM, XRD, PL, or XPS experimental results all show that high-temperature annealing can significantly improve the crystal quality and reduce the defect density, thereby resulting in noticeable improvement in carrier mobility, carrier lifetime, resistivity reduction, and electrical performance enhancement of the GZO film.

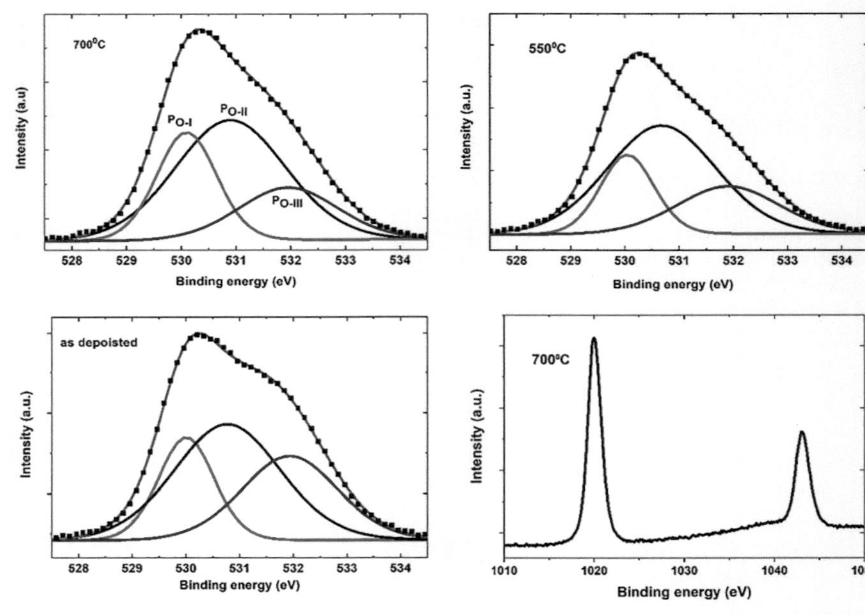

Fig. 9.7 High-resolution XPS O 1 s and Zn 2p curves of GZO films treated at different annealing temperatures

II. The Role of Annealing Atmosphere

The annealing atmosphere environment can be divided into three categories: (1) vacuum conditions, (2) inert gas environments such as argon, and (3) active reaction atmospheres such as oxygen, hydrogen, nitrogen, and Zn. Besides annealing temperature as discussed above, GZO thin film materials are also very sensitive to atmosphere environment. Amorphous or polycrystalline GZO thin films are composed of many tiny crystal particles. Interface of each tiny particle has many dangling bonds. There exist gaps between the grains, inevitably adsorbing oxygen, hydroxyl, hydrogen, carbon, and other substances. When thermal annealing at high temperatures, whether it is in a vacuum environment or an atmosphere environment, there will always be oxidation reactions: $Zn \overset{O}{\rightarrow} ZnO_{1-x} \overset{O}{\rightarrow} ZnO, Ga \overset{O}{\rightarrow} Ga(Zn)O_{1-x}$, and $ZnO_{1-x} \overset{O}{\rightarrow} ZnO$. In addition, some reduction reactions like $ZnO \overset{H}{\rightarrow} ZnO_{1-x} + H_2O$, $ZnO_{1-x} \overset{H}{\rightarrow} Zn + H_2O$, and $Ga(Zn)O_{1-x} \overset{H}{\rightarrow} Ga + H_2O$ may also take place. Various chemical reactions and microstructural adjustments have been shown to change the intrinsic donor-type defects like oxygen vacancies V_o, zinc interstitials Zn_i. and other defects [16]. For example, the carrier concentration of GZO thin film annealed at vacuum is significantly increased from 4×10^{16} cm^{-3} to 7×10^{18} cm^{-3}. It is important that the high-temperature annealing under the atmosphere provides a more abundant reaction environment for the thin film, and the chemical reaction between the elements inside the thin film and the gas substances provided by the atmosphere will inevitably be further strengthened. For example, the high-temperature treatment in a sufficient oxygen environment can effectively prompt oxidation reactions, causing metal atoms (Zn_i, Ga_i) in the interstitial position to oxidize. The whole or local oxygen-deficient phase is transformed into a standard coordination oxide phase. And the intrinsic acceptor-type defect like zinc vacancy V_{Zn} will also be partially filled with oxygen, causing the acceptor-type defects to be annihilated to some extent. In addition, the carrier concentration exhibits a significant decrease from 2.29×10^{19} cm^{-3} to 4.48×10^{17} cm^{-3}. Similarly, when hydrogen and zinc are supplied from the environment, the reduction reactions of Zn and other substances can be enhanced. When excess Zn is incorporated, the activation rate of Ga as the donor can be strengthened. Meanwhile, they can fill the zinc vacancy V_{Zn} to reduce the intrinsic acceptor-type defect zinc and even lead to the formation of the interstitial position donor-type defect Zn_i [39]. However, the average transmittance of the film annealed under Zn environment decreases from >85% to 70%–80%. If hydrogen gas is added, the $H_2 + O_2 \rightarrow H_2O$ reaction occurs, which causes the standard coordination oxide phase to lose oxygen, producing an oxygen-deficient phase and affecting the V_o^+/ V_o^{++} quantity. At the same time, interstitial atoms Zn_i and Ga_i may be introduced for producing donor defects and adjusting the electron concentration of the thin film. It is reported that the resistance value of the thin film annealed in the H_2/Ar mixed gas environment changes by more than 40%. The resistivity of annealed GZO thin films in the N_2/H_2 environment decreases from 10 $\Omega \cdot$ cm to 7×10^{-4} $\Omega \cdot$ cm, showing several orders of magnitude change [39].

Of course, the effectiveness of annealing is closely related to the preparation conditions of the GZO thin film in the early stage. For example, after using argon sputtering to deposit GZO thin films and then performing an annealing process in a vacuum, researchers have shown that the resistivity of the thin film can be improved from $1.13 \times 10^{-3}\ \Omega \cdot \text{cm}$ to $5.4 \times 10^{-4}\ \Omega \cdot \text{cm}$, reducing by an order of magnitude. There are also reports that the resistivity of the GZO thin film drops from the unannealed $8.85 \times 10^{-4}\ \Omega \cdot \text{cm}$ to the annealed $2.17 \times 10^{-4}\ \Omega \cdot \text{cm}$ [39], showing the limited effect of annealing treatment on the resistivity of the film.

9.4 Applications of GZO Thin Films

GZO thin film materials have two main applications: First, as transparent electrodes, they are widely used in flat panel displays, touch screens, solar cells, light-emitting devices (LEDs, lasers), photodetectors, and other optoelectronic devices. Generally, a transparency greater than 80% is required. And the lower the square resistance of the thin film, the better for applications. Second, as an ultraviolet absorption layer, they are used to fabricate ultraviolet detection devices.

9.4.1 Transparent Electrode Applications

a. Application in Flat Panel Displays

In liquid crystal flat panel displays, GZO thin film plays an important role in control units and color display. Using transparent conductive GZO as the transparent electrodes of source and drain, a saturated mobility of $13.5\ \text{cm}^2/\text{V} \cdot \text{s}$, a switch ratio of 3.56×10^7, and a subthreshold voltage slope of 0.43 V/decade are achieved in the control unit of thin film transistor (TFT) array. The threshold voltage of organic field–effect transistors (OTFTs) with GZO electrodes can be improved from −12.4 to −7.7 V. In the red–green–blue color filter display array, the opening and closing times of the lateral electric-field-type liquid crystal with the GZO thin film electrodes are 17.5 ms and 18.0 ms, respectively, with a contrast ratio of 1100:1. GZO thin film is also a powerful low-cost transparent conductive electrode for application in 20-inch large-size television screens. Therefore, GZO thin film has great competitive potential in the display field.

b. Application in Light-Emitting Devices

For semiconductor light-emitting diodes (LEDs), to improve the electro-optical conversion efficiency of LEDs, both optical and electrical factors must be considered. Using GZO as an electrode for LEDs can increase the total flux of light in and out and helps to reduce the lateral expansion resistance, reaching the goals of the working voltage and power loss both reductions. The light output of GZO-based LED can reach 4700 A/cm^2 which is twice that of traditional metal Ni/Au LED and 25% higher than that of ITO LED. The electro-optical conversion efficiency of the GZO device is thus significantly improved. In addition, various methods can be used to prepare and optimize GZO transparent electrodes in semiconductor LEDs, such as MOCVD-prepared GZO LEDs. Their output power under an injection current of 350 mA reaches 256.51 mW, which is 21.5% higher than that of ITO LEDs. Compared with commercialized ITO LEDs, the light output of GZO LEDs prepared by atomic layer deposition increased by 15% under the action of 20 mA driving current. If combined with surface-patterned GZO, the light output was further increased by 30%. Furthermore, by adjusting the sputtering conditions of GZO thin film electrodes, the GZO LED's threshold voltage was reduced from 4.9 to 3.0 V.

GZO film also shows excellent performance in organic LED (OLED) devices. A luminous efficiency of 0.2 lm/W at 100 cd/m^2 was achieved. By optimizing the structure, the performance of the device was further improved. For example, the luminance of an OLED with a double-layer transparent electrode can reach 3640 cd/m^2 under 10 mA/cm, which is 1.5 times higher than that of an OLED with single-layer transparent electrode. Threshold voltage of an inverted structure organic light-emitting diode (iOLED) is 1.9 V, and the luminance at 1000 mA/cm^2 is 7800 cd/m^2, with the highest current efficiency of 0.78 cd/A. By using photonic crystal, the current efficiency of the GZO OLED with a spin-coated positive electrode was enhanced to approximately 120%, and the power efficiency was approximately enhanced by 136%.

c. Application in Solar Cells

GZO film can be used as transparent conductive electrode in various types of thin-film solar cells, including solar cells of amorphous silicon, polycrystalline silicon, $CuIn_xGa_{1-x}Se_2$ (CIGS), CdTe, perovskite, dye-sensitized (DSSC), and organic materials. In amorphous silicon solar cells, the short-circuit current J_{SC} of the planar and roughened solar cells with GZO as the back contact reached 14.47 mA and 14.92 mA, respectively. The conversion efficiency of the roughened amorphous silicon solar cell with GZO back contact was 9.36%, which is 4.5% higher than that of the amorphous silicon solar cells. The efficiency of the CIGS thin-film solar cells with GZO as the antireflection window can also reach 10.8%. The efficiency of the n-i-p perovskite thin-film solar cells with a GZO–ITO composite electrode was 9.67%, which is much higher than the 5.25% efficiency of using only an ITO electrode. The conversion efficiency of the N719 DSSC under one sun was demonstrated to be 0.77%.

d. Application in Detectors

Highly transparent GZO films can be used as transparent conductive electrodes in various types of detectors whose performance can be significantly improved. For example, the highest light detection responsivity of the back-to-back metal-semiconductor-metal (MSM) structure ultraviolet detectors with the traditional metal (Ni/Au) electrodes was 0.11 A/W at a wavelength of 350 nm and gradually decreased as the light wavelength became shorter, as shown in the right chart of Fig. 9.8. However, the detectors with the same size GZO electrodes exhibited a much broader

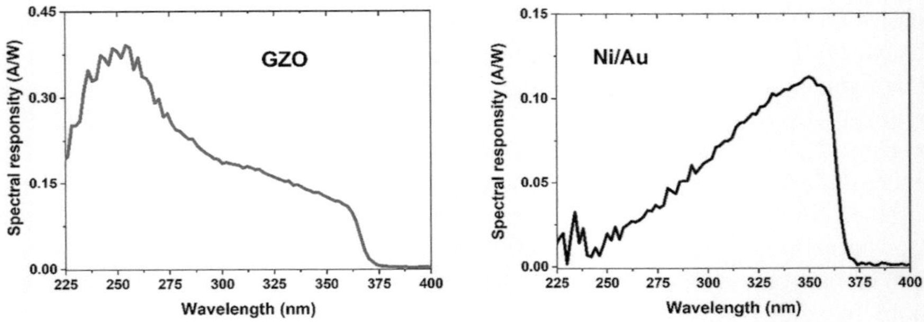

Fig. 9.8 Comparison of the photoresponse characteristics of MSM ultraviolet detectors with GZO transparent electrodes and Ni/Au metal electrodes on GaN substrates

spectral response. When a response value of 0.12 A/W at 350 nm was achieved, the light detection responsivity gradually increased as the light wavelength became shorter. Such that a highest light detection responsivity of 0.38 A/W was obtained at 250 nm, showing an enhanced deep ultraviolet photoresponse, as seen in the left chart of Fig. 9.8.

9.4.2 Detector Applications

There are a lot of examples of using GZO thin films as photoresponse active layers in preparation of ultraviolet detectors. For example, a switch ratio of 6 was achieved in the GZO thin film ultraviolet detectors prepared by spin coating, and internal and external quantum efficiencies of 22.5% and 37.4%, respectively, in the GZO thin film ultraviolet detectors with asymmetric electrode structures prepared by sputtering. The photoresponse of the sputtered GZO photodetector at zero bias was 0.058 A/W. When GZO thin film deposited on a glass substrate was processed into photodetector, its photoresponse can reach 0.23 A/W. By adding a ZnO buffer layer, the photoresponse of GZO photodetector can be further increased from 406 A/W to 1125 A/W. In the photodetector prepared by using GZO nanorods on p-GaN, a photoresponse of 46.2 A/W at 365 nm was obtained. Using a graphitized silicon substrate, the GZO thin film detector was prepared, showing a detection response 0.18 A/W at 370 nm, which is 2.5 times higher than that of the GZO thin film detector on a flat silicon substrate [40]. By depositing GZO on sapphire for preparing ultraviolet photodetectors and after annealing at 900 °C, a novel voltage modulation effect was found. Figure 9.9 shows the tested photoresponse curves. At low bias voltages, the highest photoresponse was measured at 370 nm. As the applied voltage increased, the device's peak photoresponse shifted toward short-wavelength direction. At the same time, the photoresponse also gradually increased with increasing the bias voltage. When the voltage applied to the device was greater than 3 V, the device's photoresponse at deep ultraviolet wavelengths (i.e., 290 nm) exceeded its near-ultraviolet

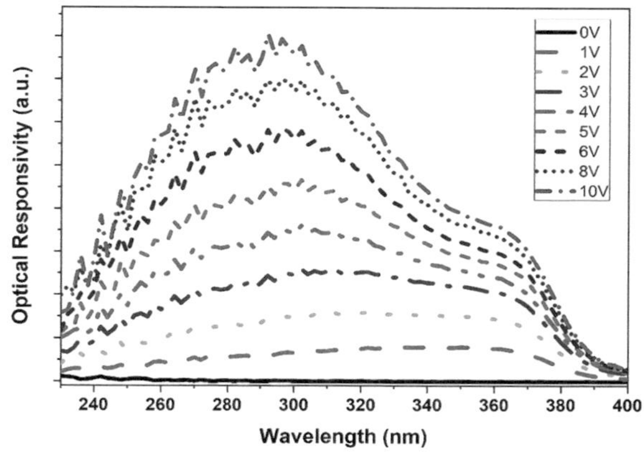

Fig. 9.9 Test results of the photoresponse characteristics of the voltage-modulated GZO ultraviolet detector prepared on sapphire [41]

(~370 nm) photoresponse. Eventually, a dual-band detection of near-ultraviolet and deep ultraviolet light was achieved in the detector [41].

9.5 Summary

This chapter describes the electrical and optical properties of GZO thin film materials. It particularly focuses on the optoelectronic characteristics of GZO thin films prepared by physical gas deposition and liquid chemical preparation. It provides a detailed introduction to the correlations between preparation conditions such as substrate materials, sputtering modes, sputtering gases, and substrate temperatures and characteristics of prepared GZO thin films. The raw materials and process steps for preparing GZO films by liquid-phase method are also documented. The effects and some underlying mechanisms of thermal treatment conditions such as annealing temperature and atmosphere are discussed. Finally, application examples of GZO thin films as transparent electrodes and ultraviolet detection layers are presented.

Acknowledgements This research was supported by the National Natural Science Foundation of China (Project No. 11974378) and the Vacuum Interconnected Station, Nanofabrication Platform, and Testing Platform of the Suzhou Institute of Nanotech and Nanobionics, Chinese Academy of Sciences.

References

1. Minami T (2005) Transparent conducting oxide semiconductors for transparent electrodes. Semicond Sci Technol 20(4):S35–S44
2. Özgür Ü, Alivov YI, Liu C, Teke A, Reshchikov MA, Doğan S, Avrutin V, Cho S-J, Morkoc H (2005) A comprehensive review of ZnO materials and devices. J Appl Phys 98(4):041301

3. Minami T, Sato H, Nanto H, Takata S (1985) Group III impurity doped zinc oxide thin films prepared by RF magnetron sputtering. Jpn J Appl Phys 24(10):L781–L784
4. Bhosle V, Tiwari A, Narayan J (2006) Electrical properties of transparent and conducting Ga doped ZnO. J Appl Phys 100(3):033713
5. Li Y, Huang Q, Bi X (2013) The change of electrical transport characterizations in Ga doped ZnO films with various thicknesses. J Appl Phys 113(5):053702
6. Lee D-H, Kim KN, Chun YS, Kim S, Lee SY (2012) Substitution mechanism of Ga for Zn site depending on deposition temperature for transparent conducting oxides. Curr Appl Phys 12(6):1586–1590
7. Mandalapu LJ, Xiu FX, Yang Z, Liu JL (2007) Ultraviolet photoconductive detectors based on Ga-doped ZnO films grown by molecular-beam epitaxy. Solid-State Electron 51(7):1014–1017
8. Park S-M, Ikegami T, Ebihara K (2006) Effects of substrate temperature on the properties of Ga-doped ZnO by pulsed laser deposition. Thin Solid Films 513(1–2):90–94
9. Walsh A, Da Silva JLF, Wei SH (2011) Multi-component transparent conducting oxides: progress in materials modelling. J Phys Condens Matter 23(33):334210
10. Kaul AR, Gorbenko OY, Botev AN, Burova LI (2005) MOCVD of pure and Ga-doped epitaxial ZnO. Superlattice Microst 38(4–6):272–282
11. Look DC, Hemsky JW, Sizelove JR (1999) Residual native shallow donor in ZnO. Phys Rev B 82(12):2552–2555
12. Zhu DL, Wang GJ, Jia F, Xu WY, Han S, Cao PJ, Liu WJ, Zeng YX, Liu XK, Lu YM (2017) Study of H-related defects in Ga-doped ZnO thin films deposited by RF magnetron sputtering in Ar+H_2 ambient. Mat Tech 32(7):424–429
13. Yamamoto I, Sakemi T, Awai K, Shirakata S (2004) Dependence of carrier concentrations on oxygen pressure for Ga-doped ZnO prepared by ion plating method. Thin Solid Films 451:439–442
14. Honda S, Watamori M, Oura K (1996) The effects of oxygen content on electrical and optical properties of indium tin oxide films fabricated by reactive sputtering. Thin Solid Films 281:206–209
15. Banerjee R, Das D, Batabyal AK, Barua AK (1986) Characterization of tin doped indium oxide-films prepared by electron-beam evaporation. Sol Energy Mater 13(1):11–23
16. Look DC, Leedy KD, Tomich DH, Bayraktaroglu B (2010) Mobility analysis of highly conducting thin films: application to ZnO. Appl Phys Lett 96(6):062102
17. Look DC, Leedy KD, Vines L, Svensson BG, Zubiaga A, Tuomisto F, Doutt DR, Brillson LJ (2011) Self-compensation in semiconductors: the Zn vacancy in Ga-doped ZnO. Phys Rev B 84(11):115202
18. Sernelius BE, Berggren KF, Jin ZC, Hamberg I, Granqvist CG (1988) Band-gap tailoring of ZnO by means of heavy Al doping. Phys Rev B 37(17):10244–10248
19. Yamada T, Makino H, Yamamoto N, Yamamoto T (2010) "Ingrain and grain boundary scattering effects on electron mobility of transparent conducting polycrystalline Ga-doped ZnO films. J Appl Phys 107(12):123534
20. Seto J (1975) The electrical properties of polycrystalline silicon films. J Appl Phys 46(12):5247–5254
21. Yamada Y, Kadowaki K, Kikuchi H, Funaki S, Kubo S (2016) Positional variation and annealing effect in magnetron sputtered Ga-doped ZnO films. Thin Solid Films 609:25–29
22. Yamamoto T, Yoshida HK (2000) Control of valence states in ZnO by codoping method. Mater Res Soc Symp Proc 623:223–234
23. Joseph M, Tabata H, Kawai T (1999) P-type electrical conduction in ZnO thin films by Ga and N codoping. Jpn J Appl Phys Part 2 38(11A):L1205–L1207
24. Ryu YR, Zhu S, Look DC, Wrobel JM, Jeong HM, White HW (2000) Synthesis of p-type ZnO films. J. Crystal Growth 216(1–4):330–334
25. Lu JG, Fujita S, Kawaharamura T, Nishinaka H, Kamada Y, Ohshima T, Ye ZZ, Zeng YJ, Zhang YZ, Zhu LP, He HP, Zhao BH (2007) Carrier concentration dependence of band gap shift in n-type ZnO:Al films. J Appl Phys 101(8):083705
26. Burstein E (1954) Anomalous optical absorption limit in InSb. Phys Rev 93(3):632–633

27. Moss TS (1954) The interpretation of the properties of indium antimonide. Proc Phys Soc London Sect B 67(418):775–782
28. Jain SC, McGregor JM, Roulston DJ (1990) Band-gap narrowing in novel III–V semiconductors. J Appl Phys 68(7):3747–3749
29. Ratcliff EL, Sigdel AK, Macech MR, Nebesny K, Lee PA, Ginley DS, Armstrong NR, Berry JJ (2012) Surface composition, work function, and electrochemical characteristics of gallium-doped zinc oxide. Thin Solid Films 520:5652–5663
30. Gabás M, Ochoa-Martinez E, Navarrete-Astorga E, Landa-Canovas AR, Herrero P, Agullo-Rueda F, Palanco S, Martinez-Serrano JJ, Ramos-Barrado JR (2017) Characterization of the interface between highly conductive Ga:ZnO films and the silicon substrate. Appl Surf Sci 419:595–602
31. Pugalenthi AS, Balasundaraprabhu R, Gunasekaran V, Muthukumarasamy N, Prasanna S, Jayakumar S (2015) Effect of thickness on the structural, optical and electrical properties of RF magnetron sputtered GZO thin films. Mat Sci Semicon Proc 29:176–182
32. Shinde SS, Shinde PS, Oh YW, Haranath D, Bhosale CH, Rajpure KY (2012) Structural, optoelectronic, luminescence and thermal properties of Ga-doped zinc oxide thin films. Appl Surf Sci 258(24):9969–9976
33. Kon M, Song PK, Mitsui A, Shigesato Y (2002) Crystallinity of gallium-doped zinc oxide films deposited by DC magnetron sputtering using Ar, Ne or Kr Gas. Jpn J Appl Phys 41:6174–6179
34. Cuong HB, Lee C-S, Jeong S-H, Lee B-T (2017) Realization of highly conductive Ga-doped ZnO Film with abnormally wide band-gap using magnetron sputtering by simply lowering working pressure. Acta Mater 130:47–55
35. Assunção V, Fortunato E, Marques A, Águas H, Ferreira I, Costa MEV, Martins R (2003) Influence of the deposition pressure on the properties of transparent and conductive ZnO:Ga thin-film produced by r.f. sputtering at room temperature. Thin Solid Films 427(1–2):401–405
36. Clatot J, Campet G, Zeinert A, Labrugere C, Rougier A (2011) Room temperature transparent conducting oxides based on zinc oxide thin films. Appl Surf Sci 257(12):5181–5184
37. Wu F, Fang L, Pan YJ, Zhou K, Huang QL, Kong CY (2010) Effect of substrate temperature on the structural, electrical and optical properties of ZnO:Ga thin films prepared by RF magnetron sputtering. Physica E 43:228–234
38. Chen WK, Huang JC, Chen YC, Lee MT, Juang JY (2019) Deposition of highly transparent and conductive Ga-doped zinc oxide films on tilted substrates by atmospheric pressure plasma jet. J Alloy Compd 802:458–466
39. Lee CS, Lee BT, Jeong SH (2020) In-depth study on defect behavior and electrical properties in Ga-doped ZnO films by thermal-treatment under different chemical equilibrium. J Alloy Compd 818:152892
40. Wang RX, Wu CY, Peng Q, Ge XT, Ning JQ, Sun Q, Huang R, Huang ZL, Zhu WQ, Yang H (2020) Multi-Effects on the characteristics of Ga-doped ZnO film metal-semiconductor-metal ultraviolet photodetectors. Semicond Sci Technol 35(1):015007
41. Wang RX, Yang LC, Xu SJ, Zhang XD, Dong X, Zhao YC, Fu K, Zhang BS, Yang H (2013) Bias-voltage dependent ultraviolet photodetectors prepared by GaO_x+ZnO mixture phase nanocrystalline thin films. J Alloy Compd 566:201–205

Chapter 10
Characterization of Optoelectronic Processes of Localized Carriers in Semiconductors and Photovoltaic Devices

Zhicheng Su and Shijie Xu

10.1 Introduction

Over the past few decades, the efficiency records of semiconductor photovoltaic devices have been continuously broken. Among them, gallium indium phosphide (GaInP) ternary compound, as an important direct bandgap semiconductor material, has greatly promoted the development of high-efficiency photovoltaic industry. Currently, the efficiency of multi-junction solar cells based on gallium indium phosphide has exceeded 32%, and it has been widely used in environments such as providing electric power for systems operating in space.

Understanding the physical mechanisms of carrier generation, transport, and recombination in photovoltaic devices is crucial for further improving device efficiency. The lifetime of carriers, especially minority carriers, is an important parameter determining the efficiency of photovoltaic devices. Therefore, in-depth study of the physical mechanisms affecting carrier lifetime is crucial for improving device design and efficiency.

However, due to the complex multilayer structure of photovoltaic devices, it has always been a challenging task to non-destructively measure the carrier dynamics inside the devices. In recent years, some photoelectric and spectroscopic characterization methods have been introduced into the research of photovoltaic devices, and many new research results have been achieved.

Z. Su (✉) · S. Xu
School of Electronic Science and Engineering, Southeast University, Nanjing, China
e-mail: suzhicheng@seu.edu.cn

S. Xu
Department of Optical Science and Engineering, Fudan University, Shanghai, China

S. Xu (ed.), *Optical Characterization of Microstructures and Optoelectronic Devices Based on Wide Band Gap Semiconductors*, Wide Bandgap Semiconductors,
https://doi.org/10.1007/978-981-95-1928-6_10

This chapter will introduce some novel photoelectric characterization methods, including electroluminescence (EL), time-resolved photoluminescence (TRPL), and time-resolved photocurrent (TRPC), and discuss the methods of using these techniques to study carrier recombination dynamics in photovoltaic devices. At the same time, we have established corresponding physical models for the motion mechanism of carriers, especially transient dynamic processes, and used them to analyze and discuss the carrier recombination dynamics in photovoltaic devices. These characterization methods, physical models, and studies of carrier dynamics and mechanisms will provide important support for the future development of photovoltaic devices based on gallium indium phosphide and other semiconductor materials.

The development of semiconductor materials and nanostructures is changing with each passing day. From solar cells in spacecraft to displays in mobile phones, semiconductor materials have a wide range of applications [1–10]. However, compared with the rapid development of semiconductor materials and their applications in devices, the basic understanding of photoelectric phenomena in these new material systems is still insufficient [11–14]. Thanks to the progress of experimental technology, noncontact photoelectric characterization methods also provide great potential for exploring physics in novel semiconductor materials [15–24].

Gallium indium phosphide ($Ga_xIn_{1-x}P$) is an important semiconductor material, usually used to manufacture solar cells and light-emitting devices [25, 26]. At room temperature, the bandgaps of GaP and InP are 2.25 and 1.27 eV, respectively. By changing the content of Ga/In components, the bandgap of $Ga_xIn_{1-x}P$ can be adjusted between 1.27 and 2.25 eV, which gives this material great flexibility, and it is therefore widely used in various semiconductor devices [26–29].

In the alloy of III-V semiconductors, local Ga and In atoms are distributed on alternating {111} lattice planes, forming a local superlattice structure, also known as a "domain" structure. This CuPt-type ordering reduces the lattice symmetry order and may strongly change the electronic and optical properties of the alloy. The obvious consequences include a reduction in the bandgap, valence band splitting, and optical anisotropy [30]. It is well known that the CuPt-type order in Group III-V alloys is significantly affected by the V/III ratio, substrate temperature during growth, and growth rate. By changing the growth conditions, different degrees of long-range order can be obtained. Under specific growth conditions, the alloy can only obtain a certain degree of order, which means that the entire alloy lattice does not have the same order. One result of this is that GaInP alloys contain a statistical distribution of domains with different degrees of order, which has been confirmed by spatially resolved photoluminescence measurements [31]. Specifically, due to the uneven distribution of Ga/In components, some areas are enriched in InP(GaP) components, while GaP(InP) components are reduced, resulting in a long-range disordered structure of $Ga_xIn_{1-x}P$. In smaller areas, Ga/In may grow evenly in proportion, forming an ordered domain structure, resulting in a local ordered structure of $Ga_xIn_{1-x}P$.

In the locally ordered $Ga_xIn_{1-x}P$ structure, the ordered structure is often surrounded by randomly disordered areas. The random distribution of long-range ordered Ga/In components introduces Anderson localization [32]. In the ordered domain structure, carriers can transmit freely. Our study of the polarization luminescence spectrum of locally ordered GaInP shows that localized and nonlocalized states coexist in this material system [33]. The Mott transition may occur between the two electronic states [34]. Therefore, the locally ordered GaInP material provides us with an excellent platform to verify the Mott transition using optical methods.

Carrier localization in actual crystal solids caused by various defects (such as defects, impurities, component fluctuations, and lattice distortions) is a common phenomenon [35–39]. Carrier localization and related phenomena, due to their scientific significance and profound impact on the electrical, magnetic, and optical properties of material systems, are still widely concerned topics. With the rapid development of In/Ga alloy-based light-emitting diodes, the carrier localization effect caused by structural defects has received more and more attention [36, 40, 41]. For example, local carriers caused by alloy disorder can produce effective luminescence and abnormal thermodynamic behavior [42].

In order to explain these abnormal luminescence behaviors related to carrier localization, many attempts have been made. Among them, the localized-state ensemble (LSE) fluorescence model developed by our research group (Professor Xu Shijie's research group) has been widely used [43–45]; for example, it has been cited in two famous classic textbooks published by Springer in Germany: "*Semiconductor Optics*" and the revised edition of the book "*The Physics of Semiconductor*." This model is rooted in a newly derived distribution function for distinguishable localized carriers. It not only quantitatively reproduces the S-shaped temperature dependence of the fluorescence peak of collective fluorescence of localized carriers in different material systems but also explains the V-shaped temperature dependence of the fluorescence peak width, deeply elucidating the characteristics of collective fluorescence of localized carriers. However, the early version of this model is only applicable to steady-state luminescence. To analyze and interpret the transient or time-resolved luminescence of localized carriers, we have further successfully extended the LSE model to quantitatively analyze the temperature-dependent time-resolved photoluminescence (TRPL) of localized carriers [46].

At the same time, we have also developed a universal model for the temperature-related time-resolved photocurrent (TRPC) of minority carriers in doped semiconductors [47]. This model first establishes the analytical relationship between TRPC and TRPL lifetime and can quantitatively analyze the temperature-related TRPC data measured by GaInP-based solar cells. It is expected that these physical models and the study of carrier recombination dynamics may have significant and far-reaching impacts on the design and device physics of semiconductor optoelectronic devices.

10.2 Photoluminescence and Electroluminescence

Photoluminescence (PL) is an important characterization method in spectroscopy. It generally uses a light source with energy greater than the bandgap to excite semiconductors or other luminescent solid materials, causing electrons in the valence band to be excited to the conduction band by absorbing photon energy. When the excited electrons transite back to the valence band, the changed energy of the electrons is released in the form of photons, forming the so-called photoluminescence.

The principle of electroluminescence (EL) is similar to PL. The main difference is that EL uses electrical energy to excite the sample to emit light. The electroluminescence we are going to discuss here is the light emission caused by charge injection in the semiconductor p–n junction. Under the drive of an external voltage, electrons and holes will be injected into the p–n junction from both electrodes, and some of these injected electrons and holes can undergo radiative recombination in the depletion zone of the p–n junction, thus causing electroluminescence.

Here, we take the p–n junction in GaInP solar cells as an example to discuss the application of electroluminescence characterization in photovoltaic devices [42].

In this study, the GaInP photovoltaic device used for testing was grown on a p-type GaAs substrate using low-pressure metal–organic chemical vapor deposition (MOCVD) technology. The detailed layer structure of this photovoltaic device is shown in Fig. 10.1a. The thickness of the GaInP base layer is 756 nm, and the thickness of the emission layer is 70 nm. In previous research, we have shown that the red light mainly originates from the GaInP base layer.

Under a constant injection current of 15 mA, EL measurements were performed on the GaInP cell at different temperatures. Figure 10.1b shows the measured EL spectra. As shown in the figure, when the temperature is below 20 K, only a broad peak centered at 641 nm (marked as peak 1) is observed. When the temperature is above 20 K, another peak (marked as peak 2) on the high energy side of peak 1 becomes visible. At 30 K, the center wavelength of peak 2 is at 630 nm, and its peak position gradually redshifts with increasing temperature. In the temperature range of 10 K to 50 K, peak 1 dominates in the EL spectrum, but its intensity rapidly decreases with increasing temperature and quenches at about 80 K. Instead, peak 2 becomes the dominant emission at higher temperatures.

In addition to the significant changes in the relative intensities of the two luminescent peaks, their line shapes also show interesting changes with increasing temperature. For example, peak 1 presents an asymmetric line shape at low temperatures, with a long low-energy tail, showing the typical characteristics of collective luminescence of localized states. In contrast, peak 2 shows an asymmetric line shape at high temperatures, with a long high-energy tail, and a symmetric line shape at temperatures < 60 K. To further study the spectra of the two emission peaks, their peak positions in relation to temperature are shown in Fig. 10.1c. The temperature change of the peak position of peak 2 follows the redshift trend described by the Varshni empirical formula, and the red solid line is the redshift rule described by the Varshni empirical formula. However, the peak position of peak 1 shows a V-shaped evolution, first redshifting, and then blueshifting. These huge differences in spectral characteristics indicate that the two luminescent peaks may originate from two different electronic states of different properties. As we discussed earlier, spontaneous local ordering in GaInP alloys can lead to locally ordered domains embedded in the remaining disordered areas. These locally ordered domains, composed of alternating InP/GaP monolayer superlattice structures, may have different degrees of order and sizes,

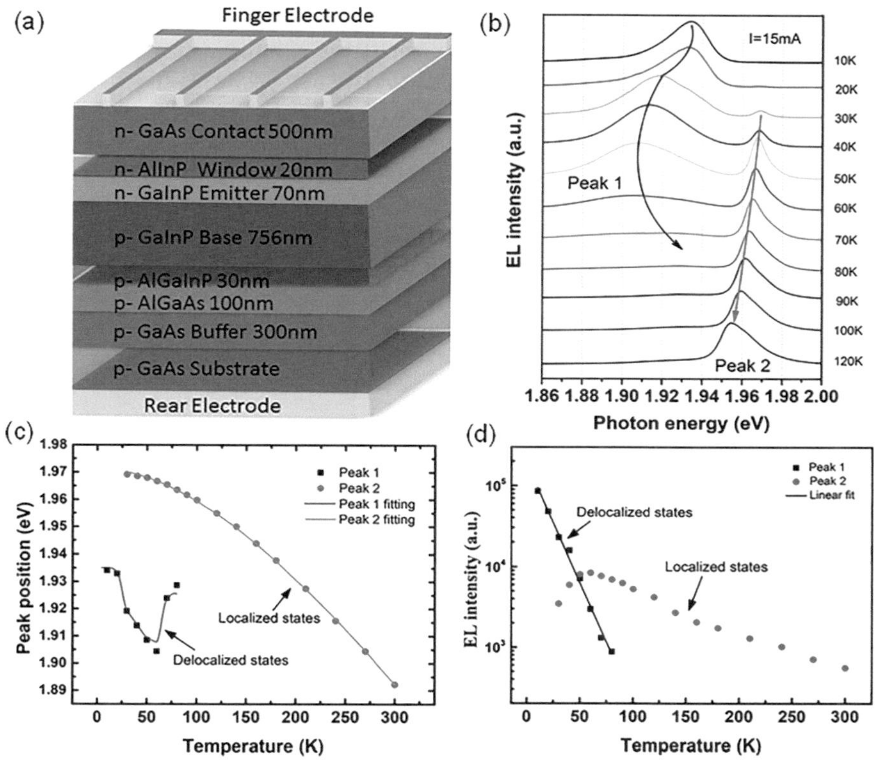

Fig. 10.1 Temperature dependence of electroluminescence. **a** Schematic diagram of the structural details of GaInP photovoltaic devices. **b** The measured EL spectra of the samples under different temperatures and a constant injection current of 15 mA. **c** Temperature dependence of the two EL emission peak positions. **d** Relationship between the integrated intensity of peak 1 and peak 2 and temperature (semi-logarithmic scale)

leading to locally extended states with a wide energy distribution. It should be noted that compared to the bandgap of the completely disordered area, this local extended state has a lower bandgap. On the other hand, another type of electronic state may form in the remaining disordered areas where Ga and In atoms fluctuate completely randomly.

According to the concept of Anderson localization, the second type of electronic state formed in the disordered area should be a localized state [32, 39, 42, 48, 49]. In GaInP alloys with a direct bandgap, radiative recombination of carriers occupying two different electronic states may produce two separate luminescent bands, which may have very different characteristics, including different peak energies (central wavelengths), spectral line shapes, and even polarization states. As mentioned in the introduction, we have measured the polarization of the photoluminescence signal from the GaInP epitaxial layer, clearly showing that the emission spectrum contains fully polarized and fully unpolarized parts, which come from the radiative recombination of carriers in the ordered domains and in the remaining completely disordered areas, respectively. In addition to having different peak positions and line shapes, these two fluorescence polarization components also exhibit markedly different temperature dependencies. In this section, we will explore the evidence of electroluminescence from the transition of carriers from superlattice-like extended states to Anderson-type localized states in GaInP photovoltaic cells.

To quantitatively explain the abnormal temperature dependence of the luminescence of localized carriers often observed, we (Xu's research group) have developed a general model, the localized-state ensemble (LSE) luminescence model [43–45]. According to this model, the luminescence spectrum of localized state carriers with Gaussian density of state (DOS) is as follows:

$$I \propto \rho(E) \cdot f(E, T) = \rho_0 e^{-(E-E_0)^2/2\sigma^2} \cdot \frac{1}{e^{(E-E_a)/k_BT} + \tau_{tr}/\tau_r}, \tag{10.1}$$

where $\rho(E) = \rho_0 e^{-(E-E_0)^2/2\sigma^2}$ represents the Gaussian DOS due to the distribution of localized states in the energy space, E_0 is the center energy level position of the Gaussian DOS, E_a is the characteristic energy level, k_B is the Boltzmann constant, and τ_{tr} (τ_r) represents the time constant for carriers to escape (radiative recombination) from localized states.

The second part of the above equation can perhaps be understood as a distribution function for distinguishable localized carriers; when $\frac{\tau_{tr}}{\tau_r} = 1$, the distribution function is formally identical to the Fermi–Dirac distribution function. By solving for the maximum value of the right side of Eq. (10.1) with respect to temperature, we obtain the peak energy of the localized state collective fluorescence. Considering the Varshni empirical formula, which monotonically redshifts with increasing temperature for the ideal semiconductor bandgap width, we can finally write the peak energy of the localized state carrier collective fluorescence as follows:

$$E(T) = \left(E_0 - \frac{\alpha T^2}{\theta + T}\right) - x \cdot k_BT, \tag{10.2}$$

where θ is the Debye temperature, α is the Varshni parameter, and x is a dimensionless temperature-dependent parameter, which can be obtained by numerically solving the following formula:

$$xe^x = \left[\left(\frac{\sigma}{k_BT}\right)^2 - x\right]\left(\frac{\tau_r}{\tau_{tr}}\right)e^{(E_0-E_a)/k_BT}, \tag{10.3}$$

The above LSE model is used to fit the V-shaped temperature dependence of the peak position of peak 1, and the fitting curve is shown in Fig. 10.1c. The parameters used for the best fit are $E_0 = 1.935$ eV, $\alpha = 6.989 \times 10^{-2}$meV/K, $\theta = 450$ K, $E_a = 1.956$ eV, $\tau_{tr}/\tau_r = 3.684 \times 10^{-2}$, and $\sigma = 26.0$ meV. The temperature dependence of the peak position of peak 2 follows the trend described by the Varshni Formula [45], as shown by the solid line in Fig. 10.1c. The best fit parameters obtained with the Varshni formula are $E_0 = 1.971$ eV and $\alpha = 0.659$ meV/K.

It is clear that the transition of carriers from delocalized states (local superlattices or ordered domains) to Anderson localized states is a dynamic process related to temperature. At least within a certain temperature range, this carrier transition will lead to a rapid decay of peak 1 intensity and a corresponding increase in peak 2 intensity. As shown in Fig. 10.1d, the intensity of peak 1 exhibits a rapid exponential decay, while the intensity of peak 2 first increases, reaching a maximum at 60 K, and then undergoes a slower exponential decay as the temperature continues to rise.

As shown in Fig. 10.2a, the EL spectrum of the photovoltaic device measured at a constant temperature of 50 K with the injection current ranging from 0.6 to 15.0 mA to further investigate the two luminescent bands. As expected and observed, the intensities of peaks 1 and 2 both increase with the increase in injection current, reflecting the physical mechanism of increasing carrier concentration involved in luminescence. However, the dependence of the two peaks on the injection current is very different. For peak 1, while its intensity grows, there is a significant blueshift in the peak position, indicating the presence of a typical electronic state-filling effect. This state-filling effect may cause a significant Burstein–Moss shift in the absorption edge and luminescent peak [50]. Unlike the performance of peak 1, the position and linewidth of peak 2 unchanged with the increase in injection current. The unusual behavior of peak 2 and its much narrower linewidth than peak 1 indicate that the electronic-state energy distribution of peak 2 is narrower and the density of state is higher.

We know that changing the injection current will cause changes in the carrier concentration in the device, which will affect the quasi-Fermi level of localized carriers, that is, E_a in the LSE model. From the intensity distribution of the two luminescent bands and their evolution with injection current as shown in Fig. 10.2a, we demonstrate that the majority of the electrically injected carriers are first localized in the first type of electronic state (local atomic ordered superlattice region) and then undergo radiative recombination, producing luminescent peak 1. As the injection current increases, more and more carriers will be injected into the device and occupy more electronic states. According to the well-known Pauli exclusion principle, this will lead to an increase in the quasi-Fermi level of carriers, thereby pushing up the luminescent peak position, which is the physical reason for the blueshift of the peak position of luminescent peak 1 with the increase in injection current that we observed. The LSE model can be used to analyze the change of peak 1 position with injection current very well. As shown on the right side of Fig. 10.2b, the solid line represents the fitting curve of the LSE model for the peak 1 position, and the fit between the theoretical fitting curve and the experimental data is quite good. The adjustable parameter E_a used in the fit is also drawn in short solid lines in the right side of Fig. 10.2b. It is evident that as the injection current increases, E_a continues to increase, representing that the concentration of carriers in the device continues to increase with the increase in the injection current. The left side of Fig. 10.2b also shows the DOS distribution curve of peak 1 (black solid line). To better understand

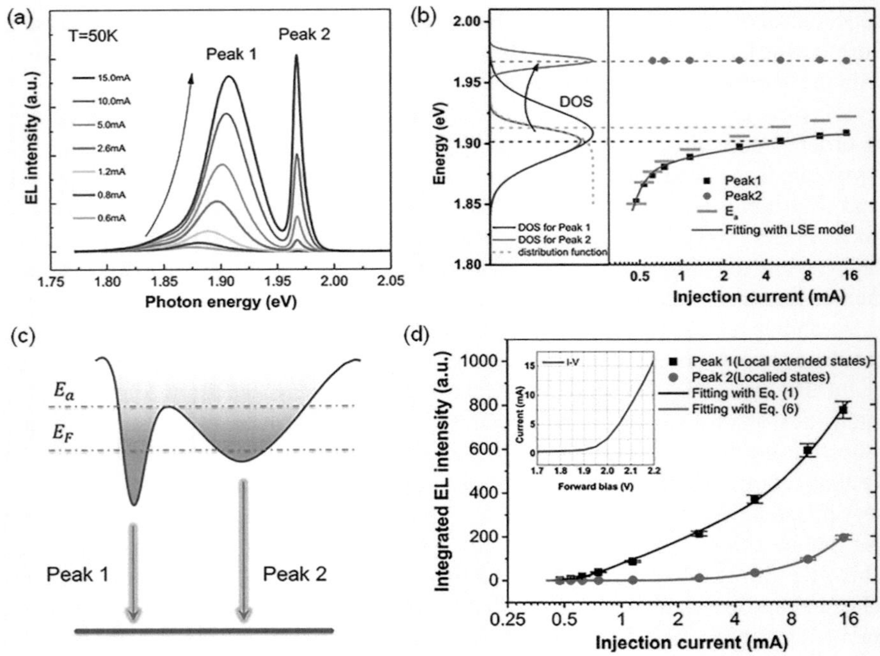

Fig. 10.2 Evolution of the two peak positions of electroluminescence. **a** EL spectra of the battery measured at different injection currents at 50 K. **b** Calculated state density (DOS) (left) and changes in the two peak positions in the EL spectrum under different injection currents (right). For example, the distribution function of the LSE model (dashed line) and the calculated luminescent spectrum (solid line) are marked with an injection current of 5 mA. **c** Energy band diagram of the two luminescent peaks. **d** Relationship between the integrated intensity of peaks 1 and 2 and the injection current (semi-logarithmic scale)

the application of the LSE luminescence model, the distribution function in the LSE luminescence model (green dashed line) is also shown. In addition, we also plotted the theoretical luminescence spectrum of localized carriers under an injection current of 5 mA (blue solid line) in the left column of Fig. 10.2b.

To further verify that most of the injected carriers are indeed captured by the first type of electronic state under the injection current of our interest, we measured the I-V characteristics of the photovoltaic cell under forward bias, as shown in Fig. 10.2d. Obviously, the I-V curve of the cell shows the typical rectifying characteristics of the p–n junction. At the same time, the integrated EL intensities of peaks 1 and 2 are also shown in Fig. 10.2d. Within the measurement range, peak 1 always dominates over peak 2. Moreover, its intensity changes closely follow the trend of the injection current. These data further prove that the first type of electronic state (localized extended state formed by local atomic ordered domains) that produces peak 1 luminescence captures most of the injected carriers.

After deriving some key parameters of peak 1 with the LSE model, we can have a more detailed discussion of peak 2. As shown in Fig. 10.2a, although the intensity of peak 2 increases with the increase in the injection current, it has an almost constant peak position (1.968 eV) and maintains a narrow linewidth (about 9 meV). As pointed out earlier, peak 2 originates from the radiative recombination of Anderson-type localized states related to disordered domains. Compared with peak 1, the integrated intensity of peak 2 is relatively weakly correlated with the injection current, because the luminescence of peak 2 is mainly dominated by the thermal transfer of carriers from the first type of electronic state to the second type of electronic state. According to the LSE model, the thermal transfer process of carriers can be expressed as $\frac{\rho(E)\cdot f(E,T)}{\tau_{tr}} e^{(E-E_c)/k_B T}$ to quantitatively describe. Therefore, the integral intensity of peak 2 can be described as follows:

$$I_2 \propto N_{\text{Activation}} \propto \int_{-\infty}^{\infty} \rho(E) \cdot f(E,T) \cdot e^{(E-E_c)/k_B T} dE, \qquad (10.4)$$

where $E_c = 1.931\ eV$ is obtained from the fit at 50 K mobility edge.

According to Mott's theory, the mobility edge is defined as the energy value that separates local and nonlocal states in amorphous materials or doped semiconductors [34]. Generally speaking, carriers with lower energy are spatially fixed and only migrate through thermal activation; carriers with higher energy have a certain mobility. Therefore, the mobility edge is considered to be the characteristic energy value that separates the two types of local electronic states. As shown in Fig. 10.2d, the theoretical fitting results are in good agreement with the experimental data.

Based on the above experimental data and the simulation results of the LSE model, we can now establish a fairly accurate physical picture to describe the localization distribution of carriers, radiative recombination, and thermal migration in partially ordered GaInP alloys. As shown in Fig. 10.2c, there are two different electronic states coexisting in lightly doped p-type GaInP alloys. At low temperatures, most of the injected carriers are captured by the electronic states caused by local atomic ordered domains, some of which produce EL peak 1 through radiative recombination. As the injection current increases, the quasi-Fermi level occupied by carriers in these electronic states will rise, resulting in a noticeable Burstein–Moss blueshift for peak 1. Moreover, more carriers will thermally transfer from these locally extended electronic states to Anderson-type localized states, leading to an enhancement of EL peak 2.

10.3 Time-Resolved Photoluminescence Model and Application

The photoluminescence (PL) introduced in the first section of this chapter actually refers to time-integrated photoluminescence (TIPL), also known as steady-state photoluminescence. Since the sample is luminescent under constant light excitation conditions, TIPL is a quasi-equilibrium spectrum and does not reflect information related to the time evolution of carriers.

Unlike TIPL, time-resolved photoluminescence (TRPL), also known as transient photoluminescence, provides carries information about the evolution of carriers over time and can be used to study the dynamic process of carriers in luminescent materials. To observe the evolution of photoluminescence over time, we must use a pulsed light source or electrical pulse to excite the sample, so that we can measure the evolution curve of a certain wavelength over time. Currently, there are several techniques for TRPL or TREL measurements, the latter using electrical pulses to excite luminescent samples. The streak camera stands out as the most potent tool for TRPL measurements, achieving sub-picosecond time resolution when paired with a femtosecond laser. Other techniques include Kerr gating and upconversion fluorescence.

For the steady-state spectrum of localized carrier luminescence, including exciton luminescence, we have previously analyzed and calculated its behavior using the LSE model. This analysis has provided a deep understanding of the changes in peak position, peak shape, intensity, etc., of localized carrier steady-state luminescence with temperature. Additionally, it allowed us to obtained some quantitative values of key physical factors (parameters) that determine localized carrier steady-state luminescence. However, there is a lack of quantitative understanding of the dependence of the transient process of localized carrier luminescence on parameters such as temperature and energy (wavelength). The main reason is that there was no suitable theoretical model to quantitatively describe the time-resolved spectrum of localized carrier luminescence. Unlike identical fermions or bosons in the general statistical sense, localized carriers are distinguishable, they carry different energies and occupy different localized electronic states, and each contributes differently to luminescence. In the steady-state LSE model, a new distribution function related to the ratio of nonradiative and radiative recombination time constants of localized carriers plays a decisive role. In the transient process of localized carrier luminescence, we can imagine that its fluorescence lifetime constant or lifetime must be a function of energy or wavelength and also a function of temperature. As shown in Fig. 10.3a, the two-dimensional image of the TRPL spectrum of InGaN/GaN quantum wells measured experimentally using a streak camera technique clearly shows a strong dependence of its fluorescence lifetime on energy [51]. Below, we will describe the universal model of time-resolved spectroscopy that we have recently developed based on the steady-state LSE model [46]. In this model, we not only prove that the luminescence intensity of localized carriers at a specific energy still follows a single exponential decay with time but also derive the analytical expressions for the changes in effective luminescence lifetime and nonradiative recombination lifetime with temperature and energy. By applying this model, we quantitatively explain the time-resolved luminescence experimental data of several material systems published in the literature, demonstrating its universality and accuracy.

For a luminescent system with a density of state (DOS) of $\rho(E)$, we can use the following partial differential equation to describe the time evolution of the excited carrier concentration [43–45]:

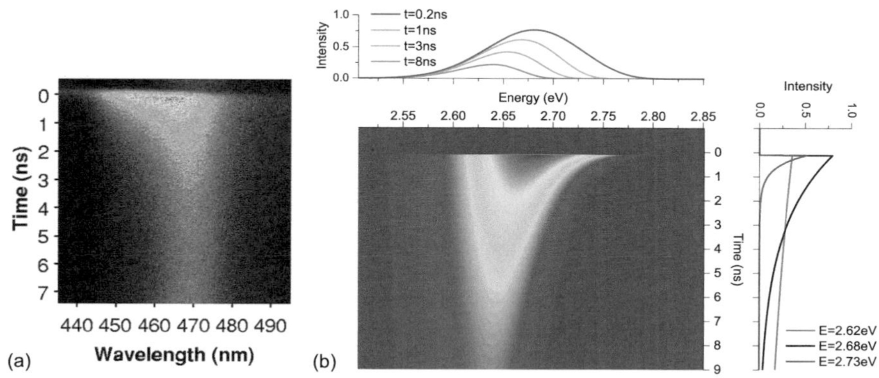

Fig. 10.3 TRPL two-dimensional spectrum. **a** The experimentally measured TRPL spectrum of InGaN/GaN quantum wells [51]. **b** The calculated TRPL spectrum image of the localized state system

$$\frac{\partial N(E,T,t)}{\partial t} = G + \gamma_c N'\eta - \frac{N(E,T,t)}{\tau_{tr}} e^{\frac{(E-E_a)}{k_B T}} - \frac{N(E,T,t)}{\tau_r}, \tag{10.5}$$

where G, N', γ_c, and η represent the generation rate of carriers due to photoexcitation, electrical injection, etc.; the total number of thermally activated carriers; the recombination coefficient of thermally activated carriers; and the recombination efficiency, respectively. E_a still represents a unique energy of the material, depending on the specific material, and τ_{tr} and τ_r are two time constants characterizing the thermal activation and radiative recombination of localized carriers; the reciprocal of the latter constant represents the probability of generating fluorescence.

Let us assume that the recombination efficiency is proportional to the number of unoccupied localized states [52, 53], that is:

$$\eta = \frac{N(E,T,t)e^{(E-E_a)/k_B T}}{\int_{-\infty}^{+\infty} N(E',T,t)e^{(E'-E_a)/k_B T} dE'}. \tag{10.6}$$

Under such assumptions and pulse excitation conditions, the rate Eq. (10.6) can be rewritten as

$$\frac{\partial N(E,T,t)}{\partial t} = G(E,t) - (1-\gamma_c)\frac{N(E,T,t)}{\tau_{tr}} e^{\frac{(E-E_a)}{k_B T}} - \frac{N(E,T,t)}{\tau_r}. \tag{10.7}$$

Simplify Eq. (10.7) to

$$\frac{\partial N(E,T,t)}{\partial t} + P(E,T)N(E,T,t) = G(E,t), \tag{10.8}$$

where

$$P(E,T)=\frac{(1-\gamma_c)e^{\beta(E-E_\alpha)}}{\tau_{tr}}+\frac{1}{\tau_r}, \tag{10.9}$$

and in the above equation, $\beta = 1/k_BT$.

One solution to the partial differential Eq. (10.8) is as follows:

$$N(E,T,t)=\rho(E)e^{-P(E,T)\cdot t}\int g(t)e^{P(E,T)\cdot t}dt. \tag{10.10}$$

For pulse excitation, the carrier generation term can be written as $G(E,t) = g(t)\rho(E)$, where $g(t)$ is expressed as follows:

$$g(t)=g_0e^{\frac{-(t-t_0)^2}{2\sigma_t^2}}. \tag{10.11}$$

In the above equation, g_0, is a constant, t_0 represents the center time of the Gaussian pulse, and σ_t is an important parameter controlling the carrier excitation process. The final localized carrier luminescence intensity as a function of time can be expressed as

$$I_L(E,T,t)=\frac{N(E,T,t)}{\tau_r}\propto\rho(E)\cdot\left\{\text{erf}\left[\frac{(t-t_0)-\sigma_t^2/\tau_L}{\sqrt{2}\sigma_t}\right]+1\right\}e^{-(t-t_0)/\tau_L}, \tag{10.12}$$

where

$$\tau_L(E,T)=\frac{1}{P(E,T)}=\frac{\tau_r}{1+\frac{\tau_r}{\tau_{tr}}(1-\gamma_c)e^{\frac{(E-E_a)}{k_BT}}}, \tag{10.13}$$

or written as

$$\frac{1}{\tau_L(E,T)}=\frac{1}{\tau_r}+\frac{1}{\tau_{tr}}(1-\gamma_c)e^{\frac{(E-E_a)}{k_BT}}. \tag{10.14}$$

Equations (10.13) or (10.14) provides the analytical expressions for the variation of the time constant of localized carrier collective fluorescence with energy and temperature, etc. Obviously, when the recapture rate of thermally activated localized carriers $\gamma_c = 1$, the fluorescence lifetime of localized carriers τ_L is equivalent to its radiative recombination lifetime τ_r. This is because in this model, the thermal activation escape of localized carriers is the only channel for nonradiative recombination of carriers. $\gamma_c = 1$ means that all localized carriers eventually participate in radiative recombination.

In fact, the expression (10.13) can be further simplified to

$$\tau_L=\frac{\tau_r}{1+e^{\alpha(E-E_m)}}. \tag{10.15}$$

This concise expression has been widely used in the literature, where E_m is a specific energy defined by Oueslati et al., at which the carrier radiative recombination rate equals the transmission rate [54]. α is a model parameter, in units of reciprocal energy. Please note that our model not only considers the variation of fluorescence time constant with luminescent energy but also considers the temperature effect. Therefore, Eqs. (10.13) or (10.14) provide a universal quantitative description for the collective fluorescence time constant of localized state carrier.

As shown in Fig. 10.3b, we used this model to simulate the time-resolved PL spectra of localized carriers. The calculated TRPL spectra were plotted in a two-dimensional image. The image on the top shows the theoretical PL spectra at different delay times, noting the redshift of the fluorescence peak over time, which is due to the higher thermal escape of localized carriers at higher energy levels. The image on the right shows the decay curves of luminescence intensity at three different photon energies over time. Obviously, the theoretical TRPL spectra show very unique and interesting trends in their evolution over time. For example, the luminescence lifetime shows a clear dependence on energy, and the luminescence spectrum shows an interesting dependence on delay time, that is, the rapid redshift displayed by the peak position in the early delay time and the rapid decay at higher energies. These theoretically predicted trends in evolution are in good agreement with the experimental spectral measurements.

In addition, we know that the fluorescence lifetime is not solely determined by the radiative recombination time constant of the carriers, it is also constrained by the nonradiative recombination time constant, which can usually be written as $\tau_L^{-1} = \tau_{nr}^{-1} + \tau_r^{-1}$. Using this equation, we can derive the following expression:

$$\tau_{nr}(E, T) = \frac{\tau_{tr}}{1 - \gamma_c} e^{-(E-E_\alpha)/k_B T}. \tag{10.16}$$

The above physical quantity can be considered as the effective nonradiative recombination time of localized carriers. This formula shows that the effective nonradiative recombination time of localized carriers exhibits a clear exponential dependence on temperature and energy, playing a significant role in the effective fluorescence lifetime. As shown in Fig. 10.3b, the relationship between temperature and the nonradiative lifetime of localized excitons in InGaN multi-quantum wells measured by Narukawa et al. can be quantitatively explained by this formula [55].

Now, we apply this model to the time-resolved photoluminescence data of single-layer InGaN alloy measured by Satake et al. [56]. In this InGaN ternary alloy, the spontaneous radiation of localized excitons is the main luminescence mechanism. It can be observed from Fig. 10.4a that a good agreement has been achieved between theory and experiment.

In addition to GaInP and InGaN ternary alloys, the InGaN/GaN core quantum wells, which are commonly used for manufacturing blue-green LEDs, also have a strong localized luminescence effect. Okamoto et al. measured the photoluminescence lifetime of localized carriers at different energies and temperatures in InGaN/GaN quantum wells [51]. The squares and circles in Fig. 10.4c and d are their

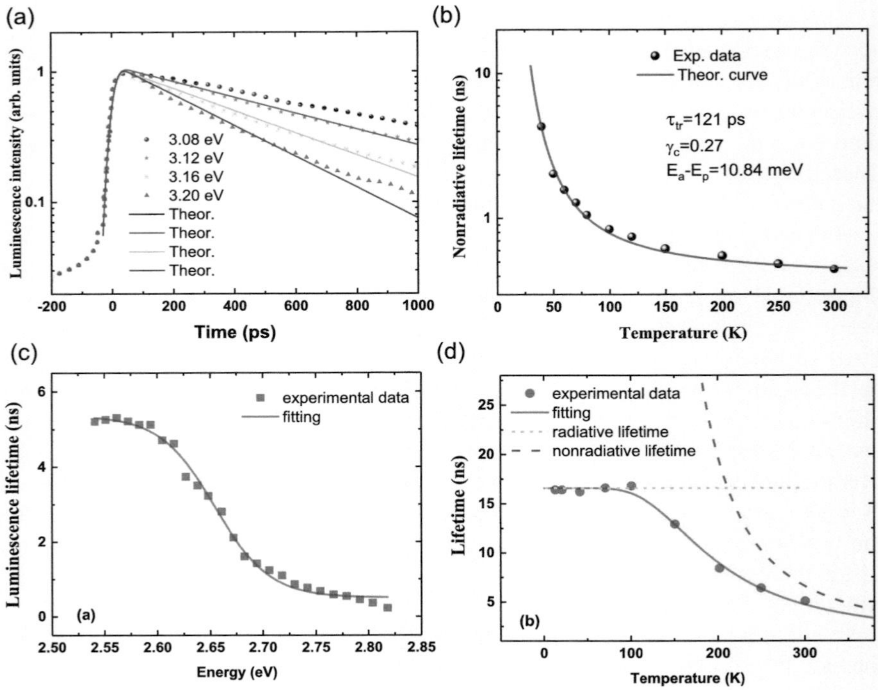

Fig. 10.4 Energy and temperature dependence of TRPL spectrum. **a** Time-resolved photoluminescence intensity at different luminescence energies of a single layer of InGaN alloy measured by Satake et al. [56]. **b** Temperature dependence of the nonradiative lifetime (circles) of localized excitons in InGaN multiple quantum wells [55]. Fluorescence lifetime of localized carriers in InGaN/GaN quantum wells coated with Ag thin film measured by Okamoto et al. [51]. **c** Fluorescence lifetime at different energies at room temperature. **d** Fluorescence lifetime at fixed energy but different temperatures

experimental data. From the figures, we can see that the fluorescence lifetime shows good agreement between theory and experiment with respect to its variation with both energy and temperature. This once again demonstrates the universality and accuracy of the theoretical model we developed. In Fig. 10.4d, we can also clearly see the explanation of the temperature dependence of the nonradiative recombination lifetime of localized carriers measured by the experiment by the theoretical model. The calculated curve in the figure uses $\tau_r = 16.557$ ns, the dashed line is the theoretical effective nonradiative lifetime τ_{nr}, and the solid line is the theoretical luminescence lifetime. For localized carriers, their radiative recombination lifetime τ_r usually does not change with temperature in the low-temperature region. In the high-temperature region, τ_{nr} significantly shortens and mainly determines the overall lifetime of luminescence.

In summary, we have established an analytical model of time-resolved luminescence of localized carriers, successfully extending the LSE model to time-resolved fluorescence spectra. This model can quantitatively describe the dispersive thermodynamic behavior of localized carriers, mainly composed of analytical expressions for the effective luminescence lifetime and nonradiative recombination lifetime of localized carriers. The establishment and application of this model have given us an unprecedented deep understanding of the time-resolved luminescence process of localized carriers, especially the temperature and energy dependence of these dynamic processes occurring in semiconductors and devices.

10.4 Time-Resolved Photocurrent Model and Application

Similar to TRPL, time-resolved photocurrent (TRPC) is also a function of time and can be used to study the time evolution dynamics of photo-generated carriers in semiconductors and optoelectronic devices. The difference from TRPL is that TRPC measures the evolution curve of current over time. Unlike luminescence, photo-generated carriers must reach the electrode to detect the current signal. Therefore, it can be expected that even for the same sample, TRPL and TRPC may show different time constants. Short pulse light is also used as an excitation source in actual TRPC measurements.

Currently, both TRPL and TRPC experimental techniques are often used to detect the lifetime of carriers in semiconductors and optoelectronic devices [15, 57]. Experiments show that in a given semiconductor or device, the measured TRPL and TRPC signals may show different time and temperature dependencies [58]. Luminescence is a negative impact process for photovoltaic cells, as it represents the radiative recombination loss process of carriers in transit. However, photocurrent (PC) is positively correlated with the electrical output power of photovoltaic devices. Therefore, the PC process is a positive influence for photovoltaic cells. For the study of time-resolved carrier dynamics in photovoltaic devices, TRPC may represent a more direct and important technique, as its time constant is most directly related to the photoelectric conversion efficiency of photovoltaic devices. However, to the best of our knowledge, there seems to be no report on the theoretical model of time-resolved photocurrent signals in semiconductors, which makes us not yet have a deep understanding of the physical reasons affecting the lifetime of photocurrent in photoelectric devices, which also hinders the further design and development of photovoltaic and other photoelectric devices to some extent.

Here, we attempt to establish a general model for the temperature-related TRPC of minority carriers in doped semiconductors and seriously discuss the recombination and transport processes of carriers in photovoltaic devices [47].

We start by analyzing the space- and time-related minority carrier concentration in p-type doped semiconductors. For simplicity, the spatial distribution of minority carrier concentration is only considered in one dimension, such as the x direction, and it is not difficult to extend to three dimensions after obtaining the results. We assume that the semiconductor is uniformly excited and the electric field is uniformly distributed in the x direction. This concentration distribution of carriers $N(x, t)$ can be described by the continuity partial differential equation [59]:

$$\frac{\partial N(x,t)}{\partial t} = D\frac{\partial^2 N}{\partial x^2} + \mu E\frac{\partial N}{\partial x} + G - \frac{N}{\tau}, \tag{10.17}$$

In the above equation, D represents the diffusion coefficient of carriers (electrons), μ is the mobility of electrons, E is the applied electric field, τ is the lifetime of minority carriers (electrons), and G is the carrier generation term produced by external excitation. It is assumed that the generation term can be written as the following expression: $G(x, t) = X(x) \cdot g(t)$, a possible solution to the above equation can be written as the product of two independent functions of spatial coordinate x and time t:

$$N(x,t) = AX(x) \cdot n(t), \tag{10.18}$$

$$X(x) \propto e^{-ax}. \tag{10.19}$$

In Formula (10.18), A is a dimensionless constant, while a in Formula (10.19) is a constant with the unit of inverse length, which characterizes the vertical spatial distribution of minority carriers. The constants A and a can be determined by specific spatial boundary conditions. Here, we only consider the time function $n(t)$, which can be solved by the following formula:

$$\frac{dn(t)}{dt} = a^2 Dn - a\mu En + g - \frac{n}{\tau}. \tag{10.20}$$

A solution to the above equation can be written as

$$n(t) = e^{-Pt} \cdot \left(\int ge^{Pt}dt + C\right), \tag{10.21}$$

where

$$P = \frac{1}{\tau} + a\mu E - a^2 D. \tag{10.22}$$

We know that in time-resolved measurements, pulse excitation is used. The time-dependent function of the carriers generated by pulse excitation can be expressed as $g(t) = g_0 e^{\frac{-(t-t_0)^2}{2\sigma_t^2}}$. Under this Gaussian pulse excitation, the solution of minority carrier concentration can be written as

$$n(t) = \sqrt{\frac{\pi}{2}}\sigma_t g_0 \left\{ \mathrm{erf}\left[\frac{t - \left(t_0 + \sigma_t^2 P\right)}{\sqrt{2}\sigma_t}\right] + 1 \right\} e^{\sigma_t^2 P^2/2 + (t_0 - t)P}, \tag{10.23}$$

where erf represents the Gaussian error function.

According to the definition of current density [59]

$$J_{pc}(t) = \rho(t) \cdot v_d, \tag{10.24}$$

where $\rho(t) = qn(t)$ represents the charge density and v_d is the average drift velocity of the carriers.

Assuming that the average drift velocity remains constant, the change in photocurrent density over time can be represented as

$$J_{pc}(t) \propto \left\{ \mathrm{erf}\left[\frac{(t - t_0) - \sigma_t^2/\tau_{pc}}{\sqrt{2}\sigma_t} \right] + 1 \right\} e^{-(t-t_0)/\tau_{pc}}, \tag{10.25}$$

Here, σ_t is the parameter controlling the rising process, which is related to the material and excitation conditions, while τ_{pc} is the parameter controlling the decay process of the photocurrent. t_0 is the reference time of the system. The decay lifetime of the photocurrent density in the above equation can be expressed as

$$\frac{1}{\tau_{pc}} = \frac{1}{\tau} + a\mu E - a^2 D. \tag{10.26}$$

Using the Einstein relation $D = \frac{\mu k_B T}{q}$, Eq. (10.26) can be rewritten as

$$\frac{1}{\tau_{pc}} = \frac{1}{\tau} + a\mu E - a^2 \frac{\mu k_B T}{q}. \tag{10.27}$$

In the above formula, τ can be the PL lifetime of minority carriers. If so, we can establish a clear relationship between PL and PC lifetimes:

$$\frac{1}{\tau_{pc}} = \frac{1}{\tau_{pl}} + a\mu E - a^2 \frac{\mu k_B T}{q}. \tag{10.28}$$

The PL lifetime of local state carriers has been derived in previous chapters, as expressed by Formulas (10.13) or (10.14). From Formula (10.28), it can be seen that the effective lifetime of the photocurrent is related to not only the fluorescence lifetime of the carrier but also a function of the minority carrier diffusion coefficient or mobility, as well as a function of the electric field. These dependencies are predictable because the photocurrent of minority carriers is related to not only their fluorescence lifetime but also the parameters related to control transport. When the electric field $E = 0$ or very small, the diffusion coefficient and mobility of minority carriers can be expressed as

$$D = \frac{1}{a^2}\left(\frac{1}{\tau_{pl}} - \frac{1}{\tau_{pc}} \right), \tag{10.29}$$

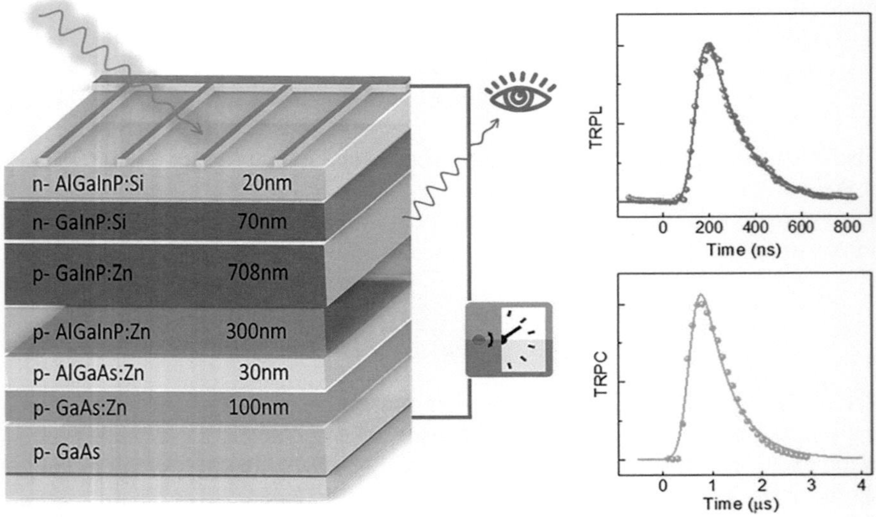

Fig. 10.5 The structure of the studied GaInP solar cell and the schematic diagram of time-resolved fluorescence and time-resolved photocurrent measurements

$$\mu = \frac{q}{a^2 k_B T}\left(\frac{1}{\tau_{pl}} - \frac{1}{\tau_{pc}}\right). \tag{10.30}$$

These formulas suggest that when we can measure the fluorescence lifetime and photocurrent lifetime of minority carriers, we may be able to determine the diffusion coefficient and mobility of minority carriers. Another important conclusion is that when the electric field is zero or very small, the time constant of TRPC is longer than that of TRPL. Such theoretical predictions have been confirmed by our following experiments, as shown in Fig. 10.5.

The effective lifetime of the photocurrent is much longer than the lifetime of photoluminescence. Moreover, the two lifetimes have different dependencies on temperature. For example, for T < 100 K, the effective lifetime of the photocurrent will rapidly decrease with increasing temperature. But when the temperature exceeds 100 K, the effective lifetime of the photocurrent tends to saturate at ~550 ns. The lifetime of photoluminescence changes little when the temperature is below 100 K, but it decreases rapidly when the temperature exceeds 100 K.

As shown in Fig. 10.6, we used the obtained PL and PC effective lifetime data to plot $\ln\left(\frac{1}{\tau_{pl}}\right)$ and $\ln\left(\frac{1}{\tau_{pl}} - \frac{1}{\tau_{pc}}\right)$ or $\ln\left(a^2 D\right)$ regarding $1000/T$ Arrhenius plot. Clearly, 100 K is a critical temperature where the trend of PL lifetime and PC lifetime of minority carriers' changes. For the diffusion coefficient, 100 K is also a critical temperature. Below this temperature, the diffusion coefficient slowly decreases as the temperature rises, but when it is above this temperature, the diffusion coefficient significantly increases with the rise in temperature.

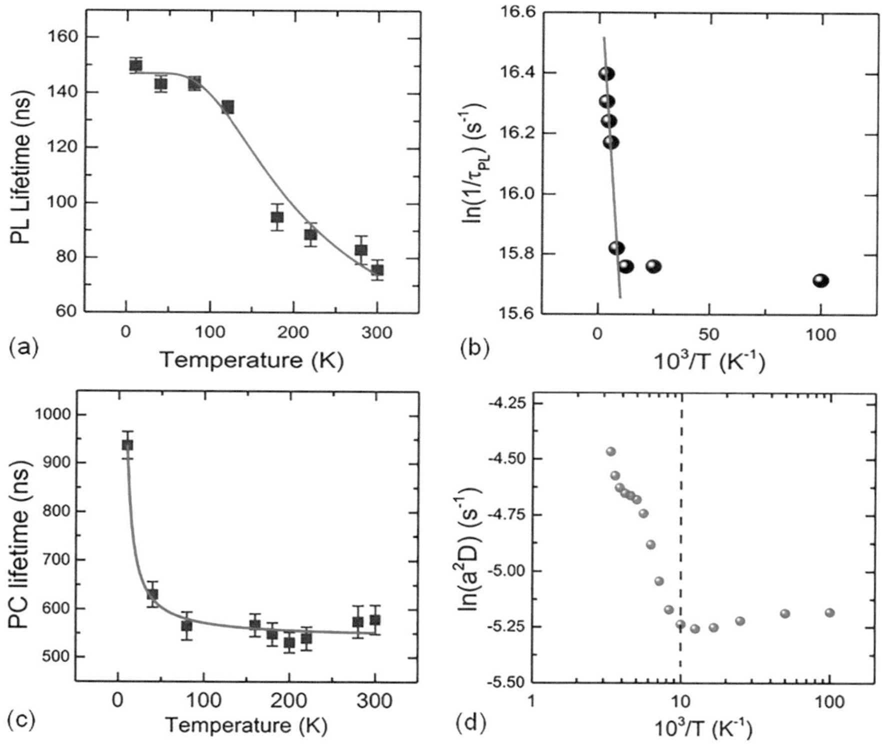

Fig. 10.6 Time-resolved fluorescence and time-resolved photocurrent analysis. **a** Relationship between GaInP photovoltaic cell PL lifetime and temperature. **b** $\ln\left(\frac{1}{\tau_{pl}}\right)$ Arrhenius plot. **c** Relationship between photovoltaic cell PC lifetime and temperature under zero bias. **d** $\ln(a^2D)$ Arrhenius plot

According to related studies, the change in majority carrier concentration with temperature may be a key factor in the abovementioned minority carrier phenomena. In the low-temperature region of <100 K, thermal ionization of impurities will be a major physical process. In the GaInP field samples we studied, the base area is p-type doped. Therefore, in this low-temperature region, the hole concentration may increase exponentially with the rise in temperature. The temperature dependence of photocurrent lifetime can be simply expressed as [60]

$$\tau_c \propto \frac{1}{n_h} \propto e^{(E_f - E_v)/kT}, \tag{10.31}$$

where E_v and E_f are the valence band and conduction band quasi-Fermi levels, respectively, and n is the majority carrier concentration.

As shown in Fig. 10.6c, the temperature dependence of the photocurrent lifetime can be well reproduced by Eq. (10.31), where the solid line is the theoretical fitting line. A relatively small energy difference $E_f - E_v = 0.47meV$ is obtained in the fitting, which means that the hole concentration is very sensitive to the temperature change in the low-temperature ionization zone. When the temperature rises above 100 K, most of the p-type impurities are ionized. When the temperature rises further, the hole concentration basically does not change anymore, showing no longer sensitivity to temperature. The result is shown in Fig. 10.6c, where the photocurrent lifetime of minority carrier electrons tends to be stable.

10.5 Summary

In summary, we have extended the steady-state LSE model for localized carrier collective fluorescence to transient situations under pulse excitation, obtained the analytical theoretical expression for time-resolved fluorescence, and successfully applied it to quantitatively analyze the experimental data of various materials published in the literature, gaining unprecedented deep understanding of the related time-resolved photoelectron processes. In addition, we have developed a temperature- and electric-field-dependent model for minority carrier transient photocurrent in semiconductors and successfully used it to quantitatively explain the temperature- and reverse-bias-related time-resolved photocurrent of GaInP single-junction photovoltaic cells. We have also derived the analytical relationship between the photocurrent and photoluminescence lifetime of minority carriers in semiconductors and obtained the photocurrent lifetime and other key parameters, such as diffusion coefficients. This general model may have a significant impact on the deep understanding of the internal photoelectron physical processes of semiconductor optoelectronic devices and the design and performance improvement of semiconductor optoelectronic devices

References

1. Barrigón E, Heurlin M, Bi Z, Monemar B, Samuelson L (2019) Synthesis and applications of III–V nanowires. Chem Rev 119(15):9170–9220
2. Green MA, Dunlop ED, Hohl-Ebinger J, Yoshita M, Kopidakis N, Ho-Baillie AWY (2020) Solar cell efficiency tables (version 55). Prog Photovoltaics Res Appl 28(1):3–15
3. Yang Q, Kruk S, Xu Y, Wang Q, Srivastava YK, Koshelev K, Kravchenko I, Singh R, Han J, Kivshar Y, Shadrivov I (2020) Mie-resonant membrane huygens' metasurfaces. Adv Func Mater 30(4):1–7
4. Arab H, MohammadNejad S, KhodadadKashi A, Ahadzadeh S (2020) Recent advances in nanowire quantum dot (NWQD) single-photon emitters. Quantum Inf Process 19(2):44
5. Kwak D-H, Ramasamy P, Lee Y-S, Jeong MH, Lee J-S (2019) High-performance hybrid InP QDs/black phosphorus photodetector. ACS Appl Mater Interfaces 11(32):29041–29046

6. Siyushev P, Nesladek M, Bourgeois E, Gulka M, Hruby J, Yamamoto T, Trupke M, Teraji T, Isoya J, Jelezko F (2019) Photoelectrical imaging and coherent spin-state readout of single nitrogen-vacancy centers in diamond. Science 363(6428):728–731
7. Wang L, Zhou H, Hu J, Huang B, Sun M, Dong B, Zheng G, Huang Y, Chen Y, Li L, Xu Z, Li N, Liu Z, Chen Q, Sun L-D, Yan C-H (2019) A Eu^{3+}-Eu^{2+} ion redox shuttle imparts operational durability to Pb-I perovskite solar cells. Science 363(6424):265–270
8. Yang Z, Albrow-Owen T, Cui H, Alexander-Webber J, Gu F, Wang X, Wu TC, Zhuge M, Williams C, Wang P, Zayats AV, Cai W, Dai L, Hofmann S, Overend M, Tong L, Yang Q, Sun Z, Hasan T (2019) Single-nanowire spectrometers. Science 365(6457):1017–1020
9. Rong X, Wang M, Geng J, Qin X, Guo M, Jiao M, Xie Y, Wang P, Huang P, Shi F, Cai YF, Zou C, Du J (2018) Searching for an exotic spin-dependent interaction with a single electron-spin quantum sensor. Nat Commun 9(1):739
10. Long G, Jiang C, Sabatini R, Yang Z, Wei M, Quan LN, Liang Q, Rasmita A, Askerka M, Walters G, Gong X, Xing J, Wen X, Quintero-Bermudez R, Yuan H, Xing G, Wang XR, Song D, Voznyy O, Zhang M, Hoogland S, Gao W, Xiong Q, Sargent EH (2018) Spin control in reduced-dimensional chiral perovskites. Nat Photonics 12:528–533
11. Li X, Rui M, Song J, Shen Z, Zeng H (2015) Carbon and graphene quantum dots for optoelectronic and energy devices: a review. Adv Func Mater 25(31):4929–4947
12. Nagl A, Hemelaar SR, Schirhagl R (2015) Improving surface and defect center chemistry of fluorescent nanodiamonds for imaging purposes—a review. Anal Bioanal Chem 407(25):7521–7536
13. Veldhuis SA, Boix PP, Yantara N, Li M, Sum TC, Mathews N, Mhaisalkar SG (2016) Perovskite materials for light-emitting diodes and lasers. Adv Mater 28(32):6804–6834
14. Kang X, Chong H, Zhu M (2018) Au 25 (SR) 18: the captain of the great nanocluster ship. Nanoscale 10(23):10758–10834
15. Wang YT, Chen MH, Lin CT, Fang JJ, Chang CJ, Luo CW, Yabushita A, Wu KH, Kobayashi T (2015) Use of ultrafast time-resolved spectroscopy to demonstrate the effect of annealing on the performance of P3HT:PCBM solar cells. ACS Appl Mater Interfaces 7(8):4457–4462
16. Hoffman DP, Ellis SR, Mathies RA (2014) Characterization of a conical intersection in a charge-transfer dimer with two-dimensional time-resolved stimulated Raman spectroscopy. J Phys Chem A 118(27):4955–4965
17. Huxter VM, Oliver TA, Budker D, Fleming GR (2013) Vibrational and electronic dynamics of nitrogen-vacancy centres in diamond revealed by two-dimensional ultrafast spectroscopy. Nat Phys 9(11):744–749
18. Mukamel S (1995) Principles of nonlinear optical spectroscopy. Oxford University Press
19. Shah J (1999) Ultrafast spectroscopy of semiconductors and semiconductor nanostructures. Springer-Verlag, Berlin Heidelberg
20. Ulbricht R, Hendry E, Shan J, Heinz TF, Bonn M (2011) Carrier dynamics in semiconductors studied with time-resolved terahertz spectroscopy. Rev Mod Phys 83(2):543–586
21. Richter JM, Branchi F, de Almeida V, Camargo F, Zhao B, Friend RH, Cerullo G, Deschler F (2017) Ultrafast carrier thermalization in lead iodide perovskite probed with two-dimensional electronic spectroscopy. Nat Commun 8(1):1–7
22. Okuda T (2017) Recent trends in spin-resolved photoelectron spectroscopy. J Phys Condens Matter 29(48)
23. Grischkowsky D, Keiding S, Van Exter M, Fattinger C (1990) Far-infrared time-domain spectroscopy with terahertz beams of dielectrics and semiconductors. J Opt Soc Am B 7(10):2006
24. Joyce HJ, Wong-Leung J, Yong CK, Docherty CJ, Paiman S, Gao Q, Tan HH, Jagadish C, Lloyd-Hughes J, Herz LM, Johnston MB (2012) Ultralow surface recombination velocity in InP nanowires probed by terahertz spectroscopy. Nano Lett 12(10):5325–5330
25. Gu J, Yan Y, Young JL, Steirer KX, Neale NR, Turner JA (2016) Water reduction by a p-$GaInP_2$ photoelectrode stabilized by an amorphous TiO_2 coating and a molecular cobalt catalyst. Nat Mater 15(4):456–460

26. Cordoba C, Zeng X, Wolf D, Lubk A, Barrigón E, Borgstrom MT, Kavanagh KL (2019) Three-dimensional imaging of beam-induced biasing of InP/GaInP tunnel diodes. Nano Lett 19(6):3490–3497
27. King RR, Law A, Edmondson KM, Fetzer CM, Kinsey GS, Yoon H, Sherif RA, Karam NH (2007) 40% Efficient metamorphic GaInP/GaInAs/Ge multijunction solar cells. Appl Phys Lett 90(18):183513–183516
28. Yuan X, Li L, Li Z, Wang F, Wang N, Fu L, He J, Tan HH, Jagadish C (2019) Unexpected benefits of stacking faults on the electronic structure and optical emission in wurtzite GaAs/GaInP core/shell nanowires. Nanoscale 11(18):9207–9215
29. Polman A, Knight M, Garnett EC, Ehrler B, Sinke WC (2016) Photovoltaic materials: present efficiencies and future challenges. Science 352(6283):aad4424–aad4424
30. Horner GS, Mascarenhas A, Froyen S, Alonso RG, Bertness K, Olson JM (1993) Photoluminescence-excitation-spectroscopy studies in spontaneously ordered GaInP2. Phys Rev B 47(7):4041–4043
31. Smith S, Cheong HM, Fluegel BD, Geisz JF, Olson JM, Kazmerski LL, Mascarenhas A (1999) Spatially resolved photoluminescence in partially ordered GaInP2. Appl Phys Lett 74(5):706–708
32. Su ZC, Ning JQ, Deng Z, Wang XH, Xu SJ, Wang RX, Lu SL, Dong JR, Yang H (2016) Transition of radiative recombination channels from delocalized states to localized states in a GaInP alloy with partial atomic ordering: a direct optical signature of Mott transition? Nanoscale 8(13):7113–7118
33. Ning JQ, Xu SJ, Deng Z, Su ZC (2014) Polarized and non-polarized photoluminescence of GaInP2 alloy with partial CuPt-type atomic ordering: ordered domains vs. disordered regions. J Mater Chem C 2(30):6119
34. Mott N (1987) The mobility edge since 1967. J Phys C Solid State Phys 20(21):3075–3102
35. Chomette A, Deveaud B, Regreny A, Bastard G (1986) Observation of carrier localization in intentionally disordered Gaas/Gaalas superlattices. Phys Rev Lett 57(12):1464–1467
36. Schömig H, Halm S, Forchel A, Bacher G, Off J, Scholz F (2004) Probing individual localization centers in an InGaN/GaN quantum well. Phys Rev Lett 92(10):106802
37. Glusac K, Köse ME, Jiang H, Schanze KS (2007) Triplet excited state in platinum−acetylide oligomers: triplet localization and effects of conformation. J Phys Chem B 111(5):929–940
38. Roati G, D'Errico C, Fallani L, Fattori M, Fort C, Zaccanti M, Modugno G, Modugno M, Inguscio M (2008) Anderson localization of a non-interacting Bose-Einstein condensate. Nature 453(7197):895–898
39. Schwartz T, Bartal G, Fishman S, Segev M (2007) Transport and Anderson localization in disordered two-dimensional photonic lattices. Nature 446(7131):52–55
40. Wang YJ, Xu SJ, Li Q, Zhao DG, Yang H (2006) Band gap renormalization and carrier localization effects in InGaN/GaN quantum-wells light emitting diodes with Si doped barriers. Appl Phys Lett 88(4):041903
41. Li Z, Kang J, Wei Wang B, Li H, Hsiang Weng Y, Lee YC, Liu Z, Yi X, Chuan Feng Z, Wang G (2014) Two distinct carrier localization in green light-emitting diodes with InGaN/GaN multiple quantum wells. J Appl Phys 115(8):083112
42. Su ZC, Xu SJ, Wang RX, Ning JQ, Dong JR, Lu SL, Yang H (2017) Electroluminescence probe of internal processes of carriers in GaInP single junction solar cell. Sol Energy Mater Sol Cells 168:201–206
43. Li Q, Xu SJ, Cheng WC, Xie MH, Tong SY, Che CM, Yang H (2001) Thermal redistribution of localized excitons and its effect on the luminescence band in InGaN ternary alloys. Appl Phys Lett 79(12):1810–1812
44. Li Q, Xu SJ, Xie MH, Tong SY (2005) A model for steady-state luminescence of localized-state ensemble. Eur Lett (EPL) 71(6):994–1000
45. Li Q, Xu SJ, Xie MH, Tong SY (2005) Origin of the 'S-Shaped' temperature dependence of luminescent peaks from semiconductors. J Phys Condens Matter 17(30):4853–4858
46. Su ZC, Xu SJ (2017) A generalized model for time-resolved luminescence of localized carriers and applications: dispersive thermodynamics of localized carriers. Sci Rep 7(1):13

47. Su ZC, Xu SJ (2019) Effective lifetimes of minority carriers in time-resolved photocurrent and photoluminescence of a doped semiconductor: modelling of a GaInP solar cell. Sol Energy Mater Sol Cells 193(December 2018):292–297
48. Soukoulis CM, Economou EN (1999) Electronic localization in disordered systems. Waves Random Media 9(2):255–269
49. Lagendijk A, van Tiggelen B, Wiersma DS (2009) Fifty years of Anderson localization. Phys Today 62(8):24–29
50. Chakraborty P, Datta G, Ghatak K (2003) The simple analysis of the burstein-moss shift in degenerate n-Type semiconductors. Physica B 339(4):198–203
51. Okamoto K, Niki I, Scherer A, Narukawa Y, Mukai T, Kawakami Y (2005) Surface plasmon enhanced spontaneous emission rate of InGaN/GaN quantum wells probed by time-resolved photoluminescence spectroscopy. Appl Phys Lett 87(7):071102
52. Zwiller V, Pistol ME, Hessman D, Cederström R, Seifert W, Samuelson L (1999) Time-resolved studies of single semiconductor quantum dots. Phys Rev B 59(7):5021–5025
53. Mair RA, Lin JY, Jiang HX, Jones ED, Allerman AA, Kurtz SR (2000) Time-resolved photoluminescence studies of InxGa1−xAs1−yNy. Appl Phys Lett 76(2):188–190
54. Oueslati M, Benoit C, Zouaghi M (1988) Resonant Raman scattering on localized states due to disorder in GaAs1-xPx. Phys Rev B 37(6):3037–3041
55. Narukawa Y, Kawakami Y, Fujita S, Nakamura S (1999) Dimensionality of excitons in laser-diode structures composed of InGaN. Phys Rev B 59(15):10283–10288
56. Satake A, Masumoto Y, Miyajima T, Asatsuma T, Nakamura F, Ikeda M (1998) Localized exciton and its stimulated emission in surface mode from single-layer InxGa1-xN. Phys Rev B 57(4):R2041–R2044
57. Prechtel L, Song L, Manus S, Schuh D, Wegscheider W, Holleitner AW (2011) Time-resolved picosecond photocurrents in contacted carbon nanotubes. Nano Lett 11(1):269–272
58. Ahrenkiel RK, Johnston SW, Kuciauskas D, Tynan J (2014) Dual-sensor technique for characterization of carrier lifetime decay transients in semiconductors. J Appl Phys 116(21):214510
59. Yuan JS, Liou JJ (1998) Semiconductor Device Physics and Simulation. Springer, US, Boston, MA
60. Wolpert D, Ampadu P (2012) Temperature effects in semiconductors. In: Managing temperature effects in nanoscale adaptive systems, pp 15–33. Springer New York, New York

Chapter 11
Luminescence Properties and Characterization Techniques of InGaN/GaN QDs in Nanowires

Zilan Wang and Zhibiao Hao

11.1 Introduction

Compared to traditional planar structure quantum dots (QDs), indium gallium nitride (InGaN)/gallium nitride (GaN) QDs in nanowires effectively reduce lateral stress and greatly decrease dislocation density and polarization electric field, resulting in superior luminescence performance. By controlling the composition and size of the QDs in nanowires, the direct bandgap can be altered, enabling its luminescence spectrum to span the entire visible light range. Optoelectronic devices based on InGaN/GaN QDs in nanowires have excellent performance, such as high quantum efficiency and low threshold current [1, 2]. In recent years, with the deepening exploration of QDs in nanowires, significant progress has been made in understanding their properties and applications.

Firstly, the growth process of high-quality, small-size, and controllable growth point InGaN/GaN QDs in nanowires is the prerequisite for preparing high-performance devices. In 2016, Professor Deng Hui's team achieved effective control over the position and size of InGaN/GaN QDs in nanowires and manufactured a 90 × 90 array of InGaN/GaN QDs with a diameter of 30 nm and a length of 230 nm using a top-down method. They then used this array to prepare an electrically pumped InGaN/GaN QDs in nanowire single-photon emission diode, achieving chip-level integration of InGaN/GaN QDs in nanowires [3]. In the same year, Puchtler Tim J. and others used a self-assembly method to achieve the growth of m-plane QDs on the sidewalls of GaN nanowires for the first time, alleviating planar strain, reducing defects, and

Z. Wang (✉)
School of Optoelectronic Engineering and Instrumentation Science, Dalian University of Technology, Dalian, China
e-mail: wangzilan@dlut.edu.cn

Z. Hao
Department of Electronic Engineering, Tsinghua University, Beijing, China

S. Xu (ed.), *Optical Characterization of Microstructures and Optoelectronic Devices Based on Wide Band Gap Semiconductors*, Wide Bandgap Semiconductors,
https://doi.org/10.1007/978-981-95-1928-6_11

improving the efficiency of extracting light from the end of the nanowires. In the study of single-photon emission from InGaN QDs, these QDs showed very short recombination lifetimes (as low as 170 ps), high linear polarization, and anti-bunching effects [4].

Loading different QDs on a single nanowire to achieve white light emission and integrating full-color LEDs into chips to make smart lighting, projection, display, and other devices have become a reality [5]. In 2016, Professor Zetian Mi's team prepared an InGaN/GaN QD in nanowire light-emitting diode integrated with an Al tunnel junction, which solved the problems of undesirable light absorption and high voltage loss in traditional GaN-based tunnel junction designs. This device has low resistance, high hole injection rate, and an operating voltage of only 2.9 V and can achieve low-resistance, high-brightness light-emitting diodes in the visible and deep ultraviolet spectral range [6]. In the same year, the team conducted research on full-color InGaN/GaN QDs in nanowire light-emitting diodes. By selectively epitaxially growing nanowires, they made an array of InGaN/GaN QDs in nanowires of different diameters, demonstrating a multicolor single-nanowire LED array that covers almost the entire visible spectrum. The device's operating voltage was further reduced to 2 V, and the leakage current under reverse bias was negligible [5]. In 2020, Professor Nguyen Hieu P T's team demonstrated a high-efficiency white LED using InGaN/GaN QDs in nanowires plus two additional InGaN quantum wells. Using electrically coupled InGaN quantum wells can reduce electron leakage into the p-GaN section, enhance the ability of the active region to capture electrons, and eliminate light generated outside the active region. The resulting QDs in nanowire white light-emitting diode has an internal quantum efficiency (IQE) of about 58.5%, with a fairly stable white light emission [7].

In addition to micro-/nanostructured LEDs, InGaN/GaN QDs in nanowires also show excellent characteristics in multiple fields such as laser diodes [8], photovoltaic devices [9], and detectors [10]. In 2019, Masaya Professor Notomi's team has meticulously controlled the position, geometric shape, size, and macroscale of InGaN/GaN QDs in nanowires to create a 10×10 array of InGaN/GaN QDs in nanowires with varying geometric shapes, each with a diameter of 300 nm and a length of 4 μm. They have achieved a nanowire array on the same chip that individually supports room-temperature lasers based on Fabry–Perot resonance, paving the way for integrated arrays of photon and polariton nanolasers based on nanowires emitting in the ultraviolet and visible range [11]. In the same year, Aiello and others measured and analyzed the characteristics of a visible light ($\lambda \sim 550$ nm) InGaN/GaN nanowire disk array photodetector. The nanowire array was grown on a (001) silicon substrate using plasma-assisted molecular beam epitaxy, and a single elongated QDs was formed through strain relaxation. Under excitation conditions of 300 W/cm^2, a photoluminescence (PL) peak was observed at 565 nm at room temperature, with a linewidth of about 280 meV. The device also showed a very large gain (up to 10^{3} under a 3 V bias), and this device is expected to be used for silicon-based infrared (IR) detectors based on InGaN sub-bandgap transitions [10]. As the level of technology and device performance improves, research on the luminescence mechanism of QDs is also advancing, extending the application of luminescence and providing theoretical guidance for process and device improvements. In 2017, Professor Luo Yi's team used ultraviolet photoelectron spectroscopy (UPS) to describe the band bending effect in InGaN/GaN QD in nanowires coated with different dielectric layers. They also studied their passivation mechanism through time-resolved PL, proving that the deposition of the dielectric layer effectively changed the surface state of the nanowires, thereby weakening or enhancing the band bending near the surface, resulting in a significant increase in the luminescence intensity of the passivated nanowires [12]. In the same year, Cameron Nelson and others used frequency-domain nonlinear spectroscopy, combined with low-temperature and room-temperature conditions, to conduct a series of measurements on the red-light-emitting InGaN disks in GaN nanowires (DINW) grown along the polar c-axis on a silicon substrate. They found a strong resonance in the nonlinear absorption spectrum of the InGaN disks in the GaN nanowires, with a decay rate at low temperatures on the microsecond scale, indicating that the dynamics related to the nonlinear signal are mainly dominated by metastable traps. After mitigating the combined effects of the nonresonant background and metastable traps, this structure can contribute to future optical applications [13]. In recent years, due to the outstanding advantages of InGaN/GaN QDs in nanowire heterostructures, applications based on quantum optics have made considerable progress. QDs in nanowire heterostructures have a high photon collection rate and can be designed to prepare entangled photon pair emitters, which are widely used in quantum information processing and quantum cryptography schemes [14], promoting rapid development in the field of quantum optics [15]. In 2020, Md Rezaul Karim Nishat and others used the full configuration interaction (FCI) method and the 10-band (sp^3s^*-spin) tight-binding model to calculate the exciton energy and fine structure splitting (FSS) of InGaN/GaN QDs in nanowire devices. Since the FSS of the biexciton state will destroy the entanglement characteristics of the photon pair, it needs to be eliminated; the nonpolar m-plane structure has a lower built-in electric field and shows a very small FSS (within 10< (in the μeV range), making it an ideal candidate for the development of entangled photon pair generators (EPPGs) for quantum key distribution (QKD) [16].

This chapter systematically introduces the photoluminescence characteristics and characterization methods of InGaN/GaN QDs in nanowires. The first section introduces the basic luminescence characteristics of QDs in nanowires, including the influence of different growth conditions on the wavelength of PL and the decay characteristics at different measurement temperatures. The second section discusses the luminescence characteristics of QDs in nanowires affected by surface states and introduces the band bending model, surface-state passivation methods, and characterization techniques. The third section explores the localized luminescence characteristics of QDs in nanowires and compares the effects of different surface treatments on the localized distribution. Finally, the future development prospects of QDs in nanowires are proposed.

11.2 Luminescence Characteristics of InGaN/GaN QDss in Nanowires

This section focuses on the luminescence characteristics of InGaN/GaN QDs in nanowires. They are grown on Si(111) substrates using plasma-assisted molecular beam epitaxy (PA-MBE) technology. The scanning electron microscope (SEM) image shown in Fig. 11.1a shows evenly arranged nanowires on the silicon wafer substrate, with an average height of about 300 nm and a density of about 10^{10} cm^{-2}. The diameter distribution (b) shows that the diameter of most nanowires is 25–35 nm. Figure 11.1c and d shows the high-angle annular dark field scanning transmission electron microscope (HAADF-STEM) and in situ energy dispersive spectrometer (EDS) images, respectively. The highlighted areas represent InGaN QDs, with a thickness of about 2 nm, and the In component is evenly distributed as shown in the figure [17].

11.2.1 Relationship Between the Spectrum of InGaN/GaN QDss in Nanowires and Growth Conditions

During the growth process, the emission band of InGaN/GaN QDs in nanowires can be controlled by adjusting conditions such as the In component and growth temperature. A variable-temperature PL system can be used to measure the luminescence spectrum of InGaN/GaN QDs in nanowires, where the excitation wavelength is 325-nm laser, the focal length of the spectrometer is 550 mm, and the minimum resolution is 0.5 nm. Figure 11.2 compares the PL spectra of QDs in nanowires prepared

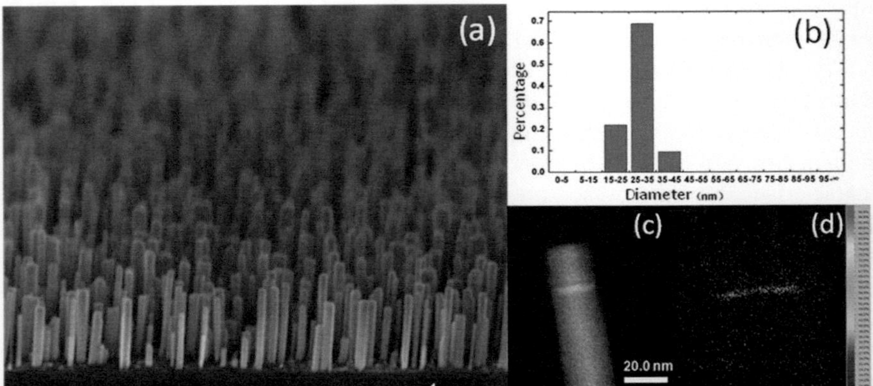

Fig. 11.1 Cross-sectional SEM view and single-nanowire HAADF-STEM image of InGaN/GaN QDs in nanowires [17]

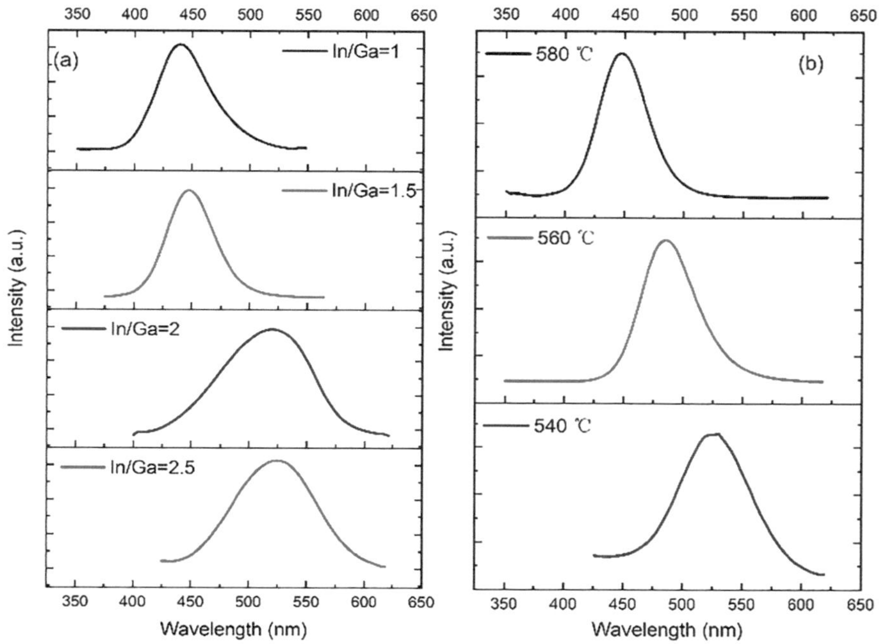

Fig. 11.2 Effects of different In components (**a**) and different growth temperatures (**b**) on the emission of QDs in nanowires

under different In:Ga component ratios (a) and different growth temperatures (b). The diameters of the nanowires shown in the figure are basically the same, and the spectral measurement temperature is a low temperature condition of 10 K. It can be seen that by controlling the material composition and growth temperature, InGaN/GaN QDs in nanowires exhibit different emission wavelengths, which is convenient for different applications.

11.2.2 Relationship Between the Spectrum of InGaN/GaN QDss in Nanowires and Measurement Temperature

Temperature-dependent PL spectra provide insights into the impact of thermal effects on carrier recombination at the microscopic level. As shown in Fig. 11.3, during the process of increasing the temperature of InGaN/GaN QDs in nanowires from 10 to 300 K, the PL intensity gradually decreases, the spectral linewidth gradually broadens, and the peak position gradually redshifts. The PL peak position is 448 nm at 10 K and gradually moves to 458 nm at 300 K. This characteristic of the peak position monotonically redshifting with increasing temperature conforms to the band contraction rule described by the Varshni formula.

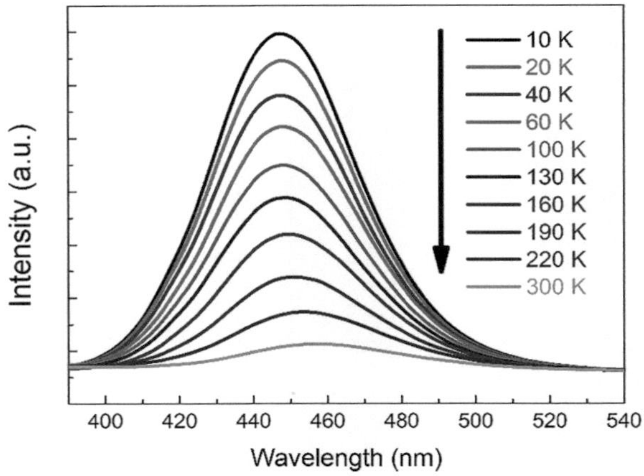

Fig. 11.3 Variable-temperature PL spectrum of QDs in nanowires

By statistically analyzing the decay law of the integral intensity of PL, that is, the Arrhenius decay curve, the change law of the carrier recombination process affected by thermal energy can be analyzed. According to the trend of the integral intensity of the PL peak with temperature in Fig. 11.3, the decay curve of the spectral integral intensity with temperature is drawn, as shown below.

As shown in Fig. 11.4, the black scatter points are the integral intensities of the PL peaks obtained at various temperatures. As the sample temperature increases, the integral intensity of PL (data points from right to left) gradually decreases. It is generally believed that the radiation efficiency is close to 100% at low temperatures. As the sample temperature increases, nonradiative recombination centers will be thermally activated, causing the PL intensity to decrease. The thermal activation energy of the nonradiative recombination center can be obtained through Arrhenius fitting (red line), and the source of the nonradiative recombination center can be analyzed. The fitting formula is as follows:

$$I(T) = \frac{I_0}{1 + A_1 \exp\left(-\frac{E_1}{k_B T}\right) + A_2 \exp\left(-\frac{E_2}{k_B T}\right)} \tag{11.1}$$

The coefficient of determination for the fit is 0.998, indicating that the experimental results are consistent with the model. According to the Arrhenius fit, as the temperature increases, the decay of the luminescence intensity includes two channels, A_1 and A_2 represent the pre-exponential factors of the two exponential processes, E_1 and E_2 represent the thermal activation energies of the two channels, I_0 represents the integrated PL intensity at $T = 0$ K, and k_B is the Boltzmann constant. The sources of the two nonradiative centers can be analyzed as follows: From the perspective of energy size, E_1 is 7.8 meV, E_2 is 48.2 meV, and both activation energies are much smaller than the potential barrier energy of the quantum well and may be related to

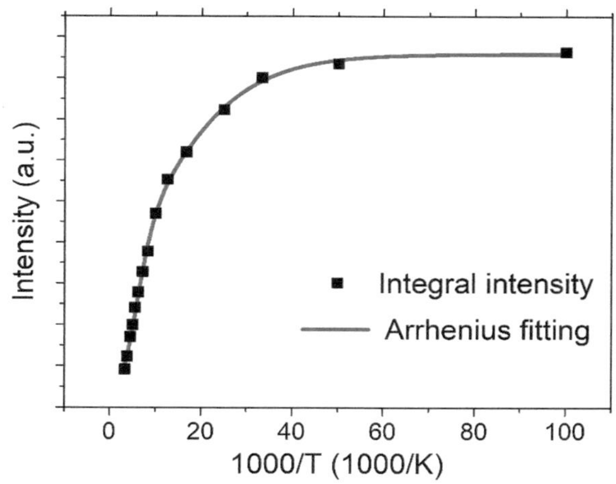

Fig. 11.4 Decay curve and Arrhenius fitting of the integral intensity of variable-temperature PL spectrum of QDs in nanowires

defects, potential fluctuations, composition fluctuations, and interface effects. The smaller E_1 is usually considered to originate from the interface effects of the sample [18], while the larger E_2 is often attributed to the effect of dislocations[19, 20]. However, due to the structural characteristics of QDs in nanowires, there are no dislocation defects [21], and considering the large number of surface states on the surface of the nanowires, the source of E_2 may be surface defects of the nanowires. IQE is defined as the ratio of the integrated emission intensity at room temperature to that at low temperature. The IQE of this QDs in nanowire is about 10%. Therefore, reducing the number of surface states of QDs in nanowires is the key to improving the IQE of QDs emission. The characterization and manipulation mechanism of surface defects of QDs in nanowires will be discussed in detail in the next section.

11.3 Luminescence Characteristics of InGaN/GaN QDss in Nanowires Affected by Surface States

11.3.1 Mechanism of Surface States on Luminescence Characteristics

Due to the small size of the nanowires and the large surface area/volume ratio, there are a large number of dangling bonds and various unintentional doping defects, as well as surface damage caused during the preparation process. These energy states present on the surface of nanomaterials can be called surface states. Surface states usually act as nonradiative recombination centers in the emission process, leading to a decrease in PL intensity and a decrease in thermal stability, causing PL quenching at higher temperatures. In addition, the Fermi-level pinning effect of surface states

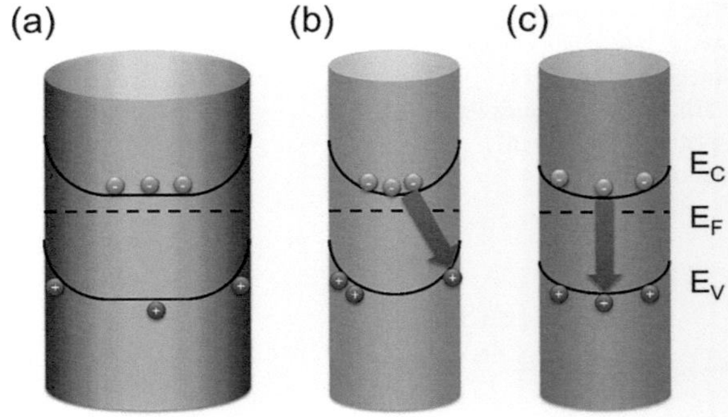

Fig. 11.5 Schematic diagram of band bending in GaN-based nanowires

can also cause the surface of the nanowire to be in a depletion zone within a range of tens of nanometers, thereby causing the energy band of the nanowire surface to bend [12], as shown in Fig. 11.5.

Figure 11.5 shows the band bending of nanowire surfaces under different conditions, where (a) represents a nanowire with a larger diameter and (b) and (c) represent nanowires with smaller diameters and different degrees of band bending. The figure indicates the distribution of negatively charged electrons and positively charged holes on the band, with arrows representing the recombination transition process. The positions of the conduction band (E_C), valence band (E_V), and Fermi level (E_F) are not absolute. The presence of a large number of surface states leads to a series of changes in surface characteristics, such as the depletion of surface free charges [22], spatial charge separation, and Fermi-level pinning at the surface [23]. Due to the charged characteristics of GaN-based nanowires, surface states cause the band to bend upward. As shown in Fig. 11.5a, for nanowires with diameters larger than 80 nm, band bending only occurs near the surface, while the position near the center is similar to normal bulk materials, and the position of the Fermi level is determined by the concentration of carriers. For nanowires with smaller diameters, the depletion region occupies the entire nanowire, as shown in Fig. 11.5b. In this study, the diameter of InGaN/GaN QDs in nanowires is less than 30 nm, the band bending phenomenon is significant, and the entire nanowire is in the depletion region. Under such circumstances, in the photo-generated carriers, electrons tend to stay at the bottom of the conduction band in the center of the nanowire, while holes tend to gather at the top of the valence band on the outside of the nanowire, causing the electron and hole wave functions to separate in space. This ultimately leads to a decrease in carrier recombination luminescence efficiency, a decrease in transition energy (slanted arrow), and a redshift in the PL spectrum. When the surface states decrease, as shown in Fig. 11.5c, the band bending phenomenon weakens, the carriers are redistributed, the overlap integral of the electron and hole wave functions increases, the radiation efficiency improves, the transition energy increases (vertical downward arrow), and the PL spectrum shows a blueshift. The next section will discuss common surface-state passivation methods and their mechanisms of action.

11.3.2 Passivation Methods and Impact on Luminescence Characteristics

Common nanowire surface passivation methods usually require changing the surface morphology of the nanowires. In order to ensure that the morphological features of the QDs in nanowires are not affected by passivation, an ultrathin dielectric film coating method is chosen here, selecting commonly used GaN-based materials such as silicon nitride (SiN_x), silicon dioxide (SiO_2), and aluminum oxide (Al_2O_3) as passivation materials, to coat the nanowire surface and compare the effects. After multiple experiments, the growth conditions and film thickness of the passivation layer were optimized, and the deposition thickness of the passivation material was finally set to about 20 nm, with the radial thickness of the passivation layer on the nanowire less than 10 nm, to ensure that the morphological features of the QDs in nanowires are not changed. The Al_2O_3 film deposition used the atomic layer deposition (ALD) method, with a substrate temperature of 200 °C, using trimethyl aluminum (TMA) and water as precursors for 200 growth cycles. SiN_x and SiO_2 thin-film deposition uses plasma-enhanced chemical vapor deposition (PECVD) with a substrate temperature of 300 °C. The following figure compares the morphology of unpassivated and passivated QDs in nanowires observed under a scanning electron microscope. To clearly show the edges of the nanowires, we specifically selected parts where the nanowires are relatively sparse.

Figure 11.6 shows the morphology comparison of cluster nanowires and single nanowires before and after passivation. Among them, (a) and (d) are cluster nanowires, (b) and (e) are single nanowires, (c) and (f) are backscattered images of (b) and (e), and images (b), (c), (e), and (f) share a scale. The passivation layer is a 20-nm alumina thin film. Comparing the two images, the passivation effect can be seen. In the backscattered images, the brightness difference corresponds to different materials, the darker surface part is the coated passivation layer, and the brighter internal part is the GaN nanowire. From the surface coating effect, the radial thickness of the passivation layer is less than 10 nm, and the diameter of the nanowires does not change significantly in the electron microscope, ensuring that the original three-dimensional confinement effect of the QDs and the size effect of the nanowires remain unchanged.

The PL characterization method before and after passivation is the same as the basic characterization of QDs in nanowires. In order to accurately compare the PL intensity, it is necessary to ensure that the optical path conditions are completely consistent during the measurement process. The PL excitation wavelength used here is 405 nm, and the photoluminescence spectra at low temperature (10 K) and room temperature (300 K) are compared under the same optical path, as shown in Fig. 11.7.

For detailed spectral parameter information, see the table below.

Figures 11.7 and Table 11.1 show the PL of QDs in nanowire clusters under the action of different passivation materials, including spectral graphs under different measurement conditions at room temperature and low temperature. The blue line

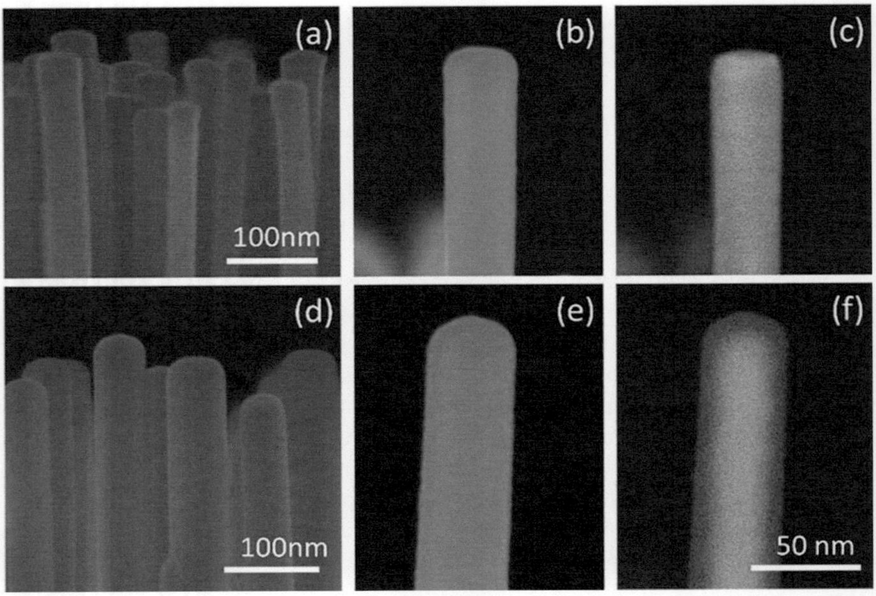

Fig. 11.6 Comparison of scanning electron microscope images of unpassivated nanowires (**a**–**c**) and alumina-passivated nanowires (**d**–**f**)

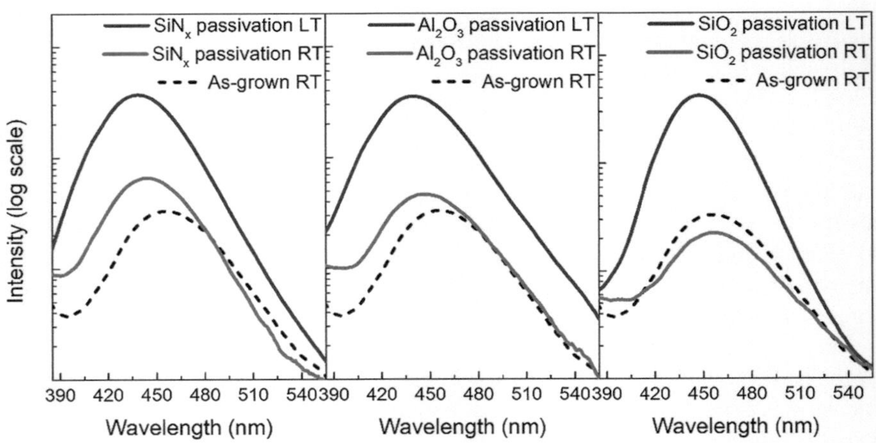

Fig. 11.7 Comparison of PL spectra before and after passivation of SiN_x, Al_2O_3, and SiO_2 [12]

represents the spectrum at low temperature (LT), the red line represents the spectrum at room temperature (RT), and the black dashed line represents the spectrum of the unpassivated sample at room temperature. For easy comparison of intensity, the vertical axis uses a logarithmic scale. Using the PL spectrum of unpassivated (as-grown RT) at room temperature as a standard, it can be clearly seen that the emission intensity of SiN_x and Al_2O_3 increases after passivation, and there is a significant blueshift. However, the luminescence intensity of the SiO_2-passivated sample did not increase, and there was no blueshift. Considering the changes in refractive index and light extraction efficiency introduced by the coating material, we used the transfer matrix method to calculate the transmittance under different passivation conditions. When the passivation film thickness is below 20 nm, the impact of different passivation layers on light extraction efficiency is within 1%. Therefore, the change in emission intensity shown in the figure should primarily be attributed to the surface passivation effect of different materials. Moreover, this experiment has been repeatedly verified, and QDs in nanowires of different luminescent wavelengths have obtained consistent results.

The IQE in Table 11.1 is calculated according to the currently commonly used method, which assumes that nonradiative centers are frozen at low temperatures, carriers do not participate in nonradiative recombination, and the efficiency is approximately 100%. Therefore, IQE is the ratio of room-temperature emission integral intensity to low temperature, and IQE can be used as an important indicator to measure the luminescent performance of QDs in nanowires. Using SiN_x-passivated QDs, the IQE has doubled, indicating that the passivation method effectively reduces nonradiative recombination and improves luminescence performance.

In addition, the passivation process through surface coating only affects the surface area of the nanowires. According to the changes in PL, the surface state and surface manipulation can be further analyzed in the role of luminescent recombination. From the comparison of emission before and after passivation, it can be seen that the QDs in nanowires passivated with Al_2O_3 and SiN_x not only increases in intensity but also exhibits a significant blueshift. This is not only due to the reduction in the density of surface states but also related to the reduction in the degree of band bending. As shown in Fig. 11.5, the repair of band bending is accompanied by an increase in transition energy, and in this experiment, the blueshift can reach up to 70 meV. At the same time, the overlap integral of the electron and hole wave functions increases, which is conducive to radiative recombination and enhances luminescence. The sample passivated with SiO_2 does not show a significant change in the luminescent peak position, indicating that the passivation material has no obvious repair effect on the band, so the PL shows the enhancement of luminescence and peak shift completely match the trend of band bending changes. The next section

Table 11.1 Peak position and internal quantum efficiency of QDs in nanowires before and after passivation

	Peak position (RT) (nm)	Peak position (10K) (nm)	IQE (%)
As-grown	457.5	447.5	10.23
SiNx passivation	445	439	19.23
Al_2O_3 passivation	449	440	14.88
SiO_2 passivation	458	447	7.39

will validate and quantitatively characterize the theory of surface band modification and the mechanism of surface passivation based on the experiment.

11.3.3 Characterization of Surface Band Bending of QDs in Nanowires

To study the surface of nanostructures, surface-sensitive characterization methods are needed. Therefore, to characterize the band bending characteristics of nanowire surfaces, this study uses ultraviolet photoemission spectroscopy (UPS) as a detection method. UPS uses ultraviolet light to excite electrons inside the material to emit, and the valence electron structure inside the material can be inferred by collecting and scanning the electron energy spectrum. Its detection depth is between 1 and 3 nm, which is extremely sensitive to the sample surface. The UPS used in this experiment is integrated on the Kratos Axis Ultra delay-line detector (DLD) X-ray photoelectron spectroscopy, the excitation source energy is 21.22 eV He I spectral line, the background vacuum is below 10^{-10} Torr, gold substrate is used for Fermi energy level position calibration, scanning range is $-2 \sim 23$ eV, and resolution is 0.025 eV. In order to compare the regulatory effects of different passivation materials on the surface of nanowires, the UPS test samples selected in this experiment include unpassivated nanowires and nanowire samples coated with ultrathin passivation layers [12], and the UPS test results of different surface band structures are shown in the following figure:

Figure 11.8 shows the energy distribution of emitted electrons from different nanowires under He light source irradiation, with the spectral lines being translated longitudinally. To clearly show the cutoff position on the left, the intensity of the left cutoff edge has been amplified five times. The horizontal axis corresponds to the binding energy of different valence electrons, where the energy of the cutoff edge of the left spectral line, that is, the intersection of the reverse extension line (dashed line) of the rising section in the figure and the horizontal line (dashed line), corresponds to the energy difference (E_F–E_V) from the Fermi level to the valence band [24]. Due to different surface treatments, the degree of band bending varies, so the energy difference of the Fermi level relative to the valence band also differs. As shown in Fig. 11.5, due to the Fermi-level pinning effect, nanowires with larger band bending have the Fermi level closer to the top of the valence band; that is, the value of E_F–E_V is smaller; while for nanowires with smaller band bending, the value of E_F–E_V is larger. Therefore, the band bending situation of the nanowire surface can be inferred from the low-energy cutoff edge of the UPS spectrum. For unpassivated samples, the energy from the Fermi level to the valence band is approximately 2.099 ± 0.050 eV. For nanowire samples coated with different passivation films, the relative position of the energy levels has shifted compared to the unpassivated samples. The values of E_F–E_V for samples coated with SiN_x and Al_2O_3 have increased, indicating that the degree of band bending on the nanowire surface has decreased; that is, the band has

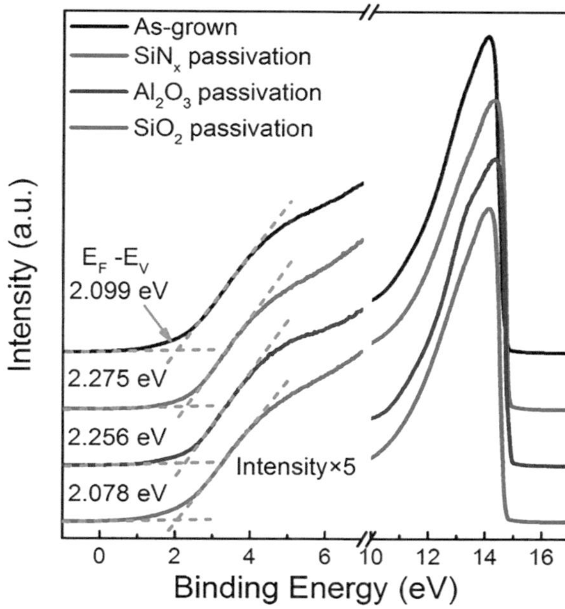

Fig. 11.8 UPS spectra of unpassivated and passivated nanowire samples[12]

become "flatter." For samples coated with SiO_2, the value of E_F–E_V has decreased, indicating an increase in the degree of band bending.

11.3.4 The Impact of Surface Passivation on PL Lifetime

The density of surface states and the issue of band bending in QDs in nanowires also affect the dynamic characteristics of emission. Time-resolved photoluminescence (TRPL) can reflect the decay curve of PL lifetime over time under ultrafast pulse excitation of QDs, corresponding to the recombination lifetime of different surface-controlled download carriers. The excitation source for the TRPL used in this experiment is a Ti:Sapphire laser, with a pulse width of 170 fs and a repetition rate of 80 MHz. The excitation wavelength is chosen to be 405 nm, with the aim of only exciting InGaN QDs luminescence, while avoiding exciting the intrinsic luminescence of GaN nanowires. A single-photon avalanche detector is used to collect signals, with a resolution of 50 ps. Figure 11.9 shows the decay curve of PL intensity over time. To clearly show the differences, the spectral lines have been translated on the vertical axis [12]. Due to the effects of polarization and other factors in InGaN/GaN QDs in nanowires, the decay trend does not conform to a single exponential decay [25]. To obtain the PL lifetime value, the stretched exponential model is adopted here, with the expression as follows:

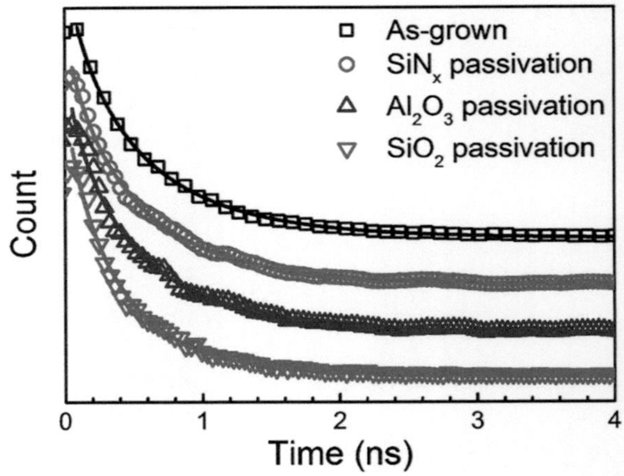

Fig. 11.9 TRPL spectra of unpassivated and differently passivated QDs in nanowires

$$\mathrm{I}(t) = I_0 \exp\left[-\left(\frac{t}{\tau}\right)^{\beta}\right] \tag{11.2}$$

where I_0 is the photon count at $t = 0$ and τ is the PL lifetime of the carrier. A curvature factor β is introduced into the expression to explain the non-exponential decay, which is related to the size of the polarization electric field in the QDs [25]. According to the fitting results, the value of β is around 0.8, indicating a smaller polarization electric field in the QDs. This is because the axial stress is released during the growth of the nanowire. The fitting results are shown in the table.

The τ listed in Table 11.2, obtained by fitting with the extended exponential model, is the PL lifetime of the QDs in nanowires obtained by fitting the TRPL results with the extended exponential model. The radiative recombination lifetime τ_r and nonradiative recombination τ_{nr} can be obtained from the following formulas:

$$\frac{1}{\tau} = \frac{1}{\tau_r} + \frac{1}{\tau_{nr}},\ \eta_{in} = \frac{\tau_{nr}}{(\tau_r + \tau_{nr})} \tag{11.3}$$

where η_{in} is the IQE, refer to the experimental data in Table 11.1. From the calculation results, it can be seen that compared with the unpassivated QDs in nanowires, the PL radiative recombination lifetime of SiN_x– and Al_2O_3-passivated becomes shorter

Table 11.2 Carrier lifetime, radiative recombination lifetime, and nonradiative recombination lifetime of QDs in nanowires with different surface treatments

	τ_{nr} (ps)	τ_r (ps)	τ_{nr} (ps)
As-grown	439	4294	489
SiNx passivation	460	2392	570
Al_2O_3 passivation	454	3065	533
SiO_2 passivation	341	4618	368

and nonradiative recombination becomes longer. However, SiO_2 passivation shows the opposite trend. This once again verifies the band bending theory. The surface treatment of SiN_x and Al_2O_3 makes the energy band of the nanowire surface flat, the overlap integral of the electron–hole wave function increases, the recombination efficiency increases, so the radiative recombination lifetime becomes shorter. At the same time, due to the reduction of surface states, the concentration of nonradiative centers decreases, leading to a longer nonradiative recombination lifetime. Among them, SiN_x passivation is more effective in reducing the density of surface states. Conversely, the SiO_2 passivation process may introduce different nonradiative centers, not achieving the desired effect.

11.4 Localized Luminescence Characteristics of InGaN/GaN QDss in Nanowires

The localized distribution will affect the recombination mode of carriers in QDs in nanowires, PL lifetime, quantum efficiency, etc. One of the obvious signs is the special line shape change of the PL peak with temperature, which includes S-type movement, V-type movement, etc. And after the surface treatment of the QDs in nanowires, the localized-state luminescence of the QDs has changed significantly, showing a significant surface-dependent characteristic of the localized-state distribution. This section selects QDs in nanowires prepared by the top-down etching method from a single-layer quantum well structure. During the etching process, due to the enhancement of quantum confinement effect and the release of stress, the emission wavelength of the QDs is different from the original quantum well, but both are affected by the localized-state distribution. This section will elaborate on the localized PL characteristics of QDs in nanowires.

11.4.1 The Impact of Surface Manipulation on the Luminescence of QDss in Nanowire

To demonstrate the luminescent characteristics of the localized states of QDs in nanowires, we selected QDs in nanowires prepared by the top-down method and those treated with different surface treatments. By comparing the trend of PL spectrum peak position with temperature, we analyzed the differences in the distribution of localized states. As shown in the figure below, Fig. 11.10a is an electron microscope image of QDs in nanowires, with a height of about 120 nm, a diameter of 30 nm, and a well layer of 2 nm. Figure 11.10b is a schematic diagram of the preparation of QDs in nanowires and different surface treatments. The method for fabricating QDs in nanowires through quantum well etching is detailed in reference [26]. The samples treated with a strong alkaline etching solution (etched QD [EQD]) and those coated with an aluminum oxide passivation film (passivated QD [PQD]) are described in the surface treatment methods in reference [27].

Figure 11.11 shows the PL spectra of top-down prepared QDs, QDs in nanowires after alkali solution corrosion (EQD), and passivated QDs in nanowires (PQD), with PL measurement temperature at 10 K. According to the position of the emission peak, the spectrum can be divided into two parts: The ultraviolet emission peak near 3.263 eV [28] comes from the intrinsic emission of GaN, and the two smaller peaks with lower energy are the phonon replicas of intrinsic emission; the blue PL band near 2.8–2.9 eV comes from the emission of InGaN QDs, which is the focus of this section.

For the untreated QDs, their emission peak (black curve) is at 2.901 eV, which is blueshifted compared to the single quantum well of 2.737 eV during growth. This shift in peak energy is mainly due to the increase in quantum confinement effect and

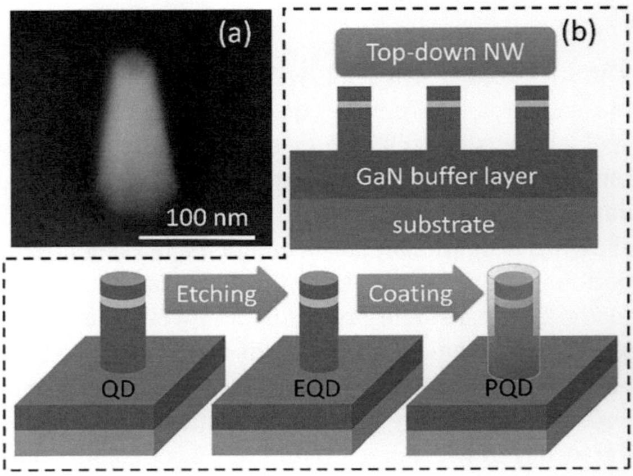

Fig. 11.10 Schematic diagram of the preparation of QDs in nanowires [27]

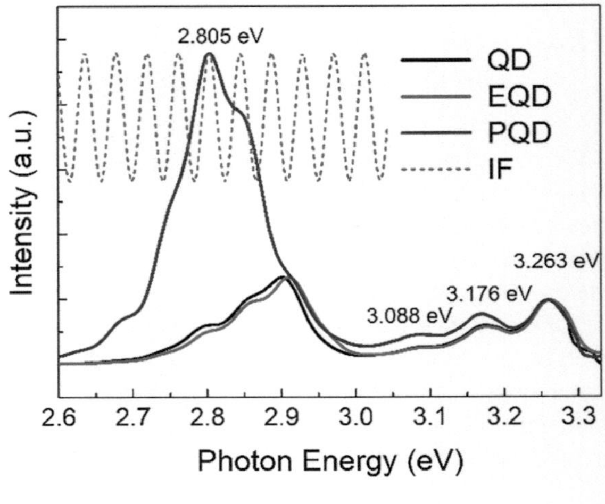

Fig. 11.11 PL spectra of QDs in nanowires (QD, EQD, and PQD)

the release of stress caused by top-down etching [26]. After further wet corrosion in KOH solution (EQD), its emission intensity has hardly changed, only a slight blueshift in the PL peak position, which is consistent with the further increase in quantum confinement effect. Finally, the spectrum of QDs passivated with Al_2O_3 (PQD) is significantly different from the other two spectra, with an increase in emission intensity by five times and a significant redshift in the emission peak, which may be related to the stress change in the coating layer. The fine structure appearing in the emission peaks shown in Fig. 11.11 is attributed to the interference fringes of PL emission and substrate reflection [29, 30]. The dashed line in the figure represents the interference fringe distribution calculated from the thickness of the buffer layer and the contour of the PQD peak, which can be removed by fitting for more accurate positioning of the emission peak.

Surface passivation not only affects the PL characteristics at low temperatures but also changes the way carriers are localized. The carriers in InGaN QDs are driven by thermal energy to redistribute among different energy states, which is manifested as an abnormal change trend of the PL peak position with temperature in the variable-temperature PL spectrum. Here, the InGaN QDs with different surface treatments at 10–300 K PL measurements were conducted at different temperature ranges, and Fig. 11.12 shows the temperature dependence of four quantum structures: quantum wells (QW), QDs in nanowires (QD), etched QDs (EQD), and passivated QDs (PQD) within the temperature range of 10–300 K. The x-axis represents the measurement temperature, and the y-axis represents the energy range near the PL peak. All PL spectra have been normalized, with different brightness corresponding to the relative intensity of the spectrum. White represents the intensity of the PL peak (normalized intensity >0.98), and the movement trend of the white band with temperature shows the change trend of the PL peak with temperature. Here, in order to more accurately calibrate the position of the PL peak, the interference fringes of each original spectral line have been fitted and removed.

As shown in Fig. 11.12, the trend of the movement of the four PL peaks with temperature does not conform to the traditional semiconductor band shrinkage theory (Varshni), indicating that the carrier recombination characteristics of these quantum structures are mainly based on localized-state luminescence. First, the PL peak of the quantum well (QW) shows an S-shaped shift with temperature; that is, it first redshifts (photon energy decreases) and then blueshifts (photon energy increases), which is consistent with the previously reported GaN-based materials [29]; this is a typical result of localization effects. Although the PL emission wavelength of QD is different from that of QW, its PL peak still shows an S-shaped movement trend, indicating that the carriers of QD are still affected by similar localization distribution despite being subject to stronger quantum restriction effects. However, the peak shift of EQD does not follow the S-shaped trend, the PL peak of EQD in the temperature range of 100–200 K does not blueshift but further redshifts, which means that the carriers in EQD have a different localized-state distribution, even if they have very similar PL spectra at low temperatures (Fig. 11.11). Finally, after Al_2O_3 passivation, PQD shows another new type of peak shift form, which begins to significantly blueshift at a lower temperature stage and then gradually redshifts.

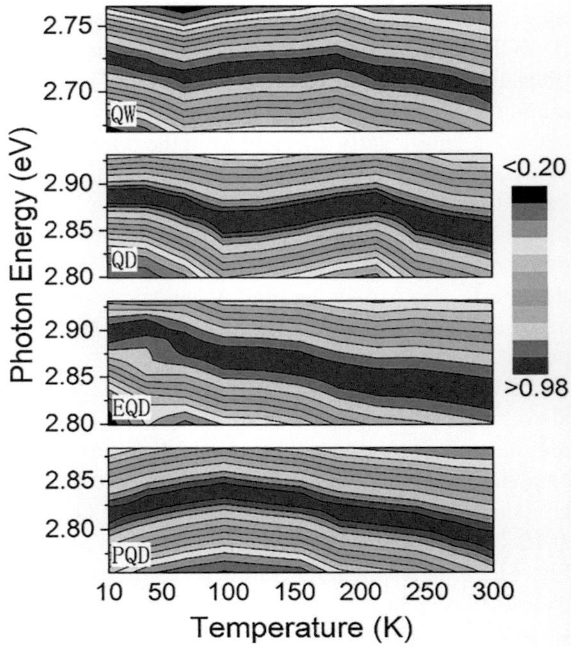

Fig. 11.12 Trend of PL peak shift of four quantum structures [27]

11.4.2 The Model for Emission Characteristics of Localized States

In order to quantitatively explain the special phenomena in the localized-state PL spectrum, Professor Xu Shijie and others proposed the localized state ensemble (LSE) model [31]. Considering an arbitrary localized-state collection system, its state density distribution $\rho(E)$ is Gaussian, and the energy corresponding to the peak of the state density distribution is represented by E_0. When this system is subjected to external excitation such as light irradiation or electrical injection, there will be carriers occupying the localized states, and the occupied energy level is represented by E_a. Therefore, at absolute zero, all localized states below the energy level E_a are occupied by carriers, leading to radiative recombination and thermal escape. If the escaped carriers do not return to their original localized states or are recaptured by other localized states, a nonradiative recombination process is formed, leading to the weakening or even quenching. If the carriers escaping from the localized states are recaptured by other localized states, the carriers can transfer between different localized states through such a process, which will cause the carriers to redistribute in the localized states.

The probability of carrier escape from a localized state is closely related to the energy of the localized state where the carrier is located, the energy difference with the energy level E_a, and the environmental temperature. According to the rate equation, it can be deduced that the distribution of carriers with energy in the localized-state ensemble (LSE), $n(E, T)$, is equal to the product of the localized-state density $\rho(E)$ and a distribution function $f(E, T)$.

$$n(E,T) = \rho(E) \cdot f(E,T) \tag{11.4}$$

In essence, $n(E, T)$ describes the spectral shape of the PL spectrum. This distribution function is

$$f(E,T) = \frac{1}{e^{(E-E_a)/k_B T} + \tau_{tr}/\tau_r} \tag{11.5}$$

where $(E_a–E)$ is the potential barrier that carriers need to overcome to escape from the localized state, k_B is the Boltzmann constant, $1/\tau_r$ is the carrier radiative recombination rate, and $1/\tau_{tr}$ is the attempted escape rate of localized carriers; this distribution is similar to the Fermi–Dirac distribution.

The PL peak can be determined by $n(E, T)$, considering that the bandgap width of the perfect semiconductor varies with temperature can be expressed by the Varshni empirical formula; the final relationship between the peak position of the PL spectrum and temperature is as follows:

$$E(T) = E_0 - \frac{\alpha T^2}{\theta + T} - x \cdot k_B T \tag{11.6}$$

where E_0 is the center energy of the localized-state distribution at 0 K, the second term is the effect of temperature on the bandgap width of the semiconductor material, α is the Varshni coefficient, θ is the Debye temperature of the material, and the third term is the correction brought by the redistribution of the carrier ensemble in the localized state, where x is a dimensionless coefficient, and its specific value is obtained by numerically solving the following nonlinear equation:

$$xe^x = \left[\left(\frac{\sigma}{k_B T}\right)^2 - x\right]\left(\frac{\tau_r}{\tau_{tr}}\right)e^{(E_0-E_a)/k_B T} \tag{11.7}$$

The above equation has exactly one solution, and the solution lies within the range of $0 < x < (\sigma/k_B T)^2$. In the region of higher temperature, the approximate solution is $x \approx (\sigma/k_B T)^2$; then, the formula for the peak position changing with temperature becomes

$$E(T) = E_0 - \frac{\alpha T^2}{\theta + T} - \frac{\sigma^2}{k_B T} \tag{11.8}$$

The above formula is the tail state model proposed by Eliseev and others [32], which is an approximation of this model in the high-temperature region. Therefore, the LSE model has a wider range of applicability and can fit the changes in the

PL peak position in the full-temperature range. In addition to fitting the changes in the PL peak position, this model can also obtain the changes in the spectral half-peak width, light integral intensity, and lifetime parameters with temperature. It has achieved great success in analyzing the spectral characteristics of localized states, quantifying the distribution parameters, escape rate, and carrier filling of localized states. It provides quantitative data support for understanding the abstract concept of localization.

It is worth noting that E_a–E_0 is a crucial parameter in the model, and its size and sign substantially determine the relationship between the change in the PL spectrum peak position and temperature. We divide E_a–$E_0 > 0$ and E_a–$E_0 < 0$ into two cases, corresponding to two physical systems.

Figure 11.13 shows the Gaussian-shaped localized-state density distribution $\rho(E)$, distribution function $f(E, T)$, and carrier distribution $n(E, T)$ under two conditions. The shaded part $n(E, T)$ corresponds to the carrier emission situation, and the temperature change causes the distribution function $f(E, T)$ in the figure to change. In the case of E_a–$E_0 > 0$, as the temperature rises, the carrier distribution $n(E, T)$ moves to a smaller state density. If the band contraction and phonon scattering are not considered, the localized-state emission will cause the peak to redshift, accompanied by a narrowing of the high-energy side spectral line shape. In the other case of E_a–$E_0 < 0$, at 0 K, the shaded part of the carrier distribution is at a smaller state density. As the temperature rises, the carrier distribution $n(E, T)$ moves to a smaller state density, corresponding to a blueshift of the localized-state emission peak, and the high-energy side spectral line will significantly broaden. In this study, we respectively call the localized-state center that causes the redshift under the condition of E_a–$E_0 > 0$ and the localized-state center that causes the blueshift under the condition of E_a–$E_0 < 0$, the red center, and blue center.

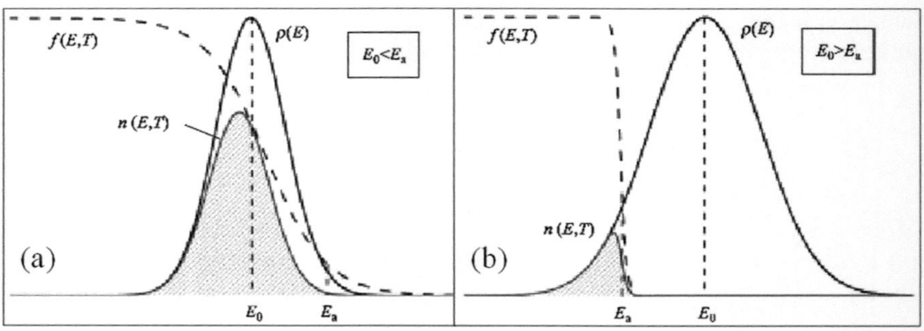

Fig. 11.13 State density distribution under the influence of localized states [31]

11.4.3 The Mechanism of Surface Manipulation on Localized Carrier Distribution

To explain the differences in localized emission exhibited by different QDs in Fig. 11.13 and to quantitatively illustrate the role of various surface treatments on the localized distribution of QDs, we apply the LSE model to systematically analyze the above results to clarify the physical mechanism behind the localization effect. By fitting the experimental data, a series of important parameters such as the carrier distribution of density of state (DOS) can be obtained, as shown in the figure below.

Figure 11.14 a–d shows the fitting results of the PL peak position moving trend with temperature calculated by the LSE model, where the black squares represent the peak energy at each temperature and the solid line represents the fitting result of the LSE model. The dashed line in (c) represents the traditional Varshni peak shift trend. The different moving trends of the four quantum structures are highly consistent with the calculation results of the model, and Table 11.3 lists the involved fitting parameters. Figure 11.14 e–h is carrier localization distribution diagrams drawn according to the fitting results, where the localization distribution function (DOS distribution function) is Gaussian, E_0 represents the center value of the Gaussian line shape, σ represents the line shape's broadening width, and the shaded area below E_a represents the energy position filled by localized-state carriers. According to the relative relationship between the filling level (E_a) value and the distribution function center (E_0), the PL peak shift can be divided into two types: One is $E_a < E_0$, represented by QW, QD, and PQD (Fig. 11.14a, b, and d), and the other is $E_a > E_0$, such as EQD (Fig. 11.14c). Firstly, analyzing the first situation, in the initial stage of temperature $T < 80$ K, due to bandgap shrinkage, carriers follow the traditional bandgap shrinkage theory and show a redshift. However, as the temperature rises from 80 to 180 K, the carriers in the localized state are activated by thermal energy, gradually moving away from their original energy level position; that is, the carriers undergo a de-localization process. The escaped carriers will be redistributed according to the state density distribution (DOS), and in the case of $E_a < E_0$, the center of the state density distribution E_0 is at a higher energy state, causing the carriers to distribute more to higher energy levels, corresponding to a blueshift in the PL spectrum, until all carriers are thermally activated and leave their original localized-state energy level. When the temperature continues to rise to near room temperature, the PL peak position is once again dominated by the bandgap shrinkage effect, showing a redshift. These three stages constitute the S-shaped change trend of the peak position with temperature unique to localized luminescence. However, for the carrier distribution situation of EQD, $E_a > E_0$; that is, in the second stage, thermally activated carriers are redistributed to lower energy states according to DOS, corresponding to a redshift in the PL spectrum. Therefore, in addition to the redshift effect caused by bandgap shrinkage (as shown by the dashed line in Fig. 11.14c), the EQD PL peak position has further redshifted until all carriers are activated. Finally, analyzing the peak shift mode of PQD, although the fitting result shows that it is consistent with the situation of $E_a < E_0$ (Fig. 11.14h), the carrier distribution of PQD is more concentrated, which leads to rapid thermal activation of carriers even at low temperatures. The first stage of the S-shaped movement is directly skipped, and the movement of the PL peak position mainly includes two stages, which is slightly different from the conventional S-shaped movement.

After a deep understanding of the distribution functions and luminescent characteristics of localized-state carriers in the four types, we can infer the mechanism of different surface treatments on the PL of QDs. The QD prepared by the top-down dry etching method, as its size decreases, stress is released and the quantum confinement

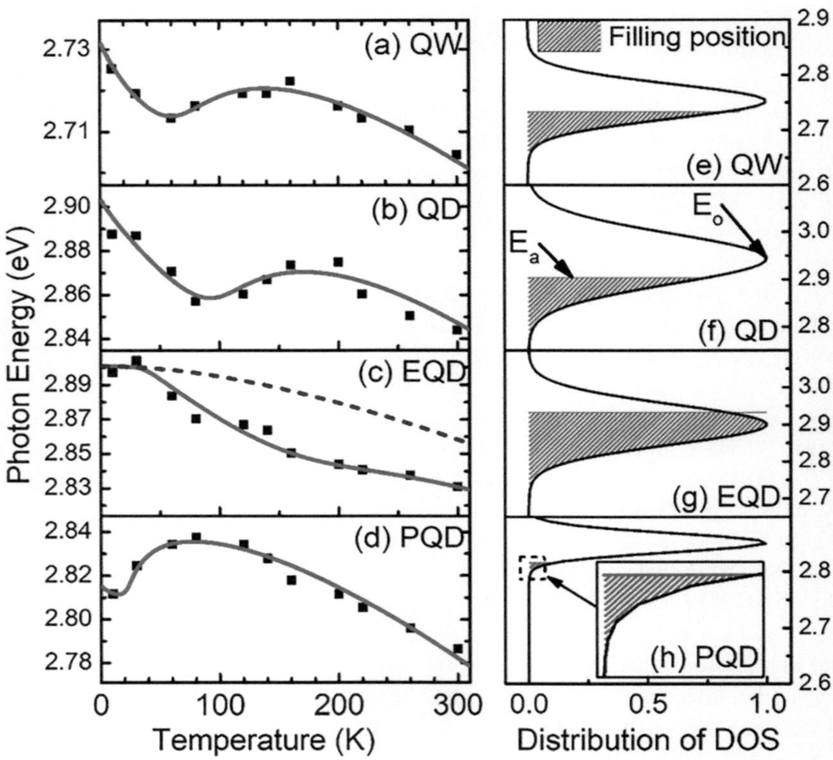

Fig. 11.14 PL peak fitting and localized-state density distribution function

Table 11.3 LSE fitting parameters of four localized PL spectra

	E_0 (eV)	σ (meV)	E_a (eV)	E_0–E_a (meV)	τ_r/τ_{tr}
QW	2.752	15.7	2.733	19.2	100
QD	2.944	26.3	2.904	39.8	735
EQD	2.901	27.1	2.934	−33.2	332
PQD	2.852	8.6	2.816	37.0	42

effect increases. Both of these factors will lead to an increase in the bandgap width, and E_a and E_0 significantly increase. At the same time, a large number of surface etching damages and defects are introduced during the preparation of QD, which constitutes new carrier "capture centers" in the luminescent recombination process. In the localized model, they can be considered as generating new localized energy levels, so the DOS distribution of QD samples is wider (σ, 26.3 meV). In addition, under normal circumstances, the GaN substrate material will cause the surface energy band to bend upward due to the influence of surface states [33–35]. This has been discussed in detail in the second section of this article. Under this influence, E_a and E_0 also respond to changes, and the distribution range of localized carriers is extended to higher energy levels.

For EQD, which has been corroded by alkali solution, some of the etching damage on its surface is removed or repaired, and the density of surface states is reduced based on QD, and the corresponding upward bending energy band structure gradually recovers and becomes flat [12]. Therefore, the center of the localized-state DOS distribution of carriers (E_0) moves down from 2.944 to 2.901 eV. At the same time, the corrosion of the alkali solution reduces the size of the QDs, and the stress is further released, and the position of E_a is slightly higher than that of QD. The combined result of these two aspects leads to a decrease in E_0 and an increase in E_a in EQD, resulting in a localized-state distribution form of $E_a > E_0$, which is the main difference between EQD and the other three samples.

Finally, the density of surface states of PQD passivated by Al_2O_3 is further reduced, and the band bending phenomenon is greatly weakened, which can refer to the content of the second section of this article. This causes the energy states originally at higher energy levels to disappear, and most of the carrier distribution returns to lower energy levels. Therefore, the broadening σ of the carrier localization distribution function becomes smaller (from 27.1 meV reduced to 8.6 meV), and the localized center E_0 value decreases (from 2.901 eV down to 2.852 eV). According to the spectral results, the Al_2O_3 coating layer increases the stress of the nanocolumns, and E_a moves to a lower energy level, causing the relationship between E_0 and E_a to return to the situation where $E_a < E_0$.

The last parameter in Table 11.3, τ_r/τ_{tr}, is a time constant related to the recombination efficiency. In the LSE model, τ_r represents the carrier radiative recombination lifetime, while τ_{tr} represents the lifetime of carrier thermally activated escape, which is proportional to the nonradiative recombination lifetime τ_{nr} [36]; that is, τ_r/τ_{tr} is proportional to τ_r/τ_{nr}. Therefore, from the formula for the quantum efficiency within the QDs, $\eta = 1/(1 + \tau_r/\tau_{nr})$, it can be seen that a smaller τ_r/τ_{tr} corresponds to a larger quantum efficiency. According to the fitting results in Table 11.3, the smallest τ_r/τ_{tr} value of PQD corresponds to the highest quantum efficiency η, which is completely consistent with the PL intensity experimental results shown in Fig. 11.12. The physical mechanism is the reduction in the density of surface states, leading to a longer τ_{nr}; at the same time, the band bending weakens, the overlap integral of the electron–hole wave function increases, and the carrier radiative recombination lifetime τ_r becomes shorter.

11.5 Summary

QDs in nanowires with microcavity-coupled structures are key research subjects in the fields of micro-/nano-optoelectronics, quantum optics, quantum communication, and biomedical applications. To realize these applications, it is essential to enhance the luminescence efficiency of QDs to optimize device performance. However, the manipulation of surface states in nanostructures remains a common challenge, particularly regarding the unclear passivation mechanisms. Section 11.3 presents a systematic study on surface passivation mechanism of InGaN/GaN QDs in nanowires, utilizing several different dielectric materials for ultrathin surface coatings while preserving the size advantages and quantum confinement effects of the QDs in nanowires. Due to the Fermi-level pinning of surface states, the energy band of InGaN/GaN QDs in nanowires bends upward, placing the entire nanowire in the depletion region and causing spatial separation of carriers. Surface passivation not only changes the quantity and nature surface states on the nanowire but also induces changes in band bending accordingly. Through systematic spectral experimental comparison, it was found that after SiN_x and Al_2O_3 passivation, the luminescence emission of QDs in nanowires became stronger, and the IQE increased by 87.98 and 45.45%, respectively. Additionally, the emission peak exhibited a blueshift, and the radiative recombination lifetime decreased. These all correspond to the weakening of the nanowire surface band bending, which means that the overlap integral of the electron–hole wave function increases, the transition energy increases, and the radiative recombination efficiency improves. In order to quantitatively characterize the degree of surface band bending, this chapter uses ultraviolet photoelectron spectroscopy to characterize the nanowire surface after different surface treatments, proving that surface manipulation effectively repairs the band bending phenomenon, and this result also verifies the physical theoretical model of surface band modification. Therefore, effective surface passivation treatment can not only reduce the density of surface states but also improve the band structure of the nanowire surface. This finding reveals the mechanism of surface passivation and provides a quantitative evaluation method for improving the performance of nanowire devices. The role of traditional passivation is also expanded; in addition to reducing surface defects, it can also serve to modify the surface of nanostructures and regulate their energy bands.

Section 11.4 discusses the localization effects of emission in InGaN/GaN QDs in nanowires, starting with the widely applicable LSE theoretical model, and compares the emission characteristics of QDs with different surface treatments. A systematic analysis of the InGaN/GaN QDs prepared by the top-down etching method, as well as the QDs treated with alkali solution etching (EQD) and Al_2O_3 passivation (PQD), reveals differences in emission. With the help of the LSE model, key physical parameters such as the carrier localization DOS distribution function reflected in the relationship between emission peak and temperature variation can be accurately calculated. The difference between E_a and E_0 determines the redistribution of carriers, influencing their recombination behavior in temperature-dependent PL spectra. In addition to the content mentioned in this chapter, the LSE model can also quantitatively describe the trend of half-peak width (FWHM) and intensity variation with temperature [37]. Therefore, the localized luminescence characteristics of carriers in QDs can be accurately quantified and used to analyze the luminescence performance of QD devices. On the other hand, this chapter discusses the impact of different surface treatments on the localized-state luminescence of QDs, finding that different surface treatments have minimal effects on the surface morphology of QDs in nanowires, but significantly change the emission behavior. These common surface treatments, including dry etching, wet etching, and surface thin-film coating techniques, affect the performance of nanostructured devices. Investigating the mechanisms of these treatments will be key to breaking through device performance bottlenecks in such devices.

InGaN/GaN QDs in nanowires, due to their excellent luminescence performance, have achieved great success in the preparation of high-efficiency LEDs, lighting, and displays. In addition, their unique structure makes them have great potential in the preparation of micro-/nanolasers and quantum lasers, where nanowires can serve as both microcavities and gain media. At the same time, due to the discrete energy-level characteristics of QDs, they can serve as ideal single-photon sources, and GaN-based materials have ultrahigh quantum recombination efficiency, with the brightness of their single-photon sources exceeding 10^6 counts per second [14, 38]. Their larger exciton binding energy facilitates stable room-temperature emission and even high-temperature emission [39, 40]. The GaN/AlGaN QDs in nanowire structure prepared by the Arakawa team at the University of Tokyo has raised the working temperature of single-photon emission to 350 K [39]. By adjusting the elemental composition, the emission wavelength can cover from ultraviolet to infrared, greatly expanding the wavelength range of single-photon emission [41, 42]. Shorter wavelengths are beneficial for reducing the size of the transmitter and receiver telescopes [43]. On the other hand, the GaN-based QD positioning growth process ensures that QDs are easy to implement electrical injection operations [44], as well as integration with other optoelectronic devices [45], to prepare on-chip integrated micro-/nano-active optoelectronic devices.

It should be noted that the realization of the above optoelectronic applications puts extremely high requirements on the optical and electrical performance of the device. However, the current preparation process of GaN-based QDs is not mature, and defects, dislocations, and surface states still need to be controlled. There are still some unresolved issues at the level of luminescence mechanism, so in order to more effectively improve the optoelectronic performance of QDs, further in-depth research on carrier recombination luminescence characteristics is needed.

In terms of luminescence characterization, the optical measurement of individual QDs is facing challenges. Various practical applications require accurate positioning and imaging of QDs with sub-wavelength dimensions, analysis of emission spectra, and PL lifetime. Therefore, ultrahigh resolution microscopic optics, high-precision displacement stages, and sensitive single-photon detection systems are essential tools for future QD luminescence characterization. More sophisticated optical systems and advanced characterization methods will also drive faster progress in the field of QD luminescence.

References

1. Wu Y, Liu B, Li Z et al (2020) Synthesis and properties of InGaN/GaN multiple quantum well nanowires on Si(111) by molecular beam epitaxy. Phys Status Solidi (A) 217(7):1900729
2. Barrigón E, Heurlin M, Bi Z et al (2019) Synthesis and applications of III–V nanowires. Chem Rev 119(15):9170–9220
3. Zhang L, Teng CH, Ku PC et al (2016) Site-controlled InGaN/GaN single-photon-emitting diode. Appl Phys Lett 108(15):153102
4. Puchtler TJ, Wang T, Ren CX et al (2016) Ultrafast, polarized, single-photon emission from m-plane InGaN QDs on GaN nanowires. Nano Lett 16(7):4608–4615
5. Ra YH, Wang R, Woo SY et al (2016) Full-color single nanowire pixels for projection displays. Nano Lett 16(7):4608–4615
6. Sadaf SM, Ra YH, Szkopek T et al (2016) Monolithically integrated metal/semiconductor tunnel junction nanowire light-emitting diodes. Nano Lett 16(2):1076–1080
7. Jain B, Velpula RT, Bui HQT et al (2020) High performance electron blocking layer-free InGaN/GaN nanowire white-light-emitting diodes. Opt Express 28(1):665–675
8. Zhao C, Alfaraj N, Subedi RC et al (2018) III-nitride nanowires on unconventional substrates: from materials to optoelectronic device applications. Prog Quantum Electron 61:1–31
9. Chatterjee U, Park JH, Um DY et al (2017) III-nitride nanowires for solar light harvesting: a review. Renew Sustain Energy Rev 79:1002–1015
10. Aiello A, Hoque AKMH, Baten MZ et al (2019) A high-gain silicon-based InGaN/GaN dot-in-nanowire array photodetector. ACS Photonics 6(7):1796–1804
11. Sergent S, Damilano B, Vézian S et al (2019) Subliming GaN into ordered nanowire arrays for ultraviolet and visible nanophotonics. ACS Photonics 6(12):3321–3330
12. Wang ZL, Hao ZB, Yu JD et al (2017) Manipulating the band bending of InGaN/GaN QDss in nanowires by surface passivation. J Phys Chem C 121(11):6380–6385
13. Nelson C, Deshpande S, Liu A et al (2017) High-resolution nonlinear optical spectroscopy of InGaN QDss in GaN nanowires. J Opt Soc Am B 34(6):1206
14. Wang X, Xu L, Jiang Y et al (2019) III–V compounds as single photon emitters. J Semicond 40(7):071906
15. Wang T, Oliver R, Taylor R (2020) Non-polar nitride single-photon sources. J Opt 22(7):073001
16. Nishat MRK, Ahmed SS (2019) Atomistic modeling of fine structure splitting in InGaN/GaN dot-in-nanowire structures for use in entangled photon pair generation. IEEE J Quantum Electron 55(1):1–1
17. Yanxiong E, Hao Z, Yu J et al (2015) MBE growth of AlN nanowires on Si substrates by aluminizing nucleation. Nanoscale Res Lett 10(1):1–7
18. Zheng XH, Chen H, Yan ZB et al (2004) Influence of the deposition time of barrier layers on optical and structural properties of high-efficiency green-light-emitting InGaN/GaN multiple quantum wells. J Appl Phys 96(4):1899–1903
19. Cho YH, Gainer GH, Fischer AJ et al (1998) "S-shaped" temperature-dependent emission shift and carrier dynamics in InGaN/GaN multiple quantum wells. Appl Phys Lett 73(10):1370–1372
20. Hao M, Zhang J, Zhang XH et al (2002) Photoluminescence studies on InGaN/GaN multiple quantum wells with different degree of localization. Appl Phys Lett 81(27):5129–5131
21. Bourret A, Barski A, Rouviere JL et al (1998) Growth of aluminum nitride on (111) silicon: microstructure and interface structure. J Appl Phys 83(4):2003–2009
22. Calarco R, Marso M, Richter T et al (2005) Size-dependent photoconductivity in MBE-grown GaN−nanowires. Nano Lett 5(5):981–984
23. Lüth H (2001) Solid surfaces, interfaces and thin films, vol 4. Springer, Berlin
24. Klein A, Säuberlich F, Späth B et al (2007) Non-stoichiometry and electronic properties of interfaces. J Mater Sci 42(6):1890–1900
25. Jahangir S, Mandl M, Strassburg M et al (2013) Molecular beam epitaxial growth and optical properties of red-emitting ($\lambda = 650$ nm) InGaN/GaN disks-in-nanowires on silicon. Appl Phys Lett 102(7):071101

26. Hu Y, Hao Z, Lai W et al (2015) Nano-fabrication and related optical properties of InGaN/GaN nanopillars. Nanotechnology 26(7):075302
27. Wang ZL, Hao ZB, Yu JD et al (2019) Surface-induced carrier localization and recombination characteristics in InGaN/GaN QDss in nanopillars. J Phys Chem C 123(9):5699–5704
28. Morkoç H (2007) Handbook of GaN materials and devices. Springer, Berlin
29. Holm RT, McKnight SW, Palik ED et al (1982) Interference effects in luminescence studies of thin films. Appl Opt 21(14):251x–2519
30. Namvar E, Fattahi M (2008) Interference effects on the photoluminescence spectrum of GaN/$In_xGa_{1-x}N$ single quantum well structures. J Lumin 128(1):155–160
31. Li Q, Xu SJ, Xie MH et al (2005) A model for steady-state luminescence of localized-state ensemble. EPL (Eur Lett) 71(6):994–1000
32. Eliseev PG (2003) The red σ 2/kT spectral shift in partially disordered semiconductors. J Appl Phys 93(9):5404–5415
33. Sabuktagin S, Reshchikov MA, Johnstone DK et al (2003) Band bending near the surface in GaN as detected by a charge sensitive probe. In: MRS online proceedings library archive, p 798
34. Chevtchenko S, Ni X, Fan Q et al (2006) Surface band bending of a-plane GaN studied by scanning Kelvin probe microscopy. Appl Phys Lett 88(12):122104
35. Peng LH, Shih CW, Lai CM et al (2003) Surface band-bending effects on the optical properties of indium gallium nitride multiple quantum wells. Appl Phys Lett 82(24):4268–4270
36. Su Z, Xu S (2017) A generalized model for time-resolved luminescence of localized carriers and applications: dispersive thermodynamics of localized carriers. Sci Rep 7(1):13
37. Bao W, Su Z, Zheng C et al (2016) Carrier localization effects in InGaN/GaN multiple-quantum-wells LED nanowires: luminescence quantum efficiency improvement and "negative" thermal activation energy. Sci Rep 6:34545
38. Holmes MJ, Arita M, Arakawa Y (2019) III-Nitride QDss as single photon emitters. Semicond Sci Technol 34(3):033001
39. Berhane AM, Jeong KY, Bodrog Z et al (2017) Bright room-temperature single-photon emission from defects in gallium nitride. Adv Mater 29(12):1605092
40. Zhou Y, Wang Z, Rasmita A et al (2018) Room temperature solid-state quantum emitters in the telecom range. Sci Adv 4 (3):eaar3580
41. Holmes MJ, Kako S, Choi K, Arita M, Arakawa Y (2016) Single photons from a hot solid-state emitter at 350 K. ACS Photonics 3(4):543–546
42. Wang T, Puchtler TJ, Zhu T et al (2017) Polarisation-controlled single photon emission at high temperatures from InGaN QDss. Nanoscale 9(27):9421–9427
43. Tsao JY, Chowdhury S, Hollis MA et al (2018) Ultrawide-bandgap semiconductors: research opportunities and challenges. Adv Electron Mater 4(1):1600501
44. Chen Y, Zhang J, Zopf M et al (2016) Wavelength-tunable entangled photons from silicon-integrated III–V QDs. Nat Commun 7(1):1–7
45. Holmes MJ, Arita M, Arakawa Y (2019) III-Nitride QDs as single photon emitters. Semicond Sci Technol 34(3):033001

Chapter 12
Characterization of Nonlinear Optical Properties of Zinc Oxide Single Crystals and Micro-Nanostructures

Changcheng Zheng and Shijie Xu

12.1 Overview

In this chapter, we introduce the nonlinear optical properties of zinc oxide (ZnO) and discuss their behavior across the optical spectrum. The discussion primarily covers the second harmonic generation (SHG) in ZnO single crystals (undoped samples and ion-implanted samples) and micro-nanostructures (submillimeter-sized hollow rod samples and nano-tetrapod samples), two-photon absorption characteristics, and several associated optical phenomena, such as the weakening or even disappearance of band-edge luminescence caused by the self-absorption effect and changes in carrier recombination lifetime. The experimental methods used are mainly steady-state and time-resolved photoluminescence spectra at variable and room temperatures. The principal experimental results mentioned in this chapter were completed in the laser spectroscopy laboratory of Professor Shijie Xu at the Department of Physics, University of Hong Kong, and also referred to the relevant achievements of other research groups in this field (results reported from around 2000 to recent). Because of space limit, this chapter does not involve intentionally doped ZnO during the growth process, ZnO composite structures, functional nanostructures of ZnO (such as potential applications in medical imaging) and other fields. Nevertheless, research in these adjacent areas is also closely related to the nonlinear optical properties of ZnO. Of course, the high-quality ZnO single crystals are ideal model systems for investigating their basic physical properties and therefore the main object of discussion in

C. Zheng (✉)
Division of Natural and Applied Sciences, Duke Kunshan University, Suzhou, China
e-mail: changcheng.zheng@dukekunshan.edu.cn

S. Xu
Department of Physics, University of Hong Kong, Hong Kong, China

Department of Optical Science and Engineering, School of Information Science and Technology, Fudan University, Shanghai, China

S. Xu (ed.), *Optical Characterization of Microstructures and Optoelectronic Devices Based on Wide Band Gap Semiconductors*, Wide Bandgap Semiconductors,
https://doi.org/10.1007/978-981-95-1928-6_12

this chapter. At the end of this chapter, we will briefly outline future prospects for the research on the nonlinear optical properties of ZnO, aiming to provide reference material for researchers and students who are interested in this field.

12.2 Introduction to Nonlinear Optical Properties

The nonlinear optical properties discussed in this chapter include optical harmonic generation (second harmonic/third harmonic, etc.), frequency mixing (sum frequency/difference frequency; harmonic is a special sum frequency), frequency up-conversion, multiphoton absorption, and stimulated Raman scattering. They are all manifestations of the interaction between light and matter. A semiclassical theoretical description, combined with the concept of photons and the electric field vector of the light wave and its induced polarization field in the medium, will be used to describe several typical nonlinear optical phenomena such as SHG and multiphoton absorption.

12.2.1 Second Harmonic Generation (SHG)

As a basic nonlinear optical phenomenon, SHG was first observed soon after the invention of the laser [1]. Simply put, SHG occurs when light interacts with matter, two coherent photons (with identical frequency and phase) annihilate to produce a double-frequency photon, as shown in Fig. 12.1, where the intermediate states 1 and 2 are virtual energy levels not associated with any real atomic or electronic states.

In this simplified physical picture, the time the atom/molecule in the medium stays at the intermediate state is infinitesimally short. The figure has already considered energy conservation, and we also need to consider momentum conservation, that is, the phase matching condition:

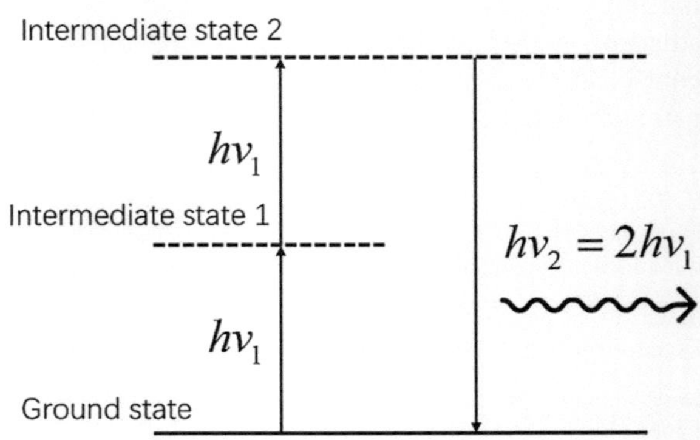

Fig. 12.1 Schematic diagram of the SHG process

$$\vec{k}_2 = \vec{k}_1 + \vec{k}_1.$$

Phase matching has a decisive influence on the efficiency of SHG. In practical applications, considering that the refractive indices of the fundamental light (ν_1) and the second harmonic light (ν_2) in the medium are different, a specific incident angle and the birefringence phenomenon in the medium are generally adopted to achieve efficient second harmonic output. In experimental research, SHG can sometimes be observed even under conditions that do not strictly satisfy phase matching. This mainly refers to the experimental design that does not strictly follow a transmissive or reflective optical path configuration. We will discuss the phenomenon of SHG in ZnO under these experimental conditions in the following text.

It is relatively complex to describe nonlinear optical behaviors such as SHG using a complete quantum theory treatment. Here, we will start from the classical polarization model which offers a simplified quantitative/semiquantitative description of the generation of the second harmonic. The relationship between the polarization intensity induced in the medium under the irradiation of electromagnetic waves (light field) and the electric field vector can be written as follows:

$$\begin{aligned} P(\vec{r},t) &= \varepsilon_0\chi^{(1)}E(\vec{r},t) + \varepsilon_0\chi^{(2)}E^2(\vec{r},t) + \varepsilon_0\chi^{(3)}E^3(\vec{r},t) + \cdots \\ &= P^{(1)}(\vec{r},t) + P^{(2)}(\vec{r},t) + P^{(3)}(\vec{r},t) + \cdots, \end{aligned},$$

where ε_0 is the dielectric constant, $E(\vec{r},t)$ represents the electric field vector of the incident light, $\vec{r}$ is the position vector, and $\chi^{(n)}$ is the nth order polarizability; for example, $\chi^{(1)}$ is the first order linear polarizability, $\chi^{(2)}$ is the second order nonlinear polarizability, etc., in which $\chi^{(2)}$ is a third order tensor, with $3 \times 3 \times 3$ components. The above formula can also be written in integral form and transformed to $\left(\vec{k},\omega\right)$ [2] through the Fourier transform. In some literature, the $\boldsymbol{d}$ tensor is also often used, and their relationship is $d_{ijk} = d_{xyz} = \frac{1}{2}\chi_{ijk}^{(2)}$. For crystals that lack central symmetry, $\chi^{(2)}$ is not zero, and the material can generate second harmonics. This chapter focuses on ZnO crystals with hexagonal wurtzite structure, which belong to C_{6v}^4 (*6mm*, or *P6₃mc*) space group/point group (as shown in Fig. 12.2). Different notations come from different classification methods, but their corresponding crystal structures are the same, all lacking central symmetry, so ZnO with a hexagonal wurtzite structure can generate effective second harmonics.

For ZnO materials grown along the c axis, according to the symmetry of the structure, its second-order polarizability tensor can be simplified to a 3×6 matrix, which has four independent nonzero elements [4] (also seen as five independent nonzero elements [5]), that is, $\chi_{xzx}^{(2)} = \chi_{yzy}^{(2)}$, $\chi_{xxz}^{(2)} = \chi_{yyz}^{(2)}$, $\chi_{zxx}^{(2)} = \chi_{zyy}^{(2)}$ (or consider $\chi_{zxx}^{(2)}$ and $\chi_{zyy}^{(2)}$ not equal), and $\chi_{zzz}^{(2)}$. However, in general, both the fundamental and harmonic light waves are far from the resonance frequency of ZnO (bandgap energy); under these circumstances, the Kleinman symmetry condition can further reduce the number of independent nonzero elements to 2, namely $\chi_{xzx}^{(2)} = \chi_{yzy}^{(2)} = \chi_{xxz}^{(2)} = \chi_{yyz}^{(2)} = \chi_{zxx}^{(2)} = \chi_{zyy}^{(2)}$ and $\chi_{zzz}^{(2)}$ [4]. Sometimes, to simplify the notation, some literature replaces

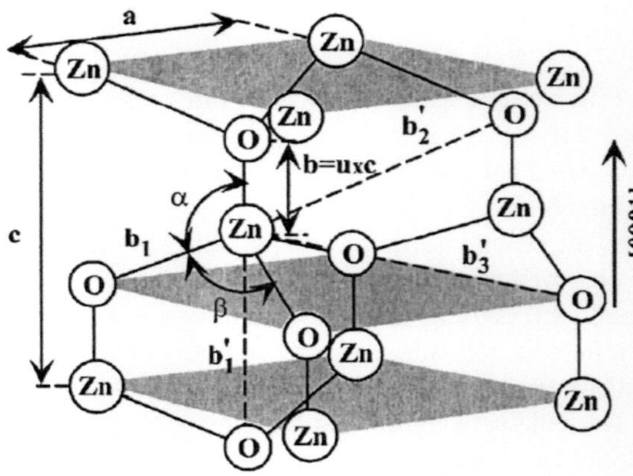

Fig. 12.2 Schematic diagram of the structure of hexagonal wurtzite ZnO [3]

the last two subscripts of $\chi^{(2)}_{xyz}$ or $\chi^{(2)}_{ijk}$ according to $xx = 1$, $yy = 2$, $zz = 3$, $yz = zy = 4$, $xz = zx = 5$, $xy = yx = 6$, to make the notation more concise. Here, to retain all structural information, we will use a notation with three subscripts. In the following sections, we will discuss the SHG in ZnO single crystals and the impact of ion implantation on it, as well as introduce the SHG in ZnO micro-nanostructures, especially the role played by the surface/interface.

12.2.2 *Two-Photon/Multiphoton Absorption-Induced Photoluminescence*

Another noteworthy nonlinear optical phenomenon is the two-photon/multiphoton absorption in ZnO and the resulting luminescence. The absorption process of two photons is similar to the SHG shown in Figure 1, both of which are carried out with the participation of two coherent incident photons. The main difference is that the result of two-photon absorption is that the electron is promoted to a real excited state (as shown in Fig. 12.3). The figure does not consider momentum conservation, impurity/defect energy levels, exciton effects, etc., in these processes.

Because the excitation photon is far from the resonance energy of the material during two-photon/multiphoton absorption, the absorption efficiency is often very low. Therefore, in order to observe two-photon/multiphoton absorption and the resulting luminescence, it is usually necessary to use a high-energy/power laser for excitation, and two-photon/multiphoton absorption often occurs simultaneously with the SHG. When the energy of two photons is close to the resonance energy of the material, a resonance enhancement of SHG may occur [6, 7]. In this chapter, we primarily examine the off-resonant regime, where the photon energy is well below half of the ZnO bandgap. In this case, the attenuation (i.e., absorption process) of the laser light (or light from other sources) as it propagates in the medium can be

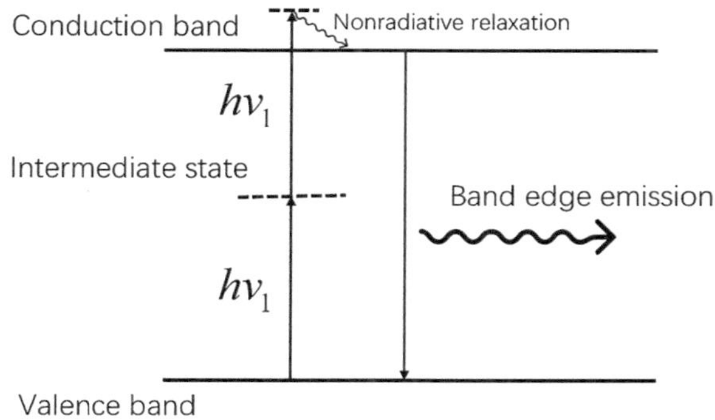

Fig. 12.3 Schematic diagram of two-photon absorption and its induced band-edge luminescence

empirically described by the following formula [8]:

$$\frac{dI(z)}{dz} = -\alpha I(z) - \beta I^2(z) - \gamma I^3(z) - \cdots,$$

where $I(z)$ is the light intensity propagating along the z-axis and α, β, γ are the single-photon, two-photon, and three-photon absorption coefficients of the medium, respectively. Generally speaking, the high-order nonlinear absorption coefficient is significantly smaller than the low-order ones, which leads to an interesting phenomenon that two-photon/multiphoton absorption tends to occur deeper inside the sample, away from the surface, thus providing a nondestructive method for optically detecting the internal features of the sample, such as impurity/defect energy levels inside the sample and electron–phonon interaction processes. In the following, we will discuss the luminescence behavior of ZnO under two-photon/multiphoton excitation and explore how these nonlinear interactions depend on structural form-single crystal, thin film, or nanostructure.

12.3 SHG in ZnO Single Crystals and Thin Films

12.3.1 Second-Order Nonlinear Polarizability of ZnO Single Crystals

In experiments of SHG from ZnO single crystals or thin film samples, there are usually two types of experimental optical configurations employed, namely transmission [4, 9, 10] and reflection [11–16]. Both geometrical configurations can be understood in terms of standard ray-optics setups. Considering the phase-matching condition, experiments with variable incidence angles or polarization control can obtain the corresponding second-order nonlinear polarizability through fitting the SHG intensities. For example, H. Cao et al. examined the SHG phenomenon of ZnO thin films under different incident angles of a pulsed laser (1064 nm, 10 Hz) in transmission mode (see Fig. 12.4) and obtained the second-order nonlinear polarizability

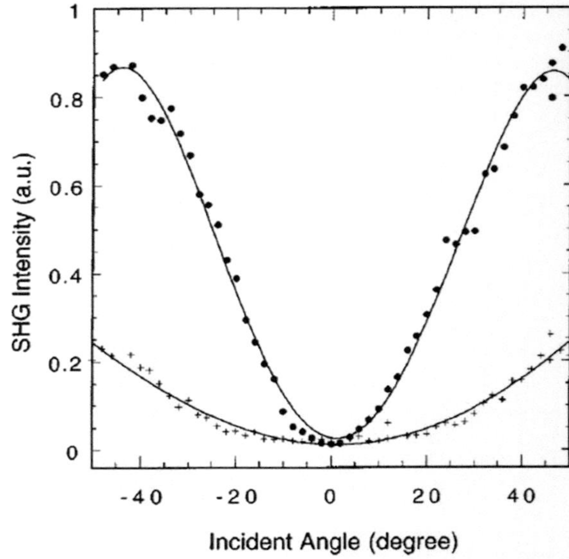

Fig. 12.4 Relationship between the intensity of the SHG in ZnO thin films and the incident angle [4]

of ZnO thin films (without considering the sign) by fitting the relationship between the relative intensity of the SHG and the incident angle as follows: $\chi_{zxx}^{(2)} \approx 3.6\ pm/V$ and $\chi_{zzz}^{(2)} \approx 13.4\ pm/V$ [4]. The crosses in the figure represent the *s-p* polarization mode, while the dots represent the *p-p* polarization mode (the first letter represents the polarization direction of the fundamental light, and the second letter represents the polarization direction of the SHG signal, where *s* represents the polarization direction perpendicular to the incident plane and *p* represents the polarization direction parallel to the incident plane).

It should be noted that the values of the second-order nonlinear polarizability obtained from different samples are often different even when identical fitting procedures are used. These differences reflect the influence of growth methods, crystallinity, impurity content, and defect density. For instance, G. Wang et al. also measured the second-order nonlinear polarizability of ZnO single crystals under comparable conditions and obtained: $\chi_{zxx}^{(2)} = 1.36\ pm/V$ and $\chi_{zzz}^{(2)} = -14.31\ pm/V$. Here, the sign determination is obtained by fitting the polarization dependence (changing the polarization direction of the incident light) on the relative intensity of the SHG [10]. On the other hand, the reflection measurement method is often used to probe the surface characteristics of ZnO thin films and the information embedded in grain boundaries [12–14]. For example, K. Y. Lo et al. studied the *3-mm* symmetry structure of the thin film surface by measuring and analyzing the SHG signal of ZnO thin films under specific polarization settings [13]. As mentioned earlier, ZnO crystals usually have a *6-mm* symmetry structure, and the SHG of this symmetry structure in the *s-s* polarization configuration is zero. However, the ideal polar surface of ZnO has a *3-mm* symmetry structure, and this surface contribution will make the SHG in the *s-s* polarization observable. Y. J. Huang and K. Y. Lo et al. used the same method

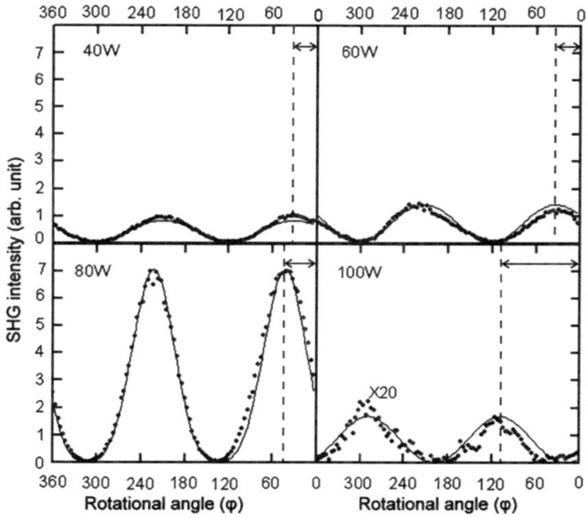

Fig. 12.5 Relative intensity of SHG in s-s polarization setup in ZnO thin films grown by sputtering at different radio frequency powers [14]

to invesigate the grain orientation in ZnO thin films, producing the SHG intensity data as follows (Fig. 12.5).

These results demonstrate that ZnO thin films grown on the (111) oriented silicon substrate via radio frequency sputtering exhibit a $[\bar{1}10]$ orientation, which is consistent with X-ray diffraction observations [14].

12.3.2 The Impact of Surface/Interface on SHG

Regarding the influence of interface effects on the SHG, we observed a significant enhancement of SHG in ZnO hollow rods at the (sub)millimeter level and attributed it to the enhanced surface/interface contribution due to multiple reflections of the fundamental pump light on the outer and inner surfaces [16]. In this work, the optical path configuration and comparative spectra are shown in Fig. 12.6.

Here, a femtosecond pulsed laser with a wavelength of 774 nm and a repetition rate of 82 MHz is used as the excitation/fundamental light source. The objective lens is placed in front of the sample to collect the light signal predomonantly emitted in the *x* direction (within a finite divergence). We found that the usual two-photon excitation-induced band-edge emission signal (~397 nm) is generated in the solid ZnO rod, while a strong SHG signal (387 nm) is observed from the hollow rod (phonon effect, and self-absorption effect, discussed in the subsequent two-photon absorption section). We confirmed its physical origin through measuring the power dependence characteristics and polarization dependence behavior of the SHG signal. For example, its excitation light power dependence characteristics are as follows:

By fitting the power dependence of the integrated intensity of the SHG signal (the inset in Fig. 12.7), we found that its power dependence coefficient is very close to

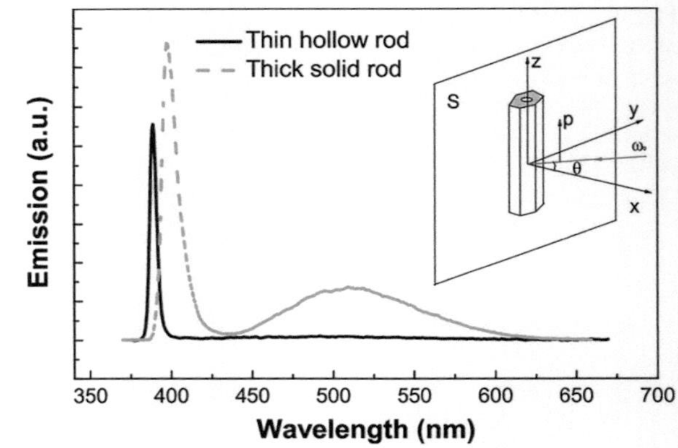

Fig. 12.6 Spectra of ZnO hollow rod (solid line) and solid rod (dashed line) under two-photon excitation at room temperature [16]

2, matching the theoretical expectation for a second-order process. Compared with the spectral signal dominated by band-edge emission in the solid rod sample, the enhancement effect of the SHG in the hollow rod arises mainly from the multiple reflections of the fundamental light between the outer and inner surfaces of the tube structure in the hollow rod and the continuous generation of the second harmonic signal during propagation, and part of the signal propagates along the *x* direction (or nearby direction), which is emitted from the sample and gets collected and detected. However, there is no such multiple reflection phenomenon in the solid rod sample because the absence of an internal cavity prevents multiple reflections. Furthermore, since the energy of the second harmonic is higher than the energy of the band-edge luminescence peak, the second harmonic may be reabsorbed after it is generated inside the sample, but in the hollow column sample, the enhanced SHG intensity and reduced optical path length minimize this reabsorption effect, making it negligible in the experiment. We also conducted polarization-dependent measurements and fittings on the photoluminescence data from the sample; the results are as follows:

Here, we used a simplified semiquantitative model to fit the polarization-dependent results of the SHG signal [17, 18]:

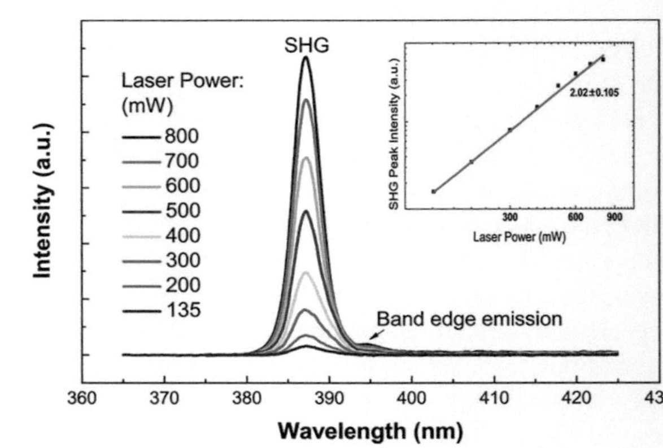

Fig. 12.7 Excitation power dependence of SHG in ZnO hollow rod [16]

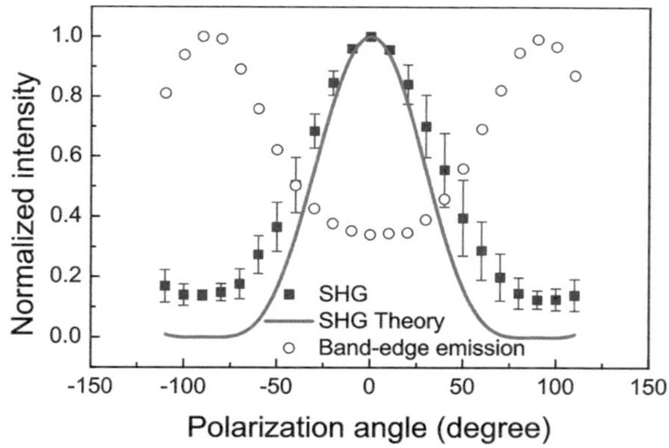

Fig. 12.8 Polarization-dependent measurements and fitting results of SHG signal and band-edge luminescence signal in ZnO hollow rod at room temperature [16]

$$I_{SHG} \propto \left[P^{(2)}(\omega = 2\omega_0)\right]^2,$$

where $P^{(2)}(\omega = 2\omega_0) = \chi^{(2)}_{ijk} E_j E_k$, $\chi^{(2)}_{ijk}$ is the second-order nonlinear polarizability of ZnO, and $\boldsymbol{E}$ is the electric field vector. Applying *Kleinman* symmetry condition leaves two independent nonzero tensor elements, $\chi^{(2)}_{zxx} = 1.36pm/V$ and $\chi^{(2)}_{zzz} = -14.31pm/V$ [10]. Considering the incident angle (~26°) in the experiment and the geometric relationship among the incident direction, polarization direction (φ), and crystal axis direction in the upper right corner of Fig. 12.6, we can write:

$$P^{(2)}(\omega = 2\omega_0) \propto \left(0.26\sin^2\varphi - 14.3\cos^2\varphi\right).$$

The fitting result of the SHG signal using this expression is shown as the solid curve in Fig. 12.8, which agrees the experimental data well, confirming the model's validity. From the results in Fig. 12.8, one also observes a competition between the SHG and band-edge luminescence in ZnO, and similar effects have also been observed in ZnO nanorods [19].

12.3.3 The Impact of Defects/Impurities on SHG

Beyond surface and interface effects, the SHG in ZnO also provides insight into defects and impurities within the crystal. As mentioned earlier, the distribution and orientation of grain boundaries in ZnO thin films have been studied through the SHG [12, 14]. SHG has also been used to analyze the impact of growth conditions/methods on the crystal quality of thin film samples, often in agreement with the results of X-ray diffraction analyses [20]. We have studied the impact of ion implantation on the

SHG in ZnO single crystal samples and analyzed the quasi-interface/grain boundary conditions induced by ion implantation in the samples [16]. The experimental method used is similar to that in Fig. 12.6, except that the single crystal sample here is a bulk sample with a thickness of 0.5 mm, and the collected light signal is along the *c*-axis of the sample. We compared the samples implanted with He, Zn, and Cu ions, respectively, and the untreated one, and evaluated the effect of annealing on the SHG from these samples. Typical SHG signals obtained are shown in Fig. 12.9, with an excitation/fundamental light wavelength of 840 nm. It can be seen that the band-edge emission remains evident when excited at this wavelength ($h\nu_2 = 2h\nu_1 < E_g$). If the energy of the excitation/fundamental light led to a SHG photon energy that was greater than the energy of the band edge emission peak (near 396 nm), in our experimental setup, due to the reabsorption effect, the SHG signal would be almost unobservable. Therefore, when utilizing the SHG signal from the sample to analyze its properties, an appropriate excitation/fundamental light wavelength or photon energy should be chosen. Another point to mention is that the band-edge emission signal observed here may also originate from the strong Rabi oscillation process during strong light-matter coupling driven by two-photon absorption [21].

From the figure, it is evident that the implantation of He ions and Zn ions, respectively, will significantly enhance the SHG in ZnO single crystals, while the implantation of Cu ions has almost no impact on the SHG (the ion diffusion in matter model,

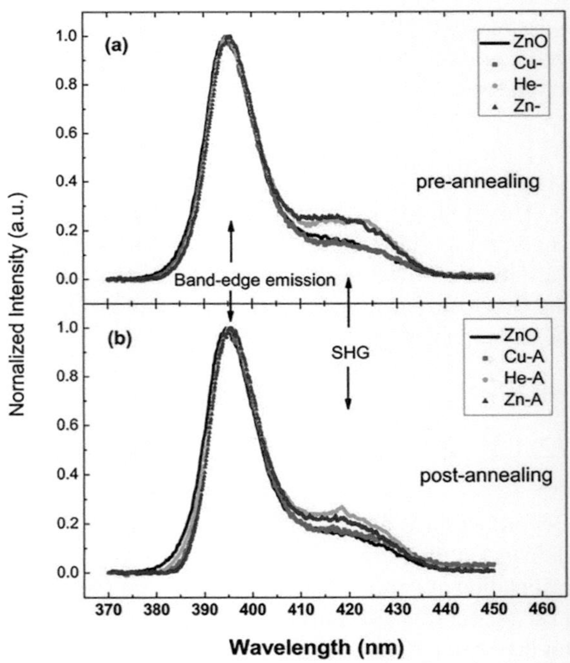

Fig. 12.9 Spectra of band-edge emission and SHG of ZnO single crystal samples with/without ion implantation under two-photon/multiphoton excitation, (**a**) before annealing and (**b**) after annealing [15]

TRIM, reveals that Cu ions only accumulate within a few tens of nanometers from the surface of ZnO, while Zn ions and He ions exist within a few hundred nanometers beneath the surface). At the same time, annealing treatment partially reduces this enhancement effect. Based on these experimental results, and in conjunction with X-ray diffraction and rocking curve measurements and fitting, we propose that effective ion implantation will introduce one or more quasi-interfaces near the crystal surface, thereby providing additional interface contributions to enhance the efficiency of SHG. We confirmed the formation of quasi-interfaces near the surface by fitting the rocking curve of the sample (the main peak and multiple satellite peaks), considering the stress distribution in the ion-implanted region [15].

We briefly summarize here the physical process and experimental results of SHG in ZnO single crystals and discuss some achievements in studying the surface, interface, grain boundaries, and other important properties in the sample using SHG. Simply put, the SHG in ZnO not only comes from its lack of central symmetry in the bulk structure and the disruption of the crystal structure by impurities/defects but also needs to consider various boundary effects, and such boundary effects become even more pronounced in ZnO micro-nanostructures.

12.4 SHG Phenomenon in ZnO Micro-Nanostructures

12.4.1 SHG in ZnO Nanowires

Among the existing and potential applications of ZnO, the nonlinear optical properties of ZnO micro-nanostructures are crucial for its application in nanostructured optoelectronic devices, solar cells, two-photon imaging, etc. [5, 22–24]. Due to their large surface area/volume ratio, ZnO micro-nanostructures often show significantly enhanced SHG signals, which are comparable to or stronger than those from ZnO thin films or single crystal samples [5, 6, 25–32]. These structures fully demonstrate the important role of the surface/interface in SHG. For example, C. F. Zhang et al. observed a markedly enhanced SHG signal from ZnO nanowires, as shown in Fig. 12.10 [21].

Peak A is the band edge emission, peak B is the SHG signal, and the broad peak C is the impurity level luminescence. Compared with the results of ZnO single crystals in Fig. 12.9, the relative intensity of the SHG signal in nanowires is considerably greater than that of band edge emission, even though they both originate from two-photon processes (see detailed discussion in the literature). This is also evidenced by the spectra under different excitation wavelengths (Fig. 12.11).

As shown in Fig. 12.11, across the tested range of excitation wavelengths, the intensity of the SHG signal remains higher than that of band-edge emission. This offers guidance for the application of ZnO nanowire materials in the field of nonlinear optics.

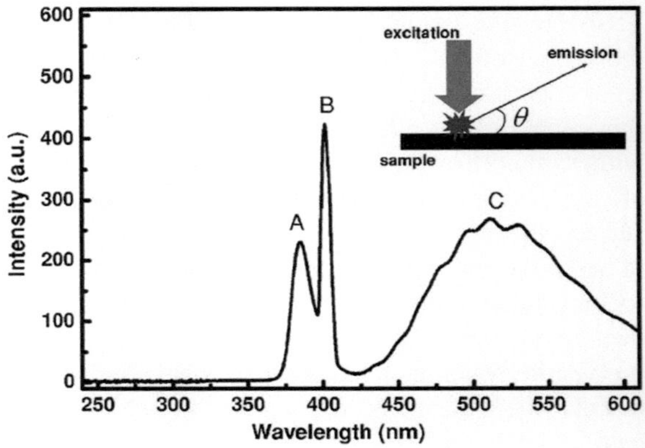

Fig. 12.10 Spectrum of ZnO nanowires under 806 nm femtosecond laser excitation [21]

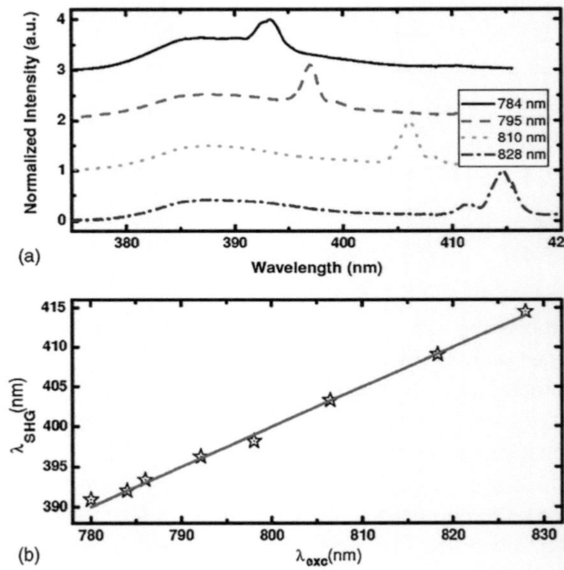

Fig. 12.11 SHG signal of ZnO nanowires under femtosecond light excitation at different wavelengths: **a** spectra and **b** wavelength relation [21]

12.4.2 SHG in ZnO Nano-Tetrapod Structures

An unusual SHG reponse also appears in another type of ZnO nanostructure, namely ZnO nano-tetrapod structures [26, 28, 31], as shown in Fig. 12.12.

The spectrum includes SHG signals (excitation-wavelength-dependent), band-edge emission signals (around 395 nm), and broad peaks of impurity/defect luminescence. As can be clearly seen from Fig. 12.12, even under two-photon excitation conditions $\left(\lambda_{ex} = 730\,nm,\ E_g > h\nu_1 > \frac{E_g}{2}\right)$, a significant SHG signal (365 nm) can be observed in the ZnO nano-tetrapod structure. When the excitation/fundamental

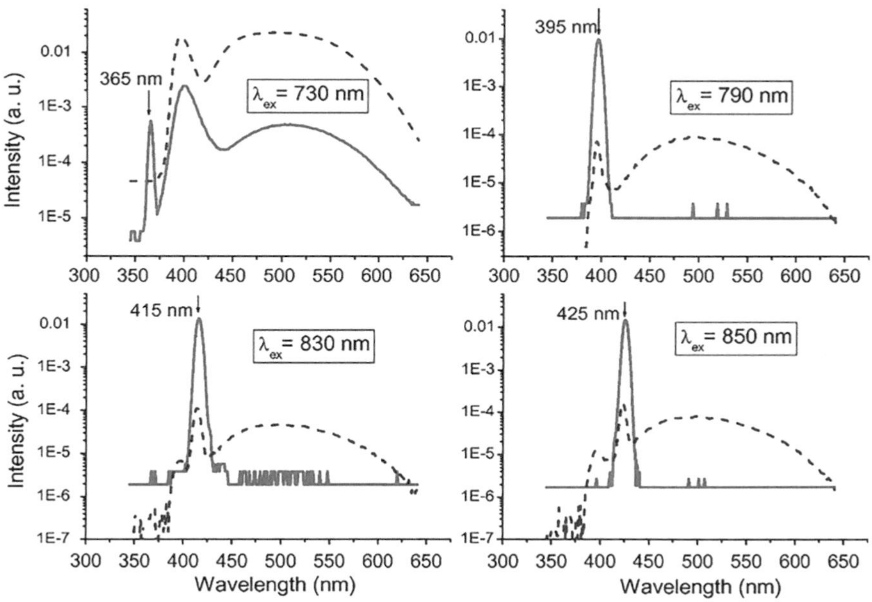

Fig. 12.12 Luminescence spectra of ZnO nano-tetrapod structure (solid line) and bulk material (dashed line) under two-photon/multiphoton excitation [26]

light energy is close to or lower than $E_g/2$, the spectral signal of the ZnO nano-tetrapod structure is dominated by the SHG, while other signals almost completely disappear. More detailed results of changing the excitation/fundamental light wavelength or photon energy are shown in Fig. 12.13.

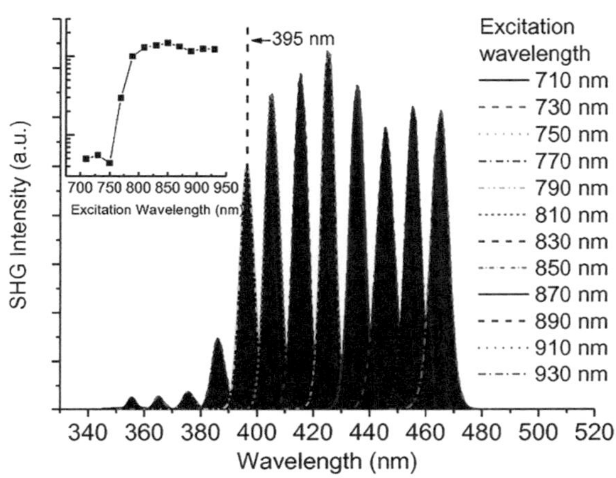

Fig. 12.13 SHG in ZnO nano-tetrapod structures under different excitation/fundamental light wavelength irradiation [26]

From the inset in the upper left corner, it can be more clearly seen that the intensity of the SHG significantly increases with the increase in the excitation/fundamental light wavelength (or the decrease of photon energy) and jumps near the band-edge luminescence energy. There are mainly two reasons for this phenomenon: the localized surface (second-order) nonlinear polarizability caused by the structural asymmetry and discontinuity (abrupt change) of the light field at numerious branch surfaces of the nano-tetrapod structure; on the other hand, when the photon energy of the excitation light/fundamental light is lower than $E_g/2$, the band-edge luminescence significantly weakens due to multiphoton absorption, and the self-absorption effect of the SHG signal also reduces.

From the above discussion, it can be concluded that the significantly increased surface structure in ZnO micro-nanostructures will introduce additional (second-order) nonlinear polarizabilities, thereby enhancing the efficiency of SHG. In ZnO nano-/micron-thick films or even polycrystalline structures, similar surface/interface enhancement effects also occur at the grain boundaries, which is why these types of samples can still exhibit considerable SHG signals after the bulk contribution weakens [5]. Meanwhile, when experimentally probing the SHG in ZnO, one inevitably encounter luminescence phenomena caused by two-photon/multiphoton absorption, which we will discuss in the next section.

12.5 Spectral Performance and Self-Absorption Phenomenon in ZnO under Two-Photon Excitation

12.5.1 Two-/Multiphoton Absorption-Induced Photoluminescence in ZnO

From the previous two sections, we have seen that under intense femtosecond laser excitation, ZnO not only generates second harmonics but also emits photoluminescence arising from two-photon/multiphoton absorption [15, 16, 21, 26]. Here, we discuss in detail the photoluminescence produced by two-photon/multiphoton absorption in ZnO and consider the possible occurrence of self-absorption phenomena. Fig. 12.3 in the first section shows the simplified two-photon absorption and its resulting luminescence, while the multiphoton absorption process is similar. In actual measurements and data analysis, we must also consider the strong exciton effect in ZnO (large exciton binding energy of 60 meV), exciton–phonon interaction, and the promounced self-absorption effect due to the deep penetration depth of two-photon/multiphoton excitation. Of course, this self-absorption effect can also be observed under single-photon excitation [33]. Let us first examine the photoluminescence spectrum of a ZnO single crystal under two-photon/multiphoton excitation, as shown in Fig. 12.14.

Unlike the ZnO micro-nanostructures discussed in the previous section, in ZnO single crystals, when the excitation photon energy (or corresponding wavelength)

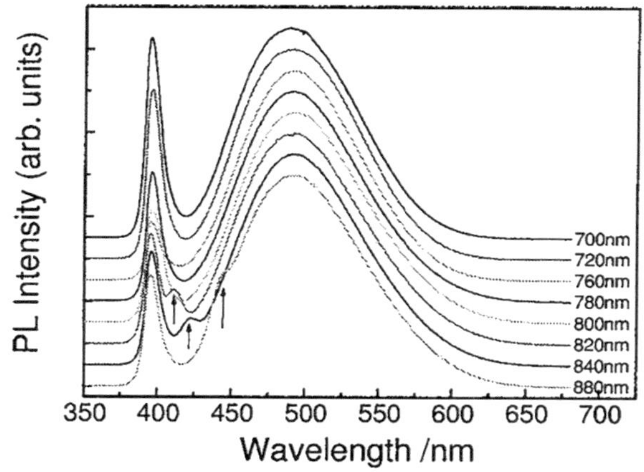

Fig. 12.14 Photoluminescence spectra of ZnO single crystal under femtosecond laser excitation at different wavelengths at room temperature [34]

is greater than $E_g/2$ (less than E_g), almost no SHG signal was observed. However when the excitation photon energy decreases (wavelength increases) further, the SHG signal gradually becomes noticeable (as indicated by the arrows in Fig. 12.14). By analyzing the dependence of the intensity of the band-edge luminescence peak near 395 nm on the excitation laser power/energy, we can determine the actual multiphoton process, as shown in Fig. 12.15:

Theoretically, the excitation-power dependence coefficients of the luminescence processes caused by two-photon and three-photon absorption should be 2 and 3, respectively. The deviation of the power dependence coefficients obtained by fitting from the theoretical values may result from the influence of impurity/defect energy

Fig. 12.15 Dependence of the integrated intensity of the band-edge luminescence peak in ZnO under femtosecond light excitation at different wavelengths against the excitation power [34]

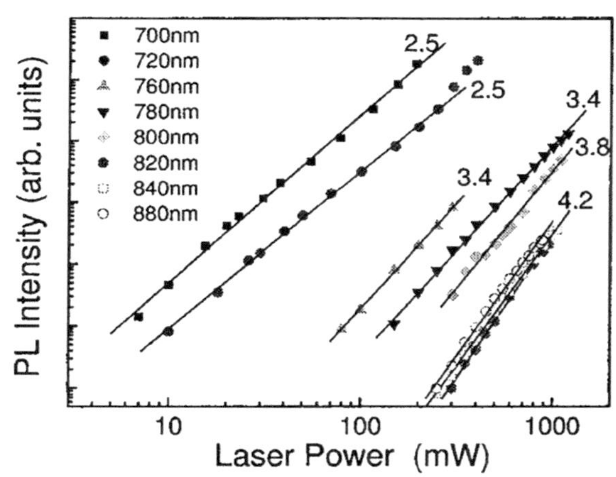

levels, as well as the impact of phonon interactions and self-absorption on band-edge luminescence at room temperature. In this work, we also conducted interferometric autocorrelation measurements of the band-edge luminescence under different excitation wavelengths. By analyzing the interference pattern, we found that the signal-to-background ratio under 720 nm excitation is nearly consistent with the theoretical value of 8:1 expected for a two-photon absorption process. Similarly, under 778 nm excitation, the signal-to-background ratio is close to the theoretical value of 32:1 for the three-photon absorption process. Between these two excitation wavelengths, both two-photon and three-photon processes coexist [34].

12.5.2 Self-Absorption Phenomenon Under Two-/Multiphoton Excitation in ZnO

We know that the bandgap of ZnO single crystals at room temperature is 3.37 eV. Considering the exciton binding energy of 60 meV, the corresponding luminescence peak should appear around 375 nm. The photoluminescence peak of ZnO at room temperature that we observe (i.e., band-edge luminescence, under single-photon excitation, such as 325 nm excitation) tends to shift slightly toward longer wavelengths. This shift is primarily due to the strong electron/exciton–phonon interaction at room temperature (e.g., a first-order phonon sideband, redshift of about 72 meV relative to the exciton peak energy). However, this phenomenon becomes more pronounced under two-photon/multiphoton excitation, suggesting that the physical origin of the emission may also change fundamentally [33, 35]. As shown in Fig. 12.16, the difference between photoluminescence spectra of ZnO single crystals under single-photon and two-photon excitation at low temperature is not very obvious. The self-absorption effect mainly occurs near the resonance energy of the exciton (both bound exciton and free exciton), i.e., around 369 nm or 3.36 eV. As the temperature increases, this effect becomes progressively stronger, and the self-absorption boundary shifts from the exciton region toward the phonon sideband (i.e., the low-energy end, as shown in Fig. 12.16c). The main reason for this behavior is that with increasing temperature, the luminescence of free excitons gradually dominates the tested spectral range. More importantly, the exciton–phonon interaction also strengthens significantly with rising temperature. This allows the photoluminescence photons generated inside the sample under two-photon excitation (the excitation photons can effectively penetrate into the deep region of the sample due to their small absorption coefficient) to be unable to reach the sample surface due to the self-absorption effect and thus cannot be detected. This comprehensive process is clearly reflected in the temperature-dependent evolution of the spectra shown in Figure 12.16b.

To further verify the self-absorption effect, we conducted photoluminescence spectral measurements under single-photon excitation (325 nm) using front excitation and side detection while varying the distance between the excitation light focus point and the sample surface. The experimental results obtained under varying temperature conditions are as expected: the self-absorption boundary gradually redshifts toward

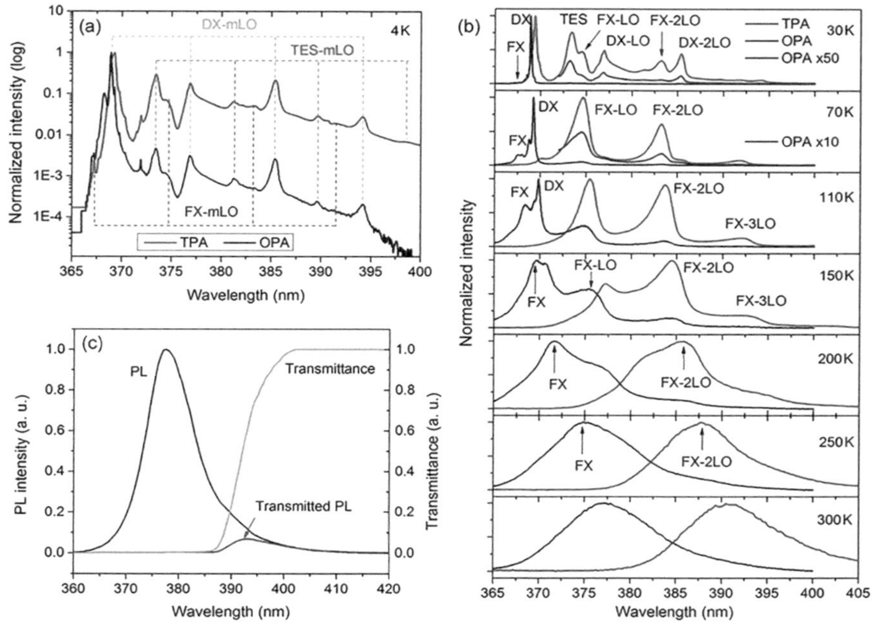

Fig. 12.16 Photoluminescence spectra of ZnO single crystals under one-photon (OPA) and two-photon (TPA) excitation: (**a**) PL at 4 K, (**b**) PL at various temperatures, and (**c**) PL and transmission at room temperature [33]

longer wavelengths side as the temperature increases [33]. As mentioned earlier, when the two-photon energy approaches the exciton (or band-edge luminescence/self-absorption boundary) energy, a significant enhancement of SHG occurs [6, 7, 24]. This enhancement mainly originates from the resonant coupling between two-photon fluorescence and SHG. We also measured the photoluminescence spectra of a ZnO solid rod under two-photon excitation (730 nm) with side excitation and front detection, again varying the distance between the excitation light focus point and the front surface):

From Fig. 12.17, we can clearly observe that as the excitation location gradually moves away from the surface of the sample, the high-energy side of the observed photoluminescence spectra gradually undergoes a redshift, which indicates that photons in this energy range experience increasingly self-absorption, preventing them from escaping the sample and being detected. It can therefore be predicted that as the temperature rises, the enhanced electron/exciton–phonon interactions will further strength this self-absorption effect, affecting the luminescence near the band edge and even in the phonon sideband. Consequently, the spectra observed under two-photon/multiphoton excitation differ significantly from the expected band-edge luminescence (i.e., the typical photoluminescence spectrum under single-photon excitation) as shown in Fig. 12.16b. Furthermore, due to the different luminescence mechanisms of ZnO under single-/two-photon excitation, the lifetimes of the corresponding (non)radiative recombination may also differ. Y.C. Zhong et al. measured the time-resolved photoluminescence (TRPL) spectra of the phonon sideband of ZnO tetrapod nanostructures under both single- and two-photon excitations. By fitting the decay curves with a single-exponential model, they found that the lifetimes under the two excitation conditions at room temperature were very similar (17.3 ns and 16.7 ns), while at low temperature (10 K), they were 2.1 ns and 3.0 ns, respectively [28]. The main reason for this result is that the ZnO tetrapod nanostructure sample contained very few surface defect states, so no obvious nonradiative recombination was observed. In our own TRPL spectrum measurements of the phonon sideband of ZnO single crystal samples under single- and two-photon excitations (at 25 K), the lifetimes extracted from the decay curves were ~0.7 and ~1.4 ns, respectively. These results reveal that different recombination mechanisms occur in ZnO bulk materials under different excitation conditions. Similar results have also been observed in free-standing gallium nitride (GaN) samples [36].

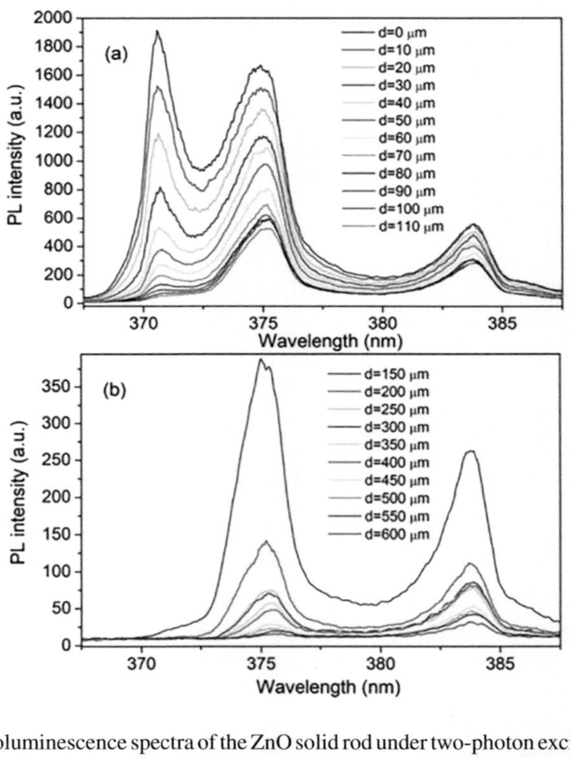

Fig. 12.17 Photoluminescence spectra of the ZnO solid rod under two-photon excitation at different excitation depths at low temperature (4.15 K): (**a**) 0–110 μm and (**b**) 150–600 μm [35]

12.6 Prospects for the Study of Nonlinear Optical Properties of ZnO and Other Wide Bandgap Semiconductor Materials

The SHG and spectral properties of ZnO under two-photon/multiphoton excitation discussed above will also occur in other wide bandgap semiconductors (such as GaN) [18, 36–42]. For example, the spectra of GaN thin films under two-photon/multiphoton excitation are shown in Fig. 12.18.

Under two-photon excitation $\left(E_g > h\nu_1 > \frac{E_g}{2}\right)$, a clear SHG signal can be observed in the GaN thin film, and its band-edge luminescence peak appears around 367 nm. Compared with the photoluminescence spectrum under single-photon excitation, this band-edge luminescence also exhibits a redshift. The reason for this phenomenon is similar to that illustrated in Fig. 12.16b, mainly arising from the self-absorption effect of GaN on band-edge luminescence [36]. As we increase the wavelength of the excitation laser to the energy where the SHG begins to resonate with the band-edge luminescence, the SHG is significantly enhanced (see the inset in the upper right corner of Fig. 12.18). If we further increase the wavelength of the excitation light $\left(h\nu_1 < \frac{E_g}{2}\right)$, the SHG can still be observed, but its efficiency decreases. However, because of the sharp reduction in three-photon absorption efficiency, the band-edge luminescence almost disappears from the spectrum. We also performed polarization-dependent measurements on the SHG and fitted the results (details similar to the discussion following Fig. 12.8). The experimental results are consistent with the theoretical prediction of the SHG [18]. In addition, we observed strong two-photon fluorescence in InGaN/GaN quantum wells, as shown in Fig. 12.19.

From the figure, we can clearly observe a luminescence peak at 425 nm in the InGaN/GaN quantum well, originating from two-photon absorption. The inset in the upper-right corner of the figure shows the variation of the integrated intensity

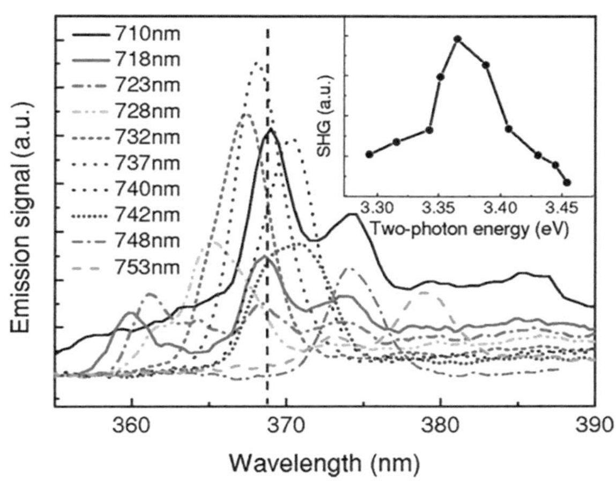

Fig. 12.18 Band-edge luminescence and SHG signals of GaN thin films under femtosecond laser excitation at different wavelengths at room temperature [18]

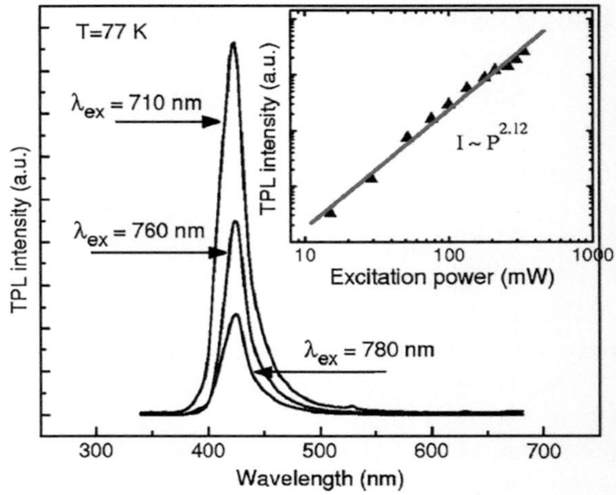

Fig. 12.19 Photoluminescence spectra of InGaN/GaN quantum wells under femtosecond laser excitation at different wavelengths at low temperature (77 K) [40]

of the 425 nm luminescence peak under 760 nm femtosecond laser excitation as a function of the excitation laser power. The power-dependence coefficient obtained from the power law fitting is 2.12, which is very close to the ideal theoretical value of 2. To further confirm the two-photon absorption phenomenon in the sample, we also conducted an interferometric autocorrelation measurement. The signal-to-background ratio, determined from the interference pattern, is ~6.6, which is close to the theoretical value of 8 [40].

12.7 Summary

From the above discussions and analysis, we can see that two-photon/multiphoton absorption phenomena are common in wide bandgap semiconductors such as ZnO and GaN, including their bulk materials, thin films, and various forms of micro- and nanostructures. When measuring the luminescence caused by two-photon/multiphoton absorption, because of the low absorption coefficient or low absorption efficiency (compared with single-photon absorption), a strong coherent light source, typically a femtosecond laser, is often used for excitation. At the same time, the low absorption coefficient also means that the excitation light penetrates deeply into the sample, so the region where two-photon/multiphoton absorption occurs is relatively large. The photons generated by radiative recombination within this region may not be detected due to the reabsorption/self-absorption within the sample. On the other hand, significant SHG often accompanies two-photon/multiphoton absorption, and these two mechanisms compete by consuming the same excitation photons. Furthermore, when their energies resonate ($2h\nu_1 \approx E_g$ or $2h\nu_1 \approx E_{edge}$, i.e., band-edge luminescence energy), the SHG is strongly enhanced. This resonance enhancement effect is particularly evident in micro- and nanostructures, or in structures with additional surfaces/interfaces, where surface/interface contributions generate much stronger second harmonics than bulk materials.

ZnO, as a wide bandgap semiconductor with a large exciton binding energy, has broad potential in optoelectronic devices, semiconductor lasers, solar cells, two-photon imaging, and other relevant fields. However, some applications require stable and low-cost p-type doped ZnO, and the progress in this area remains limited, restricting its use in heterojunctions, diodes, transistors, and other related devices. Nevertheless, ZnO can still be effectively integrated into electronic and optoelectronic devices by combining it with other semiconductor materials. The nonlinear optical properties of ZnO discussed in this chapter also have promising applications, such as two-photon imaging and the use of ZnO nanostructures in medical diagnosis [43–45]. Expanding these application areas requires a comprehensive and in-depth understanding of the nonlinear optical properties of ZnO materials, especially the enhancement mechanism of SHG and two-photon absorption near resonance, the identification of two-photon/multiphoton processes in micro/nanostructures, and the polarization characteristics of two-photon/multiphoton luminescence, all of which remain important directions for further research.

Furthermore, the experimental methods and techniques described in this chapter are also applicable to the study of the nonlinear optical properties of other materials, for example, two-photon fluorescence in perovskites [46] and SHG in two-dimensional materials [47]. Nonlinear optics, as an important branch of optics, plays a crucial role in physics, materials science, and other interdisciplinary fields.

References

1. Franken PA, Weinreich G, Peters CW, Hill AE (1961) Generation of optical harmonics. Phys Rev Lett 7(4):118
2. Shixiong Q, Gongming W (2002) Nonlinear optics. Fudan University Press, Shanghai
3. Ozgur U, Alivov YI, Liu C et al (2005) A comprehensive review of ZnO materials and devices. J Appl Phys 98(4):041301
4. Cao H, Wu JY, Ong HC et al (1998) Second harmonic generation in laser ablated zinc oxide thin films. Appl Phys Lett 73(5):572–574
5. Larciprete MC, Centini M (2015) Second harmonic generation from ZnO films and nanostructures. Appl Phys Rev 2(3):031302
6. Prasanth R, van Vugt LK, Vanmaekelbergh DAM, Gerritsen HC (2006) Resonance enhancement of optical second harmonic generation in a ZnO nanowire. Appl Phys Lett 88(18):181501
7. Lafrentz M, Brunne D, Rodina AV et al (2013) Second-harmonic generation spectroscopy of excitons in ZnO. Phys Rev B 88(23):235207
8. Guangsheng H (2018) Nonlinear optics and photonics. Shanghai Science and Technology Press, Shanghai
9. Wang G, Kiehne GT, Wong GKL et al (2002) Large second harmonic response in ZnO thin films. Appl Phys Lett 80(3):401–403
10. Wang G, Wong GKL, Ketterson JB (2001) Redetermination of second-order susceptibility of zinc oxide single crystals. Appl Optics 40(30):5436–5438
11. Dai DC, Xu SJ, Shi SL et al (2006) Observation of both second-harmonic and multiphoton-absorption-induced luminescence in ZnO. IEEE Photonic Tech Lett 18(13–16):1533–1535

12. Lo KY, Lo SC, Chu SY et al (2006) Analysis of the growth of RF sputtered ZnO thin films using the optical reflective second harmonic generation. J Cryst Growth 290(2):532–538
13. Lo KY, Huang YJ, Huang JY et al (2007) Reflective second harmonic generation from ZnO thin films: a study on the Zn-O bonding. Appl Phys Lett 90(16):161904
14. Huang YJ, Lo KY, Liu CW et al (2009) Characterization of the quality of ZnO thin films using reflective second harmonic generation. Appl Phys Lett 95(9):091904
15. Zheng CC, Xu SJ, Ning JQ et al (2011) Ion-implantation induced nano distortion layer and its influence on nonlinear optical properties of ZnO single crystals. J Appl Phys 110(8):083102
16. Zheng CC, Xu SJ, Ning JQ et al (2011) Inner surface enhanced femtosecond second harmonic generation in thin ZnO crystal tubes. J Appl Phys 109(1):013528
17. Shen YR (2003) The principles of nonlinear optics. Wiley-Interscience, Hoboken, N.J.
18. Yang H, Xu SJ, Li Q et al (2006) Resonantly enhanced femtosecond second-harmonic generation and nonlinear luminescence in GaN film grown on sapphire. Appl Phys Lett 88(16):161113
19. Dai J, Zeng JH, Lan S et al (2013) Competition between second harmonic generation and two-photon-induced luminescence in single, double and multiple ZnO nanorods. Opt Express 21(8):10025–10038
20. Larciprete MC, Passeri D, Michelotti F et al (2005) Second order nonlinear optical properties of zinc oxide films deposited by low temperature dual ion beam sputtering. J Appl Phys 97(2):023501
21. Zhang CF, Dong ZW, You GJ et al (2006) Femtosecond pulse excited two-photon photoluminescence and second harmonic generation in ZnO nanowires. Appl Phys Lett 89(4):042117
22. Look DC (2001) Recent advances in ZnO materials and devices. Mat. Sci. Eng. B 80(1–3):383–387
23. Ma T, Wang Y, Tang R et al (2013) Pre-patterned ZnO nanoribbons on soft substrates for stretchable energy harvesting applications. J Appl Phys 113(20):204503
24. Protte M, Weber N, Golla C et al (2019) Strong nonlinear optical response from ZnO by coupled and lattice-matched nanoantennas. J Appl Phys 125(19):193104
25. Zhang XQ, Tang ZK, Kawasaki M et al (2004) Second harmonic generation in self-assembled ZnO microcrystallite thin films. Thin Solid Films 450(2):320–323
26. Shi SL, Xu SJ, Xu ZX et al (2011) Broadband second harmonic generation from ZnO nanotetrapods. Chem Phys Lett 506(4–6):226–229
27. Djurisic AB, Leung YH (2006) Optical properties of ZnO nanostructures. Small 2(8–9):944–961
28. Zhong YC, Wong KS, Djurisic AB et al (2009) Study of optical transitions in an individual ZnO tetrapod using two-photon photoluminescence excitation spectrum. Appl Phys B 97(1):125–128
29. Chen SL, Stehr J, Reddy NK et al (2012) Efficient upconversion of photoluminescence via two-photon absorption in bulk and nanorod ZnO. Appl Phys B 108(4):919–924
30. Han XB, Wang K, Long H et al (2016) Highly sensitive detection of the lattice distortion in single bent ZnO nanowires by second-harmonic generation microscopy. ACS Photonics 3(7):1308–1314
31. Zhao J, Fan JT, Liu W et al (2019) Ultra-broadband second-harmonic generation in ZnO nano-tetrapod with over-one-octave bandwidth. IEEE Photonic Tech Lett 31(3):250–252
32. Wang RQ, Wang F, Long J et al (2019) Polarized second-harmonic generation optical microscopy for laser-directed assembly of ZnO nanowires. Opt Lett 44(17):4291–4294
33. Ye HG, Su ZC, Tang F et al (2017) Extinction of the zero-phonon line and the first-order phonon sideband in excitonic luminescence of ZnO at room temperature: the self-absorption effect. Sci Bull 62(22):1525–1529
34. Dai DC, Xu SJ, Shi SL et al (2005) Efficient multiphoton-absorption-induced luminescence in single-crystalline ZnO at room temperature. Opt Lett 30(24):3377–3379
35. Wang XR, Yu DP, Xu SJ (2020) Determination of absorption coefficients and Urbach tail depth of ZnO below the bandgap with two-photon photoluminescence. Opt Express 28(9):13817–13825

36. Zhong Y, Wong KS, Zhang W et al (2006) Radiative recombination and ultralong exciton photoluminescence lifetime in GaN freestanding film via two-photon excitation. Appl Phys Lett 89(2):022108
37. Wang H, Wong KS, Foreman BA et al (1998) One- and two-photon-excited time-resolved photoluminescence investigations of bulk and surface recombination dynamics in ZnSe. J Appl Phys 83(9):4773–4776
38. Sun CK, Liang JC, Wang JC et al (2000) Two-photon absorption study of GaN. Appl Phys Lett 76(4):439–441
39. Toda Y, Matsubara T, Morita R et al (2003) Two-photon absorption and multiphoton-induced photoluminescence of bulk GaN excited below the middle of the band gap. Appl Phys Lett 82(26):4714–4716
40. Li Q, Xu SJ, Li GQ et al (2006) Two-photon photoluminescence and excitation spectra of InGaN/GaN quantum wells. Appl Phys Lett 89(1):011104
41. Noor ASM, Torizawa M, Miyakawa A et al (2008) Simultaneous observation of single- and two-photon excitation photoluminescence on optically quenched wide-gap semiconductor crystals. Appl Phys Lett 93(17):171107
42. Godiksen RH, Aunsborg TS, Kristensen PK et al (2017) Two-photon photoluminescence and second-harmonic generation from unintentionally doped and semi-insulating GaN crystals. Appl Phys B 123(11):270
43. Kachynski AV, Kuzmin AN, Nyk M et al (2008) Zinc oxide nanocrystals for nonresonant nonlinear optical microscopy in biology and medicine. J Phys Chem C 112(29):10721–10724
44. Breunig HG, Weinigel M, Konig K (2015) In vivo imaging of ZnO nanoparticles from sunscreen on human skin with a mobile multiphoton tomograph. Bionanoscience 5(1):42–47
45. Raghavendra AJ, Gregory WE, Slonechi TJ et al (2018) Three-photon imaging using defect-induced photoluminescence in biocompatible ZnO nanoparticles. Int J Nanomed 13:4283–4290
46. Cao ZL, Lv BH, Zhang HC et al (2019) Two-photon excited photoluminescence of single perovskite nanocrystals. J Chem Phys 151(15):154201
47. Stielm T, Schneider R, Kern J et al (2019) Supercontinuum second harmonic generation spectroscopy of atomically thin semiconductors. Rev Sci Instrum 90(8):083102

MIX
Papier aus verantwortungsvollen Quellen
Paper from responsible sources
FSC® C105338

If you have any concerns about our products,
you can contact us on
ProductSafety@springernature.com

In case Publisher is established outside the EU,
the EU authorized representative is:
Springer Nature Customer Service Center GmbH
Europaplatz 3, 69115 Heidelberg, Germany

Printed by Libri Plureos GmbH
in Hamburg, Germany